Sources of Construction Information

An Annotated Guide to Reports, Books, Periodicals, Standards and Codes

Volume 1: Books
by
JULES B. GODEL

The Scarecrow Press, Inc.
Metuchen, N.J. and London
1977

Library of Congress Cataloging in Publication Data

Godel, Jules B   1925-
  Sources of construction information.

  Includes index.
  CONTENTS: v. 1. Books.
  1. Engineering--Bibliography. 2. Architecture--
Bibliography. 3. Construction industry--Bibliography.
I. Title.
Z5851.G62 [TA145]016.624   77-4671
ISBN 0-8108-1030-1

Copyright © 1977 by Jules B. Godel

Manufactured in the United States of America

PREFACE

This book is one of a series devoted to construction information. It is intended to be a working tool for use by professionals who value their time. Sources of Construction Information is by no means a complete encyclopedia, but nor is it merely a book that tells where to find another book. Instead, the series is a systematic description of existing books, reports, standards, and other publications of interest to the architect and planner, engineer, contractor, manager, and building official.

The complete work is divided into five principal Parts:

- A   Books
- B   Government Reports
- C   Professional, Trade Association, and Planning Organization Reports
- D   Standards and Codes
- E   Organizations, Periodicals, Directories

Parts A, B, and E will be separate volumes, C and D will be combined in a single volume.

The uniqueness of this work lies in the fact that it covers all the major sources of construction information with the exception of magazine and journal articles. In most cases, each publication is described in a brief abstract, a feature that enables the reader not only to identify, but also to evaluate a large body of data without assistance from a librarian. In every case, enough information is given to locate or acquire a publication of interest. The value of this series is that it takes the reader past the first stage of identifying references from which he would just begin his search. The time saved by this rapid access to construction information should have a marked effect on the user's ability to be better informed on subjects, both within his specialty and outside his immediate expertise.

Two years have been spent in research, compilation and editing. More than 1000 book publishers and professional, trade, labor, and government organizations were solicited. From these efforts, about 10,000 construction-related publications were identified and processed. Many individuals made significant contributions. The conception for this series resulted from consultations with William H. Edgerton, president of CPC, and with Bradford Perkins, managing partner of Llewelyn-Davies Associates. Editing was done by Phyllis Oder, Karen Bird and Joan Kutner. Compilers were Kenneth Godel, Marian McGraw, Beverly Price and Tom Stoenner. Holly Haskell and Deborah Baron typed the manuscript for Volume 1. The efforts of all are gratefully acknowledged.

CONTENTS

Preface		iii
Introduction		vii
How the Book Is Organized		ix
Section A01	Planning--Urban and Regional	1
Section A02	Planning--Project	45
Section A03	Management	75
Section A04	Design--Architectural	115
Section A05	Design--Civil and Structural	238
Section A06	Design--Mechanical and Electrical	320
Section A07	Design--Transportation, Heavy Construction	360
Section A08	Building Materials, Components, Finishes	375
Section A09	Construction, Fabrication, Installation	423
Section A10	Construction Industry	466
Section A11	Selected Topics	482
Appendix	Book Publishers and Distributors	515
Index		525

INTRODUCTION

The one common thread of the construction industry in the United States is its fragmentation. Its major participants--design professionals, contractors, suppliers, lower union locals, owners, etc.--all number in the tens of thousands and almost none of these individual elements ever accounts for more than one per cent of the industry. Thus, in spite of the fact that it is the second largest American industry (with an annual put-in-place production value usually equal to more than 10 per cent of the GNP), construction remains a decentralized, chaotic conglomeration of small, loosely-organized elements far more akin to the mode of 19th than 20th Century industry.

Nowhere is the impact of this situation more apparent than in the lack of organized and available information about the construction industry. An individual seeking the answer to almost any question about this industry can easily be stymied by the difficulty of obtaining even a place to start his research. In fact, one inspiration for this book came from the frustrating experience my staff and I had in putting together a detailed review of the construction industry for one of our studies. This overview required research drawn from a wide variety of sources--books, magazine, government publications, university reports, phone interviews--over half of which would probably not have been found by a researcher not already familiar with the industry. Our years of construction experience had to take the place of the bibliographies available for so many other topics.

The actual idea that there was a need for a comprehensive bibliography of the construction field came from Bill Edgerton, but it was not until Jules Godel agreed to undertake the development of such a book that it actually began to become a reality. What has emerged is a document far broader in scope and concept than either Bill or I had originally envisioned. It deals with a very wide range of questions currently facing individuals in this industry. Specifically, it covers all five of the major sources of construction industry topics (books, government reports, standards, non-government reports, codes), and identifies periodicals and newsletters and such non-published sources as trade associations, unions, professional groups, etc. Moreover, in each of the above areas the search is shortened because not only is a comprehensive list presented of available information sources, but most are annotated to clarify their scope and subject matter.

These features, along with an extensive index, make this series an invaluable starting point for:

---the design professional or manufacturer wishing to review standards relevant to a particular product;

---the architect or engineer looking for a directory of potential clients of a particular type, such as hospital administrators;

---the research team assembling a bibliography;

---the individual trying to decide which books to order for his office library;

---the firm seeking information on a new building, material, or construction type; or

---the contracting firm seeking new business procedures, or estimating and cost data.

In short, any individual or organization wondering where to start in research of any major industry topic from computer applications in design to roof construction.

Future editions of this book will undoubtedly expand and refine the work of the first edition. For the present, however, it represents an invaluable addition to the far-too-limited reference material now available in this industry.

BRAD PERKINS

September 30, 1976

HOW THE BOOK IS ORGANIZED

As noted in the Preface, Sources of Construction Information is divided into five Parts, A to E, depending upon the type of source; i.e., books, reports, standards, etc. Within each of these, except for Part E, the entries are further divided into eleven broad subject sections. An index lists all individual sources by access number which identifies the section and category wherein the source can be found. For information on a specific subject, the reader should refer directly to the index. On the other hand, he may want an overview of all publications in a general grouping, e.g., the specifications on fire safety. This information can be found quickly in Part D, Section 11.

Part E lists all the segments of the construction industry (owners, design professionals, contractors, planners, etc.), along with the trade and professional organizations to which they might affiliate, the periodicals that relate to their work, and the directories or information sources to which they would refer. The emphasis is to assemble organizational and reference data for use in developing new professional and business contacts.

Parts A through D are divided into the 11 consecutively numbered subject sections, each having a number of related topics as shown below:

-01 Planning--Urban and Regional
    Housing
    Land Use
    New Towns
    Urban Decay and Rehabilitation
    Zoning

-02 Planning--Project
    Cost Estimates
    Finance
    Site Selection
    Specific Projects (Hospital, School, Shopping Center, etc.)

-03 Management
    Contracting and Construction Management
    Contracts and Law
    Engineering Management
    Inspection, Tests, Quality Control

- 04   Design--Architectural
        Acoustics
        Architecture; by Geographic Region
        Architecture; by Building Type
        Design Details
        Historic Periods and Structures

- 05   Design--Civil and Structural
        Computer-Aided Design
        Hydraulics
        Pollution and Environmental Impact
        Seismic and Windstorm Design
        Soil and Rock Mechanics

- 06   Design--Mechanical and Electrical
        Electrical Equipment
        Heating, Ventilation, Air Conditioning
        Lighting
        Plumbing
        Vibration

- 07   Design--Transportation, Heavy Construction
        Airports
        Bridges
        Highway Engineering

- 08   Building Materials, Components, Finishes
        Adhesives, Sealants
        Concrete, Brick, Masonry
        Metals Properties, Corrosion
        Paints, Coatings, Finishes
        Plastics, Elastomers
        Wood and Wood Products

- 09   Construction, Fabrication, Installation
        Construction Equipment
        Excavation
        House Construction
        Renovation and Rehabilitation

- 10   Construction Industry
        Building Trades
        Information Sources
        Industrialized Buildings

- 11   Selected Topics
        Drafting, Illustration, Blueprint Reading
        Exams for Professional Registration
        Fire Safety
        General Safety and Security

The 11 subject sections are similar to the 16-division Uniform Construction Index (UCI), in that they represent broad generic headings, are few in number, and are repeated in each Part. However, there are important differences. The UCI is admirably suited for specifications, data filing and cost analysis. Because it is so product-oriented, it is excellent for use with manufacturers' literature. However, our subjects are heavily biased towards planning, management, design, and materials. Strict adherence to the UCI format would have bunched more than 50 per cent of our entries into UCI's Division 1--General Requirements, leaving other divisions almost devoid of information sources. To suit our special needs, we have used some UCI divisions, deleted others, and added a number of special categories.

Every information source has a unique access number that permits its rapid identification in the index and provides the means for locating it from among the 11 subject sections. The access number has two parts, a prefix containing a letter and a two-digit number, and the actual source number. The letter in the prefix, A through D, corresponds with the Parts, while the two-digit code ranges between -01 and -11 to match the Sections. An example is shown below:

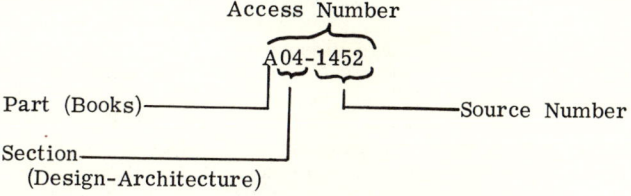

In each section, the access numbers run consecutively.

Volume 1 of Sources of Construction Information contains almost 3000 books from 300 publishers. Most were published in the United States, some in Great Britain, and a small number are from Holland and West Germany. All are in the English language.

There were some arbitrary editorial decisions in determining whether a publication was a book or a report. Generally, a book is considered to be at least 100 pages in length, written by individuals, and not the result of a specific study. However, there are numerous examples of reports that could be construed as books, and vice versa. This is particularly true of government agency publications. It is suggested that the reader refer to Volume 2, Government Reports, for an overview of book-length information sources.

The subject sections containing the most titles in Volume 1 are architecture (A04), structural design (A05), management (A03), urban planning (A01), construction techniques (A09), and building materials (A08). However, some books, especially handbooks, cross many subject lines and may not be found in the section where the reader expects it to be. Entries are listed only in one section, never repeated. For this reason, the index should be the primary place from where to begin the information search.

xi

Each entry appears in a standard bibliographical form containing the access number, author(s), title, publisher, date of publication, number of pages, cost, and special information. In most cases, there follows a brief abstract. A few words of explanation will familiarize you with this format.

Authors are listed alphabetically within each section. With knowledge of the author's name and the subject of the book, the reader has the option of finding the book without use of the index. Again, remember that some books could be listed in several sections, even though they appear in only one. Often, however, the reader will know from past experience that a book is in a particular section and can bypass the index for rapid access. Publishers' names are abbreviated to save space. Their full names and addresses are to be found in the Appendix. Book costs should serve only as a guide. Rapid escalation in the cost of some books and the discounting of others make accurate pricing difficult.

For the most part, access numbers run consecutively throughout the volume. There are gaps in the access numbers, however, purposely left at the end of each section to permit the late addition of entries. In a few cases numbers are skipped within a section. Some duplication of entries was discovered during preparation of the index and several were deleted. There are other cases of apparent duplication; that is, two different access numbers for entries bearing the same title and author. Closer inspection will reveal minor differences, such as different publishers or different editions. As long as there were two editions in print or a significant difference in cost for the same book from two publishers, we elected to list them separately.

Most abstracts stem from publishers' catalogs or pre-publication advertisements. In some cases they have been edited to reduce their length to between 50 and 100 words and to delete self-serving adjectives and phrases. The compilers and editors in no way attempt to evaluate the worth of a book. This is left to the reader's judgment.

A conscientious attempt has been made to assemble as comprehensive a list of books as was possible given the limits of time and resources. We recognize the possibility that important references may have been inadvertently omitted. Future editions of this book will be updated and we would appreciate hearing from readers about information sources that we have missed.

Section   A01

PLANNING: URBAN AND REGIONAL

           Housing
           Land Use
           New Towns
           Urban Decay and Rehabilitation
           Zoning

A01-0001
Abrams, C. The City is the Frontier. Harper and Row, 1965. 394 pp. $1.95
Appraisal of urban renewal which offers measures to answer the myriad problems.

A01-0002
Alden, J. and Morgan, R. Regional Planning: A Comprehensive View. John Wiley, 1974. 364 pp. $17.95
Synthesizes the current state of regional planning and suggests areas of improvement and lines for future development.

A01-0003
Altshuler, A. A. The City Planning Process: A Political Analysis. Cornell Univ. Press, 1966. 464 pp. $14.50 (paperback $4.95)
The author presents four detailed case studies of land-use planning in Minneapolis and St. Paul, Minnesota. The planner's goals and means of achieving them, the strategies he adopts in his dealings with politicians, and his concern with "professionalism" are all subjected to careful analysis.

A01-0004
American Public Health Assn. Planning the Neighborhood. Public Administration Service, 1948. 108 pp. $3.00

A01-0005
Anderson, M. The Federal Bulldozer: A Critical Analysis of Urban Renewal, 1949-1962. MIT Press, 1964. 272 pp. $5.95
This book represents the first major analysis of the federal urban renewal program since it started in 1949. In making this study, Professor Anderson gathered information about every urban renewal project in the country. The author explains what the federal urban renewal program is, why it was started, how it works, and what it has accomplished since 1949.

A01-0006
ASCE. The Use of Underground Space to Achieve National Goals. American Society of Civil Engineers, 355 pp. $7.00
This report evaluates the potential benefits that can accrue to society from a transfer of many urban and other life-support functions to sub-surface space. The capability to make these transfers hinges on a large R & D effort, a public policy decision to achieve the transfers, and the creation of the institutional arrangements to execute and implement the program.

Planning--Urban and Regional

A01-0007
ASTM. <u>1975 Compilation of ASTM Standards in Building Codes.</u>
American Society for Testing and Materials, 1975. 3100 pp. $89.50 ($71.60 for ASTM members)

This two-volume publication includes all of the ASTM Standards that have been adopted in building codes throughout the United States and Canada as authentic sources of test procedures and as a basis for acceptable quality for materials and constructions. This edition has over 524 specifications and methods of test, 206 new or revised standards and 5 deletions.

A01-0008
Babcock, R. F. and Bosselman, F. P. <u>Exclusionary Zoning: Land Use Regulation and Housing in the 1970s.</u> Praeger, 1973. 350 pp. $17.50

Reviews land use regulation and its relationship to housing production. Includes specific recommendations for new policies and techniques at local and regional levels. The history and legal status, impact on cost, quantity, and location of the present system are reviewed with emphasis on low and moderate income housing.

A01-0009
Bagby, G. <u>Housing Rehabilitation Costs.</u> Lexington Books, 1973. 81 pp. $9.50

Using data on rehabilitated row houses in Philadelphia and comparable new row housing, the author shows that rehabilitation costs somewhat less, but this approach often produces housing less desirable in terms of salability than new construction. This work investigates and evaluates federal administration of rehabilitation programs.

A01-0010
Bailey, J. <u>New Towns in America: The Design and Development Process.</u> American Institute of Architects or John Wiley, 1973. $23.75 (for AIA members, less 10%)

The creation of new towns in America as outlined by 22 experts plus comparative land-use plans of 32 historic and current new towns.

A01-0011
Bannon, J. <u>Leisure Resources: Its Comprehensive Planning.</u> Prentice-Hall, 1976. 512 pp. $12.95

For recreation and parks administrators, landscape architects, urban planners, and physical education teachers this book provides a complete analysis of leisure planning resources, including administration of the planning process, population trends, survey of the community, open space and land analysis, and evaluation of services.

A01-0012
Banz, G. <u>Elements of Urban Form.</u> McGraw-Hill, 1970. 220 pp. $16.50

Presents the contemporary urban environment as one continuous artifact serving a multiplicity of functions simultaneously.

A01-0013
Barnett, J.   Urban Design as Public Policy: Practical Methods for Improving Cities. McGraw-Hill, 1974. 200 pp. $15.00
   This book demonstrates the possibilities of making meaningful physical improvements in cities, within the framework of present social and governmental systems. Architects, urban planners, and local, state, and federal administrators will find this book useful and thoughtprovoking.

A01-0014
Batty, M.   Urban Modelling. Cambridge Univ. Press, 1976. 384 pp. $36.00
   This book introduces a series of mathematical models of urban phenomena which are relevant to scientists and planners attempting to understand and predict the future form of cities and regions. It describes the models both as urban theories and as computer algorithms, and thoroughly explores the process of calibrating the models to reflect reality.

A01-0015
Becker, H. J.   New Housing in Finland. International Publications, 1964. 182 pp. $20.00
   A study of regional planning and social housing in an architecturally leading country. 130 photos, 12 scale models, 120 floor plans, 65 site plans, maps, etc.

A01-0016
Bell, G. D., et al., eds.   Urban Environments and Human Behavior. Dowden, Hutchinson and Ross, 1973. 288 pp. $15.00
   Provides a working tool for architects, urban designers, landscape architects and other planners who need to be aware of the human requirements of environmental design. It will also serve as a reference for public housing and urban planning administrators, the social and behavioral scientists who provide the information which forms the basis for administrative decisions, and builders.

A01-0017
Benevolo, L.   The Origins of Modern Town Planning. MIT Press, 1967. 147 pp. $7.95

A01-0018
Beresford, M.   New Towns of the Middle Ages. Praeger, 1967. 690 pp. $18.50
   A history and reference of town plantations in England, Wales, and Gascony.

A01-0019
Berry, B. J. L.   Growth Centers in the American Urban System. Vol. 1, Community Development and Regional Growth in the Sixties and Seventies. Ballinger Publishing Co., 1973. 208 pp. $9.50
   This two-volume study offers an investigation of the U.S. urban hierarchy on a consistent nationwide basis, together with an

analysis of both metropolitan and non-metropolitan growth centers since 1960. Volume 1 reviews the theory of urban-regional growth, and analyzes the theoretical basis of growth center strategies. Berry studies the possibilities of channeling development into peripheral areas of present development, and outlines a nationwide application of growth center strategies.

A01-0020
Berry, B. J. L. Growth Centers in the American Urban System. Vol. II, Working Materials on the U. S. Urban Hierarchy Organized by Economic Regions. Ballinger Publishing Co., 1973. 600 pp. $19.50
 This second volume contains the data from a nationwide study of the U.S. urban hierarchy and growth centers within the American urban system organized by functional economic areas. It provides the supporting evidence for Volume 1 and a set of working materials for the development of local growth center policies.

A01-0021
Berry, B. J. L. Race and Housing: The Chicago Experience 1960-1975. Ballinger Publishing Co., 1976. 448 pp. $17.50
 The author evaluates a series of HUD-financed experimental fair housing programs that he monitored between 1967 and 1972. After outlining the housing programs in question and examining their essential inability to reach stated goals, Professor Berry looks at the broader questions of attitudes to integration, both white and black. He then considers questions related to the operation of metropolitan housing markets, and relates these issues back to the social-psychological and community issues and to the dilemmas of the open housing experiments.

A01-0022
Berry, F. Housing: the Great British Failure. Charles Knight and Co., 1974. £4.00
 Examines the requirements which should be taken into consideration for good housing, and argues that perhaps 40% of all British housing falls short of standards accepted as commonplace in many other countries. Looks at the history of state intervention in some detail, and describes the effects of policies in the housing sector.

A01-0023
Bestor, G. C. City Planning Bibliography. American Society of Civil Engineers, 1972. 518 pp. $9.00
 This publication is of value to all who are concerned with recent developments in the growth and improvement of urban centers. About 75% of the 1,837 entries are annotated. The book covers the nature and form of cities, the history of cities and city planning, and contemporary planning.

A01-0024
Binns, G. D., et al. Knight's Building Regulations. (Metric

Edition) Charles Knight and Co., 1973. 347 pp. £11.25
This new edition is an annotation of the Building Regulations 1972, the statutory conditions and standards for building work in England and Wales. Contents include materials, site preparation, structural stability, fire precautions, sound and thermal insulation, chimneys, heating and sanitary engineering.

A01-0025
Branch, M. C. Comprehensive Urban Planning: A Selective Annotated Bibliography with Related Materials. Sage Publications, 1970. 480 pp. $20.00 (Acquire from American Society of Planning Officials) ($18.50 for ASPO members)
An annotated bibliography on comprehensive planning.

A01-0026
Branch, M. C. Planning Urban Environment. John Wiley, 1974. 254 pp. $22.00
Focuses on environmental pollution and the kind of city planning required to alleviate the problems of pollution. Drawn from a diversity of disciplines: engineering, economics, biology, chemistry, zoology, geology, planetary and space science, and urban and regional planning.

A01-0027
Branch, M. C., ed. Urban Planning Theory. John Wiley, 1975. 596 pp. $35.00
Describes the theoretical bases from which planning is being advanced. Papers are included on such subjects as the purpose of zoning, the city image and its elements, spatial organization theory as a basis for regional planning, information systems and public planning, continuous city planning, systems theory and management, model design, gaming, the local planning agency, the city planner and the public interest, and crowding.

A01-0028
Breese, G., et al. The Impact of Large Installations on Near-by Areas: Accelerated Urban Growth. Sage Publications, 1969. 640 pp. $24.00
Studies the changes in land use, government and economy caused by the establishment of large government or industrial plants in previously rural or semi-urban locales.

A01-0029
Brooks, R. O. New Towns and Communal Values: Case Study of Columbia, Maryland. Praeger, 1974. 250 pp. $16.50
Evaluates a "new town" to which the development process has achieved the re-creation of a community with specific communal goals--economic self-sufficiency, a degree of political autonomy, equality within the community. "Development episodes" survey housing, social planning, mass transit, education facilities, and "village associations."

Planning--Urban and Regional

A01-0030
Brown, A. J. and Sherrard, H. M. An Introduction of Town and Country Planning. American Elsevier, 1969. 423 pp. $25.00

This revised edition provides a general introduction to town planning and relates practice overseas, notably in Britain and the U.S.A., to Australian conditions. The range and proportion of Australian material covered has been appreciably increased.

A01-0031
Browne, K. West End: Renewal of a Metropolitan Centre. Architectural Press, 1971. 176 pp. £6.50

Taking the west end of London as a case history, this book illustrates principles of renewal with maps, drawing, photographs and text.

A01-0032
Burchell, R. W. and Hughes, J. W. Planned Unit Development: New Communities American Style. Center for Urban Policy Research, 256 pp. $12.95

In recent years, PUD construction in the United States has doubled--an indication that this may be the land use technique of the future. This book traces PUD's origins and its current and future impact on residential patterns, including an analysis of the socioeconomic characteristics of PUD residents and an examination of the local costs and revenues associated with planned unit development.

A01-0033
Calsat, J. H. and Sydler, J. P., eds. International Vocabulary of Town Planning and Architecture. International Union of Architects, Paris, France. 1970. 368 pp. 138 francs

Four thousand architectural and town planning terms are defined in French, German, and English.

A01-0034
Caminos, H., et al. Urban Dwelling Environments: An Elementary Survey of Settlements for the Study of Design Determinants. MIT Press, 1969. 224 pp. $17.50 (loose sheets, $11.00)

A comparative method for studying the urban environment is attempted in this book, whereby the total environment is successively scaled down to smaller and smaller segments, so that at each stage the segment under study is already placed in its larger physical and social context. Sixteen localities (eight each in North and South America) are examined.

A01-0035
Campbell, C. New Towns: Another Way to Live. Prentice-Hall, 1976. 384 pp. $11.95

This book provides a comprehensive analysis of new towns, or planned communities, based on actual cases of residents' experiences, transportation, governance, citizen self-help efforts in education, and the new towns of Reston, Virginia; Columbia, Maryland; Soul City, North Carolina; Park Forest South, Illinois; Jonathan and Cedar-Riverside, Minnesota; and Irvine, California.

A01-0036
Cantacuzino, S., ed. New Uses for Old Buildings. Architectural Press, 1974. £6.00
This book is intended to show the many possibilities of adapting old buildings to new uses at a time when many people are becoming concerned with the widespread destruction of existing environments. Each job illustrated is backed up by a technical account of the problems encountered and how they were solved.

A01-0037
Carver, H. Cities in the Suburbs. Univ. of Toronto Press, 1968. 128 pp. $7.50 (paperback, $1.50)
Mr. Carver examines what has provided a sense of community for city groupings of the past and how leading planners of our day (Ebenezer Howard, Clarence Stein, Le Corbusier, Frank Lloyd Wright) have suggested it be found for modern cities. His own proposal for achieving this goal is simple, and originates in the earliest views of a city as a place in which an urban society achieves its individual character by congregating around its own social institutions.

A01-0038
Case, F. E., ed. Inner-city Housing and Private Enterprise. Praeger, 1972. 248 pp. $15.00
Based on studies in nine cities.

A01-0039
Catanese, A. J. Scientific Methods of Urban Analysis. Univ. of Illinois Press, 1972. $12.50
The author brings together techniques from mathematics, operations research, systems analysis, information science, and planning and develops a framework for their application to urban problems. Complex scientific methods are literally translated into an urban studies perspective.

A01-0040
Chadwick, G. A Systems View of Planning. Pergamon, 1971. 384 pp. $11.25
Sets out a theory of the process known as town and regional planning. This theory is based on more general texts in systems theory, cybernetics and related fields; but the author provides both a philosophy and a methodology for the planning field.

A01-0041
Chaplin, F. S., Jr. Urban Land Use Planning. Univ. of Illinois Press, 1965. 498 pp. $7.95 (acquire from American Society of Planning Officials)
The second edition reflects the shift in emphasis from the use of "craftsman's" methods of planning analysis to more scientific

approaches. New chapter on developments in theory and a section on the use of models in planning.

A01-0042
Chapman, S. D., ed. The History of Working-class Housing: A Symposium. Rowman and Littlefield, 300 pp. $16.50
A selection of Britain's largest urban centers was chosen. The wide regional variations in standards of living called for assessment of each by a different specialist.

A01-0043
Chartrand, R. L. Systems Technology Applied to Social and Community Problems. Hayden, 1971. 496 pp. $12.50
Reference work that traces the local, state, and federal legislative efforts to utilize innovative technology to non-defense, non-space problems. The book analyzes use of systems tools and techniques in environmental pollution, transportation planning, housing, law enforcement, education, health services.

A01-0044
Chartrand, R. L., ed. Hope for the Cities: A Systems Approach to Human Needs. Hayden, 1971. 224 pp. $10.50
In the belief that better communication must be established between legislators and technologists, a series of seminars was developed by Mr. Chartrand under the aegis of the Washington Operations Research Council. Each session featured a member of Congress with responsibilities and a personal interest in one of the problem areas. Following the presentation of the problem from the congressional vantage point, a senior member of the operations research systems analysis community set forth the possible role of systems technology in meeting the subject problem. This book is the culmination of idea-action exchange between congressmen and technologists.

A01-0045
Cherry, G. E. Urban Change and Planning: A History of Urban Development in Britain since 1750. G. T. Foulis, 264 pp. $15.00
This urban history describes the process of urban change against a social and economic background. The author focuses on the planning movement itself, the acceptance of its ideas and their effect on town development. The book offers a general survey of urban development in all major British towns.

A01-0046
Chung, H. C. The Economics of Residential Rehabilitation: Social Life of Housing in Harlem. Praeger, 1973. 246 pp. $15.00
Defines the concept of social life of housing as the period during which a structure continues to function as standard housing, once it is rehabilitated with some capital investment. Formulates a 20-year renewal program which encourages short-term rehabilitation of "deteriorating" buildings and immediate replacement of "dilapidated" buildings.

A01-0047
Claire, W. H., ed. Handbook on Urban Planning. Van Nostrand Reinhold, 1973. 416 pp. $27.95
A reference that shows how to implement urban plans. Explains the roles of architects and engineers in planning, and covers housing, community, and transportation planning.

A01-0048
Clapp, J. A. New Towns & Urban Policy: Planning Metropolitan Growth. Kennikat Press, 1971. $12.50
An evaluation of the new-town concept as a mechanism for introducing order into urbanization and achieving the objectives of a variety of public programs which have not been previously linked to any unified national urban policy.

A01-0049
Clawson, M. America's Land and Its Uses. Johns Hopkins Univ. Press, 1972. 178 pp. $8.50 (paperback $2.45)

A01-0050
Clawson, M., ed. Modernizing Urban Land Policy. Johns Hopkins Univ. Press, 1973. 256 pp. $11.00
Economists, planners, and lawyers look at the history of land use controls, the economic costs of housing discrimination, legal attacks on exclusionary controls, and conflicting housing and environmental concerns.

A01-0051
Clawson, M. and Hall, P. Planning and Urban Growth: An Anglo-American Comparison. Johns Hopkins Univ. Press, 1973. 296 pp. $12.50
U.S. and English urban land development is compared by the authors. They analyze the customs, legal bases, and planning practices and present quantitative information on the results of planning processes in terms of land use changes, dwelling construction, land prices, and densities of residential development.

A01-0052
Clawson, M. and Stewart, C. L. Land Use Information: A Critical Survey of U.S. Statistics, Including Possibilities for Greater Uniformity. Resources for the Future, Inc. 1966. 422 pp. $6.00

A01-0053
Cooper, J. R. and Guntermann, K. L. Real Estate and Urban Land Analysis. Lexington Books, 1974. 544 pp. $14.00
The book emphasizes the urban milieu in which the real estate investment decision is made. It provides the basic tools of housing market analysis and in the three traditional approaches to value.

A01-0054
Crawford, D., ed. A Decade of British Housing 1963-73.

Architectural Press, 1974. 200 pp. £2.50
This is a selection of important building studies of housing
projects which have appeared in the Architects' Journal and the
Architectural Review over the last ten years. Each project is appraised socially and technically, and a cost analysis given.

A01-0055
Crecine, J. P., ed. Financing the Metropolis: Public Policy in
 Urban Economies. Sage Publications, 1970. 640 pp.
 $20.00 (paperback $7.50)

A01-0056
Crosby, T. Architecture: City Sense. Van Nostrand Reinhold,
 1965. 96 pp. $2.45
This book attempts to present a coherent approach to the
environment and particularly to the regeneration of existing urban
areas.

A01-0057
Cullen, G. The Concise Townscape. Van Nostrand Reinhold,
 1971. 192 pp. $3.95
"Townscape" is the art of giving visual coherence and organization to the jumble of buildings, streets and spaces that make
up the urban environment. Its concepts were embodied in the book
Townscape (1961), which established itself as a major influence on
architects, planners and others concerned with what cities should
look like. In this shortened version, the studies of specific towns
have been left out and instead Cullen sets his ideas in the context
of the seventies.

A01-0058
Cullingworth, J. B. Town and Country Planning in England and
 Wales. Univ. of Toronto Press, 1965. 341 pp. $9.50
This work provides a picture of the planning system and
how it is changing. An historical introduction leads into an account of the machinery of planning and the major new provisions of
the 1968 Town and Country Planning Act. The problems of land
values, amenity, derelict land, planning for leisure, new and expanding towns, urban renewal and the search for an adequate means
of regional planning are treated.

A01-0059
Dahinden, J. Urban Structures for the Future. Praeger, 1972.
 220 pp. $25.00
Presents new concepts of the way we can and probably will
live-in three-dimensional grid cities, biotectures, even massive
clip-on and plug-in structures. Evaluates these proposed plans
according to standards of flexibility, controlled economic growth,
and social environment.

A01-0060
Davidovich, V. G. Town Planning in Industrial Districts. Israel
 Program for Scientific Translations, 1960. 320 pp.

This book deals with theoretical and practical problems of town planning for centers of heavy industry in the USSR. It describes analytical methods and suggests solutions to practical problems of town planning in industrial districts.

A01-0061
DeChiara, J. and Koppelman, L. Planning Design Criteria. Van Nostrand Reinhold, 1969. 250 pp. $25.00
This handbook of standards of city planning and urban design brings together information found previously in scattered literature. Topics include master plans, land use, housing, air and rail transportation, parks and recreation, new towns, zoning and codes.

A01-0062
DeChiara, J. and Koppelman, L. Planning Design Criteria. Reinhold Books, 396 pp. $26.50
A graphic reference book of current urban design standards. Coverage includes neighborhood and new town concepts: industrial development and economic base; working with and within special government programs.

A01-0063
deWolfe, I., ed. Civilia: The End of Sub Urban Man. Crane, Russak, 1971. 156 pp. $20.50
This book offers a solution to the urban decay. In it, a group of architects and planners provides an original approach to how a new city of one million inhabitants can be built. They take a site of near-derelict land in the British midlands and, with a series of imaginative photomontages, show how a new city might be created.

A01-0064
Dickey, J., et al. Metropolitan Transportation Planning. McGraw-Hill, 1974. 512 pp. $19.50
This volume emphasizes an interdisciplinary approach to solving metropolitan transportation problems. Impact problems, land use control mechanisms, and different types of transportation services are among the topics covered.

A01-0065
Dietz, A. G. H., et al. Housing in Latin America. MIT Press, 1965. 259 pp. $10.00
This is one in a series of studies undertaken by the Inter-American Program of the Civil Engineering Department at MIT. It is based on the belief that the combined efforts of sociologists, economists, political scientists, architects, engineers, and builders are required to find solutions for the housing problems in Latin America. Problems are examined from the demographic and sociological standpoints as well as the technological aspects of materials, methods of construction, and research.

Planning--Urban and Regional

A01-0066
Downs, A. Federal Housing Subsidies: How are They Working?
    Lexington Books, 1973. 160 pp. $11.00
    The book investigates the effectiveness of federal housing subsidies program and concludes that it is effective, as far as it goes. The author recommends high priority in attacking the social, "nonhousing" aspects of housing.

A01-0067
Doxiadis, C. A. Urban Renewal and the Future of the American
    City. Public Administration Service, 1966. 174 pp. $6.00
    Develops principles and criteria to use in formulating programs and appraising progress. Suggests a policy and program for reconstruction of urban areas to make them dynamic and habitable.

A01-0068
Ducsik, D. W., ed. Shoreline for the Public. A Handbook of Social, Economic, and Legal Considerations Regarding Public Recreational Use of the Nation's Coastal Shoreline. MIT Press, 1974. 280 pp. $12.50
    The Coastal Zone Management Act of 1972 proclaims a national interest in the problem of decreasing public recreational space in the face of mushrooming demand. This study discusses the social significance of the problem, outlines the causes of coastal mismanagement in terms of the organization of economic and political activity, and examines in detail the legal issues pertinent to the formation of public policy.

A01-0069
Eberhard, J. P. The Performance Concept: A Study of its Application to Housing. Vol. 2. National Technical Information Service, 1969. 194 pp. $3.00 Order No. PB-184876
    A National Bureau of Standards report, investigated the hypothesis that, if adequate performance standards for low-cost housing could be developed, and used, cost-reducing innovations would be introduced into the design of low-cost housing. The report finds this hypothesis to be generally correct, but emphasizes the need for the development of the necessary knowledge to implement the performance concept. This volume contains appendices titled as follows: Needs of the rural poor in low cost housing; Housing of Indians on reservations; Conceptual structure of low cost/low income housing; Implementation of the performance concept in regulatory and acceptance systems such as the minimum property standards.

A01-0070
Eckbo, G. The Landscape We See. McGraw-Hill, 1969. 256 pp.
    $16.00
    A reference book covering all the aspects of total environmental design.

A01-0071
Erber, E., ed. Urban Planning in Transition. Viking Press, 352 pp. $4.95
Based on papers commissioned for the American Institute of Planners.

A01-0072
Evans, H., ed. New Towns: The British Experience. Halsted Press, 1972. 208 pp. $11.95
Discusses administrative framework, regional and economic planning, and new town planning. Includes bibliography and statistics on new towns.

A01-0073
Everett, Robinson O. and Johnston, John D., Jr. Housing Series: Law and Contemporary Problems. Oceana, 1968. 384 pp. $14.00
Part I of this symposium deals with perspectives and problems of housing. Part II deals with the federal government's role in housing.

A01-0074
Feagin, J. R., et al. Subsidizing the Poor: An Evaluation of the Boston Housing Authority Program. Lexington Books, 1972. 192 pp. $15.00
Focusing on data from the Boston Housing Authority's low-income rent supplements, the study defines the characteristics of Boston's low-income groups and examines the demography of the people receiving subsidies. Measures the program's social impact and its success in terms of management, personal relations, housing markets, costs and family preferences.

A01-0075
Federal Highway Administration. Coordination of Urban Development and the Planning and Development of Transportation Facilities. U.S. Gov't. Printing Office, 1974. 132 pp. $2.50 Order No. TD 2.2:Ur 1/6 S/N 5001-00076
One problem facing American cities is providing transportation to, from, and within the central business district and the expanding suburban areas. This report summarizes practices used to deal with this problem in England, Scotland, Spain, Switzerland, France, Germany, Denmark, Sweden, Australia, and Canada.

A01-0076
Field, C. G. and Rivkin, S. R. The Building Code Burden. Lexington Books, 1975. 240 pp. $14.50
Providing a decent home for every American family is a vital national goal, but some of the most powerful tools to meet this need--technological innovation and housing industrialization--are still tangled in restrictive and outdated government regulations. The pattern of building codes and their content act as a burden upon the American consumer and the ability of government to efficiently achieve its housing goals for lower income families.

Planning--Urban and Regional 15

A01-0077
Flawn, P. T. Environmental Geology Conservation, Land-Use Planning, and Resource Management. Harper & Row, 1970. 313 pp. $14.50 ($13.25 to ASPO members)
   Best described as a textbook on geology for planners. Explains the use of geologic maps for planning and methods for incorporating geologic principles into zoning and subdivision regulations.

A01-0078
Frieden, B. J. The Future of Old Neighborhoods: Rebuilding for a Changing Population. MIT Press, 1964. 272 pp. $12.50
   This book challenges current interpretations of declining urban neighborhoods that stress the decay of old communities and the need for extensive clearance and proposes a policy of gradual and continuous rebuilding of the old areas, keeping pace with the abandonment of housing and replacing only surplus houses.

A01-0079
Frieden, B. J. and Kaplan, M. The Politics of Neglect: Urban Air from Model Cities to Revenue Sharing. MIT Press, 1975. 312 pp. $14.95
   The book traces the Model Cities Program from its origins. The authors chart the subsequent inability of both the Johnson and Nixon administrations to implement the program effectively, and the reasons why results failed to measure up to rhetorical goals and early over-optimism. By analyzing the performance of the federal bureaucracy, Congress, and the White House, this study explains why officials in Washington were unable to meet the priorities of the cities and why the cities in turn were unable to use federal resources to make significant improvements in their poverty neighborhoods.

A01-0080
Friedmann, J. and Alonso, W., eds. Regional Development and Planning: A Reader. MIT Press, 1964. 772 pp. $15.00
   A large number of topics are considered, including: location theory, theory of spatial organization, the role of resources and migration in regional development, the role of the city in regional development, problems of peripheral rural areas, the definition of regions, the concept of planning regions, objectives and measures of regional development, regional investment criteria, and institutional aspects of regional development planning.

A01-0081
Friendly, P. H., et al. Benefit-Cost Applications in Urban Renewal: A Feasibility Study. National Technical Information Service, 1969. 279 pp. $3.00 Order No. PB-182969
   The study assesses the applicability of benefit-cost analysis to three specific types of urban-renewal decisions, identifies the factors and variables that must be included in a benefit-cost analysis of renewal undertakings, and examines the extent to which these variables and factors can be accurately measured or predicted.

A01-0082
Gallion, A. B. and Eisner, S. *The Urban Pattern: City Planning and Design.* Van Nostrand Reinhold, 1963. 435 pp. $11.95

A01-0083
Geddes, P. *City Development: A Study of Parks, Gardens, and Culture-Institutes.* Rutgers Univ. Press, 1973. 268 pp. $22.50
   A complete study of the past, present, and possible future of the small Scottish manufacturing town of Dunfermline, incorporating every aspect of town life from its parks to its sewers, from historic preservation to new construction. Dunfermline was the birthplace of Andrew Carnegie, and, upon his death in 1903, he bequeathed half a million pounds for city improvements. Patrick Geddes' study is the outgrowth of this endowment.

A01-0084
George, C. J. and McKinley, D. *Urban Ecology: In Search of an Asphalt Rose.* McGraw-Hill, 1974. 168 pp. $5.50 (paperback $3.95)
   This book emphasizes the application of ecological principles to human affairs. Beginning with a definition of the city and some fundamental ecological principles, it proceeds to discuss the development of cities, air and noise pollution, city planning, density and sanity.

A01-0085
Gerson, W. *Patterns of Urban Living.* Univ. of Toronto Press, 1970. 113 pp. $10.00
   This study attempts to determine the types of people, places and spaces that should be near one's home to provide the best possible life in modern urban society.

A01-0086
Gibberd, F. *Town Design.* Praeger, 1967. 372 pp. $22.50
   This study of urban planning deals with town design as an art, with architecture, landscape, and road design as integral parts that blend to form the "urban scene." Using examples from the Renaissance to the present, the author treats civic spaces, shopping centers, industrial buildings and estates, and residential neighborhoods.

A01-0087
Godwin, G. *Directory of Official Architecture & Planning.* George Godwin Ltd. £5.00
   Senior officials concerned with architecture and planning matters are given for each of the new counties and districts, and maps and indexes show how the old local government areas have been absorbed into the new.

A01-0088
Goetze, R. *Building Neighborhood Confidence: A Humanistic*

Planning--Urban and Regional 17

Strategy for Urban Housing. Ballinger Publishing Co.,
1976. 128 pp. $13.50
As availability of federal subsidies winds down and America
rediscovers the values of its existing neighborhoods, this book develops new approaches for neighborhood revitalization and evaluates
their application in Boston. By focusing on neighborhood dynamics
and observing how many communities regenerate, policy makers
can often tap the resources within many neighborhoods. This
volume emphasizes the expanded opportunities open to central cities
under the new community development block grants.

A01-0089
Golany, G. New Towns Planning and Development: A Worldwide
Bibliography. Urban Land Institute, 1973. 256 pp.
$10.00 ($8.00 for ULI members)
A reference work on new towns. Its 4,551 entries cover
all aspects of new towns planning and development throughout the
world since the beginning of the 20th century. The book's classification system makes it a useful working document. Includes a
detailed table of contents, a separate section for each country, and
identification of 600 sources as the most informative entries. A
chapter offers sources that concern future urban patterns and possible effects of new towns on future environments.

A01-0090
Golany, G., ed. Strategy for New Community Development in the
United States. John Wiley, 1975. 293 pp. $28.00
Examines and evaluates the recent New Communities movement in the United States; the need for delineation of national and
state policy on future urbanization of the U.S.; the need for a new
urban pattern; new communities as a response to contemporary
problems; location, physical structure, social, economic, financial,
legislative, and other aspects of new communities as a unique feature of the United States.

A01-0091
Goldstein, S., et al. Residential Mobility, Migration, and Metropolitan Change. Ballinger Publishing Co., 1974. 224 pp.
$15.00
Focusing on Rhode Island, the authors compare area data
and individual residence histories to ascertain the social, economic,
and demographic correlates of migration, the determinants of mobility, and the extent to which residential satisfaction and other
factors account for changes in area characteristics of various metropolitan sectors. The research findings are used to test a variety
of models designed to explain and predict residential mobility.

A01-0092
Gottmann, J. Megalopolis: The Urbanized Northeastern Seaboard
of the United States. MIT Press, 1964. 810 pp. $4.95
This geographical region is the most active crossroads on
earth. Professor Gottmann's aim is to give the reader a fundamental understanding of its basic problems and the interrelationships

among the social and economic processes at work in this region. He analyzes the complex forces that have created Megalopolis.

A01-0093
Gottschalk, S. S. Communities and Alternatives: An Exploration of the Limits of Planning. John Wiley, 1975. 169 pp. $9.95 (paperback $5.95)
A unique analysis of how a new community is and is not planned. The author presents a systematic conceptual analysis, carefully constructed, which puts to further use the important distinction between formal and communal organizations developed by George Hillery.

A01-0094
Groberg, R. P. Urban Renewal Programs Assisted by Title 1 of the Housing Act of 1949. National Technical Information Service, 1969. 285 pp. $3.00 Order No. PB-185277
Contents include introduction and review of accomplishments, premises and summary of recommendations, federal goals and objectives for urban renewal, forms of federal financial assistance, federal, state, and local agencies concerned with Title 1 programs, data on Title 1 urban renewal and related federally-aided programs.

A01-0095
Group for Environmental Education. The Process of Choice. MIT Press, 1974. 168 pp. $10.00
Five separate booklets relate different phases of the choice and change process; an overall Introduction to the process and its separate parts; What You Want, which is about how to decide what goals to aim for; Your Resources, both natural and human; What You Are Allowed to Do, dealing with such constraints as building and zoning laws; and How You Make Choices, which ties everything together. New concepts are introduced through stories and poems, drawings and photographs, puzzles and cartoons.

A01-0096
Gruen, N. J. and Gruen C. Low and Moderate Income Housing In the Suburbs: An Analysis for the Dayton, Ohio Region. Praeger, 1972. 262 pp. $15.00
Analyzes economic, social, and sociocultural attitudes of those to be affected by implementation of "The Dayton Plan," designed to provide improved suburban housing for low and moderate income groups. Recommends that affected communities be funded for increased service costs to meet the non-housing needs of the new residents.

A01-0097
Haar, C. M. and Iatridis, D. S. Housing the Poor in Suburbia: Public Policy at the Grass Roots. Ballinger Publishing Co., 1976. 458 pp. $14.50
The subject of the conflict of zoning to exclude the poor from the suburbs is examined from the views of public policy, land use patterns, politics, and social aspects.

Planning--Urban and Regional

A01-0098
Hall, P. London 2000. Praeger, 1970. 282 pp. $10.50
This updated edition provides answers to questions that can be asked about every growing city. London is used as a model for future economic growth, traffic conditions and housing requirements.

A01-0099
Hall, P. The World Cities. McGraw-Hill, 1966. 256 pp. $4.95 (paperback $2.45)
A study of seven leading metropolitan cities--their growth and future plans: London, New York, Paris, Tokyo, Moscow, and the great complexes of Holland and Rhine-Ruhr.

A01-0100
Harvey, J. Conservation of Buildings. Univ. of Toronto Press, 1973. 240 pp. $15.00
Mr. Harvey takes the view that human needs are important and that buildings or groups of buildings must be conserved to accommodate some part of their original function, and must also be regarded in the context of their surroundings. He divides his book into three sections: What to Save and How to Save It, Craftsmanship and Materials, and Continuing Conservation.

A01-0101
Heberseimer, L. The Nature of Cities: Its Origin, Growth and Decline. Paul Theobald, 285 pp. $8.75
A discussion of all problems of city planning from early settlements down to the present day industrial city.

A01-0102
Heilbrun, J. Real Estate Taxes and Urban Housing. Columbia Univ. Press, 1966. 195 pp. $8.50
This book provides an analysis of the impact of real estate taxes on the quality of urban housing in an attempt to discover whether real estate tax reform would help reach the goal of adequate housing for every citizen. The author describes the way in which the house-operating firm and the rental-housing industry function, with emphasis on the consumer demand for rental-housing services, and the operating and investment decisions by which building concerns direct the supply of such services.

A01-0103
Heroux, R. L. and Wallace, W. A. Financial Analysis and the New Community Development Process. Praeger, 1973. 184 pp. $13.50
Uses single period linear programming models to define interrelationships and identify social and environmental restraints in the development of "new towns." A multiperiod linear programming model aids in the large-scale land development process.

A01-0104
Holliday, J., ed. City Centre Redevelopment. Halsted Press,

1974.   244 pp. $19.50

This book is a record of changes which have taken place in five city centres since the war, and particularly during the last ten years. The cities, Birmingham, Coventry, Leicester, Liverpool and Newcastle upon Tyne, are sufficiently diverse in their characters and planning to illustrate differences which arise from traditions, fortunes and personalities.

A01-0105
Hoover, E. M.   An Introduction to Regional Economics.   Alfred Knopf, 1971.   416 pp. $10.50

Relates theory to economic problems through examples based, in general, on the American experience. The theme is a concern for the economics of spatial relationships. The book discusses such subjects as locational selection and the behavior of large groups of areas, and such urban problems as downtown obsolescence, transportation, slums, and the financing of public services.

A01-0106
Houghton-Evans, W.   Planning Cities: Legacy and Portent.   Beekman Publishers, 1975. $24.00

Comprehensive survey and critique of town planning, with emphasis on socioeconomic objectives and creative design. Presented in the light of worldwide debate concerning the urban environment and its future.

A01-0107
Housing and Urban Development Dept.   Guidelines for Urban Renewal Land Disposition.   U.S. Gov't. Printing Office, 1975.   458 pp. $5.05   Order No. HH 1.6/3:Ur 1/4 S/N 023-000-00288-4

The purpose of this manual is to provide local urban planning agencies with theories, suggestions, and related data that can help them in the final disposition of land parcels. Each chapter discusses problems that have been known to impede the progress of community projects, and suggests how they may be remedied.

A01-0108
Housing and Urban Development Dept.   Housing in the Seventies.   U.S. Gov't. Printing Office, 1974.   275 pp. $5.10   Order No. HH 1.2:H 81/47 S/N 023-000-00269-8

This book recounts the history of Federal involvement in housing, explains the programs that have evolved, assesses the cost-effectiveness of those programs, describes the housing activities of State and local governments, and outlines patterns of housing production and finance and the structure and technology of the housing industry.

A01-0109
Housing and Urban Development Dept.   1972 HUD Statistical Yearbook.   U.S. Gov't. Printing Office, 1974.   376 pp. $4.20   Order No. HH 1.38:972 S/N 2300-00257

Planning--Urban and Regional

Contains comprehensive data on program and financial operations of the Department of Housing and Urban Development and statistical information related to housing and urban activities.

A01-0110
Housing and Urban Development Dept. A Study of Property Taxes and Urban Blight. Part 1 and Part 2. U.S. Gov't. Printing Office, 1973. 329 pp. $3.45 Order No. HH 1.2:T 19/2 S/N 2300-00239
The findings of this study are based on interviews with 228 property owners in ten cities; Atlanta, Baltimore, Chicago, Detroit, Nashville, Oklahoma City, Philadelphia, Portland, Providence, and San Francisco. The study examines the effects of the property tax within the central cities of each of the ten different metropolitan areas. No comparisons were made with taxation in the surrounding suburban areas.

A01-0111
Housing and Urban Development Dept. A Study of the Effects of Real Estate Property Tax Incentive Programs Upon Property Rehabilitation and New Construction. U.S. Gov't. Printing Office, 1973. 376 pp. $3.70 Order No. HH 1.2:R 22 S/N 2300-00238
The findings of this study are based on interviews with 164 property owners and data on 398 of their properties, located in nine cities (New York, Newark, Boston, St. Louis, Honolulu, Pittsburgh, Southfield, Mi., Arlington, Va., and Fairhope, Al.).

A01-0112
Howard, E. Garden Cities of To-Morrow. MIT Press, 1965. 168 pp. $2.45
The book led directly to two experiments in town-founding which have had profound influence on practical urban development throughout the world.

A01-0113
Hughes, J. W., ed. New Dimensions of Urban Planning: Growth Controls. Center for Urban Policy Research, 246 pp. $8.95
The papers gathered in this volume--from two conferences on urban planning--outline the dimensions of the emerging growth control movement. A case study of the basic forces of suburban expansion explores the rationale for using a regional approach to determine housing needs. Current efforts to control growth are viewed through a series of somewhat conflicting court decisions. Other discussions focus on assumptions about the cost of providing community services to new residents and the impact of housing modes, price, and supply.

A01-0114
Hutchinson, B. Principles of Urban Transport Planning. McGraw-Hill, 1974. 500 pp. $19.50
A systems framework is used to provide an analysis of the

strategic processes involved in urban transport planning. Trip
distribution analysis, traffic assignment analysis, and various concepts of urban structure useful for generating alternative arrangement of land use and transport networks are among the topics
covered.

A01-0115
Ingram, G. K., et al. The Detroit Prototype of the NBER Urban
 Simulation Model. Columbia Univ. Press, 1973. 233 pp.
 $12.50
 This urban simulation model draws heavily on economic
theory and depicts the collective effects that utility-maximizing
households and profit-maximizing firms have upon urban housing
markets over time.

A01-0116
Jacobs, J. The Death and Life of Great American Cities. Random House, 480 pp. $1.95

A01-0117
James, F. J. and Hughes, J. W. Economic Growth and Residential Patterns: A Methodological Investigation. Center for Urban Policy Research, 216 pp. $8.00
 The book analyzes various economic models determining
job/residential growth patterns.

A01-0118
Jedd, D. R. and Mendelson, R. E. The Politics of Urban Planning: The East St. Louis Experience. Univ. of Illinois
 Press, 1973. $7.95
 From 1960 to 1969, more than 125 studies were made on
problems in East St. Louis. Yet few signs of physical improvement or social renewal appeared. In an attempt to account for
this wide disparity between planning and implementation, the authors scrutinize politics within the planning profession. They find
that urban planning in the United States has been sustained as much
to provide professional employment and security as to reach social
objectives.

A01-0119
Jensen, R. Cities of Vision. John Wiley, 1974. 382 pp.
 $18.50
 This book is concerned with the creation or regeneration of
the cities of the future. Chapters on city planning and design,
urban structure, economics and social factors, land use, and urban renewal are interwoven with the politics and programs of urban planning.

A01-0120
Kaplan, H. Urban Renewal Politics: Slum Clearance in Newark.
 Columbia Univ. Press, 1963. 219 pp. $7.50
 Between 1949 and 1960, Newark, New Jersey launched nine
slum clearance and urban renewal projects. In this book, Dr.

Kaplan describes the problems and successes of the city's housing authority and its relationships with the various groups whose support it needed.

A01-0121
Kaplan, M.   Urban Planning in the 1960s: A Design for Irrelevancy. Praeger, 1973. 144 pp. $14.00 (paperback available from MIT Press $2.95)
A critical evaluation of the functions and activities of city planners during the 1960s, including an appraisal of "new towns" in terms of their contribution to residents of suburban areas, rural enclaves, and cities. Recommends a set of ground rules for planning and an agenda for action in the 1970s.

A01-0122
Keeble, L.   Principles and Practice of Town and Country Planning. American Society of Planning Officials, 1969. 415 pp. $15.00 (for ASPO members $14.00)
Outlines principles and standards for land use and circulation planning, based on admixture of scientific determinants and humanistic values.

A01-0123
Koppelman, L. E., et al.   The Urban Sea: Long Island Sound. Praeger, 1974. 180 pp. $16.50
Outlines planning and administrative mechanisms for dealing with the new problems arising from the fact that more than 90% of the U.S. population now lives in concentrated urban centers adjacent to the Atlantic, Pacific, and Great Lakes shores.

A01-0124
Lansing, J. B., et al.   New Homes and Poor People: A Study of Chains of Moves.   Univ. of Michigan--Inst. for Social Research, 1969.   136 pp. $5.00
The construction of new, relatively expensive housing has a chain reaction which trickles down to the lower income groups. The methodology employed and the nationwide sample that comprises 3,000 interviews provide the authors with a reliable basis for evaluating this phenomenon, assessing the working of the housing market, and analyzing the extent to which the critical needs of the poor are met.

A01-0125
Le Corbusier.   The Athens Charter.   Viking Press, 1973.   128 pp. $10.00
This is the first English translation of this plea for urban reform--for architects, planners, technologists, and government officials to work together for a more humane urban environment.

A01-0126
Lent, S.   Plumbing Code of New York City--Guide and Interpretation.   Industrial Press, 1971.   239 pp. $14.00
Clarifies and interprets a Code that even the most

experienced engineer finds difficult to understand in its original form. The book is useful in all geographical areas where inspection delays and construction problems can result from a lack of correct information.

A01-0127
Lenz-Romeiss, F. The City: New Town or Home Town? Praeger, 1973. 176 pp. $8.95
Presents proposals and guidelines for approach to contemporary problems. Discusses the dehumanizing effect modern cities have as the feeling of belonging to a community is destroyed. Examines the medieval town in relation to the needs of contemporary urban planning.

A01-0128
Leonard, C. N. Cities of the Industrial Age: Dream and Reality in Historical Perspective. Schenkman, 1973. 250 pp. $8.95 (paperback $3.95)
The focus of this study is on the nineteenth century, and a comparison of that formative period for the development of urban concepts which have influenced twentieth-century urban planning.

A01-0129
Levin, M. R., et al. New Approaches to State Land Use Policies. Lexington Books, 1974. 160 pp. $12.50
This study examines the problem of the peripheral land use problems related to Planned Unit Development and focuses on existing New Jersey land use legislation. It concludes by proposing an administrative approach tailored to New Jersey's special needs and also offers proposed "development rights" legislation, a technique whereby states can encourage municipalities to set aside large amounts of permanent open space without requiring governmental expenditure for costly land use acquisition.

A01-0130
Lewis, D., ed. The Growth of Cities. John Wiley, 1971. 256 pp. $24.95
Illustrated collection of articles examining the organic growth and change of cities and showing their modification as a result of the impact of identifiable forces--economics, social change, technological change, etc. Includes articles on Greek, ancient Saharan, and medieval cities. Describes the suburban phenomenon and the problem of the in-migration of rural peoples (South American barrios, the inner cities of major U.S. cities) and then examines planning and architectural solutions.

A01-0131
Lewis, D., ed. Urban Structure. John Wiley, 1968. 283 pp. $18.50
Chapters include town design, planning, concern for pedestrians, form and the work of a number of architects and urban planners.

Planning--Urban and Regional

A01-0132
Liblit, J. *Housing--The Cooperative Way.* Twayne Publishers, 1964. 300 pp. $6.00

A compendium of readings for anyone interested in cooperatives. Reviews the background of cooperative housing, how it works, and what is needed to make it work better. Of help to those concerned with housing, slum clearance and urban renewal.

A01-0133
Linowes, R. R. and Allensworth, D. T. *The Politics of Land Use: Planning, Zoning, and the Private Developer.* Praeger, 1973. 178 pp. $13.50

Shows how current master planning processes most often encourage economic segregation and prevent needed commercial, industrial, and higher-density residential development. Proposes sweeping changes in government powers over land use.

A01-0134
Listokin, D. *The Dynamics of Housing Rehabilitation: Macro and Micro Analyses.* Center for Urban Policy Research, 240 pp. $10.00

Rehabilitation of the existent housing stock in urban areas is a housing strategy that could alleviate the housing shortage in our cities. In practice, however, rehabilitation has proven disappointing to its sponsors and to the people it has housed. This study analyzes the reasons for past failures and proposes now strategies for the implementation of a successful housing rehabilitation program. Among the topics discussed are financing, acquisition of properties, maintenance, management, housing code enforcement, tax incentives, the rehabilitation technology, means of obtaining local community support.

A01-0135
Listokin, D. *Fair Share Housing Allocation.* Center for Urban Policy Research, 180 pp. $8.00

Many courts and state and local governments have turned to "fair share" plans to determine the equitable distribution of housing in a region according to such criteria as broadening the socioeconomic mix of communities and protecting the environment. This monograph examines the emerging fair share strategy, discusses its theoretical impact, and analyzes its track record to date.

A01-0136
MacFadyen, D. *Sir Ebenezer Howard and the Town Planning Movement.* MIT Press, 216 pp. $8.95

Sir Ebenezer Howard is recognized as the father of the new towns (or garden cities) movement. This biography, originally published in 1933, is being reissued now to serve the revival of interest in new towns. The book presents the personal aspects of Howard's life and a detailed account of the planning and building of the new towns of Letchworth and Welwyn.

A01-0137
McFarland, M. C. and Vivrett, W. K. Residential Rehabilitation.
    National Technical Information Service, 1968. 323 pp.
    $3.00 Order No. PB-176246
        An attempt to trace the development of the concept of residential rehabilitation over the past 30 years, to report the progress which has been made, to identify the current conceptions and their deficiencies, and to suggest the shape of the future.

A01-0138
McKown, R. Comprehensive Guide to Town Planning Law and Procedures. George Godwin Ltd., 1973. 160 pp. £5.00
        With more than a dozen Acts of Parliament specifically concerned with one aspect or another of town and country planning, as well as statutory orders and regulations, there is now a mass of interrelated legislation which is often difficult for the layman to follow. This book offers guidance to this subject by bringing together the key legal requirements from all relevant legislation and by classifying these according to subject.

A01-0139
McLoughlin, J. B. Urban and Regional Planning: A Systems Approach. Praeger, 1970. 340 pp. $9.00
        This book provides a practical framework for planning by applying theories of systems and control, which have proved successful for industry and defense needs, to the creation of human environments. The author shows how these theories can relate advances in geography, economics, and the social sciences to each other and to more traditional planning techniques. He also describes how planners can simulate the behavior of cities through cybernetics.

A01-0140
McNulty, T. F. and Stevens, M. O. World of Variation. Braziller, 1971. 147 pp. $6.95 (paperback $2.95)
        Using models and illustrations, new principles of physical and social organization are offered as alternatives to more traditional planning processes.

A01-0141
Masotti, L. H. and Hadden, J. K., eds. The Urbanization of the Suburbs. Sage Publications, 1973. 640 pp. $20.00
        Suburbia is undergoing a significant transition from its traditional role as dependent "urban fringe" to independent "neo-city." This volume confronts the challenge to the cities presented by unplanned and ungoverned suburban development.

A01-0142
Masser, I. Analytical Models for Urban and Regional Planning.
    John Wiley, 1972. 157 pp. $12.50
        Presents analytical models and techniques that have been developed for population growth, the prospects of economic activity,

Planning--Urban and Regional

and the spatial organization of population and economic activity in cities and towns. Gives computer programs.

A01-0143
Mayer, A. The Urgent Future. McGraw-Hill, 1967. 184 pp. $16.50
   Covers the skills and powers involved in the design and creation of buildings, cities, and regions.

A01-0144
Meier, R. L. Planning for an Urban World: The Design of Resource-Conserving Cities. MIT Press, 1975. 352 pp. $27.50
   The author has chosen a biological analogy as the most appropriate way to describe the dynamics of the worldwide growth of urban concentrations during a period marked by inevitably decreasing resources. Cities are modeled as living systems in varying states of internal health and disease, in shifting ecological balance with other populations at the rural, urban, and national scale.

A01-0145
Meyerson, M., et al. Housing, People and Cities. McGraw-Hill, 1962. 386 pp. $9.95
   This study combines the findings of outstanding experts on the crucial problems of urban housing.

A01-0146
Mields, H. Federally Assisted New Communities: New Dimensions in Urban Development. Urban Land Institute, 1973. 288 pp. $16.00 (for ULI members, $12.00)
   Provides information on new towns, focusing on government assistance--a guide through the mazes of local, state, and federal rules and regulations. This publication includes: Title IV and Title VII program provisions; complete application process; detailed descriptions of new towns in the USA; new towns economics and financing; HUD suggested format for Title VII cash flow analysis; sample agreements, and appendix.

A01-0147
Mills, E. S. Urban Economics. Scott, Foresman, 1972. 278 pp.
   Begins with theoretical foundations and historic trends of urban areas and covers urban problems of poverty, slum housing, urban renewal, transportation and pollution.

A01-0148
Morris, A. E. J. History of Urban Form: Prehistory to the Renaissance. George Godwin Ltd. or John Wiley, 1972. 280 pp. $19.75
   Illustrated with over 300 photographs, maps and plans, this volume traces the development of the form of cities from the first ancient settlements to the full close of the Renaissance in Europe in the early nineteenth century. For the USA, the Civil War is taken as a comparable date. The author has included many

significant historical plans and has attempted to establish examples of detail planning in their contemporary urban contexts.

A01-0149
Morris, P. State Housing Finance Agencies: An Entrepreneurial Approach to Subsidized Housing. Lexington Books, 1974. 179 pp. $13.50
This book provides a comprehensive review of the activities of state housing finance agencies and analyzes the legal and constitutional impediments to more creative programming to meet the needs of low and moderate income families. It presents a series of housing programs and designs, intended as models for thought rather than as blueprints, that effectively utilize underlying real estate values as subsidies to supplement federal and state assistance.

A01-0150
Multiple Dwelling Law of New York State. Gould Publications, 1972. 417 pp. $7.50
The book contains abstracts of rules, regulations, etc., applicable to the maintenance and alteration of structures and buildings in the City of New York.

A01-0151
Murphy, R. E. The Central Business District. Aldine Publishing Co., 1972. 193 pp. $8.95
An examination of the central business district, this volume describes its general characteristics and recent trends. Specific types of businesses are examined for their movements away from the central business districts.

A01-0152
Muth, R. F. Cities and Housing. Univ. of Chicago Press, 1969. 356 pp. $10.75
The author presents an analysis of the price system in urban housing and residential land markets, and investigates alternative theories concerning the determinants of housing quality in inner-city areas. He examines the major arguments now current concerning slums and their creation, and what the various government bodies can do about them.

A01-0153
Newman, O. Defensible Space: People and Design in the Violent City. Architectural Press, 1973. 260 pp. £4.95
The author believes that law enforcement, by itself, is not the answer to violence in cities. He has demonstrated through studies of neighboring buildings and their inhabitants that more imaginative design configurations in both the public and private housing sectors can prevent the criminal encounter.

A01-0154
Nourse, H. O. The Effects of Public Policy on Housing Markets. Lexington Books, 1973. 140 pp. $11.00

In this overview of contemporary housing policy, the author analyzes factors which combine to create housing problems. The papers examine housing markets and the author designs a real property value index to evaluate the standards presently used for property tax policy formulation.

A01-0155
Osborn, F. J. Green-Belt Cities. Schocken Books, 1971. 203 pp. $2.45

A summary of the essential ideas behind the new towns movement. This book is considered a classic in new town planning.

A01-0156
Osborn, F. J. and Whittick, A. The New Towns: The Answer to Megalopolis. MIT Press, 1970. 456 pp. $25.00

The work is divided into two parts: the first, by Osborn, concerns the background of the new towns movement and the questions of policy and planning that apply to new towns generally. Part Two, by Whittick, considers various new towns one by one. Descriptions are given of the first twenty-three of the thirty towns authorized in Great Britain up to 1968.

A01-0157
Page, A. N. and Seyfried, W. R. Urban Analysis. Scott, Foresman, 1970. 437 pp.

This survey of the current problems in urban development and housing brings together thirty-six articles that focus on various aspects of the role of housing in the urban scene. While the articles emphasize economic factors, they also give insight into the social aspects of the urban environment.

A01-0158
Papageorgiu, A. Integration of the Past: Preservation in City Planning. Praeger, 1971. 350 pp. $29.50

This book is a study of the preservation of historic city centers. The author emphasizes the importance of avoiding a museum approach to preserving historic buildings and areas; he proposes instead a search for new architectural values that can be combined with the traditional structures of an area. Among the problems he treats are the creation of new roads, underground railways, highrise buildings, and even the changes wrought in cities by the growth of new suburbs. He also shows the techniques used by a number of European cities.

A01-0159
Pass, D. Vällingby and Farsta--From Idea to Reality: The New Community Development Process in Stockholm. MIT Press, 1972. 200 pp. $15.00

Popular and professional literature through the world has spotlighted Stockholm's accomplishments in solving its urban problems. Particular attention has been given to Vällingby and Farsta, two of the more than twenty suburban communities developed by the city since the Second World War.

A01-0160
Passonneau, J. R. and Wurman, R. S. Urban Atlas: 20 American Cities. MIT Press, 1967. 160 pp. $25.00

This book provides an approach to structuring and representation of information in graphic form. Ten plates each are provided for the following cities: New York, Chicago, Detroit, San Francisco, and Los Angeles. Five plates each for the remaining fifteen cities. These provide information on population density; residential, industrial-commercial, park, and institutional land usages; income per residential area; condition of housing vs. income; population density vs. income. On several pages variables of all 20 cities are displayed, allowing for easy comparison.

A01-0161
Perloff, H. S. and Sandberg, N. C., eds. New Towns: Why-- And for Whom? Praeger, 1973. 270 pp. $16.50

Edited papers by participants in the New Towns Symposium held at UCLA, June 1972. Explores problems and opportunities inherent in new town development and its relationship to a national urban growth policy. Discusses new towns in Britain, Sweden, Holland, Canada, Israel, and the U.S.

A01-0162
Perloff, H. S., et al. Modernizing the Central City: New Towns Intown ... and Beyond. Ballinger Publishing Co., 1976. 464 pp. $16.50

Shortly after the New Town Intown program was launched in 1970, an interdisciplinary research team, headed by Dean Harvey S. Perloff undertook early monitoring of this latest federal effort to help revitalize central cities. After surveying the current status of the few New Town Intown projects launched and planned, their report discusses what can be learned from urban renewal and model cities experience. The authors then explore the benefits of a truly programmatic approach to central city modernization.

A01-0163
Peterson, G. E., et al. Property Taxes, Housing, and the Cities. Lexington Books, 1973. 160 pp. $13.00

The property tax, state the authors of this study, is not only unfair as a system, it is also unfairly administered. This critical appraisal of property taxation analyzes the effects of urban blight on America's cities and the effect of the property tax on urban housing.

A01-0164
Pitt, P. and Dufton, J. The Guide to The London Building (Constructional) Bylaws 1972 and Building Acts. Architectural Press, 1973. 88 pp. £2.75

The London Building Acts and Bylaws regulate all building work in Inner London. Apart from governing a substantial slice

Planning—Urban and Regional

of the total construction in this country, they are also taken as a model by other municipalities in Britain and the Commonwealth.

A01-0165
Pratt Guide to Planning and Renewal for New Yorkers. Harper and Row, 504 pp. $15.00 (paperback $5.95)
This is a guide to planning and operations in the City of New York. Covers city-wide planning; special area programs; regional, state and federal actions affecting the city; comprehensive plans; capital and expense budgets; site selection and project initiation; zoning; landmarks preservation.

A01-0166
Prichard, R. M. Housing and the Spatial Structure of the City: Residential Mobility and the Housing Market in an English City Since the Industrial Revolution. Cambridge Univ. Press, 1976. 408 pp. $19.95
One English city, Leicester, is used as a case study to investigate the manner in which the provision and operation of the housing market in Britain has influenced the spatial evolution of urban areas. Specifically, the study uses the pattern of residential mobility and intra-urban migration to demonstrate the way in which changes in the housing market have produced changes in the social geography of cities in general.

A01-0167
Pushkarev, B. S. and Zupan, J. M. Urban Space for Pedestrians: A Quantitative Approach. MIT Press, 1975. 208 pp. $17.50
The book opens with a chapter on the various forms of urban space and then defines the issue of pedestrian space in particular, relating walkway space to buildings. The second chapter presents two analytical models, as well as a discussion of the costs of walking and the prices paid to avoid walking. This is followed by an examination of pedestrian space requirements. A final chapter presents design implications and includes recommendations on side-walk width, autofree zones, grade separation, underground spaces, and amenities.

A01-0168
Pynoos, J., et al., eds. Housing Urban America. Aldine Publishing, 630 pp. $25.00
Fifty-one articles focusing on the multidimensional aspects of housing. The book analyzes housing markets, the production of housing, and the distribution of its services over the population.

A01-0169
Redstone, L. G. The New Downtowns: Rebuilding Business Districts. McGraw-Hill, 1976. 352 pp. $21.95
Fifty case histories are given of actual renewal projects which have been completed or are now underway in the U.S., Canada, and elsewhere. These case histories detail the financial and

political backgrounds of the projects as well as their architectural design concepts and execution.

A01-0170
Regional Plan Association. Second Regional Plan Publications.
MIT Press, $10.00 each; $55.00 for all six.

The Lower Hudson. The potential of the Hudson River Valley below the George Washington Bridge as a public amenity--one example of how to meet the conflict between preserving nature and the growth of a metropolis. (1966, 78 pp.) The Region's Growth. Projections of jobs by type, population, households and income for the New York Metropolitan Region, 1965-2000, with a section on world urbanization and the Atlantic Urban Region. (1967, 143 pp.) Public Participation in Regional Planning. The importance and difficulties of involving the public in planning, and results of RPA's pioneering 1963 Goals Project. (1967, 72 pp.) Waste Management. How to organize a metropolitan area for efficient waste management: what information is needed, where controls can be applied, how to project waste generation of the future and how the pattern of urbanization affects waste generation and handling. (1968, 106 pp.) Jamaica Center. Prototype study of a major urban center: the possibilities, design, transportation and process of developing Jamaica, Queens, with a summary of arguments for large centers, particularly in the old Core of the Region. (1968, 72 pp.) Public Services in Older Cities. Proposes a fiscal strategy by which the nation can meet its responsibilities to the poor and the old cities in our metropolitan Region and elsewhere can adequately finance essential community improvement and housekeeping programs. (1968, 56 pp.)

A01-0171
Reps, J. W. The Making of Urban America: A History of City Planning in the United States. Princeton Univ. Press, 1965. 574 pp. $35.00 (for ASPO members, $30.00) (acquire from American Society of Planning Officials)

Basic reference on the subject. Traces sources of ideas and assesses factors, events and influences affecting city form from the beginning of European settlement to the first World War.

A01-0172
Reps, J. W. Monumental Washington: The Planning and Development of the Capital Center. Princeton Univ. Press, 1967. 200 pp. $17.50

In this illustrated history of the nation's capital, Mr. Reps' emphasis is on the plan prepared for the central portion of Washington in 1901 and its subsequent development. He shows that the grand design Major L'Enfant proposed was justified by history, for it was upon L'Enfant's plan that the Senate Park Commission based its proposals for future development. How the Park Commission set about its work, major changes in the city center from the past to the present, and the influences of the Washington plan on city planning elsewhere are also recorded.

A01-0173
Richardson, D. K. The Cost of Environmental Protection: Regulating Housing Development in the Coastal Zone. Center for Urban Policy Research, 1975. 250 pp. $10.00

This monograph presents the cost of all regulatory requirements for housing developments in an environmentally sensitive area--the Coastal Zone. The author evaluates the implementation impact on local regulatory procedures and consumer housing costs of an interim land use permit system instituted by a state department of environmental protection in compliance with federally sponsored guidelines. The regulatory scenarios, based on a detailed analysis of 21 single and multifamily developments in New Jersey, are directly applicable to all states within the Coastal Zone.

A01-0174
Richardson, H. W., et al. Housing and Urban Spatial Structure. Lexington Books, 1975. 208 pp. $17.50

An analysis of the spatial structure of residential property values and a test of theories of residential location and of urban spatial structure. A product of a four-year research study on residential property values in a major British city, this study also looks at intra-urban changes in the distribution of social classes.

A01-0175
Ricks, R. B., ed. National Housing Models. Lexington Books, 1973. 208 pp. $12.50

This comprehensive study brings together current work in the development of housing-related models. Econometric analyses of residential housing and construction include models developed for housing, mortgages and housing starts.

A01-0176
Rodwin, L. Housing and Economic Progress: A Study of the Housing Experiences of Boston's Middle-Income Families. MIT Press, 1961. 228 pp. $10.00

A statistical study of middle-income housing conditions in Boston in the past century, including expenditure patterns, rent levels, standards, land-use patterns, middle-income housing movements, tenure issues, and the theory of residential growth and structure. Published for the Joint Center for Urban Studies.

A01-0177
Rodwin, L., ed. The Future Metropolis. Braziller, $6.00

City planners and social scientists deal with the phenomenon of the metropolis and offer plausible aspirations for the future.

A01-0178
Rose, A. Regent Park; A Study in Slum Clearance. Univ. of Toronto Press, 1958. 242 pp. $7.50

R. P. Stands as Canada's first extensive experiment in slum-clearance and urban re-development. Within six blocks houses and apartments were to be erected and leased at rents relative to the incomes of the tenant families. The record of this experience is given here.

A01-0179
Rose, J. G. Landlords and Tenants: A Complete Guide to the Residential Rental Relationship. Center for Urban Policy Research, 288 pp. $12.95

This book describes in detail the patterns by which myriad inequities between landlord and tenants have been woven into the law, and suggests some remedies. It provides, in plain language, a comprehensive and up-to-date manual of the rights and liabilities of landlords and tenants. A basic reference work for anyone concerned with landlord-tenant relations.

A01-0180
Rose, J. G., ed. The Transfer of Development Rights: A New Technique of Land Use Regulation. Center for Urban Policy Research, 342 pp. $12.95

The author has assembled material that explains this imaginative new technique of land use control and shows how it may be used to regulate the location and timing of community growth, preserve open space and historic landmarks, encourage construction of moderate and low income housing, and help eliminate the inequities of windfalls and wipeouts.

A01-0181
Rothblatt, D. N. Regional Planning: The Appalachian Experience. Lexington Books, 1971. 304 pp. $15.00

Examines the establishment of the semi-autonomous Appalachian Regional Commission. Traces the evolution of national policy which culminated in setting up the Commission, analyzes its behavior as a planning and allocating institution and evaluates its economic and political impact on the region.

A01-0182
Rubinowitz, L. S. Low-Income Housing: Suburban Strategies. Ballinger Publishing Co., 1974. 192 pp. $11.50

Sets forth the need for solutions to metropolitan housing crisis in terms of racial and economic residential patterns, availability of jobs and housing, land, and equal educational opportunity and housing. The author analyzes the various suburban devices for keeping low income housing out. He studies the federal, state and local governments potential and actual role in open suburban communities to low and moderate-income housing.

A01-0183
Saalman, H. Haussmann: Paris Transformed. Braziller, 128 pp. $5.95 (paperback $2.95)

The epic transformation of nineteenth-century Paris by Haussmann and Napoleon III had an impact mirrored in cities throughout the world. Yet the rebuilding of Paris has been a target for contemporary urban critics. By surveying the physical growth of Paris through the ages, Howard Saalman is able to present a fresh view defending Haussmann's accomplishments.

A01-0184
Sagalyn, L. B. and Sternlieb, G. Zoning and Housing Costs: The Impact of Land-Use Controls on Housing Price. Center for Urban Policy Research, 132 pp. $8.00

This study is an analysis of lot size and other zoning and developmental requirements in terms of their impact on the selling price of single-family tract homes. Also built into the analysis are such factors as the scale of the developer's operation and some measurement of the socioeconomic standing of the municipalities in which developments are occurring.

A01-0185
Sanoff, H. and Cohn, S. EDRA I: Proceedings of the 1st Annual Environmental Design Research Association Conference. John Wiley, 1975. 359 pp. $16.00

The major focus of the conference was the discussion of developing models and methods towards a framework of coherence and definable structure of environmental design. The papers are concerned mainly with the contributions of scientific disciplines towards the creation of improved methods of problem solving as well as understanding the nature of human responses to the environment.

A01-0186
Schafer, R. The Suburbanization of Multifamily Housing. Lexington Books, 1974. 176 pp. $14.00

Schafer examines within the context of theories of residential location the recent trend towards apartments in the suburbs, including the outer suburbs. Through his empirical analysis, Schafer shows that changes in the distribution of jobs within the metropolitan area and changes in the age structure of the population have played an important role in the suburbanization of apartments.

A01-1087
Schmid, A. A. Converting Land From Rural to Urban Uses. Johns Hopkins Univ. Press, 1968. 103 pp. $4.00

The complicated and expensive conversion process is not as yet well understood. Greater knowledge of the processes also would be useful in making public decisions on zoning, taxation, public investments, transport systems, new towns, and related questions. Dr. Schmid analyzes the relevant data on conversion of rural land to urbanuses and points out what further research is needed.

A01-0188
Schwartz, H. and Abeles, P. Planning for the Lower East Side. Praeger, 1973. 225 pp. $15.00

This study combines a profile of New York City's Lower East Side with recommendations for demography, housing, transportation, environmental protection, health, education, public services, jobs, and recreation. This revision of a study prepared for the New York City Planning Commission and Housing and Development Administration in 1971 includes new material focusing on the applicability to other cities.

A01-0189
Scott, A. J. Cities and Towns. Planning. Mathematical Models. Wittenborn, 1971. 204 pp. $11.25
Planning in cities and towns, with mathematical models.

A01-0190
Seeley, I. H. Planned Expansion of Country Towns. George Godwin Ltd., 1968. 285 pp. £3.25
An illustrated account of development work in England following the passing of the Town Development Act of 1952. Each facet of town development work is studied, from initial site selection and preparation of schemes to the problems of implementation --constructional, financial and sociological.

A01-0191
Shomon, J. J. Open Land for Urban America: Acquisition, Safekeeping, and Use. Johns Hopkins Univ. Press, 1971. 176 pp. $2.50
The author exposes the poor conditions of urban planning in America and suggests many ways by which the situation could be improved.

A01-0192
Siegan, B. H. Land Use and Real Estate. Lexington Books, 1972. 273 pp. $10.00
This study of zoning in the U.S. examines current exclusionary practices. Reports are presented on the effects of zoning, alternative solutions to land use problems, and recent court cases. As the study demonstrates, zoning regulations act to limit housing production and deny to many the opportunity to improve present housing conditions.

A01-0193
Siegan, B. H. Land Use Without Zoning. Lexington Books, 1972. 273 pp. $11.00
This thoroughly documented study by a noted authority of zoning in the U.S. examines current exclusionary practices and finds their effects destructive. Detailed reports are presented on the effects of zoning, alternative solutions to land use problems, and recent court cases.

A01-0194
Sixta, G. and Parr, A. L. Urban Structure. Wittenborn, 1970. 142 pp. $12.00
A study of long range policies which affect the physical structure of an urban area was undertaken by the Planning Department of the District of Burnaby, B.C., Canada. Its purpose was to establish urban design guidelines.

Planning--Urban and Regional 37

A01-0195
Smithson, A. and Smithson, P. Urban Structuring: Studies of Alison and Peter Smithson. Van Nostrand Reinhold, 1967. 96 pp. $5.50 (paperback $2.45)
Urban innovation and professional dialogue made available to architects and urban planners.

A01-0196
Solomon, A. P. Housing the Urban Poor: A Critical Analysis of Federal Housing Policy. MIT Press, 224 pp. $12.00
This book presents a comparative evaluation of alternative strategies for housing the urban poor. The author proposes that government policy emphasize the use and upgrading of the existing housing stock. This would be accomplished by a combination of direct rent subsidies and the leasing of private dwellings, accompanied by efforts to remove housing and capital market barriers.

A01-0197
Spiegel, E. New Towns in Israel: Urban and Regional Planning and Development. Praeger, 1967. 192 pp. $17.50
This book is a result of a research program set up to determine if the Israeli experience could serve as a model for underdeveloped countries. Its conclusions are that Israel is not so much a model as a pioneer and that its experience and knowledge merit careful attention. It discusses population, land use, construction, financing, and analyzes eight new towns in detail.

A01-0198
Spreiregen, P. D., ed. The Modern Metropolis: Its Origins, Growth, Characteristics, and Planning. MIT Press, 1971. 377 pp. $3.95
This is a collection of essays by Hans Blumenfeld on the planning and growth of modern cities.

A01-0199
Spyer, G. Architect and Community. Wittenborn, 1971. 168 pp. $11.50
Urban environment and design in England showing development of the suburb, mass building programs and industrialized buildings.

A01-0200
Stegman, M. Housing and Economics. MIT Press, 1971. 544 pp. $17.50
Housing and Economics brings together in a single volume a number of scholarly articles focused on the applied economic and financial problems of rebuilding America's housing stock.

A01-0201
Stegman, M. A. Housing Investment in the Inner City: The Dynamics of Decline. MIT Press, 1972. 320 pp. $15.00
This study seeks approaches that would combine the strengths and capacities of private investment with those of the public sector

to achieve satisfactory housing for all low-income families in a reasonable period of time. It covers the Baltimore, Md. area between 1968 and 1970.

A01-0202
Stein, C. <u>Toward New Towns for America.</u> MIT Press, 1966. 263 pp. $12.50 (paperback $4.45)
A document on the housing and planning experiments that began with Sunnyside Gardens in New York City during the nineteen-twenties and which marked the beginning of modern planning and housing in the United States. It is a study of deliberate planning and replanning of suburban towns to make them safer, more efficient, and better adapted to modern living.

A01-0203
Steinitz, C. and Rogers, P. <u>A Systems Analysis Model of Urbanization and Change: An Experiment in Interdisciplinary Education.</u> MIT Press, 1970. 120 pp. $8.95
This is a report of an experiment in interdisciplinary education for environmental planning recently carried out at the Harvard Graduate School of Design. A studio course was used as a vehicle for synthesizing the analytical data and approaches of four different disciplines: Landscape Architecture, Engineering, City and Regional Planning, and Urban Design. Aims of the study were to develop a better method of exploration and of interdisciplinary teaching which would lead to an understanding of urbanization complexities.

A01-0204
Sternberg, E. D. and Sternberg, B. E. <u>Community Centers and Student Unions.</u> Van Nostrand Reinhold, 1971. 319 pp. $27.50
An overview of the social and planning concepts which serve as a foundation for the location, programming and operation of various types of community centers and student unions. The book looks at community centers and student unions from a social, community, and architectural point of view. Analyses of selected examples of community centers, recreation centers, centers for the elderly, and school-community centers in the United States and abroad.

A01-0205
Sternlieb, G. and Burchell, R. W. <u>Residential Abandonment: The Tenement Landlord Revisited.</u> Center for Urban Policy Research. 444 pp. $15.00
The authors seek to define and analyze the process of residential abandonment over time. This study is unique in that the abandonment phenomenon is isolated within a single set of structures viewed over an eight-year period. Additional probes are made into the current solvency of the urban slumlord, and the validity of such a stereotype. The implications of homesteading, minority home-ownership, growing urban tax delinquency, and increased demands for public safety are also considered.

Planning--Urban and Regional 39

A01-0206
Sternlieb, G., et al. The Affluent Suburb: Housing Needs and Attitudes. Center for Urban Policy Research. 225 pp. $14.95

Princeton, New Jersey, a town typical of the affluent suburbs arising in so many communities today, is surveyed by a team of urban affairs experts. Housing problems are studied in light of their political and economic implications.

A01-0207
Sternlieb, G., et al. Housing Development and Municipal Costs: A Methodological Investigation. Center for Urban Policy Research. 378 pp. $12.95

The relationship of public costs to private development is the key to a host of concerns facing municipalities today: the quality of local services, the magnitude of local tax burdens, the adequacy of housing stock for future residents, and the most desirable mix of types of housing in the community. This rigorous investigation focuses on how various types of housing--garden apartments, high-rise apartments, townhouses, and single-family homes--affect a municipality's expenditures.

A01-0208
Stevens, J. L. Impact of Federal Legislation and Programs on Private Land In Urban and Metropolitan Development. Praeger, 1973. 266 pp. $16.50

This study reviews federal legislation through March 1973, as it affects private lands either by outright prohibitions or through long-range conditions for grants or loans. Paraphrases legislative texts for readability; presents the background of each federal program and explains the issues involved.

A01-0209
Stone, P. A. Urban Development in Britain. Vol. 1. Population Trends and Housing. Cambridge Univ. Press, 1973. $14.50

The structure, size and costs of urban settlements are described.

A01-0210
Strong, A. L. Planned Urban Environments: Sweden, Finland, Israel, the Netherlands, France. Johns Hopkins University Press, 1971. 416 pp. $20.00

Points to European successes in population distribution, land development, water management, housing, administrative organization, and financial incentives. The legal, economic, and physical planning characteristics vary with the development philosophy of individual nations, and each offers lessons for American planning.

A01-0211
Struyk, R. J. and Marshall, S. A. Urban Home Ownership. Lexington Books, 1975. 224 pp. $14.00

This study presents a unique investigation into the economics of home ownership in U.S. cities. The difference between renters and homeowners, types of households, tax advantages, and family composition are all part of the overall question here of how home ownership affects demand, cost of housing, and desegregation in housing. For the urban planner, finance officer, and regional scientist.

A01-0212
Tager, J. and Goist, P. D., eds. The Urban Vision: Selected Interpretations of the Modern American City. Dorsey Press, 1970. 320 pp. $4.25

This volume presents some of the major American attitudes toward the industrial city, 1890 to 1965. The readings will interest diverse students of urbanism, whether they are city planners, architects, sociologists, administrators, or historians.

A01-0213
Taggart, R. Low-Income Housing: A Critique of Federal Aid. Johns Hopkins Univ. Press, 1971. 160 pp. $2.25

Part of the series, Policy Studies in Employment and Welfare (Levitan and Mangum, eds.), this book is an analysis of federal housing subsidy programs, their design, administration, and accomplishments. The housing strategies of past and present are discussed and directions for the future suggested.

A01-0214
Tandy, C., ed. Handbook of Urban Landscape. Watson-Guptill, 275 pp. $24.95

The nature of much of the development of the modern city means that the designer has to think in terms of a total environment rather than of individual structures. This book provides a guide on current trends and techniques, covering housing, parks, open spaces, play areas, gardens.

A01-0215
Tetlow, J. and Gross, A. Homes, Towns, and Traffic. Praeger, 1968. 272 pp. $8.00

This volume explores what has been done to remedy city design problems due to transportation. There is an assessment of successes and failures in Britain's new towns and other cities around the world.

A01-0216
Thomlinson, R. Urban Structure: The Social and Spatial Character of Cities. Random House, 1969. 352 pp. $8.50

A01-0217
Thompson, W. R. A Preface to Urban Economics. Johns Hopkins Univ. Press, 1965. 432 pp. $9.00 (paperback $3.95)

The author discusses the factors that affect economic growth and the level, distribution, and stability of income and employment in the framework of the small-area open economy of the urban

Planning--Urban and Regional 41

region. He moves toward a theory of urban growth and guides for the rational management of cities.

A01-0218
Town and Country Planning Association. The New Citizen's Guide to Town and Country Planning. Charles Knight and Co., 1974. 178 pp. £3.25 (paperback £1.25)
The book deals with planning law and organisation by central and regional governments in England, the making and content of plans, the uses of planning in changing and preserving the environment, planning effects, control, enforcement, and rights of appeal, controversial aspects and future potential developments in this field.

A01-0219
Triton Foundation, Inc. Triton City: A Prototype Floating Community. National Technical Information Service, 1969. 131 pp. $3.00. Order No. PB-180051
A study was made to determine the feasibility of developing the water areas of major cities by floating entirely new communities on the water adjacent to the urban core. The report includes: locational possibilities and site environment, population statistics and organization; area and space requirements; costs and other considerations. Findings of this study indicate that it is possible to provide waterfront living for large numbers of city dwellers on floating communities at the shores of our major cities.

A01-0220
Urban Land Institute. New Towns Planning and Development: A Bibliography. American Institute of Architects, 1973. $10.00 (for AIA members, less 10%)
A bibliography of over 4,500 entries on all the information written about new towns in the 20th century.

A01-0221
Urban Structure. Elek Books, 280 pp.
Twenty-four articles exploring the current revolution in urban structure. One series of essays is concerned with systems of movement in towns and with new technologies of building, another with new approaches to planning as cities expand, and a third shows the impact of these forces on old cities.

A01-0222
Van Buskirk, P. The Resurrection of an American City. Schenkman, 1972. 256 pp. $6.95
Presents a review of the evolution of social and physical problems in American cities since the turn of the century. The study of the development of the Cohoes program--from preliminary planning through the formulation of actual budgets and forecasts--provides a framework for the study of urban planning in any industrialized society.

A01-0223
Van Cleef, E. Cities in Action. Pergamon, 1970. 250 pp.

$9.50 (paperback $5.50)

A text in urban planning covering the urban revolution, distribution of cities, urban sprawl, city patterns, shopping centers, land use, industrial decentralization, and more.

A01-0224
Von Hertzen, H. and Spreiregen, P. D. Building A New Town: Finland's New Garden City, Tapiola. MIT Press, 248 pp. $17.50 (paperback $4.95)

Tapiola is a modern city built in a rural community. It was built as both an experiment and a model--an experiment on how to create towns and make them socially as well as economically habitable; as a model to challenge many accepted ideas on residential area design and to influence the future pattern of Finland's urbanization. This illustrated book shows the importance of Tapiola's experience to other countries.

A01-0225
Wagner, R. H. Environment and Man. Norton and Co., 1971. 504 pp. $7.50

The author's theme is that man's relationship to his environment has undergone several significant changes. He is now out of control of his technology and is under the direct influence of a hostile environment--the city. The natural traumas from fire to radiation and traumas created by technology, such as chemical pollution, are examined. The problems in cities and suburbs from highway and air travel congestion, to disposal of wastes are discussed.

A01-0226
Washnis, G. J. Community Development Strategies in the Model Cities Program: Impact of the Program on Eight Major Cities. Praeger, 1974. 350 pp. $18.50

Summarizes the experience to date of Boston, Chicago, Dayton, Indianapolis, New York, Newark, Savannah, and Seattle in the Model Cities Program. Shows how the program has evolved from a limited operational strategy to a broad community development concept. Evaluates new techniques contributing to improved services, and describes the program's pitfalls and successes.

A01-0227
Weaver, R. C. Dilemmas of Urban America. Atheneum Publishers, 1967. $1.95

The first Secretary of Housing and Urban Affairs discusses such topics as trends in suburbia, urban renewal, and racial policy.

A01-0228
Wexler, H. J. and Peck, R. Housing and Local Government. Lexington Books, 1975. 236 pp. $22.00

With the shift in housing responsibility to cities, local policy-related research available. This study presents a unique evaluation of this research in four areas: planning and urban renewal, regulations, real property tax, and housing-assistance programs.

A01-0229
White, B. Sourcebook of Planning Information. Shoe String Press, 1971. 632 pp. $18.00
Planning is involved in a wide range of subjects and activities--population, regional science, urban studies, the environment, traffic and transportation planning, and others. This is a text to provide guidance to available sources of information.

A91-0230
Whitehead, C. M. E. The U.K. Housing Market: An Econometric Model. Lexington Books, 1975. $16.00
This study develops a model of the new private housing market in the U.K. for the period 1955-1970. It examines the determinants of demand and supply taking account of changes in incomes, prices, costs, and the rate of inflation, and looks at the question of whether new housing can be regarded as an equilibrium market.

A01-0231
Whittick, A. Encyclopedia of Urban Planning. McGraw-Hill, 1974. 1,204 pp. $34.00
Prepared by 70 leading authorities from all over the world, this encyclopedia covers all aspects of urban and regional planning. It offers comprehensive surveys before planning--explaining the economic and social questions that have to be considered, and showing what others are doing or have done throughout the world.

A01-0232
Wiebenson, D. Tony Garnier: The Cité Industrielle. Braziller, 128 pp. $5.95 (paperback $2.95)
A major document in the history of urbanism. The plans for an ideal industrial city are explained and clarified by a discussion of Garnier's sources in the socialist, regionalist, and utopian thought of the day. Included is a translation of Garnier's description of the Cité.

A01-0233
Wilson, J. Q. Urban Renewal: The Record and the Controversy. MIT Press, 1966. 683 pp. $14.95 (paperback $4.45)
A collection of articles on the background, workings, and problems of the federal urban renewal program. It contains contributions from government officials administering these programs and some of their critics; it takes up the major economic, legal, social, political, planning, and design issues surrounding this program.

A01-0234
Wingo, L., ed. Cities and Space: The Future Use of Urban Land. Johns Hopkins Univ. Press, 1963. 268 pp. $8.00 (paperback $2.95)
Essays based on the RFF Forum lectures of 1962.

A01-0235
Wolf, P. The Future of the City: New Direction in Urban Planning. Watson-Guptill, 208 pp. $18.50
   Surveys American city planning and analyzes social, economic, administrative, and design issues. Presents over 80 American and European projects and proposals to show possible directions for cities in the future.

A01-0236
Wolff, A. The Unreal Estate. Sierra Club, 176 pp. $5.95
   This volume documents how vast parcels of the American earth have been carved up and sold to thousands of consumers in flagrant abuse of the most common principles of sound land-use planning. Its case histories range from California and New Mexico, where speculators have staked out enough lots to accommodate twice those states' populations, to the hillsides of Vermont, where "second-home" developments proliferate.

A01-0237
Woodbury, C., ed. The Future of Cities and Urban Redevelopment. Univ. of Chicago Press, 1953. 764 pp. $12.50
   Discusses long term objectives and basic problems that precede, accompany, and follow the redevelopment program. The contributors analyze the social ills that must be cured, and define what should be provided for the city dweller's health, comfort, and convenience.

A01-0238
Worskett, R. The Character of Towns. Architectural Press, 1969. 272 pp. £4.00
   The author begins by explaining the principles of a sound conservation policy. He describes the qualities of design that make up the character of a town, a neighbourhood or a group of buildings and gives some technical hints on how to formulate the results of such work in maps and drawings.

A01-0239
Yearwood, R. M. Land Subdivision Regulation: Policy and Legal Considerations for Urban Planning. Praeger, 1971. 315 pp. $16.50 (for ASPO members, $14.50)
   Discusses the rationale for regulation, park and school land dedication, timing and location controls, and administration and enforcement of subdivision regulations.

Section   A02

PLANNING: PROJECT

    Cost Estimates
    Finance
    Site Selection
    Specific Projects (Hospitals, Schools, Shopping Centers, etc.)
    Real Estate and Appraisal

A02-0250
Adie, D. W. Marinas. Architectural Press, 1974. £16.00
This illustrated book gives design guidance on the problems encountered in the construction of inland and marine boating facilities.

A02-0251
American Appraisal Co. Boeckh Building Valuation Manual. Vol. 1, Residential and Agricultural. American Appraisal Associates, 1974. 274 pp. $38.00
A costing tool for U.S. and Canadian building. Permits rapid valuation on basis of square feet of area, base building costs, and in-place costs of components. This volume includes single and multi-family residences, mobile homes, vacation homes and agricultural structures.

A02-0252
American Appraisal Co. Boeckh Building Valuation Manual. Vol. 2, Commercial. American Appraisal Associates, 1974. 401 pp. $44.00
A costing tool for U.S. and Canadian buildings. This volume covers apartments, hotels, motels, hospitals and health facilities, theatres, office buildings, terminals, garages and supermarkets.

A02-0253
American Appraisal Co. Boeckh Building Valuation Manual. Vol. 3, Industrial and Institutional. American Appraisal Associates, 1974. 416 pp. $44.00
A costing tool for U.S. and Canadian buildings. This volume covers industrial buildings, warehouses, industrialized buildings, schools and dormitories, churches, and land improvements.

A02-0254
American Assn. of Hospital Consultants. Functional Planning of General Hospitals. McGraw-Hill, 1969. 352 pp. $21.00
Developed from workshops conducted by the A.A.H.C. and prepared by over 25 experts, this work serves as a guide in remodeling facilities as well as the development of new hospitals. It covers such topics as outpatient care services, computers in hospitals, long-term care, size and organization of nursing units, and financing.

Planning--Project   47

A02-0255
American Assn. of Port Authorities. Port Planning Design and
    Construction. American Assn. of Port Authorities, 514 pp.
    $30.00
    Covers all phases of marine terminal construction, planning,
and design including future ships sizes, environmental management,
conceptual engineering, containership terminals, LASH terminals,
offshore terminals, passenger terminals, LNG terminals, contract
administration.

A02-0256
American Hospital Assn. Guide Issue-Hospitals. American Hospi-
    tal Association, $12.50
    Directory of more than 7000 U.S. hospitals, administrative
personnel connected with each and other management data is given.

A02-0257
American Institute of Real Estate Appraisers. The Appraisal of
    Real Estate. Ballinger Publishing Co., 1973. 569 pp.
    $14.50
    The standard introduction to real estate appraisal. Re-
quired in AIREA courses.

A02-0258
American Institute of Real Estate Appraisers. Condemnation Ap-
    praisal Practice. Vol. 1. Ballinger Publishing Co., 1961.
    554 pp. $12.50
    Selected articles from The Appraisal Journal from the years
1951 to 1960, covering the various aspects of condemnation appraising and court testimony.

A02-0259
American Institute of Real Estate Appraisers. Condemnation Ap-
    praisal Practice. Vol. II. Ballinger Publishing Co., 1973.
    717 pp. $12.50
    Companion to Volume 1. Contains articles from The Ap-
praisal Journal (1962-1972) covering various aspects of condemnation appraising and court testimony.

A02-0260
American Institute of Real Estate Appraisers. Problems in Urban
    Real Estate Appraisal. Ballinger Publishing Co., 1968.
    220 pp. $7.50
    Contains practical urban appraising problems including Ell-
wood and cash-flow situations, with suggested solutions.

A02-0261
Anderson, E., ed. Basic Real Estate Tax Manual. Exposition
    Press, $9.00
    This book enables the reader to make a proper tax analysis
of most real property transactions. It is addressed to students
and teachers of basic tax and real estate tax courses, real estate

brokers, attorneys, title company executives, builders, C.P.A.'s, escrow officers, licensees, investors and property owners.

A02-0262
Architectural Record. Apartments, Townhouses, and Condominiums. McGraw-Hill, 1975. 228 pp. $17.50
This revision reflects the shift in emphasis in multi-unit housing from apartments and dormitories to apartments, townhouses, and condominiums. In-depth coverage of garden complexes on suburban sites, designing for low and moderate income groups, conversions, large-scale developments, apartments and condominiums for resort areas, and interiors round out this volume.

A02-0263
Architectural Record. Hospitals, Clinics, and Health Centers. McGraw-Hill, 1960. 256 pp. $18.50
The most modern, effective ideas for planning hospitals and other medical facilities are presented with full technical details. More than 700 photographs, plans, and diagrams are included.

A02-0264
Aries, R. S. and Newton, R. D. Chemical Engineering Cost Estimation. McGraw-Hill, 1955. 280 pp. $13.00
Includes material on capital investment, physical plant costs, equipment, components other than equipment, manufacturing cost, general expense, sales and profits. Also, the effect of variable conditions upon costs and profits, cost factors in plant location, economic evaluation and nomenclature.

A02-0265
Ashley, R. Electrical Estimating. McGraw-Hill, 1961. 437 pp. $22.50
A manual which contains all the information required to estimate the cost of any electrical construction job. Covers everything from fast methods of take-off to checking the final bid, from training men and using proper tools to fabricating labor costs for special work.

A02-0266
Atton, W. Estimating Applied to Building. George Godwin Ltd., 1969. 104 pp. £2.80
In metric throughout, this guide is used in the industry and by students. The sequence of the Standard Method of Measurement is followed and labor constants are given accordingly. A wide range of practical examples is given. At the end of each section examples show how the unit rates are built up. As far as possible these have been based on the more important items in the trades concerned.

A02-0267
Atton, W. Introduction to Estimating. George Godwin, Ltd., 1973. 110 pp. £1.25
This practical guide to estimating in the construction

Planning--Project

industry, using metric throughout, is widely used both by students and by many in the industry.

A02-0268
Auditorium/Arena/Stadium Guide and International Directory. Billboard Publications, $20.00 (no cost to qualified readers)
Listing of U.S. and Canadian auditoriums, arenas and stadiums along with their descriptions.

A02-0269
Barton, J. J. Estimating for Heating and Ventilating. Butterworth, 1972. 304 pp. £4.25
This volume covers all aspects of the heating and ventilating industry, from the basic principles through the various stages up to the preparation of the tender for contract work. The text uses metric (SI) units throughout, although the equivalent imperial units are quoted in most cases and conversion tables are included.

A02-0270
Bauman, C. H. Fundamentals of Cost Engineering in the Chemical Industry. Van Nostrand Reinhold, 1964. 416 pp. $15.00
Emphasis is placed on development of short cut techniques for accurate estimation of fixed capital, operating costs, and rapid evaluation of the profitability of ventures. Includes a glossary of cost engineering terms as well as tables of conversion data, sizes and weights. Charts cover the most frequently used inflation indexes.

A02-0271
Beasley, M. R. Fell's Guide to Buying, Building and Financing a Home. Frederick Fell, 1963. 143 pp. $4.95
A glossary of real estate terms is included in a practical guide that contains material on building, buying, and selling a home, financing and mortgages, reducing building costs, moving, and settling into a new house.

A02-0272
Benson, B. Building Contractor's and Home Builder's Handbook of Bidding, Surveying, and Estimating. Prentice-Hall, 1968. 190 pp. $16.95
Guide to the intricacies of preparing a bid. Provides a systematic basis for the inclusion of job waste in estimating the cost of a job. Part I is a guide for preparing the physical take-off, Part II presents methods of procedure, Part III presents a sample take-off of the quantities, together with the plans for a small library; a sample take-off of a small residence with its plans; some typical schedules used in developing quantities; and some useful tables.

A02-0273
Bifulco, J. How to Estimate Construction Costs of Electrical Power Substations. Construction Publishing Co., 1973. $14.95

An hour-to-hour, dollar-by-dollar approach to figuring costs for substations. Contains component and equipment costs, but is primarily a method for figuring labor costs based on average man-hours needed to install each component. Contains checklists, worksheets and tables of work elements.

A02-0274
Boyce, B. N., ed. Real Estate Appraisal Terminology. Ballinger Publishing Co., 1976. 320 pp. $12.50

This volume represents the results of a two-year collaboration between the Society of Real Estate Appraisers and the American Institute of Real Estate Appraisers. In addition to an exhaustive treatment of appraisal terminology, this book provides new and expanded terminology in the areas of investment analysis, statistics, mathematics and computers, as well as reference materials on Depreciation Methods.

A02-0275
Building Design and Construction Specifying Buying Guide and Directory. Cahners Publishing Co., Annual. $4.50

Lists practically all building products manufactured in U.S., includes names and addresses of manufacturers. Uses the 16-division CSI format for building specifications.

A02-0276
Building Supply News Purchasing File Issue. Cahners Publishing Co., Annual. $2.00

Lists over 1,000 building product manufacturers in 700 different product categories. Also lists leading brand names of these manufacturers.

A02-0277
Clurman, D. and Hebard, E. L. Condominiums and Cooperatives. John Wiley, 1971. 395 pp. $19.25

A handbook that gives data on the formation and operation of a successful condominium and explains legal requirements, of the proper legal forms to be used. Covers urban planning, brokerage, development, zoning, salesmanship, appraisal, advertising, and related subjects.

A02-0278
Cohen, I. Electrical Estimating Handbook. Construction Publishing Co., 1975. $19.95

Written for both the beginner and the experienced estimator, an estimating book devoted solely to the electrical trade. Detailed presentation of measurement, unit-pricing, materials, methods, trade practices and shortcuts. Includes an explained and priced sample take-off. A 100-page presentation of dollar costs including tables for average unit prices, crew costs, productivity figures. Adjustment index for 101 cities allows the reader to adjust prices for his area.

Planning--Project

A02-0279
Colby, E. E. Practical Legal Advice for Builders and Contractors.
    Prentice-Hall, 1972. 272 pp. $16.95
    Practical legal advice on every aspect of the construction
business. Contains necessary information on: partnerships, corporations, negotiable instruments, condominiums and co-operative apartments, foreclosures, and bankruptcy. Shows how to set up a temporary partnership, which to incorporate, where to turn for a surety bond, explains the legal relationship with the architect, which employees should be bonded and how to protect an unpaid claim.

A02-0280
Collison, K. M. The Developer's Dictionary and Handbook. Lexington Books, 1974. 160 pp. $12.50
    This work contains approximately 3,000 definitions of terms in the fields of industrial and economic development; architecture; real estate; law; construction; engineering; transportation, and related fields.

A02-0281
Construction Equipment Buyers Guide. Cahners Publishing Co.,
    Annual. $20.00
    Annual national directory of who makes what products, where to buy them with listing of trade names and financial services for the construction industry.

A02-0282
Cooper, G. and Badzinski, S., Jr. Building Construction Estimating. McGraw-Hill, 1971. 320 pp. $11.95
    Completely reorganized and updated, this book leads the reader who has a working knowledge of construction methods and blueprint reading through a well-rounded course covering the everyday work of the building contractor's estimator.

A02-0283
Corwin, R. and Hefferman, P., eds. Environmental Impact Assessment. Freeman, Cooper and Co., 1975. 277 pp. $12.00
    Using the talents of engineers, biologists, economists, attorneys and other planners, the book examines the subjects--legal, ecological, social and economic--that affect the environment. The book is non-technical and is useful in familiarizing the reader in the requirements for environmental impact statements.

A02-0284
Cost Reference Guide. Equipment Guide-Book Co., 1976. $85.00
    This book is a cost guide for estimators and equipment owners for the actual costs to operate construction equipment on an hourly or per day basis. More than 6,000 line items and 80,000 separate rates are included.

A02-0285
Costonis, John J. Space Adrift--Landmark Preservation and the

Marketplace. American Institute of Architects, 1973.
$10.00 (for AIA members, less 10%)
An ingenious approach for protecting landmark buildings.

A02-0286
Cross, F. L., Jr. and Simons, J. L., eds. Industrial Plant Siting. Technomic Publishing Co., 1975. 204 pp. $45.00
Site selection is dealt with from the vantage points of an architect, engineer, financial consultant, environmental specialist and others. Such subjects as environmental impact, labor, community attitudes, land use, pollution as well as economic and legal factors are explored.

A02-0287
Darlow, C., ed. Enclosed Shopping Centres. Architectural Press, 1972. 224 pp. £16.00
This practical book explains in detail the requirements and constraints of each party to a covered shopping centre development operation. It provides complete and up-to-date survey of current trends in design, planning and finance.

A02-0288
Davis, Belfield, et al., eds. Spon's Architects' and Builders' Price Book. John Wiley, 1974. 504 pp. $10.50
Contains the materials prices, rates for social insurance, and similar on-costs current at the time of going to press. Separate price sections give both metric and imperial units for construction in England and in other European countries.

A02-0289
Davis, Belfield, et al., eds. Spon's Mechanical and Electrical Services Price Book. John Wiley, 1974. 330 pp. $12.75
Source of cost data essential for the preparation of estimates and the pricing of quantities involving mechanical and electrical service for construction in England and in other European countries.

A02-0290
Deatherage, G. Construction Estimating and Job Preplanning. McGraw-Hill, 1965. 302 pp. $19.95
Discusses the importance of accurately predetermining the order and cost of men and machines used in a construction project.

A02-0291
Debaigts, J. New Interiors for Old Houses. Reinhold Books, 164 pp. $19.95
How to preserve, restore and maintain exteriors that reflect the past, while creating and reconstructing interiors that take maximum advantage of modern methods and materials. Includes illustrative plans and examples.

A02-0292
Deering, F. R., et al. Auditoriums and Arenas: Facts from a

Planning--Project

Survey by the International Association of Auditorium Managers. Public Administration Service, 1961. 96 pp. $5.00
Describes administrative responsibilities, construction costs, and financing, sites, building facilities, financing of operations, and personnel. Presents detailed data on the management practices of major auditoriums in the United States and Canada.

A02-0293
Dent, C. Construction Cost Appraisal. George Godwin Ltd. (distributed by Construction Publishing). $16.95
A comprehensive treatment of discounted cash flow techniques in the construction industry--of special interest to all those concerned with the profitability of building projects. Includes treatment of profitability analysis and substantial discount tables.

A02-0294
Dept. of Health, Education and Welfare. Hill-Burton Project Register. National Technical Information Service, $3.00. Order No. PB 198567
A listing of the more than 10,000 hospital projects under the aegis of Hill-Burton legislation. A description of each facility and its construction cost is given.

A02-2095
Diez, R. L. and Maloney, D. Builders' Estimating Fact Book. Cahners Publishing Co., 1974. 128 pp. $9.95
A useful guide to estimating job costs for builders, contractors, subcontractors, and remodelers in the housing and light construction industry. The book includes data for establishing and estimating building techniques, labor and material requirements, product sizes, and specifications.

A02-0296
Directory of Shopping Centers in the U.S. and Canada. National Research Bureau. $70.00
Lists more than 15,000 shopping centers, their addresses, names of management, cost to operate and construct and other physical and business information.

A02-0297
Dodge Building Cost Services. 1975 Guide for Estimating Public Works Construction Costs. McGraw-Hill Information Systems, 1975. $18.60
Provides costs for about 7,000 items used in public construction, including labor, material and equipment rates. In addition to unit prices, there are data on crew makeup and output per day. Expanded coverage includes mass transit and environmental control construction.

A02-0298
Dodge Building Cost Services. 1976 Dodge Construction Systems Costs. McGraw-Hill Information Systems, 1976. 247 pp. $33.80

This book provides the cost data on the different systems and assemblies for each functional part of a building--superstructure, floors on grade exterior walls, partitions, etc. This data allows comparison costs of alternate solutions to each building part.

A02-0299
Dodge Building Cost Services. 1976 Dodge Manual for Building Construction Pricing and Scheduling. McGraw-Hill Information Systems, 1976. 238 pp. $19.80

For estimating, scheduling and checking prices of more than 9,000 items. Updated annually. Cost indices are adjusted for 50 trades and sub-trades in over 80 U.S. cities.

A02-0300
Dombal, R. W. Residential Condominiums: A Guide to Analysis and Appraisal. Ballinger Publishing Co., 1976. 80 pp. $7.00

This book considers the valuation of entire projects, individual units, conversions, and resales. Discussions of gross sellout over a period of time, value to a single purchaser and gross sellout vs. market value are presented. An invaluable tool for appraisers, lenders and others involved in this growing concept of housing.

A02-0301
Editorial Staff of "Construction Labor Report." Construction Labor Report's 1970-1971 Wage Rate Guide. Bureau of National Affairs, 1970. $7.50

Shows the wage scales and fringe benefits for construction workers in 116 cities. Also shown are hourly rates and fringes that will become effective later under long-term collective bargaining agreements.

A02-0302
Edgerton, W. H. Edgerton Building Cost Calculator and Guide to Real Estate Valuation. Construction Publishing Co., 1976. $29.95

This guide allows the computation of quick approximations of total building cost for hundreds of different building types, in any metropolitan area. Revised annually.

A02-0303
Edgerton, W. H. How to Renovate a Brownstone. Construction Publishing Co., 1973. 373 pp. $14.95

Available in this volume are answers to many questions about urban rehabilitation. How to avoid common pitfalls and time-consuming delays; how to make accurate decisions that assure a well planned and profitable renovation. Dozens of forms and worksheets.

A02-0304
Edgerton, W. H. Row House Renaissance: A Guide to Renovation.

Planning--Projects

Construction Publishing Co., 1974. $14.95
Available in this book are answers to many questions about urban rehabilitation. How to avoid common pitfalls and time-consuming delays; how to make accurate decisions that assure a well planned and profitable renovation.

A02-0305
Edson, C. L. and Lane, B. S. *A Practical Guide to Low- and Moderate-Income Housing.* Bureau of National Affairs, 1972. $45.00
A guide to all federal and state housing programs for low- and moderate-income families. Attention is given to topics such as taxes and federally assisted housing, equity syndication, securities law problems, acceptable site selection under new HUD criteria, HUD rehabilitation programs, and the role of industrialized housing in the federal programs. A complete set of forms and exhibits is provided for each topic examined, and the entire manual is supplemented periodically.

A02-0306
Ellwood, L. W. *Ellwood Tables for Real Estate Appraising and Financing.* Part I--Explanatory Text. Ballinger Publishing Co., 1970. 134 pp. $10.00
Appraisal Institute's revised third edition of the appraisal profession's celebrated "red book" on mortgage-equity capitalization.

A02-0307
Ellwood, L. W. *Ellwood Tables for Real Estate Appraising and Financing.* Part II--Tables. Ballinger Publishing Co., 1970. 432 pp. $12.00
Separate loose-leaf volume of tables only, including expansion of all prior tables for contemporary market high interest rates, and revision of Table "Cy" to include new income adjustment factor. Prepared for those who want only expanded tables and revised "Cy" Tables.

A02-0308
Essex, D. L. *Bonding Versus Pay-as-You-Go in the Financing of School Buildings.* AMS Press, 1931. $10.00
A discussion of the relative merits of bonding and pay-as-you-go financing of school buildings. Contends that neither plan offers a perfect solution to the problem of financing, and suggests a policy that allows both pay-as-you-go and bonding to be used under the most favorable circumstances.

A02-0309
Fabian, D., ed. *Aquatic Buildings.* Hoffman Publications, 500 pp. $25.00
Photos, scale plans, details of hundreds of Europe's best pools, recreation facilities and buildings. Includes sections on hotel, indoor-outdoor, salt water, wave, therapy pools, etc. All descriptions in both English and German.

A02-0310
Foster, N. Construction Estimates from Take-Off to Bid. McGraw-Hill, 1972. 288 pp. $19.95
This second edition features three revised chapters that show full working drawings, specifications, take-off, estimate, analysis of unit prices, and explanations for an industrial building with office area. It teaches the fundamentals of good estimating, using detailed examples, valuable tips, and step-by-step procedures for developing accurate estimates.

A02-0311
Frost, M. J. Values for Money: Techniques of Cost Benefit Analysis. Cahners, 1972. 238 pp. $12.50
This book describes a technique for official decisions. It sets out the case for objective evaluation and shows how cost benefit method identifies the areas of gain and loss from a proposal, sets them in a logical order of priority and seeks an acceptable value in money terms for intangible factors.

A02-0312
Gabrielsen, M. A., ed. Swimming Pools: A Guide to their Planning, Design & Operation. Hoffman Publications, 222 pp. $12.50
Covers everything from choosing the site, pool design, water treatment, pool structures, to final staffing, administering and operating the public and semi-public pool. A project of the Council for National Cooperation in Aquatics.

A02-0313
Galeno, J. The Plumbing Estimating Handbook. Construction Publishing Co., 1976. $19.95
For standard and special building and sitework plumbing systems, it covers: take-off procedures, pricing techniques, common and uncommon estimating practices. Learn how to make preliminary plumbing estimates ... How to cross-check your final costs ... When conceptual estimating can help. It also tells how to price change orders and estimate alterations and additions.

A02-0314
Garrett, R. L., et al. The Valuation of Shopping Centers. Ballinger Publishing Co., 1976.
Prepared for the American Institute of Real Estate Appraisers, this is a guide to the elements the appraiser must consider in shopping center appraisal: building and mall layout and design, parking and landscaping, tenant mix, lease obligations, competition, and market potential. Contains analysis of the three accepted approaches to value and their relative applicability to the appraisal of shopping centers.

A02-0315
Geddes, S. and Chrystal-Smith, G. Estimating for Building and Civil Engineering Works. Butterworth, 1971. 368 pp. £6.00

Planning--Projects   57

The change to both SI units and decimal currency has affected all trades and professions involved in building and civil engineering works, and in the transition period following these changes a reliable primary reference source is needed. Estimating data for costing all types of work in each trade are given, and the different methods of tendering and of compiling estimates are covered, including the Bill of Quantities and the Schedule of Rates.

A02-0316
Gladstone, J. Mechanical Estimating Guide-Book. McGraw-Hill, 1970. 320 pp. $19.00

With the rapid rise of costs of both labor and materials in air conditioning, this book should prove useful to hold down the rate of future increases. It provides a rapid and comprehensive method of cost estimating. Topics covered include mechanical cooling and heating equipment, fans and blowers, towers, pumps, tanks, piping, ductwork, and electrical wiring.

A02-0317
Goleman, Harry A., ed. Financing Real Estate Development. American Institute of Architects, 1974. $21.50 (for AIA members, less 10%)

A primer on the essentials of the strategy and terminology of real estate financing.

A02-0318
Goodkin, L. M. When Real Estate and Home Building Become Big Business. Cahners, 1974. 625 pp. $19.95

This book focuses on mergers, acquisitions, and joint ventures in the real estate and housing field. It considers the role of inflation, the federal government, and the stock market on mergers and joint ventures and analyzes the managerial, financial, marketing, and accounting problems. In a section on joint ventures, the book gives step-by-step procedures for preparing a feasible joint venture, analyzes the key elements of a typical agreement, and tells how to identify profitable joint venture development opportunities.

A02-0319
Grant, E. L. and Ireson, W. G. Principles of Engineering Economy. Ronald Press, 1970. 640 pp. $12.00

This book discusses and analyzes the basic concepts underlying engineering economy decision-making, and explains the principles necessary for making sound determinations for the acquisition and retirement of capital goods and for choosing among various methods of financing. Discussion of the tax aspects of decisions among alternative choices and analysis of after-tax minimum rates of return.

A02-0320
Gross, J. Concise Desk Guide to Real Estate Practice and Procedure. Prentice-Hall, 1976. 300 pp. $14.95

This guide is a compendium of information on the most important aspects of real estate practice. The book includes a complete glossary of real estate terms and a guide to the terminology of real estate law, as well as over 30 model letters readily adaptable to the reader's needs. In addition, there are sample contracts, and appraisal reports.

A02-0321
Guide for Planning Educational Facilities. Council of Educational Facility Planners, 1969. 204 pp. $7.50

Completely rewritten to reflect current practices and expanded to cover finance and building program administration. A guide to facility planning for anyone responsible for planning, designing, constructing, maintaining or equipping facilities for education. A textbook for educational facility planning courses, or for the layman interested in obtaining better facilities for his community.

A02-0322
Guthrie, K. M. Process Plant Estimating, Evaluation and Control. Craftsman Book Co., 1974. 612 pp. $25.00

The estimating data cost engineers, chemical process designers and project analysts need to evaluate the cost of the next generation of refineries and chemical process plants. Based on cost data developed from field construction records of 50 major capital projects. Included: labor, material and equipment costs for every type of refinery and chemical process plant, indirect costs, pipe estimating data, project control techniques, computer cost systems and more.

A02-0323
Halperin, D. A. Construction Funding: Where the Money Comes From. John Wiley, 1974. 236 pp. $13.95

Contents include sources of money, capitalization and the importance of money, site selection and development, preparing the request for funds, draw schedules, cash forecasting, and time value of money.

A02-0324
Halsey, H. R. Borrowing Money for the Public Schools: A Study of Borrowing Practices in the Administration of Public Schools in Florida. AMS Press, 1929. $10.00

An attempt to investigate the use of public credit in connection with the administration of the public schools in Florida. Includes an overview of borrowing in general, and specific considerations for the issuance of school bonds. Covers the period from 1919 through the onset of the depression.

A02-0325
Hamer, A. M. Industrial Exodus from Central City: Public Policy and the Comparative Costs of Location. Lexington Books, 1973. 120 pp. $12.50

This study develops a framework for analyzing the relevant

Planning--Project

costs of locating manufacturing firms at different sites in an urban area through a case study of Boston. The primary conclusion is that the realization of maximum profit over space by a wide range of manufacturing firms involves very few factors of production and favors suburban locations.

A01-0326
Hanford, L. D. Feasibility Study Guidelines. Institute of Real Estate Management, 130 pp. $3.50
Offers direction and explanation of the contents of a feasibility study, beginning with developer contact, and working through the various analyses--economic, geographic, demographic, etc.-- to the final conclusions and report format. The book contains a sample feasibility study for a residential development.

A02-0327
Hanke, B. R. The Homes Association Handbook. Urban Land Institute, 1964. 422 pp. $15.00
A study conducted by ULI under the co-sponsorship of FHA and five other agencies. The Bulletin contains findings and recommendations of the Urban Land Institute resulting from its nationwide investigation of existing subdivisions and community properties maintained by homes associations.

A02-0328
Harberger, A. C., et al. Benefit Cost Analysis 1971: An Aldine Annual. Aldine Publishing Co., 1972. 532 pp. $20.00

A02-0329
Harrell, G. T. Planning Medical Center Facilities for Education, Research, and Public Service. Pennsylvania State Univ., 1974. 232 pp. $15.00
This book is a review of the experience of one new medical school with a sampling of the unique features of others. The long lead-time required for the planning of a medical educational program and the special architectural solution to the problem of housing it, in the author's opinion, are not sufficiently appreciated.

A02-0330
Harry, J. Construction Cost Guide. Research Guide Publications, 1976. 223 pp. $29.95
Cost data on materials, labor and equipment for residential, industrial and commercial construction in the U.S. and Canada. Topics include sitework, concrete, masonry, metals, carpentry, HVAC, plumbing and electrical work. Square foot building costs for over 90 building types are included, along with a percentage breakdown of electrical, plumbing and HVAC costs.

A02-0331
Heavy Construction Cost File. Construction Publishing Co., 1976. $19.95
Thousands of unit costs relating to heavy construction and public works estimating.

A02-0332
Hedden, W. P. Mission: Port Development. American Assn. of Port Authorities, $10.00 (paperback $3.95)
Discusses technical problems in port development, port planning, operations, economics, administration, staff training; case histories.

A02-0333
Hinrichs, H. H. and Taylor, G. M. Systematic Analysis: A Primer on Benefit-Cost Analysis and Program Evaluation. Goodyear Publishing, 1972. 160 pp. $3.95
Provides a thorough grounding in the principles and methods of systematic analysis and evaluation of public sector programs.

A02-0334
Hoagland, H. E. and Stone, L. D. Real Estate Finance. Richard Irwin or Dorsey Press, 1973. 611 pp. $12.00
This book provides basic orientation in commonly used instruments and institutional structures and policies and understanding of the interrelationships involved in accomplishing the real estate financing functions. Interest rates, equity participation, development of secondary markets and government involvement in subsidized housing are examined.

A02-0335
Hornung, W. J. Estimating Building Construction: Quantity Surveying. Prentice-Hall, 1970. 224 pp. $10.95
Introduces fundamental principles of taking-off quantities of building materials. Shows in the simplest form how quantities of building various materials are found from the plan, and the methods of itemizing them. Comprised of sixteen study units--each having an introduction, technical information with explanations and numerous illustrations, self-examination, assignments, and supplementary material.

A02-0336
Housing and Urban Development, Dept. of. HUD Condominium Cooperative Study. U.S. Gov't. Printing Office, 1975.
This three-volume study is a comprehensive Department of Housing and Urban Development compilation of State laws applicable to condominiums. It is an extensive analysis of problems and abuses in condominium and cooperative development and conversion.
Vol. 1, 384 pp. $5.15, Order No. HH 1.2:C 75/5/v.1 S/N 023-000-003030-1. Contains the study's findings on U.S. condominium history, economics, and problems and abuses.
Vol. 2, 572 pp. $6.50, Order No. HH 1.2:C 75/5/v.2 S/N 023-000-00304-0. An analysis of condominium and cooperative housing in selected major metropolitan housing markets.
Vol. 3, 112 pp. $1.75, Order No. HH 1.2:C 75/5/v.3 S/N 023-000-00305-8. Primarily a supplementary volume, containing the data gathered from national surveys conducted for the study.

Planning--Project

A02-0337
Hunt, W. D., Jr., ed. Creative Control of Building Costs.
American Institute of Architects, 1967. $16.00 (for AIA members, less 10%)
Pioneers the view that control of construction costs should be an intrinsic part of the entire architectural process. Examines the principles and methods of cost control. Tells how to use them to determine realistic budgets and maintain those budgets throughout construction of the building.

A02-0338
Huntoon, M. C., Jr. PUD: A Better Way for the Suburbs.
American Institute of Architects, 1971. $10.00 (for AIA members, less 10%)
An illustrated book on Planned Unit Development--72 pages of narration, site plans, and full color photographs of 12 PUD developments. History, current suburban response, and viewpoint of the community resident on PUD.

A02-0339
IREM. 1975 Income/Expense Analysis--Apartments, Condominiums & Co-operatives. Institute of Real Estate Management, 1975. 168 pp. $29.95
Shows detailed financial data on income and expenses of more than 3,400 buildings in 130 major cities. The data are grouped by building type, age and size, and by metropolitan area. Topics include: apartment rentals, garages and parking, vacancies and delinquencies, payroll, utilities and fuels, management, painting and decorating, maintenance and repairs, supplies, services, insurance, and real estate taxes.

A02-0340
IREM. 1975 Office Building Experience Exchange Report. Institute of Real Estate Management, 1975. 100 pp. $50.00
Published by the Building Owners and Managers Association International, this publication provides a source of information concerning the income and expense operation of office buildings in the United States and Canada. General averages of some 700 member buildings in 86 cities are recorded. Information is presented in a series of uniform tables based on cents per square foot per year in relation to building statistical areas, with supplemental studies by height; size and age of building; population of cities; special energy analysis; occupancy studies and market inventory.

A02-0341
Johnson, I. E. The Instant Mortgage Equity Technique. Lexington Books, 1972. 390 pp. $14.00
The book aids the appraiser in understanding and using the best method in the valuation and analysis of income-producing properties, the mortgage equity technique. This book compiles complete, totally precomputed mortgage-equity overall-rate tables which bring speed and accuracy to the appraisal of income property.

A02-0342
Johnson, I. E. Selling Real Estate by Mortgage Equity Analysis.
    Lexington Books, 1976. 544 pp. $16.00
    The mortgage-equity technique is one of the most meaningful approaches to real estate valuation and investment analysis. The streamlined technique presented in this volume employs totally precomputed overall rates and patterned procedures and clarifies and simplifies the technique for both novice and expert. It analyzes income producing property and tells how to list it at a realistic, justifiable price.

A02-0343
Kenny, M. F. Concrete Estimating Handbook. Construction Publishing Co., 1974. $19.95
    Written for both the beginner and the experienced estimator. Includes an explained and priced sample take-off. A 100-page presentation of dollar costs including tables for average unit prices, crew costs, productivity figures. Adjustment index for 101 cities allows the reader to adjust prices for his area.

A02-0344
Kenny, M. F. Masonry Estimating Handbook. Construction Publishing Co., 1973. $19.95
    Detailed presentation of: measurement, take-off, unit pricing, materials and methods, trade practices and shortcuts. Explained and priced sample take-off. Nearly 100 pages devoted to dollar costs: average unit prices, mason crew costs, productivity figures. Adjustment index for 101 cities lets the reader adjust the Handbook's prices for his area.

A02-0345
Kinnard, W. N., Jr. Income Property Valuation: Principles and Techniques of Appraising Income-Producing Real Estate.
    Lexington Books, 1971. 520 pp. $13.00
    Treats income capitalization in real estate appraisal. This volume emphasizes the mortgage-equity approach to income property valuation and covers the problems encountered by income property appraisers.

A02-0346
Kinnard, W. N., Jr. Principles and Techniques of Real Property Appraising. Lexington Books, 1976. 528 pp. $15.00
    Designated the official text of the Society of Real Estate Appraisers in its introductory courses, this book is a comprehensive work on techniques for the valuation of real property. It covers the underlying principles of urban land economics, real estate, and valuation which influence the work of the real estate appraiser.

A02-0347
Knowles, J., Jr. Single-Family Residential Appraisal Manual.
    American Institute of Real Estate Appraisers or Ballinger Pub. Co., 1967. 164 pp. $6.00

Planning--Project

A guide covering the fundamental principles and practices of appraising a single-family dwelling, including instructions for writing the appraisal report.

A02-0348
Kolstad, C. Rapid Electrical Estimating and Pricing. McGraw-Hill, 1969. 320 pp. $12.50

This work gives fast means for estimating the cost of electrical systems in all types of residential and commercial buildings. Charts permit quick determination of prices of installed equipment, allowing the estimator to get the cost of an installed wiring system without having to break it down into its various components.

A02-0349
Land, C. B. Machinery and Equipment Pricing Guide. Construction Publishing Co., 1976. $29.95

A book of prices of industrial machinery and equipment. Each entry contains a new, used and salvage price to provide the user with an independent price-check over a wide range of industrial items. Annual editions. Over 2,000 separate items are priced including motors, drives, electrical generating and transmission equipment, heating, refrigerating, material handling and more.

A02-0350
Lee Saylor, Inc. Current Construction Costs 1976. Lee Saylor, Inc., 1976. $14.95

A manual of current construction pricing and scheduling divided into the 16-Division CSI format. Costs are included for 9000 items used for building, industrial, and public works construction. Adjustment indexes for 105 U.S. metropolitan areas are featured as well as in-place costs for budget estimates.

A02-0351
LeJeune, E. G. Manual of Concrete Estimating. Concrete Construction Publications, 128 pp. $5.95

Of all aspects of construction, concrete work may be the most difficult to estimate accurately. The LeJeune method is described and illustrated in this book. It is applicable to the estimating needs of small, medium, and large firms.

A02-0352
Lowe, C. W. Critical Path Analysis by Bar Chart. Cahners, 1969. 198 pp. $10.75

Taking the critical path method as its basis, this book shows how a simple chart technique--the job progress chart--can be applied to the planning, scheduling and continuous control of all work involved in a project, however large and complex it may be.

A02-0353
McKee-Berger-Mansueto. Building Cost File. Construction Publishing Co., 1976. 302 pp. $19.95

Unit prices of labor, materials, and equipment used by architects, engineers, contractors, sub-contractors, owners, and managers to compute building costs item by item. Annually produced in four regional editions, 72-city adjustment index. 7500 entries covering all phases of general building construction.

A02-0354
McKee-Berger-Mansueto. Design Cost File. Construction Publishing Co., 1976. $29.95
  Provides composite and unit-in-place costs for all items found in building construction. Sound and thermal transmission factors also given. Intended primarily for comparison of alternatives but can also be used as a pricing guide. Revised annually.

A02-0355
McKeever, J. R., ed. Community Builders Handbook. Urban Land Institute, 1968. 526 pp. $20.00 (for ULI members, $16.00)
  A text on community development. Arranged in four sections: residential subdivisions, special land uses, shopping centers, and industrial parks.

A02-0356
McKenna, H. D. A House in the City. Van Nostrand Reinhold, 1973. 160 pp. $12.95
  A handbook offering solutions to the special problems that go with owning a brownstone house. Advice on choosing and financing a house, selecting a neighborhood, and actual renovation.

A02-0357
McNeill, T. F. and Clark, D. S. Cost Estimating and Contract Pricing. American Elsevier, 1966. 514 pp. $23.00
  A presentation of up-to-date techniques in pricing and estimating reinforced by over 100 graphs, charts, and illustrations, the book covers both the manual and computerized approaches. It is a reference and working tool for estimating and pricing, finance and accounting, purchasing and procurement, industrial engineering and production. The Estimating System is based on the "Jo Bloc Technique" which has wide acceptance in American industry. A complete glossary of contract and pricing terms is included.

A02-0358
Means, R. S. Building Construction Cost Data. R. S. Means, 1976. 344 pp. $14.50
  An estimating book for building construction estimating covering 17,000 items. Costs given on square and cubic foot basis with range and median cost of 50 building types and plumbing, HVAC and electrical costs and percentages tabulated separately. Contains crew sizes with equipment and average daily productivity for almost every item. Labor costs are available either as bare costs or as costs including subs' overhead and profit.

A02-0359
Means' Building Systems Cost Guide. R. S. Means, 1976. 200 pp. $27.50

Planning--Project

For architects, designers, planners and contractors to obtain current total building preliminary costs on a system-by-system basis. Using preliminary design parameters such as bay size, number of stories, type materials, live load, etc., as input, project costs can be developed. Price indexes are included for over 200 cities in the United States and Canada.

A02-0360
Metcalf, K. Planning Academic and Research Library Buildings. McGraw-Hill, 1965. 431 pp. $18.95

Written for architects, librarians, academic boards, city, and governmental offices, this definitive guide covers all facets of the design and planning of libraries. Topics discussed include height, accommodations for readers and staffs, lighting, ventilating, furniture and equipment, and even the width of the aisles and shelves.

A02-0361
MHMA, Land Develop. Div. Mobile Home Site Planning Kit: Basic Information Concerning Construction Park Development. Mobile Homes Manufacturers Assn., 1971. $10.00 Looseleaf.

A02-0362
Milne, J. A. Builders' Estimating Simply Explained. George Godwin Ltd., 1971. 194 pp. £2.25

This study of the techniques of builders' estimating covers contract procedure, information required for estimating, the method for determining the percentage required for profit and on cost, and the calculation of "all-in" labour rates. Pricing of a preliminaries bill is explained and there are examples of the build-up of unit rates for items in most sections of SMM. Also chapters on mechanical plant, motor transport, pro-rata rates and incentive schemes.

A02-0363
Minnesota Electric Assn. Estimating-Accounting Manual for Electrical Contractor Dealers. Minnesota Electric Association. Vol. 1, 90 pp. Vol. 2, 315 pp.

A tool for electrical contractors and others. The first volume includes business practices, finance, collections and labor relations. The second volume of technical data contains tables, formulas, etc., and information for lighting layouts. Cost estimation of a wide array of electrical work comprises much of the book. The books are offered on a lease basis only. The cost is $70.00 for the first year and $20.00 per year thereafter, including cost of updating.

A02-0364
Moselle, G. W. National Construction Estimator. Construction Book Co., 1973. 288 pp. $6.75

Accurate building costs for residential, commercial and industrial construction. Material prices for every commonly used building material, the proper labor cost associated with installation of the material. Many time-saving rules of thumb, waste and coverage factors and estimating tables are included.

A02-0365
Nierstrasz, F. H. J. Building for the Aged. North-Holland, 1961. 187 pp. $16.40
An assessment is given of the building requirements to which special dwellings for the aged, such as pension homes, nursing homes, homes for mentally infirm old people and geriatric departments in general hospitals should conform. A considerable number of ground-plans and photographs of projects already realized are included.

A02-0366
Nutt, M. C. Functional Plant Planning, Layout, and Materials Handling. Exposition Press, $15.00
Helps solve functional phases of plant site selection, plant planning, layout and materials handling. Assembles data and material which previously had been scattered among various texts and reference sources.

A02-0367
Ottaviano, V. B. National Mechanical Estimator. Ottaviano Technical Services, 1973.
Thirteen chapters covering budget estimating, estimating duct work, insulation, piping, piping hook-ups, controls, power wiring, plumbing, air pollution abatement devices are given in this looseleaf book. This data was compiled from more than 50 seminars on mechanical estimating given by the author.

A02-0368
Ottaviano, V. B. National Mechanical Estimator. R. S. Means, 1976. 200 pp. $35.00
The manual covers: sheet metal, piping, plumbing, budget estimating, air-conditioning service, computer estimating, air and water pollution, air and water balancing, fiberglass duct, control wiring, duct work and much more.

A02-0369
Page, J. S. Estimator's Construction Man-Hour Manual. Gulf Publishing Co. or R. S. Means, 1960. 241 pp. $10.95
Areas covered include demolition, site-grading, structural excavation, concrete roadways, foundation piling, formwork, reinforcing steel and mesh, concrete, masonry, structural steel, carpentry, millwork, metal doors, sash glass and glazing, special walls and ceiling, roofing, siding, ornamental metal, special partitions and painting.

A02-0370
Page, J. S. Estimator's Equipment Installation Man-Hour Manual. Gulf Publishing Co. or R. S. Means, 1964. 164 pp. $10.95

Planning--Project

This manual covers air-conditioning units, blowers, boiler units, dust collectors, steam-jet ejectors, evaporators, continuous centrifugal extractors, fans and motors, filters, flotation machines, gas holders, pumps, reactors and screens and related equipment.

A02-0371
Page, J. S. Estimator's Manual of Equipment and Installation Costs. Gulf Publishing Co. or R. S. Means, 1963. 232 pp. $10.95

In field testing by experienced estimators, this manual proved an accurate method for computing the cost of equipment and the cost to install it. Breaks costs down into graph form according to size and capacity. Areas covered are equipment, piping, instrumentation, insulation, electrical and other costs.

A02-0372
Page, J. S. Heating, Plumbing and Air-Conditioning Man-Hour Manual. Gulf Publishing Co. or R. S. Means, 1961. 145 pp. $10.95

This manual is divided into six sections. Section 1 covers service piping, plumbing and drainage. The earthwork section covers all types of machine excavation. Section 3 discusses concrete and masonry work. Section 4 covers heating, ventilating and air-conditioning. Section 5 covers equipment-installation, and Section 6 contains technical information.

A02-0373
Page, J. S. and Nation, J. G. Estimator's Electric Man-Hour Manual. Gulf Publishing Co. or R. S. Means, 1959. 231 pp. $10.95

Areas covered in 15 sections include conduit, boxes and fittings, service and feeder wiring, branch circuit items, lighting fixtures, panelboards and cabinets, safety switches, hangers and fasteners, signal systems, demolition, excavation and concrete.

A02-0374
Page, J. S. and Nation, J. G. Estimator's Piping Man-Hour Manual/Revised Edition. Gulf Publishing Co., 1968. 164 pp. $10.00

Areas covered in 10 sections include shop fabrication of pipe and fittings, field fabrication and erection, alloy and nonferrous fabrication, underground piping, hangers and supports, painting, scaffolding, insulation, sample estimate and technical information.

A02-0375
Parker, A. Planning and Estimating Dam Construction. McGraw-Hill, 1971. 320 pp. $27.50

This book will aid construction and design engineers who need data on construction equipment and plant facilities, design, river diversion procedures, placement methods, and powerhouse construction. Because planning and estimating are covered in such a comprehensive manner, this book will definitely help in improving every technique in submitting more successful, profitable bids.

A02-0376
Parker, A.   Planning and Estimating Under-Ground Construction.
     McGraw-Hill, 1970.   350 pp. $18.00
     The construction methods and equipment used to excavate and line tunnels, shafts, and under-ground chambers are described in this book which details the application of tunnel boring mechanics, protective shields, drills, muckers, transportation systems, ventilation facilities, shaft sinking equipment, and concreting plants.

A02-0377
Paxton, A.   National Repair and Remodeling Estimator.   Craftsman Book Co., 1973.   256 pp. $6.50
     A pricing guide for dwelling reconstruction costs. Specific data applicable on every remodeling job. Up-to-date material costs and labor figures based on repair and remodeling jobs across the country. Professional estimating techniques to help determine the material needed, the quantity to order, the labor required, the correct crew size and the actual labor cost for any area.

A02-0378
Peart, A. F.   Design of Project Management Systems and Records.
     Cahners, 1972.   206 pp. $12.95
     This book describes the basic systems framework and the detailed documentation necessary to define project objectives; analyze materials and finance; plan the timescale; co-ordinate parallel activities; control expenditure; report results.

A02-0379
Peurifoy, R.   Estimating Construction Costs.   McGraw-Hill, 1958.
     446 pp. $17.50
     Discusses the principles of estimating construction costs--from accurate production rates for labor and equipment to hourly costs of owning and operating such equipment.

A02-0380
Pulver, H.   Construction Estimates and Costs.   McGraw-Hill, 1969.   644 pp. $22.50
     The edition includes new sections on freight and express forwarding companies, kinds and measurement of materials; drywall conduction; air cooling systems; and other topics not previously covered. Presents each phase of construction work separately and illustrates simple, arithmetical methods for accurately estimating its cost.

A02-0381
Rabb, J. and Rabb, B.   Good Shelter: A Guide to Mobiles, Modulars, and Prefabricated Homes.   Harper and Row, 1976.
     192 pp. $6.95
     The Rabbs have researched the products of over 100 manufacturers of factory-produced homes. They give details on design,

Planning--Project

construction materials, costs, construction, and problems for each type of home; they have written a special chapter on their admitted preference: domes. In addition, the authors take the reader step by step through the decision-making process, rendering valuable advice on assessing your living needs, choosing the right home, selecting your site, dealing with brokers, lawyers, electricians, plumbers, contractors, and builders, and telling you what you need to know about zoning, building codes and, in the case of mobiles, park rules.

A02-0382
Redstone, L. New Dimensions in Shopping Centers and Stores. McGraw-Hill, 1973. 384 pp. $19.95

This book discusses some of the new concepts in shopping center design, including new trends in the design of regional centers, new approaches in store design, the reconversion of existing un-economic buildings, and others. The how-to-do-it aspects of planning design, construction, maintenance, protection, etc., are included in a special section of the book.

A02-0383
Rental Rate Blue Book. Equipment Guide-Book Co., 1976. $75.00

More than 22,000 monthly, weekly, daily and hourly rates for renting most types of construction equipment are given in this compilation. An area adjustment map permits variations in rental costs to be taken into account.

A02-0384
Richardson. Fabricators and Erectors Estimating Standards. Richardson Engineering Services, 1976.

Covers in one volume the cost of fabrication and erection of structural steel, miscellaneous steel, pressure vessels and tanks.

A02-0385
Richardson. General Construction Estimating Standards. Richardson Engineering Services, 1976. 2000 pp. $48.50

A three-volume cost book for the General Building and Heavy Construction Contractor covering commercial and industrial building construction, and public works projects.

A02-0386
Richardson. Light Construction Estimating Standards. Richardson Engineering Services, 1976.

A one-volume book that covers the cost estimates for the construction of light commercial and industrial buildings, single and multiple family housing. Intended for the small building contractor.

A02-0387
Richardson. Mechanical and Electrical Construction Estimating Standards. Richardson Engineering Services, 1976.

This single-volume work is used for estimating mechanical and electrical construction including: air conditioning, heating and ventilation, utility and plumbing piping systems for General Construction projects.

A02-0388
Richardson. Process Plant Construction Estimating Standards. Richardson Engineering Services, 1976. 3000 pp. $225.00
  A four-volume book of cost estimating data for construction of chemical plants, refineries, solids processing, water treatment, waste treatment, pollution control, and industrial facilities and buildings. Volume one includes sitework, piling and concrete. The second volume covers masonry, metals, carpentry, doors, windows, finishes and specialties. The third costs mechanical and electrical, while the fourth involves process equipment.

A02-0389
Roth, D. S. Architects, Contractors and Engineers Guide to Construction Costs. Concrete Construction Publications, 1974. 128 pp. $7.00
  Provides cost information covering every common type of material or equipment. Follows the "Uniform System for Construction Specifications, Data Filing, and Cost Accounting" outline found in most specifications as received from architects. Cost data for concrete and concrete formwork, masonry, metals, carpentry, moisture protection, finishes, doors and windows, mechanical and electrical systems, conveying systems and other special aspects of construction. Adjustment factors are provided for conversion to local cost levels for 55 major metropolitan areas. Revised editions published every July.

A02-0390
Rothschild, Bernard B. Construction Bonds and Insurance Guide. American Institute of Architects, 1973. $8.00 (for AIA members, less 20%)
  Reference on bonds and insurance related to construction projects, with glossary of insurance terms and suggested check list.

A02-0391
Siddall, J. N. Analytical Decision Making in Engineering Design. Prentice-Hall, 1972. 431 pp. $14.95
  Shows how decision-making in design is done analytically. Brings together and conceptually integrates concepts of value, probalistic decision theory, optimization analysis, and reliability. Shows how engineering analysis can be applied to get the best feasible design rather than just a feasible design.

A02-0392
Smith, H. C., et al. Real Estate and Urban Development. Dorsey Press, 1973. 450 pp.
  This book emphasizes the decision-making aspects of real estate. Included are the traditional matters of real estate courses but with less focus on legal and institutional arrangements. The macro aspects of public decisions involving real estate are included:

Planning--Project

A02-0393
Sokol, A., Jr. Contractor or Manipulator? Univ. of Miami Press, 228 pp. $12.50
This book is a guide to construction financing from beginning of construction to completion. It is aimed at those who grant or take out mortgages for interim or end-financing of construction projects. It intends to simplify and expedite all transactions between savings and loan associations or other trustees, and contractors and subcontractors. All the methods and sample forms in the book apply to the smallest home improvement job as well as to a construction job of several million dollars. They are valid in all states that have a mechanic's lien law.

A02-0394
Steinberg, J. and Stempel, M. Estimating for the Building Trades. American Technical Society, 504 pp. $9.95
This second edition teaches exact procedures for accurate, professional estimating. Detailed house plans are included in the text for use in sample and practice estimates. This revision also includes five completely revised or new chapters: two on framing, one on plumbing, one on electric wiring, and one recapitulating the entire study to reinforce what has been learned.

A02-0395
Stires, D. M. and Wenig, R. P. Pert/Cost. Cahners, 1964. 398 pp. $12.50
Designed to give reader a working knowledge of the information and procedural requirements necessary to carry out an effective PERT/COST implementation using a hypothetical case problem method. Not intended to serve as a primary source of information in describing the concepts and principles of the DOD/NASA PERT/COST system. Conforms to guidelines of DOD and NASA.

A02-0396
Strickland, W., et al., eds. Reports, Specifications and Estimates of Public Works in the United States of America. Augustus Kelley. $12.50

A02-0397
Thomas, P. I. How to Estimate Building Losses and Construction Costs. Prentice-Hall, 1975.
Gives information and examples of how to prepare estimates for repairing and replacing damaged building components and equipment. Covers damage from fire, water, smoke, explosions, flooding, backing up of sewers, earthquake and other causes. Actual estimating and development of unit costs are covered.

A02-0398
Turin, D. A., ed. Aspects of the Economics of Construction. George Godwin Ltd. (In preparation.)

Professor Turin has gathered together a team of well-known contributors to cover major aspects, both of the place of construction in the total economy and of the factors influencing the economics of building.

A02-0399
Turner, D. F. Quantity Surveying Practice and Administration. George Godwin Ltd., 1972. 230 pp. £3.40

The first section of this book discusses the general organization of the quantity surveyor's practice and its relationship with the client and with other members of the construction team. There follows consideration of the quantity surveyor's work from pre-tender activities to the settlement of final account. Attention is given to general procedures and to such variants as the prime cost contracts, the all-in service and others. Insolvency of the contractor is dealt with in one of the two case studies.

A02-0400
Urban Land Institute. PUD--A Better Way for the Suburbs. American Institute of Architects, 1971. $10.00 (for AIA members, less 10%)

History, current suburban response and viewpoint of the community resident on Planned Unit Development with 12 site plans and full-color photographs.

A02-0401
Walker, F. R. The Building Estimator's Reference Book. Frank R. Walker Co., $19.50

Conforms with the Construction Specifications Institute, Inc. format. Topics include, role of contractor, drawings and specifications, cost estimates, CPM, equipment, site work, concrete, masonry, metals, carpentry, thermal and moisture protection, doors and windows, finishes, specialties, conveying systems, electrical, mechanical, remodeling and mensuration. Explains how to figure mark-ups, and other pointers useful in solving problems of estimating, appraisal, cost accounting and job schedules.

A02-0402
Wass, A. Building Construction Estimating. Prentice-Hall, 1970. 324 pp. $12.50

A treatise on preliminary, detail, and quantity estimating, this edition includes such material as: the critical path method--The American Institute of Architects' uniform system for construction specifications, data filing, and cost accounting; concrete form hardware; and a fuller treatment of land leveling, cut and fill.

A02-0403
Weiss, J. Better Buildings for the Aged. McGraw-Hill, 1971. 286 pp. $28.50

Analyzes the major factors involved in the successful planning, financing, and construction of housing for the aged. Tells what to build and how to build.

A02-0404
Wendt, P. F. Real Estate Appraisal: Review and Outlook. Univ. of Georgia Press, 1974. 276 pp. $11.00
  Reviews developments in the theory and practice of real estate since 1950, and focusing upon the most recent innovations, predicts future developments likely to occur in the field.

A02-0405
Werbin, I. V. Practical Legal Library for Architects, Contractor and Engineers. Central Book Company. Vol. 1, Legal Phases of Construction Contracts, $10.00. Vol. 2, Legal Guide for Contractors, A. & E., $10.00. Vol. 3, Legal Cases for Contractors, A. & E., $10.00. Vol. 4, Law for Contractors, A. & E., $12.50. Vol. 5, Legal Opinions for Contractors, Architects and Engineers, $12.50. 5 volumes, 1970. 2048 pp. $49.50
  The library is comprised of five volumes which discuss 494 different legal problems that may arise in construction contracts and on the job. The material is based upon court decisions and the principles of law are presented in understandable language.

A02-0406
What Went Wrong? Council of Educational Facility Planners, 1968. 248 pp. $5.00
  A compendium of planning, construction, operational and maintenance errors to avoid in planning educational facilities. Based on the experience of the various authors, who describe the price of mistakes in terms of the problems they produce. A prescription against mistakes for those who must plan an educational facility or for those who must maintain one.

A02-0407
Wheeler, E. Hospital Modernization and Expansion. McGraw-Hill, 1971. 288 pp. $25.00
  This book discusses the problems of adding to or modernizing an existing hospital. Covers problems associated with analysis, design, and execution of various program types. A source book for architects, administrators, trustees, doctors, nurses and others.

A02-0408
Whitman, E. S. and Schmidt, W. J. Plant Relocation. American Management Assn., 1966. 158 pp. $7.50 (Order from University Microfilms, Inc.)
  A report on how General Foods moved four outmoded plants to one modern facility. Discusses all aspects of the move; the reasons for it, site selection and plant design, solutions to personnel problems, and public relations. Points out many problems and decisions facing the company that must relocate.

A02-0409
Wicks, H. How to Plan, Buy or Build Your Leisure Home. Prentice-Hall, 1976. 544 pp. $19.95

Gives information on site selection, area selection and ownership costs. There is thorough information here on getting financial help, working with brokers, lawyers, architects, and builders, building an energy-saving house, buying an older home and remodeling it, and building an expandable house. Written from the consumer's point of view.

A02-0410
Wood, C.   Mechanical Estimators Handbook.   Craftsman Book Co., 1971.   288 pp. $8.95

A guide to estimating pipe and fittings for residential and commercial construction, power plants, refineries, water works, booster stations, sewer plants and storage facilities. The labor required to handle, cut, fit and install pipe, the weight, diameter, wall thickness and test pressures needed. Information on planning and layout, figuring overhead and profit and professional cost estimating techniques.

A02-0411
Wynne, G. B.   Building Estimating.   Charles Merrill, 1973.
320 pp.

The author introduces principles, theories, and procedures involved in building and construction cost estimating, and discusses the factors governing the cost of construction materials. He looks at the controlling elements involved in overhead and profit. Topics covered include excavation, concrete, structural steel, masonry, wood and laminated construction, moisture protection, glazing, finished, prefabricated materials and mechanical equipment.

Section   A03

MANAGEMENT

                Contracting and Construction Management
                Contracts and Law
                Engineering Management
                Inspection, Tests, Quality Control

A03-0420
Abbett, R. W. Engineering Contracts and Specifications. John Wiley, 1963. 461 pp. $16.25

A03-0421
Abrahamson, M. W. Engineering Law and the I.C.E. Contracts. Applied Science, 1974. 500 pp.
A guide to the law for civil engineers, employers and contractors, particularly to their rights and duties under the ICE Contract forms.

A03-0422
Abrams, E. D. and Blackman, E. B. Managing Low and Moderate Income Housing. Praeger, 1973. 188 pp. $13.50
Details the management of low and moderate income housing from the architectural conception through relations with tenant organizations. Isolates the problems and possibilities which exist for the sponsor/developer, the manager, and the tenants at each stage of the process.

A03-0423
ACI Manual of Concrete Inspection. American Concrete Institute, 1975. 268 pp. $10.00 ($7.50 for ACI members)
This revised manual provides inspection information for concrete mixes, formwork, cold and hot weather techniques, grouting, terrazzo, mortar, stucco, lightweight aggregrates, underwater concrete work, bridge decks and other topics.

A03-0424
Acret, J. California Construction Law Manual. Shepard's Citations, 1975. $22.50
Thirteen chapters on construction contracts, arbitration, contractor's license law, real estate law, mechanics' liens, subcontractor claims, and public contracts comprise this manual. Of use to California contractors, building professionals and attorneys.

A03-0425
Adrian, J. J. Quantitative Methods in Construction Contracting. American Elsevier, 1973. 448 pp. $20.00
The book presents a discussion of the various required tasks of construction contracting from a quantitative point of view. It shows how operations research, economic analysis, and information system models can be applied to such contracting tasks as: finding and selecting projects, planning, scheduling, estimating, and bidding on projects and controlling of project and overall company

costs. Each task serves as background material for the application of appropriate quantitative models.

A03-0426
Adrian, J. J. Quantitative Methods in Construction Management. American Elsevier, 1973. 491 pp. $20.70
The book offers a thorough discussion of the various required contracting tasks, and the application of quantitative models to them. By means of modeling the information relevant to each required contracting task, and by recognition of the interdependencies that exist between them, the book provides the tools for making decisions that accomplish profit making objectives.

A03-0427
Aguilar, R. Systems Analysis and Design in Engineering, Architecture, Construction and Planning. Prentice-Hall, 1973. 448 pp. $17.95
A discussion of the systems approach to the solution of complex problems in engineering, architecture, construction management and planning. Fundamental principles of economics and deterministic and stocastic systems are discussed.

A03-0428
AIA Building Construction Legal Citator: 1971. American Institute of Architects, 1973. $35.00 (for AIA members, less 20%)
This research manual gives quick access to all cases to date concerning AIA contracts and forms. Current and past editions of each contract form reproduced and cross-referenced. Useful to the architect and architect's attorney in preventing or dealing with litigation.

A03-0429
AIA. Comprehensive Architectural Services. McGraw-Hill, 1965. 241 pp. $12.50
Describes all aspects of the practices and services, in all areas of design and construction, which architects perform. Shows how architectural services such as analysis of finance, feasibility, and real estate are performed. Discusses methods architects use to market their services.

A03-0430
AIA. Creative Control of Building Costs. McGraw-Hill, 1967. 256 pp. $18.50
Emphasizes the need for cost control, details the application of cost control techniques, and helps architects determine realistic budget and control costs in initial design stages.

A03-0431
AIChE. Engineering Construction Contracts, April 1970 Proceedings. American Institute of Chemical Engineers, 1970. $15.00. Order No. X-44
This proceedings volume recounts an exploratory conference on legal and technical problems in developing standard contract

clauses in engineering construction contracts. Presentations plus comment and exchange on the clauses found in chemical and process engineering-construction contracts. Discussion of the possibility of reaching common understandings on clauses.

A03-0432
AIChE. Engineering Construction Contracts, August 1970 Proceedings. American Institute of Chemical Engineers, 1970. $35.00. Order No. X-45

Recorded are panel discussions and questions and answers on insurance, pollution clauses, plant acceptance, guarantees and warantees. Topics are analyzed from owner, contractor, and vendor points of view.

A03-0433
AIChE. Engineering Construction Contracts. American Institute of Chemical Engineers, 1972. $50.00. Order No. X-47

Panel discussions, comments, questions and answers, dialogue carried on by representatives of contractors, owners and suppliers. Included are sessions dealing with selection of contractors, selection of contract types, insurance and indemnity, changes and change orders. From 1971 Proceedings.

A03-0434
AIChE. Engineering Construction Contracts. American Institute of Chemical Engineers, 1973. 145 pp. $35.00. Order No. X-48

Owners, contractors, suppliers from more than 77 large, medium and small organizations extend their conversations, joined by a construction labor leader and a governmental representative of OSHA. Subjects include: bidding/negotiation; construction financing; impact of government regulations, with accent on OSHA; construction costs, with accent on labor. Reading list of engineering/contracting articles included. From 1972 Proceedings.

A03-0435
Andersen, A. and Co. Financial Management for Architectural Firms--A Manual of Accounting Procedures. American Institute of Architects, 1970. 163 pp. $12.00 (for AIA members, less 20%)

This book offers new techniques for overall firm management; an improved method for calculating compensation; how to estimate project retail value before contract negotiations; how to exercise more efficient project and manpower control. There are accounting procedures for handling billing, payroll and outside payments, the reporting of time and expense, and the preparation of financial reports, along with illustrations of forms and reports showing the procedures in use.

A03-0436
Andree, R. V., et al. Computer Programming: Techniques, Analysis, and Mathematics. Prentice-Hall, 1973. 420 pp. $12.95

Management

A problem-solving approach for mathematics, science and engineering. Designed to teach efficient programming techniques rather than simple FORTRAN coding. Features a chapter on how to read FORTRAN, and a chapter on summarizing good programming practice.

A03-0437
Antill, J. M. Civil Engineering Management. American Elsevier, 1970. 349 pp. $16.00

Such topics as estimating, contracts, project development and feasibility, specifications, subcontracting, and financial and personnel management are dealt with in detail. A chapter is included on management failures in which an attempt is made, with the support of statistical evidence, to pinpoint the main reasons why construction projects go wrong.

A03-0438
Antill, J. M. and Woodhead, R. W. Critical Path Methods in Construction Practice. John Wiley, 1970. 414 pp. $19.95

Shows the application of critical path methods for the solution of the practical problems commonly encountered by all those engaged in the management, administration, and practice of construction.

A03-0439
Architectural Record. Techniques of Successful Practice for Architects and Engineers. McGraw-Hill, 1974. 172 pp. $14.95

Architects, engineers, administrative officers and staff personnel of architectural and engineering firms found this book to be invaluable in developing and improving a successful professional practice when it first appeared as a subscription premium for Architectural Record.

A03-0440
ASCE. National Conference on Construction Contracts. American Society of Civil Engineers, 1967. $8.00 (for ASCE members $4.00)

From ASCE Conference, Washington, D.C., May, 1967.

A03-0441
Ashley, R. Electrical Contracting. McGraw-Hill, 1961. 280 pp. $19.95

Shows how to manage contracting on a profitable basis. Includes data on overhead cost studies.

A03-0442
ASME. Managing for Improved Engineering Effectiveness. American Society of Mechanical Engineers, 1972. 142 pp. $5.00 (for ASME members, $4.00)

Topics covered include engineering department scheduling and control, the use of computers, engineering effectiveness, styles of project managers, systems approach to management of capital, impact of OSHA, and the growing role of independent laboratories.

A03-0443
ASTM. Manual on Quality Control of Materials. American Society
    for Testing and Materials, 1960. 152 pp. $3.00 (for ASTM
    members $2.40)
    Presents information on statistical methods and quality control methods. Part 1 discusses the application of statistical methods in condensing information. Part 2 is on the problem of presenting limits to indicate the uncertainty. Part 3 covers the control chart method of analysis of data.

A03-0444
Atkinson, I. Construction Management. Applied Science, 1971.
    150 pp. £2.00
    The author begins with the approach to management studies and a discussion of the human aspects. Marketing methods and trade unions are dealt with, followed by chapters on decision making; the history of management; company structure and finance; and policy making. The author concludes with chapters on contracts, obligations, and legal matters.

A03-0445
AWPA. Equipment Management Manual. American Public Works
    Assn., $15.00
    One of the most comprehensive compilations of information on equipment management, the manual is being published in two parts; one available now and the other to be produced in the near future. Part I, produced by the APWA Research Foundation, consists of six segments: Spec Writing-Hardware, Replacement Analysis, Parts Inventory Control, Preventive Maintenance, Management Information Systems, and APWA Data Bank Standards.

A03-0446
Ayers, C. Specifications: For Architecture, Engineering and Construction. McGraw-Hill, 1975. 448 pp. $14.50
    This introduction to the field of architectural and engineering specification writing covers legal citations by name and number, industrial specifications, value engineering, computerized specifications, and government safety regulations.

A03-0447
Azad, H. S., ed. Industrial Wastewater Management Handbook.
    McGraw-Hill, 1976. 608 pp. $29.50
    Written for management personnel and engineers of water-using industries who are concerned with environmental conservation and pollution control, this handbook deals with the water pollution control problems of six major industries: Chemical; Petroleum; Metals; Pulp and Paper; Power; and Food and Beverage. Recent legislation, specific standards, problem definition, treatment technology, and control equipment are subjects treated in this book.

A03-0448
Barrodale, I., et al. Elementary Computer Applications: In Science, Engineering and Business. John Wiley, 1971.

254 pp. $6.25
Introduces various computer techniques used in science, engineering, and business. Divided into three sections; numerical analysis, operations research and data processing. Helps to develop programming skills through exercises and flowcharts given in each chapter. Independent of any particular programming language and requires very little knowledge of mathematics.

A03-0449
Battersby, A. Network Analysis for Planning and Scheduling.
John Wiley, 1970. 332 pp. $11.25
Examples of the main features of network analysis are introduced through the medium of case studies, which are developed with increasing complexity throughout the book. Reviews fundamentals while giving a full appreciation of the potential of the technique.

A03-0450
Bayley, L. G. Building: Teamwork or Conflict. George Godwin Ltd., 1973. 128 pp. £2.50
A view of labor relations in the construction industry, with particular reference to the effects of the Industrial Relations Act. This book will be of interest to employers and trade unions and the suggested code of industrial relations, together with the model procedure agreement, will be useful on many sites.

A03-0451
Begley, F. D. Project Management for Construction Superintendents. Cahners, 1974. 140 pp. $8.95
This book is a survey of systems, procedures, and planning techniques that can be applied directly to the running of any construction project. Chapters are arranged in chronological sequence, beginning with preliminary planning and site inspection and continuing through job administration.

A03-0452
Benjamin, J. and Cornell, C. Probability, Statistics and Decisions for Civil Engineering. McGraw-Hill, 1970. 640 pp. $18.00
Realistic engineering examples involving economic decisions under conditions of uncertainty are stressed. Presents the principles of applied probability and statistics needed to optimize and understand the relation of analysis to practical engineering decisions.

A03-0453
Benson, B. Critical Path Methods in Building Construction.
Prentice-Hall, 1970. 132 pp. $8.90
Overview of planning, scheduling, and controlling methods for building construction projects. Covers pictographs, percentages of completion by cost, percentages of completion by quantities of work performed, lazy S curves, and CPM. Eighty per cent of the book focuses on the most inexpensive ways to put a project

back on the track once it has gotten into trouble. Explains how to
use a computer for project control enabling the reader to plan a
project with or without the use of a computer.

A03-0454
Berger, S. and Godel, J. B. Estimating and Project Management
 for Small Construction Firms. Construction Publishing Co.,
 1976. 400 pp. $19.95
 Book addresses itself to needs of small contractor from understanding prints and specifications to types of construction contracts, project management and closeout. Extensive backup is
given on estimating excavation, concrete, masonry, carpentry,
and steel work. Methods are suggested to check the estimates of
HVAC, plumbing and electrical subcontractors.

A03-0455
Betz, C. E. Principles of Magnetic Particle Testing. American
 Society for Nondestructive Testing, 1966. 528 pp. $8.35
 The theory, science, art and practice of all types of industrial testing with magnetic particles are covered. Recent develop-
ments--from basic physics experimentation to production systems
and portable test equipment--are discussed.

A03-0456
Betz, C. E. Principles of Penetrants. American Society for Nondestructive Testing, 1969. 506 pp. $8.35
 This volume on penetrant nondestructive testing offers information on materials, techniques, equipment and uses of penetrants to detect cracks, leaks, porosity and similar defects in
metallic and nonmetallic parts and materials.

A03-0457
BNA Editorial Staff. OSHA and the Unions: Bargaining on Job
 Safety and Health. Bureau of National Affairs, 1973. $3.50
 This report presents the views and experience of spokesmen
for management, the unions, and the government on safety and
health bargaining in the wake of the 1970 Occupational Safety and
Health Act. It reviews the background of prior bargaining on these
issues, the nature of current union demands, the recent gains made
by unions in contract bargaining, the practical aspects of major
issues, the legal status of safety and health bargaining, and the
status of union demands for company data on safety and health.

A03-0458
Bonny, J. B. and Frein, J. P. Handbook of Construction Management and Organization. Concrete Construction Publications
 or Van Nostrand Reinhold, 1973. 688 pp. $32.50
 This handbook covers every facet of construction management. Includes the basics of contracting, building and organization, bid strategy, construction contracts, financing, equipment
maintenance and repair, the functions and organization of a contractor's engineering section, cost controls, networking techniques

Management

for project planning, scheduling and control, safety procedures, public relations for contractors, and other topics.

A03-0459
Brock, D. S. Cost Accounting Manual for Highway Contractors, A System for Cost Control. American Road Builders' Assn., 1971. 230 pp. $15.00

This manual provides the basis for development of a cost control system, cost accounting system and estimating and bidding system to fit the needs of each highway contractor. The cost control system is designed to determine costs and report them to management for comparison with estimated costs. The cost accounting system, used properly over a period of several years, will yield records of the contractor's actual costs plus his overhead costs of doing business.

A03-0460
Brunton, J. et al. Management Applied to Architectural Practice. George Godwin Ltd., 1964. 140 pp. £1.25

The authors examine the theory and practice of management in general and show how their application to the everyday problems of architectural practice can help to solve problems of both private and official practices. Allied professions will find that the application of management principles to the planning and design of buildings will help them to achieve greater productivity and improved relations between all parties concerned.

A03-0461
Burman, J. Precedence Networks for Project Planning and Control. McGraw-Hill, 1973. 344 pp. $17.00

This book introduces the various aspects of precedence networks in logical order, starting with the basic forms of network analysis and working through to advanced techniques and applications. It includes two full exercises, illustrating the functioning of precedence networks. The book is a ready-reference for the experienced manager, as well as a complete self-instruction guide for the novice.

A03-0462
Calvert, R. E. Introduction to Building Management. Butterworth, 1970. 310 pp. £4.00 (paperback £2.70)

The management activities which are encountered in a typical building concern are explained in this book. The practical approach is emphasized by the inclusion of charts and diagrams. Provides an introduction to almost every aspect of contracting.

A03-0463
Canfield, D. T. and Bowman, J. H. Business, Legal, and Ethical Phases of Engineering. McGraw-Hill, 1954. 376 pp. $13.50

Business economy. Business organizations. Interest, and discount. Annuities and sinking funds. Insurance. Depreciation, maintenance, and market values. Economic selection. Cost determination. Financial reports and records. Allocation of burden and overhead. Public utility rates. Business law. Contracts of commerce. Offer and acceptance. Statute of frauds. Engineering procedures. Engineering bids and specifications. Patents. Copyrights and trade marks. Industrial hazards and employment compensation. The practice of engineering. Engineering as a profession.

A03-0464
Case and Company. The Economics of Architectural Practice. American Institute of Architects, 1968. $6.00 (for AIA members, less 20%)
The results of a survey of costs of architectural services. Based on 223 firms and 1150 completed projects, this book enables the architect to discover his real costs of operation. Numerous graphs and forms illustrate procedures each firm can apply to establish equitable compensation for professional services.

A03-0465
Case and Company. Methods of Compensation for Architectural Services. American Institute of Architects, 1969. $8.00 (for AIA members, less 20%)
The book reviews and updates compensation methods for architectural services in light of the most varied services architects now provide. Many illustrations guide the architect in improving his services and managing his business efficiently.

A03-0466
Case and Company. Profit Planning in Architectural Practice. American Institute of Architects, 1968. $5.00 (for AIA members, less 20%)
A follow-up to The Economics of Architectural Practice. Offers architects an applicable action program to remedy the current "profit squeeze." A valuable management tool for establishing billing rates, determining direct and indirect costs, cost controls, and realistic compensation and profit expectations.

A03-0467
Caudill, W. W. Architecture by Team: A New Concept for the Practice of Architecture. Van Nostrand Reinhold, 1971. 362 pp. $17.95
This volume shows how the author's firm has integrated the talents of a broad range of specialists for more effective solutions to today's complex building problems. The author provides insights into the effects of industrialization, of the interdisciplinary approach, social reforms and the user's involvement during the planning process. Advice on professional affairs, programming, job promotion, design, computers, university planning, construction management, building systems, and management.

A03-0468
Central Electricity Generating Board. Phraseology for Civil Engineering. George Godwin Ltd., 1970. 2 vols., £12.50

Management

The phraseology is derived from an analysis of the Standard Method of Measurement of Civil Engineering Quantities. It is of value in providing a ready-made format for extension and/or modification of the method of measurement, which is necessary in order to cover special constructional techniques that arise in civil engineering works.

A03-0469
Clark, F. J., et al. Business Systems and Data Processing Procedures. Prentice-Hall, 1972. 333 pp. $8.95

Provides an overview of various business systems, their organization, nature, and scope, to prepare the reader for entry occupations in data processing. Contains an overview of systems analysis and systems improvement.

A03-0470
Clough, R. H. Construction Contracting. John Wiley, 1975. 453 pp. $16.95

Discusses the management and everyday workings of a construction contracting business. The book serves as a valuable information reference on construction topics, such as contracts, bonds, insurance, cost accounting, CPM, labor law, labor relations, and safety.

A03-0471
Clough, R. H. Construction Project Management. John Wiley, 1972. 264 pp. $17.50

A treatment of a construction project management system, written as a procedural guide for a project manager, stressing application and practicality. Emphasizes time and cost control, with time-control procedures being based upon the Critical Path Method (CPM). An Example Project is used throughout to illustrate the workings of the management system.

A03-0472
Cohen, H. Public Construction Contracts and the Law. McGraw-Hill, 1961. 400 pp. $22.50

This text carefully analyzes all procedures and legal rules relating to the contract--from prequalification to penalties for delay of work.

A03-0473
Commerce Clearing House. 1972 Government Contracts Guide. Commerce Clearing House, 1972. 920 pp. $17.50

Designed for those involved in qualifying for, negotiating and performing government contracts, this 4th edition reflects all of the ASPR revisions up to publication date. Discusses and analyzes eligibility, bidding requirements, specifications, labor requirements, cost principles, patent policies and much more. Includes checklist of required and permissible contract clauses for use in each type of contract.

A03-0474
Coombs, W. Construction Accounting and Financial Management.
McGraw-Hill, 1958. 481 pp. $22.50
   This handbook of financial management in the construction industry describes proper accounting and management procedures.

A03-0475
Cowgill, C. H. and Small, B. J. Architectural Practice. Van Nostrand Reinhold, 1959. 272 pp. $12.95
   Covers the professional, business and legal aspects of architectural practice.

A03-0476
Coxe, W. Marketing Architectural and Engineering Services. Van Nostrand Reinhold. (In preparation.)
   This book examines the promotional and new business development practices of architects, engineers, landscape architects, planners, and similar professionals.

A03-0477
Critical Path Techniques for Construction. Know-How Publications, 194 pp. $10.00
   The book instructs in the basic language of arrow and precedent network planning, and use of the Activity Time Chart. The advantages of a manual methodology are outlined.

A03-0478
Cronstedt, V. Engineering Management and Administration.
McGraw-Hill, 1961. 380 pp. $11.75
   A guide which describes the art of engineering management and administration to aid in the efficient operation of an engineering department. Covers patent laws, financial controls, safeguarding industrial secrets, etc.

A03-0479
CSI. Uniform Construction Index. Construction Specifications Institute, 1972. 314 pp. $8.00. Order No. 53310
   1972 Edition replaces the Canadian "Building Construction Index," and the United States "Uniform System for Construction Specifications, Data Filing and Cost Accounting." The new UCI has four distinct parts: Specifications Format; Data Filing Format; Cost Analysis Format; and Project Filing Format. Coordinated with the four parts is a Key Word Index of construction terms, locating each item within the individual format.

A03-0480
Dand, R. and Farmer, D. Purchasing in the Construction Industry. Beekman Publishers, 1970. 198 pp. $15.00
   This book shows how to make profit in construction by using an efficient purchasing system. It provides detailed guidance on the main aspects of the purchasing function, such as: best source of supply, materials and plant, managing supplies on the work site, and organization and conduct.

Management

A03-0481
Davis, H. E., et al. The Testing and Inspection of Engineering Materials. McGraw-Hill, 1964. 475 pp. $13.95

Chapters include: The nature of the problem; General features of mechanical behavior; The problem of failure; Measurement of load, length, and deformation--common testing apparatus; Static tension and compression tests; Static shear and bending tests; Hardness tests; Impact tests; Fatigue tests and tests of metals at low and high temperatures; Nondestructive testing and experimental stress analysis; Analysis and presentation of data; Principles of inspection.

A03-0482
Deatherage, G. Construction Company Organization and Management. McGraw-Hill, 1964. 300 pp. $19.95

The author treats organizational and directional phases of construction firm management, stressing functions relating to production costs and profits.

A03-0483
Deatherage, G. Construction Office Administration. McGraw-Hill, 1964. 303 pp. $19.95

Describes general office management functions needed for effective and efficient operation of a construction company office.

A03-0484
Deatherage, G. Construction Scheduling and Control. McGraw-Hill, 1965. 309 pp. $19.95

Provides coverage of all aspects of construction management which contribute directly to the successful conclusion of a project. Demonstrates the advantages of applying the Critical Path Method on all large projects, uses tabular schedules and bar graphs, and discusses such topics as purchasing, expediting, traffic management, and methods engineering.

A03-0485
Dell'Isola, A. J. Value Engineering in the Construction Industry. Construction Publishing Co., 1973. 181 pp. $16.50

Helps the reader increase his competitive position by showing how to save money and make money with this new construction method. Actual case histories show how to isolate unnecessary costs and pinpoint blind spots and how to apply the latest problem-solving techniques from other fields.

A03-0486
deNeufville, R. and Stafford, J. Systems Analysis for Engineers and Managers. McGraw-Hill, 1971. 320 pp. $17.50

An explanation of how systems analysis can be used to optimize the design of large scale engineering projects. Shows how the individual analytic techniques fit into the larger framework of systems analysis as a design process.

A03-0487
Dent, C.   Construction Cost Appraisal.   George Godwin Ltd.,
       1974.   £4.50
       Surveyors, building managers and many other professionals
and executives within the construction industry are becoming increasingly aware of the value of discounted cash flow techniques.
This book, intended for non-accountants within the industry, treats
the application of DCF techniques to construction industry problems. It includes the requirements of Corporation Taxation and
the use of computer analysis, and a range of worked examples.

A03-0488
Dibner, D.   Joint Ventures for Architects and Engineers.   American Institute of Architects or McGraw-Hill, 1972.   192 pp.
       $19.95
       Written primarily for architects, engineers, and contractors,
this book discusses the do's and don'ts of temporary partnerships
(generally called joint ventures). This book describes the relationships involved in this type of organization and provides the matrix
for their development.

A03-0489
Douglas, C. J. and Munger, E. L.   Construction Management.
       Prentice-Hall, 1970.   201 pp. $11.80
       Wide ranging view of the concerns, activities, and objectives involved in construction. Discusses important characteristics
of contract documents. Lists all the major construction Unions and
explains how they maintain liaison with the Contractor. Introduces
the Critical Path Method.

A03-0490
Downs, J. C., Jr.   Principles of Real Estate Management.   Institute of Real Estate Management, 1975.   476 pp. $13.50
       Topics include the rise of government involvement in real
estate, the significance of turbulent market conditions in the mid
'70's, merchandising and marketing, the administrative process, and
tenant selection. The book focuses on apartments, condominiums
and co-operatives, office buildings, commercial buildings, and
others. It concludes with an appendix containing major forms utilized by the professional property manager.

A03-0491
Dressel, G.   Organisation and Management of a Construction Company.   Applied Science, 1968.   191 pp.   £8.00
       The purpose of this book is to give advice on how a firm
of civil engineering contractors should be run. It deals primarily
with engineering in which a major part of the total expenditure in
each contract occurs on the site and for which the site labour
force (as opposed to the permanent technical staff) is recruited locally for each undertaking.

A03-0492
Dubin, M. D.   Architectural Supervision of Modern Buildings.   Van

Management   89

Nostrand Reinhold, 1963.   320 pp. $12.95

Defines the separate roles and prerogatives of architect,
engineer, contractor, and owner during the construction of a given
building project. The methods of organizing such an operation are
described. Technical aspects of building construction are included.

A03-0493
Dunham, C. and Young, R. Contracts, Specifications and Law for
Engineers.   McGraw-Hill, 1971.   523 pp. $18.50

Gives the reader: a basic understanding of various legal
principles involved in the preparation of contracts and specifications,
an acquaintance with the various legal matters which may arise in
the conduct of professional practice in engineering, architecture,
and construction, and an awareness of pertinent questions which
may arise.

A03-0494
Eacott, E. C. Specification in the Construction Industry. George
Godwin, 1970.   173 pp.   £2.00

This book sets down the principles for and describes methods of producing a sound specification. The author explains what
a specification is, how it is written, its position within the building
contract and how it is used. He gives examples of typical specifications and details suggested headings for a range of specification
clauses with extensive notes for specification writing.

A03-0495
Editorial Staff of "Construction Labor Report." Construction Craft
Jurisdiction Agreements.   Bureau of National Affairs, 1971.
$6.00

Here is a collection of the building trades jurisdictional
agreements which are recognized by the National Joint Board for
the Settlement of Jurisdictional Disputes, but which are not printed
in the "Green Book." This edition contains each agreement in
large facsimile, including photographic and diagrammatic exhibits.

A03-0496
Edwards, H. G. Specifications.   Van Nostrand Reinhold, 1961.
372 pp. $8.95

The concepts and practice of specification writing are carefully laid out keeping in mind its relationship to the contract document. Actual specifications for many facets of building construction are illustrated.

A03-0497
Emerick, R. Handbook of Mechanical Specifications for Buildings
and Plants.   McGraw-Hill, 1966.   496 pp. $16.95

Here is a checklist for engineers and architects designed to
guide the specification writer. It includes items for all types of
mechanical equipment and covers steam, diesel, gas, hydroelectric,
and nuclear power plants, steam distribution systems, hot water
systems, premises heating, and air conditioning systems.

A03-0498
Farmer, R. A. What You Should Know About Contracts. Arco, 1969. 224 pp. $4.95
A guide to contract law for the layman which outlines the requirements for a valid contract.

A03-0499
Fasal, J. H. Practical Value Analysis Methods. Hayden, 1972. $11.95
This applications oriented guide offers a basic working knowledge of the latest methods and techniques in value engineering and value analysis. Beginning with a brief discussion of value theory, the book then explains traditional approaches to VA/VE as well as recent developments such as decision theory, allocation of resources, cost effectiveness, organization of VA programs, and more.

A03-0500
Faulkner, E. Project Management with CPM. R. S. Means, $10.00
Presents the "state of the art" of Project Management with the critical path method of scheduling. It goes through the basic steps of manual preparation of CPM networks and shows how this manual approach can be adapted for computer processing. It demonstrates how profitable, construction cost control and resource allocation techniques, in conjunction with CPM, allow maximum benefits for management.

A03-0501
First Annual Conference--Value Engineering Association. Peter Peregrinus, 1968. 73 pp. $14.50
The 16 papers in this volume fall into four subject areas: value-engineering techniques; low-cost design; value engineering in practice; the future of value engineering.

A03-0502
Fletcher, L. and Moore, T. Standard Phraseology for Bills of Quantities. George Godwin, 1974. Four loose-leaf vols. £5.00 ea.
These volumes provide concise descriptions of measured work. They permit immediate access to millions of descriptions by employing a systematic presentation of conventional words and phrases and a simple process of progression which automatically establishes the correct order for any selection of items. Although designed for work item descriptions in bills of quantities, the compact phrase analysis provides a comprehensive schedule of specification notes suitable for use on drawings and trade literature. It thus provides a source of standard terminology and a direct communication link for all engaged in the design and construction process.

A03-0503
Fordham, P. Non-destructive Testing Techniques. Cahners, 1968. 166 pp. $8.95

Management   91

The first book on the subject to be designed for management, using a non-technical approach.

A03-0504
Forinton, M. A. Civil Engineering Contracts and Claims. Charles Knight and Co., 1973. £2.50
  The principal purpose of this book is to assist civil engineering contractors in obtaining proper payment for the work which they carry out. It is not an exposition of the law relating to civil engineering contracts, but it describes the application of the law to the contractor's work in England.

A03-0505
Foxhall, W. B. Professional Construction Management and Project Administration, McGraw-Hill, 1976. 136 pp. $17.50
  Illustrated with substantial segments of recent agreement forms of GSA, AIA, and private practitioners, the book features expanded sections on GSA selection procedures, contractual relationships, and liability exposures. Topics discussed include: the professional approach to management, organization for professional construction management, dealing with public and private clients, contracts and proposals, the use of computers in construction, and the participating options of small professional firms.

A03-0506
Foxhall, W. B., ed. Technique of Successful Practice for Architects and Engineers. McGraw-Hill (Book Club). 176 pp. $10.95
  This book shows you how to handle efficiently the all-important business side of professional practice. It spotlights the essentials of business development, cost control, project management, taxation, liability, personnel administration, and a host of other concerns that are vital to success today.

A03-0507
Freeth, D. and Davey, P., eds. The AJ Legal Handbook. Architectural Press, 1973. 240 pp. £6.50
  This is a revised version of a series of articles that appeared over several months in The Architects' Journal. The book covers the architect's office and the law; contracts; property; arbitration; statutory authorities and consents; copyright; planning construction regulations; professional practice; architects' responsibilities; professional examination. The contributors are mostly British lawyers.

A03-0508
Fuller, D. Organizing, Planning and Scheduling Engineering Operations. Cahners, 1969. 500 pp. $7.95
  Guide to effective engineering planning. Offers techniques and procedures, pinpoints areas with greatest potentials for improvement.

A03-0509
Geddes, S. and Chrystal-Smith, G. Building and Public Works Administration, Estimating and Costing. Butterworth, 1967. 296 pp. £3.50

While the book should be of great value to practicing engineers, surveyors and contractors, it is not beyond the understanding of the young engineer or student.

A03-0510
Gibson, J. F. A. Value Analysis--The Rewarding Infection. Pergamon, 1968. 52 pp.

A statement of the basic principles and techniques of value analysis/engineering, the manner in which it operates and the results it may be expected to achieve. Particular reference is made to the fundamental concepts of value, unnecessary cost, team operation, the job plan, functional approach, creativity, supplier participation, and to the intangible as well as tangible results.

A03-0511
Gill, P. Systems Management Techniques for Builders and Contractors. McGraw-Hill, 1968. 224 pp. $17.50

Demonstrates the use of techniques for the development, management, and control of builder and contractor organizations. Uses a case study involving a 50-home building project.

A03-0512
Gobourne, J. Cost Control in the Construction Industry. Butterworth, 1973. 158 pp. £3.00 (paperback £1.60)

A descriptive analysis of the principles of cost control most applicable to the construction industry. The aim of the book is to show how these principles can be applied to measuring and controlling the utilization of labor, plant, materials and overhead on a construction site. As no two companies are likely to evolve identical costing systems, the author describes possible variations of the system.

A03-0513
Gothie, D. L. A Selected Bibliography on Applied Ethics in the Professions, 1950-1970. Univ. Press of Virginia, 1971. 176 pp.

The bibliography is arranged alphabetically into the following professional categories: business and management, engineering, general ethical philosophy, government and politics, health sciences, law, science, and social sciences. Accompanying each professional area are listings of relevant books, monographs, periodical articles, pamphlets, unpublished speeches, and other materials. Works of particular interest are annotated.

A03-0514
Granof, M. H. How to Cost Your Labor Contract. Bureau of National Affairs, 1973. $10.00

Describes and evaluates how selected major corporations determine the financial consequences of labor contract proposals,

Management 93

and demonstrates how the "discounted cash flow model"--a technique widely used to select from among alternative capital expenditure projects--can be employed by management to arrive at collective bargaining decisions. Traditional costing methods are criticized, the environment for financial analysis is discussed, and specific current procedures in contract-change assessment are considered.

A03-0515
Gray, O. S. Cases and Materials on Environmental Law. Bureau of National Affairs, 1973. (1974 supplement) 1442 pp. $22.50

This casebook encompasses statutes, administrative source materials, and court decisions pertaining to virtually every facet of environmental protection and management. Areas covered include air and water pollution control, solid waste management, historic preservation, oil spills and hazardous materials, herbicides and pesticides, noise, aesthetics, special problems of utilities, weather control, and federal impacting programs.

A03-0516
Grear, A. C. L. and Oxborough, J. Commercial Property Management: A Practical Guide to the Legal Ownership, Use, Sale, and Acquisition of Commercial Land and Property. Beekman Publishers, 1970. 511 pp. $24.00

Combines in a single volume all the information necessary for the executive responsible for company property. A reference book which explains and illustrates all the many financial, legal and administrative aspects of property management.

A03-0517
Green, R. The Architect's Guide to Running a Job. Architectural Press, 1972. 144 pp. £1.75

The new edition shortens or omits certain items that are dealt with more fully in some of the longer books on the subject, and at the same time expands features which have been incorporated into the standard practice or have come into more prominence since the 1968 edition. The terminology has also been checked and updated.

A03-0518
Green, R. Architect's Guide to Site Management. Architectural Press, 1965. 120 pp. £1.25

Makes clear the parts played by client, architect, quantity surveyor, engineer, contract manager, site agent and clerk of works; explains how their jobs interlock. This briefing is intended to show the architect where he stands, and help him make sure that his instructions are given to the right people, and successfully carried out.

A03-0519
Griffin, C. W. Development Building: The Team Approach. American Institute of Architects, 1972. 130 pp. $15.00

(for AIA members, less 20%)

Evolving beyond their traditional role as designers, architects are entering the decision and delivery stages of building--in some cases as co-owners, in others as consultants offering new client services in the crucial decision-making processes which affect a project's ultimate success. The complexities of land acquisition, mortgage financing, ethical implications, liability insurance are explored.

A03-0520
Grummitt, C. M.   The Mechanics of Construction Management.
       Pergamon, 1968. 36 p.
       Published on behalf of the Construction Industry Training Board. It is presented in an easy to read style, with numerous cartoons and illustrations.

A03-0521
Haavind, R. C. and Turmail, R. L.   The Successful Engineer-Manager: A Practical Guide to Management Skills for Engineer and Scientists. Hayden, 1971. 176 pp. $8.95
       This guide is aimed toward the engineer or scientist who lacks background in the management art. Covering six vital facets of managing: career, decisions, people, projects, finances, and communications. It places at hand a stockpile of pertinent advice selected from articles prepared for Electronic Design.

A03-0522
Hackney, J. W.   Control and Management of Capital Projects.
       John Wiley, 1965. 305 pp. $20.75
       Contents--Introduction, capital cost estimating and control, progress planning and control, value prediction and control, procedures and reports, organization and manning.

A03-0523
Hajek, V.   Project Engineering. McGraw-Hill, 1965. 200 pp.
       $14.75
       Describes phases of a procurement program, the problems and decisions a project engineer must handle to successfully complete each phase.

A03-0524
Hanford, L. D.   Analysis and Management of Investment Property.
       Institute of Real Estate Management, 178 pp. $8.00
       Written for the property owner, appraiser, mortgage lender and trust executive as well as the property management executive, to help develop awareness of the key elements in choosing the right investment and maintaining its value. Three major divisions cover: basic concepts for investment property management; techniques and applications to various types of property; and creative management services. Development of cash flow projections, feasibility and financial analyses are included.

A03-0525
Harper, G.   Computer Applications in Architecture and Engineering.

Management   95

   McGraw-Hill, 1968.   256 pp. $16.00
      This book discusses the practical applications of the computer in architectural and engineering firms. Provides an instructional guide to applications in such areas as specifications, estimating, accounting, networking, building information systems, etc.

A03-0526
Harper, N.   Financial Management Computer Users Manual.
      American Institute of Architecture, 1971.   125 pp. $10.00
      (for AIA members, less 20%)
      This is the manual for the installation and operation of the AIA Computerized Financial Management System as originally outlined in the companion Financial Management for Architectural Firms--A Manual of Accounting Procedures. A book of instructions plus samples of all print-outs and the input, updating and maintenance forms for the system available from the AIA.

A03-0527
Hauf, H. D.   Building Contracts for Design and Construction.
      John Wiley, 1968.   342 pp. $18.25
      The author describes building industry contracts and specific types of contracts for A/E services and Architect-Consultants. Specifications, bonds and bidding and award procedures make up another part of the book. The appendix contains information on the standards of professional practice, the code of ethical conduct, and industry arbitration rules.

A03-0528
Heery, G.   Time, Cost, and Architecture.   McGraw-Hill, 1975.
      192 pp. $16.50
      The efficient control of time and cost in the design and construction of large projects is the major theme of this book on construction management. By considering equally the contribution to a project of all the professional disciplines involved, the author presents an economic approach to construction management with an eye to greater profits.

A03-0529
Heller, E. D.   Value Management: Value Engineering and Cost Reduction.   Addison-Wesley, 1971.   232 pp. $12.50
      Designed for the practicing manager, this book provides a plan for setting up and carrying out a value program and cost reduction program in any business organization. The book features an appendix with an outline of a value engineering workshop for potential training managers.

A03-0530
Hilton, M. P.   Industrial Relations in Construction.   Pergamon,
      1968.   256 pp. $5.50
      Provides a course of study in industrial relations in the construction industry which includes the history of major organizations involved and deals with the machinery of the building and civil engineering joint consultative organizations. Suitable for

students taking the examinations of the U.K. Institute of Building.

A03-0531
Hollins, R. J.   Production and Planning Applied to Building.
George Godwin, Ltd. (available from Construction Publishing Co.), 1971.   141 pp. $11.95
A guide for executives and students to the application of management principles in the construction industry and in particular to the process of marshaling and directing the resources of manpower, materials and machinery to economic and specific ends.

A03-0532
Hovanessian, S. and Pipes, L.   Digital Computer Methods in Engineering.   McGraw-Hill, 1969.   320 pp. $17.50
Digital computer programs, written in FORTRAN and BASIC programming languages, are given for numerical examples to familiarize the reader with computer logic.

A03-0533
Hunt, W. D., Jr.   Total Design: Architecture of Welton Becket and Associates.   Wittenborn, 1972.   307 pp. $22.50
This study discusses and visually demonstrates how all aspects of architecture (programming, design, engineering, production, interior design, etc.) are effectively coordinated and handled.

A03-0534
Hutton, G. H. and Devonald, A. D. G., eds.   Value in Building.
Applied Science, 1973.   136 pp.   £4.00
The book brings together some individual concepts of value in the field of building and presents techniques of valuation in allied fields, which offer scope in their application. The contributors discuss some of the many approaches to value in the construction industry. They demonstrate methods of evaluation that can be used by clients and their advisers for taking decisions based on established data, or for exposing the factors affecting choice if data is lacking.

A03-0535
ICBO.   Building Department Administration.   International Conference of Building Officials, $16.50 (for ICBO members, $14.50)
A textbook reference for all building officials. Covers areas of Building Department Administration. Discusses the purpose and history of codes and standards. Elaborates on departmental requirements including personnel, supplies and equipment. Covers communications with the public; legal aspects of departmental activities; building rehabilitation programs, and many other matters.

A03-0536
ICBO.   Plan Review Manual.   International Conference of Building Officials, 1974.   316 pp. $10.00 (for ICBO members,

$9.00)
A text that will be an aid in the field of building inspection, as well as for the plan reviewer. Divided into structural and nonstructural parts. Based on the Uniform Building Code, it can be used as a teaching and learning aid.

A03-0537
ICBO. A Training Manual in Field Inspection of Buildings and Structures. International Conference of Building Officials, 1968. 174 pp. $9.20 (for ICBO members, $8.00)
Outlines tests for instructors of courses in universities or colleges; text for those studying building construction or architecture; a supplemental guide to use with the Building Code in achieving skill in inspection techniques; contains check lists, test questions, and a sketchbook for mechanical inspectors.

A03-0538
Jabine, W. Case Histories in Construction Law: A Guide for Architects, Engineers, Contractors, Builders. Cahners, 1973. 256 pp. $14.95
This book is a compilation of legal cases involving architects, engineers, contractors, builders and subcontractors and their relationships with their customers and each other. The author orginally contributed these case studies over the past decade to Actual Specifying Engineer magazine.

A03-0539
James, M. L. Analog Computer Simulation of Engineering Systems, 2nd ed. Intext, 1971. 258 pp. $5.95
Describes the analog computer and its use as a tool in the solution and analysis of modern engineering problems and systems. Examples from all engineering disciplines illustrate the programming techniques for analog simulation of both linear and non-linear systems. For each problem, background material is provided, the mathematical model is developed, and computer solutions are given.

A03-0540
James, M. L., et al. Applied Numerical Methods for Digital Computation with Fortran. Intext, 1967. 514 pp. $11.75
Describes the fundamental principles involved in using the digital computer as a tool in the solution and analysis of engineering problems. Includes discussions of the most widely used numerical methods employed by engineers in solving problems.

A03-0541
Jamison, R. Fortran IV Programming: Based on the IBM System 1130. McGraw-Hill, 1970. 288 pp. $7.95
A presentation of planning, writing and executing Fortran IV programs to solve problems in engineering and applied mathematics. Presents Fortran IV as a gradual unfolding from the simplest complete programs on the computer to the highly sophisticated.

A03-0542
Jessup, W. E., Jr. and Jessup, W. E. Law and Specifications for Engineers and Scientists. Prentice-Hall, 1963. 585 pp. $15.00
 Explains the major fields of law and the legal problems commonly encountered by engineers and scientists such as contracts, specifications, real property and patents.

A03-0543
Jones, G. How to Market Professional Design Services. R. S. Means, or McGraw-Hill, 1974. 384 pp. $15.50
 A book to help engineers, architects, planners, and consultants bring in more business. It shows how and where to find prospects ... sell them through carefully planned presentations and proven follow-up techniques ... and use public relations effectively to build your firm's image and reputation.

A03-0544
Juran, J. Quality Control Handbook. McGraw-Hill, 1974. 1600 pp. $32.50
 This edition offers the know-how developed in industry for achieving better quality at lower cost. New sections on quality and income, quality costs, motivation, upper management and quality, support operation, and service industries are included along with discussions covering such other areas as the economics and specification of quality, the organization, acceptance, assurance, and control of quality, statistical control methods, and policies and objectives.

A03-0545
Jurecka, W. Network Planning in the Construction Industry. Applied Science, 1969. 112 pp. £10.50
 The practical problems discussed are illustrated by means of examples taken from actual construction projects including domestic dwellings, roads, docks, bridges, sewer and pipeline construction, canals and industrial buildings. Includes eleven charts, which depict network planning techniques.

A03-0546
Kantor, N. Contractual Aspects of Value Engineering. Society of American Value Engrs. $8.50 (for SAVE members $7.75)

A03-0547
Kelly, L. G. Handbook of Numerical Methods and Applications. Addison-Wesley, 1967. 354 pp. $15.00
 This volume is designed as a handbook and reference for engineers and scientists programming their own problems, and for programmers and computing center personnel working with scientists and engineers.

A03-0548
LaFara, R. L. Computer Methods for Science and Engineering. Hayden, 1973. 326 pp. $12.95

Management   99

Encompassing a wide range of practical applications, this volume offers professionals or graduate students rapid access to the basic techniques needed to solve a variety of numerical problems typically encountered in the engineering and scientific disciplines. Derivations, proofs, and methods are honed down to essentials.

A03-0549
Laidlow, C. R. Engineering Law. Univ. of Toronto Press, 1958. 461 pp. $7.50

Important changes in legislation, and a mounting number of court decisions relevant to the topics discussed, have made necessary a revision and expansion of this volume which presents to engineers and engineering students a simple treatment of the legal aspects of engineering undertaking and responsibilities, using a minimum of technical legal terms and avoiding the subtler distinctions and conflicts.

A03-0550
Langdon and Every. Model Descriptions for Engineering Services. George Godwin Ltd. Vol. 1, Mechanical, 1974. £10.00. Vol. 2, Electrical, 1974. £10.00

A reference work for quantity surveyors and engineers involved in producing bills of quantities and final accounts, either manually or by computer, for engineering services in buildings. Complete item descriptions are broken down into four or five parts, and actual known available sizes are given where appropriate.

A03-0551
Lapidus, M. Architecture: A Profession and a Business. Van Nostrand Reinhold, 1967. 224 pp. $15.00

Covers every phase of the business of architecture from opening an office, establishing rapport with a client, and keeping a practice running smoothly and profitably, through public relations, methods of securing commissions, law and accounting as they relate to architecture.

A03-0552
Lawton, M. P. Planning and Managing Housing for the Elderly. John Wiley, 1975. 336 pp. $19.95

Integrates knowledge from many areas on programming, planning and managing housing of all types for elderly people. Will make members of housing authorities, city planners, architects and designers aware of the effects of housing environments on the ability of older people to lead fulfilling lives, or perform tasks for themselves, rather than be dependent on others.

A03-0553
Leech, D. J. Management of Engineering Design. John Wiley, 1972. 314 pp. $16.95

Deals with specification and design within the context of commercial, real-life engineering. Considers time, money, resources and environmental aspects of design specifications, describes systematic design methods, and the communication and management of the design process within an organization. Emphasis is on realism, on costs and competition. A major design case-study is pursued throughout the book, as a continuous illustration of the processes being described.

A03-0554
Lefkoe, M. R. The Crisis in Construction: There Is an Answer. Bureau of National Affairs, 1970. $12.50

This book discusses the major labor-related problems in the construction industry, such as shortages of skilled manpower, excessive absenteeism and turnover, escalating wage rates, loss of management control, and unethical union practices. It identifies the causes of these problems and makes three basic recommendations for corrective action by contractors.

A03-0555
Lifson, M. W. Decision and Risk Analysis for Practicing Engineers. Barnes and Noble, 1972. $6.95

A03-0556
Lock, D. Engineer's Guide to Management Techniques. Cahners, 1973. $18.50

Complete information on general management techniques designed for the engineer--the middle man in industry whose career development depends on acquiring management rather than technical skills. Covers financial management, marketing, administration, production management.

A03-0557
Lucas, P. D. Accounting Guide for Construction Contractors. Prentice-Hall, 192 pp. $21.50

This book illustrates an accounting system that produces current and accurate information resulting in a large saving of time and money. It relates general accounting principles to the construction industry, gives special attention to the percentage-of-completion method, and shows how to advise the contractor on methods best suited to his individual needs.

A03-0558
McCracken, D. D. Fortran with Engineering Applications. John Wiley, 1967. 237 pp. $8.00

Written for those who want to get a rapid grasp of the use of a computer in the solution of problems in engineering and related disciplines. Among the twenty-nine case studies are many that will fill the needs of those who want to go beyond the fundamentals and apply computer techniques to a variety of realistic engineering problems.

A03-0559
Macfarlane, A. A. Architectural Supervision on Site. Applied Science, 1973. 189 pp. £4.00

Defines the architect's duties and responsibilities on the

Management

site and gives guidance to the architect's relations with client, builder, clerk of works and craftsmen in the various trades.

A03-0560
McKaig, T. Field Inspection of Building Construction. McGraw-Hill, 1958. 337 pp. $17.50
　　A guide to the supervision which defines responsibilities, and outlines pitfalls to be avoided in owner-architect-contractor-subcontractor relations.

A03-0561
McKown, R. Comprehensive Guide to Factory Law. George Godwin Ltd., 1974. 163 pp. £5.00
　　A classified guide to the requirements of legislation affecting factories and allied premises including offices and building sites. All legal requirements affecting safety, health, welfare, hours of work, and administration matters have been classified and indexed according to subject with full references given to legislation. In addition to the requirements of the Factories Act, Offices, Shops and Railway Premises Act, Petroleum Act and other legislation, a special chapter also covers the employers liability at Common Law.

A03-0562
McMaster, R. C., ed. Nondestructive Testing Handbook. Ronald Press, 1959. 1932 pp. 2 vols. $30.00
　　Covers all major methods of detection or measurement of the significant properties or performance capabilities of materials, parts, assemblies, equipment, or structures by tests which do not impair their serviceability. Provides the means for improving the performance reliability of equipment, developing product quality or uniformity, increasing the utilization of materials, and reducing costs and accidents due to breakdowns.

A03-0563
Marks, R. J. Aspects of Civil Engineering Contract Procedure. Pergamon, 1965. 240 pp. $4.95
　　Written by a quantity surveyor, a barrister, and a civil engineer to provide a comprehensive, account of contract procedure.

A03-0564
Marsh, P. D. V. Contracting for Engineering and Construction Projects. Cahners, 1970. 227 pp. $13.50
　　The livelihood of firms depends on contracts being secured on the best possible terms. Profits are eroded or eliminated by ignorance of proper methods of contracting and this book shows how to overcome the problems and pitfalls.

A03-0565
Marston, A., et al. Engineering Valuation and Depreciation. Iowa State Univ. Press, 1953. 508 pp. $8.50
　　Presenting an approach to depreciation in which the cost, value, and accounting concepts are separated and applied within

their individual meanings, this volume details the art of industrial property appraisal.

A03-0566
Martin, E. W., Jr. and Perkins, W. C. Computers and Information Systems: An Introduction. Richard Irwin, 1973. 500 pp. $12.50
 Two basic objectives are to provide the general manager with an understanding of the capabilities and limitations of the computer, and to provide an introduction to the subject for those who aspire to specialize in the area of information systems and computer technology.

A03-0567
Martino, R. Critical Path Networks. McGraw-Hill, 1970. 157 pp. $17.95
 A presentation of network analysis techniques, explains in direct language why and how networks are constructed, and how they make possible the effective management and control of many business and industrial operations.

A03-0568
Meier, H. Construction Specifications Handbook, Vol. 1. Prentice-Hall, 1975. 555 pp. $49.95
 Four sections describe the role of specifications in construction contracts, methods of improving front end documents and new procedures for writing specifications including computer aided master specs. The last section is a sample project manual of a Type V low-rise wood-frame project in which 61 typical specifications are shown.

A03-0569
Meier, H. Construction Specifications Handbook, Vol. 2. Prentice-Hall, 1976. 512 pp. $39.95
 Containing model specifications for 64 CSI Sections, the Handbook is a complete sample project manual with more than 70 pages of Notes to the Specifier. A guide to instant specifications for write-ups applicable to offices, schools, hospitals, and virtually any other public building. Volume II will help architects, engineers, and others who do specification write-ups.

A03-0570
Meredith, D., et al. Design and Planning of Engineering Systems. Prentice-Hall, 1973. 384 pp. $16.00
 Principles and techniques of the systems approach in engineering design and planning. Although only elementary mathematics is used, a range of system techniques is presented. Combines comprehensive treatments of all major systems techniques: linear graph analysis, mathematical modeling, CPM, linear programming, decision analysis, and simulations.

A03-0571
Merritt, F. S. Building Construction Handbook. McGraw-Hill,

Management

1975. 992 pp. $32.50
The third edition of this well-known reference incorporates the latest developments in construction technology. Basic changes in design specifications, the introduction of new materials and dimensions for building products, effective techniques for environment control and energy conservation, and innovations in building practices have all been implemented since the publication of the second edition.

A03-0572
Miller, L. Successful Management for Contractors. McGraw-Hill, 1962. 216 pp. $15.50
A guide which shows how to plan, schedule, and control operations with emphasis on obtaining more and better business. Contains hundreds of recommended practices and ideas covering contracting--from designating duties to arranging discounts, from making subcontractor agreements to reducing risks, from formulating company policy to controlling costs. Includes ready-to-use charts, records, reports, work sheets, legal forms and mailing pieces.

A03-0573
Moyle, M. P. Introduction to Computers for Engineers. John Wiley, 1967. 271 pp. $10.75
A book that reflects the increasing use of high speed computers in scientific computation and data processing; and the increasing recognition of this fact in engineering courses in the universities. The book gives a foundation in digital computers, analog computers, and numerical methods.

A03-0574
Nagarajan, R. Standards in Building. John Wiley, 1975. $24.50
Presents an up-to-date review of the principles and practices of standardization in building, which have been greatly influenced recently by the principles of modular co-ordination, theory of preferred numbers, research into functional requirements, concept of performance standards, and industrialization in building.

A03-0575
NAHB. Land Development Law for the Builder and His Attorney. National Assn. of Home Builders, $10.00 (available to NAHB members only)
Useful as a guide for executive officers, builders, and their attorneys in understanding land development law and as a point of departure for further legal casework. Covers permit fees, dedication, liability, minimum area requirement, condemnation and related subjects.

A03-0576
Nedved, J. C. Builder's Accounting. Butterworth, 1973. 184 pp. £3.80 (paperback £2.40)
This is an introduction to accounting systems as applied to the building industry. A complete accounting and costing system

is outlined, with illustrations given of all the relevant costing records. The systems applicable to small and growing firms are compared and the special requirements of partnerships and companies are discussed, as well as the different forms of business investment. Accounting and costing data are assessed for managerial decisions concerning the financial aspects of business management.

A03-0577
Neidle, M. Electrical Contracting and Management. Butterworth, 192 pp. £2.80
 Contents include supply and distribution, tariffs, power factor, land, planning, estimating, costing, standing charges, accounts, labor relations and legal requirements.

A03-0578
NESCA. Liens and Claims Manual. National Environmental Systems Contractors Assn. (Write NESCA for price)
 Loose-leaf reference manual covering statutes, forms, and legal procedures governing public and private construction work. Includes checklists and procedures which can guarantee that all requirements affecting lien and claim rights are met. Available for all states except Louisiana.

A03-0579
NESCA. Service Operation Manual. National Environmental Systems Contractors Assn. $40.00 (for NESCA members, $24.00)
 A handbook of service information for dealer-contractors who are starting a service operation or desire greater profits from existing service operations--covers administrative policies, corporate feedback, accounting, credit and collections, insurance, equipment, advertising, pricing, invoicing, dispatching, communications, recruitment and training, etc.

A03-0580
Nord, M. Legal Problems in Engineering. John Wiley, 1956. 391 pp. $11.95

A03-0581
NSPE. Professional Index of Private Practice. National Society of Professional Engrs. 175 pp. $10.00 (for NSPE members, $5.00)
 A compilation of material concerning the practice of consulting engineering. Contains sections on ethics, fees, competitive bidding, contracts, NSPE policies, professional liability insurance, and federal laws.

A03-0582
O'Bannon, R. E. Building Department Administration. International Conference of Building Officials, 587 pp. $16.50 (for ICBO members, $14.50)
 A practical textbook for all building officials. Covers all

Management 105

areas of Building Department administration. Discusses the purpose and history of codes and standards. Elaborates on departmental requirements including personnel, supplies and equipment. Covers communications with the public; legal aspects of departmental activities; building rehabilitation programs, and other matters.

A03-0583
O'Brien, J. CPM in Construction Management: Project Management with CPM. McGraw-Hill, 1971. 249 pp. $18.95
This guide to applying the Critical Path Method in the construction industry is a realistic treatment of CPM. Explains what CPM is and how to use it in facilitating communication, stimulating new ideas, saving time, controlling and reducing expenses, improving job performance, and avoiding risks and errors.

A03-0584
O'Brien, J. Scheduling Handbook. McGraw-Hill, 1968. 736 pp. $24.95
A guide to scheduling methods of all types--traditional and computerized--used in business today. Emphasis is on scheduling of time, with coverage also given to scheduling of manpower, equipment, and capital. Case studies illustrate the practical application of the principles involved.

A03-0585
O'Brien, J., et al. Contractor's Management Handbook. McGraw-Hill, 1971. 800 pp. $29.75
Presents all the techniques, the pitfalls, and the strategic opportunities involved in estimating and bidding, in writing and administering contracts, in planning and scheduling jobs, in purchasing, in financial control, and in all other critical aspects of the contractor's daily operations.

A03-0586
O'Brien, J. J. Construction Inspection Handbook. Van Nostrand Reinhold, 1975. 494 pp. $17.95
The role that an inspector plays on a construction project is covered by relating his job to the various phases, from contract documents through the actual construction process to construction management. The approach of this book is a practical one. Describes duties, requirements and procedures of the construction inspector.

A03-0587
O'Brien, J. J. Value Analysis in Design and Construction. McGraw-Hill, 1976. 320 pp. $14.00
The book's principal purpose is to encourage meaningful value analysis at the project level by establishing a bridge between the discipline of value analysis and the roles of the key design and construction professionals. Separate chapters deal with functional analysis, job planning, technology, budgeting, cost estimating (including computerized studies), design, and construction. Case

studies following each chapter show the practical applications of value analysis to everyday problems.

A03-0588
Oppenheimer, S. Directing Construction for Profit: Business Aspects of Contracting. McGraw-Hill, 1971. 384 pp. $16.50
This book emphasizes how all activity on the job, from an individual viewpoint, is directed toward making a profit. Shows how and where efforts must be intelligently directed by the individual at every level in the industry to assure more rewards for all.

A03-0589
Oughton, F. Value Analysis and Value Engineering. Soccer Associates, $10.50 (Acquire from Society of American Value Engrs.)

A03-0590
Parker, H. W. and Oglesby, C. H. Methods Improvement for Construction Managers. McGraw-Hill, 1972. 352 pp. $18.00
This book covers the modern management techniques and operating procedures needed by those in construction to become better managers. It offers systematic procedures that can be applied at all levels of the industry and employs such approaches as time-lapse motion pictures, and adapts techniques developed in other fields to fit constructions' needs. Emphasis is on the human behavioral aspects of work improvement applications, and contemporary approaches to construction safety.

A03-0591
Parris, J. Law and Practice of Arbitrations. George Godwin Ltd. 1974. £4.25
A practical guide for all in the construction industry. The author has set out both the legal aspects and the practice of arbitrations in a form easily comprehended by those who have not had legal training.

A03-0592
Peurifoy, R. Construction Planning, Equipment, and Methods. McGraw-Hill, 1970. 640 pp. $18.50
The soundness and advantages of applying engineering and management methods to the design and execution of construction is demonstrated in this book which encourages logical, scientific, and engineering analyses of construction projects, prior to and during construction, as a means of determining the most satisfactory methods of accomplishing construction.

A03-0593
Pilcher, R. Appraisal and Control of Project Costs. McGraw-Hill, 1973. 320 pp. $19.50
This book discusses the entire process of appraisal of projects in all areas of construction and their subsequent control from

a cost standpoint. It is the only book to cover the vital matter of controlling the cost of construction.

A03-0594
Pilcher, R. Principles of Construction Management for Engineers and Managers. McGraw-Hill, 1968. 382 pp. $18.75
Covers the basics of network analysis, operational research, the critical-path method, and more. Special attention is paid to the economics of construction work. Brings together fundamental techniques in construction management and develops them for practical application.

A03-0595
Popper, H. Modern Technical Management Techniques: For Engineers in Management, and for Those Who Want to Get There. McGraw-Hill, 1970. 544 pp. $24.00
This book is based on articles appearing in Chemical Engineering magazine over the past several years. Engineers in all fields who are either in management or interested in moving into management will find this book a useful guide.

A03-0596
Porter, R. Site Labour Guide. George Godwin Ltd., 1972. 24 pp. £0.20
This guide sets out in question and answer form essential points from the Industrial Relations Act as it affects the construction industry. The 90 questions and answers deal with points most likely to be raised and will be welcomed by those in supervisory posts on construction sites.

A03-0597
Practical Accounting and Cost Keeping for Contractors. Frank R. Walker, 256 pp. $7.95
Subjects include payroll taxes, laying out the job, extra work and credits, classification of labor costs, accounting for labor and material costs, obtaining material quotations and sub-bids, aids in job estimating.

A03-0598
Priluck, H. M. and Hourihan, P. M. Practical CPM for Construction. R. S. Means, 1968. $7.50
In this workbook for manual-graphic CPM scheduling, the reader is taken step by step through the analysis of a typical construction project. Basic reasoning and procedures are outlined. There is a chapter on advanced CPM scheduling and computer applications to show how they relate to the basic scheduling method presented in this book.

A03-0599
Radcliffe, B. M., et al. Critical Path Method. Cahners, 1967. 292 pp. $10.95
A presentation of the principles and practical applications of the critical path method of network planning. An effective

management tool for the construction industry, providing a scientific approach to the complex problems of planning, scheduling, and controlling projects.

A03-0600
Rathbun, I. Building Construction Specifications. McGraw-Hill, 1972. 224 pp. $8.50
 This book covers every phase of building construction including legal documents and energy systems. By studying the text and completing the worksheets, a student gains personal first hand experience in specification writing.

A03-0601
Rau, J. G. Optimization and Probability in Systems Engineering. Van Nostrand Reinhold, 1970. 403 pp. $15.00
 An approach to operations research geared to the needs of the engineer rather than the economist, management scientist, or business analyst. Integrates optimization techniques and probability concepts as they relate to engineering problems, and emphasizes methods applicable during the design, development, and evaluation of systems.

A03-0602
Ray-Jones, A. and McCann, W. CI/SfB Project Manual. Architectural Press, 1971. 160 pp. £5.00
 This book is reprinted from a 20-week series in The Architects' Journal. It sets out in detail methods of using the British version of the SfB system (CI/SfB) in conjunction with other codes, for the arrangement and cross-referencing of drawing, specifications and project information of all kinds.

A03-0603
Reiner, L. Handbook of Construction Management. Prentice-Hall, 1972. 339 pp. $15.95
 Discusses every aspect of the business of construction management from the viewpoint of the small builder as well as the large construction company. Features preparation of bids, organization of a construction project from both the progress and cost viewpoints; the types of construction contracts and their effects. Also contains chapters on the lien laws, arbitration, labor relations, field inspection and many others.

A03-0604
Rosen, H. J. Construction Specifications Writing: Principles and Procedures. John Wiley, 1974. 238 pp. $14.25
 For use by architects and engineers who want to learn systematic principles and procedures of writing specifications. Gives relationship between drawings and specifications, and describes the organization of a specification, bidding requirements, and Uniform Construction Index.

A03-0605
Rosen, H. J. Principles of Specification Writing. Van Nostrand

Management 109

Reinhold, 1967. 224 pp. $16.50

The art of specification writing has progressed in recent years. The author attributes this advance largely to the work of the Construction Specifications Institute and the Specifications Committee of the AIA in bringing together many individuals having a common interest in the improvement of specification writing. The publication of several "Standards" has resulted from these meetings, and this book is a synthesis of the viewpoints expressed in these conferences.

A03-0606
Rosenfeld. Architect's Handbook of Professional Practice. 2 vols. American Institute of Architects, $21.00 (for AIA members, less 20%)

As contemporary building design and construction techniques become increasingly complex, architects must have the best collective thinking, experience and methods of administration available. It is to this end that the AIA's Handbook is directed. Ring bound, Volume 1 contains 21 chapters plus a Glossary of Construction Industry Terms. Volume 2 includes samples of the contracts and information publications, plus the office and project forms available from AIA. The loose-leaf format can be kept current through subscription to the Handbook Supplement Service.

A03-0607
Rossnagel, W. B. Checklists for Management, Engineering, Manufacturing, and Product Assurance. Hayden, 1971. 192 pp. $14.00

Checklists are an excellent way for managers, engineers and technicians to make certain that a given specification, contract design, et al. is complete and accurate. Covering all phases of the engineering profession, this volume contains the proper balance between details and the management approach to a wide range of industrial and military problems.

A03-0608
Rowe, K. Management Techniques for Civil Engineering Construction. John Wiley, 1975. 268 pp. $32.50

Provides an improved appreciation of management techniques, applicable to a wide range of individuals from the trainee to the practicing manager. Covers operational control and planning, financial management, contract control, and material and manpower resources.

A03-0609
Royer, K. Desk Book for Construction Superintendents. Prentice-Hall, 1967. 220 pp. $16.95

Provides a guide to the non-technical aspects of construction job management and the duties expected of a practicing job superintendent. Shows how the contractor can be relieved of much of the time usually required to instruct his superintendent in such matters as essential paperwork, trade jurisdiction, subcontractor payments, and others.

A03-0610
Rubey, H. and Milner, W. W. Construction and Professional Management. Univ. of Oklahoma Press, 1971. 306 pp. $9.95
This book covers construction management activities from planning, bidding and scheduling to the control of a construction project. Topics include leasing vs. ownership of equipment, statistical theory, CPM, PERT, and analytical approach to bidding. The construction business is surveyed and elements are categorized as private vs. government and by size, i.e., small and medium contractors.

A03-0611
Rubey, H., et al. The Engineer and Professional Management. Iowa State Univ. Press, 1971. 339 pp. $3.95
Covers finance, managerial accounting and auditing, quantitative controls, valuation, rate structures, engineering economy, marketing and advertising, public relations, and more.

A03-0612
Sanderson, R. L. Codes and Code Administration: An Introduction to Building Regulations in the United States. Building Officials and Code Administrators, 244 pp. $7.95 (for BOCA members, $5.95)
A new textbook, the first written on this subject. Includes discussion of the use of codes, code enforcement and the administration of building regulations. It should be helpful to mayors and city managers who have found their code enforcement responsibilities to be perplexing and frustrating. An overview of codes and code enforcement and administration in the United States.

A03-0613
SAVE. Conference Proceedings--1970. Society of American Value Engrs., 1970. 271 pp. $17.50 (for SAVE members, $15.00)
The meeting had six sessions. The first, on application, involved VE in new product development, process analysis and plant engineering. Six management papers are followed by VE interplay with other disciplines. Ten papers on techniques stress design, progress implementation, and systems concepts. A number of papers discuss philosophy and professionalism. The last session was devoted to marketing VE and profits from VECP's.

A03-0614
SAVE. Conference Proceedings--1971. Society of American Value Engrs., 1971. 349 pp. $17.00 (for SAVE members, $15.00)
The session titles give an insight into the papers presented. These are the art and science of VE/VA, VE/VA in and out of commercial, VE relationship to allied disciplines, VE in subcontracts, VE organization and application, profit oriented programs, VE and the management process, VE for DOD, VE tools and technology and an outline of two courses, a functional analysis system technique and a VE design course.

Management

A03-0615
SAVE. Conference Proceedings--1972. Society of American Value Engrs., 1972. 307 pp. $15.00 (for SAVE members, $10.00)
Some papers given at the meeting have the following titles: Creativity-Choice or Chance?, Teamwork in Function/Cost Analysis, The Challenge to Value Engineers, What Management Expects of its Value Program, Advanced Fast Diagraming, Value Engineering in HEW Construction, Recent Building Construction Innovations at the Univ. of Alaska, VE in Systems Building, Value Analysis in the Concept Stage, VE during Design, Air Force VE in Construction, etc.

A03-0616
SAVE. Conference Proceedings--1973. Society of American Value Engrs., 1973. 179 pp. $10.00 (for SAVE members, $9.00)
The meeting was devoted to value engineering in action which covered transactional analysis and management plus VE-VA by objectives and exceptions. Another session on the anatomy of change examined the function and philosophy of tolerance and the profit in value changes. Value engineering in Scandinavia and Israel as well as in the U.S. at Chevrolet and Chrysler were covered. Other topics included program selection and the forefront of value engineering.

A03-0617
SAVE. Conference Proceedings--1974. Society of American Value Engrs., 1974. 275 pp. $15.00 (for SAVE members, $13.50)
The major topics include influencing the design process, producer-manufacturer application, design-to-cost in construction, computerized cost/value applications, diversified VE applications, ASPR VE contract clauses, management of value programs and engineering/industrial applications.

A03-0618
Seelye, E., ed. Data Book for Civil Engineers. Vol. 3, Field Practice. John Wiley, 1954. 394 pp. $18.25
This volume helps to alert the inspector and field engineer to the essentials of construction management. Testing methods for construction materials are explained as are such subjects as construction equipment, concrete, masonry, steel, foundations, surveying, wooden structures, paving, sewers and drainage.

A03-0619
Shaffer, L., et al. Critical Path Method. McGraw-Hill, 1965. 224 pp. $14.75
Enables a student of any discipline or a person practicing in any field to use this method of planning and scheduling a construction project. Only a knowledge of arithmetic is a prerequisite.

A03-0620
Sheeran, F. B.   Management Essentials for Public Works Administrators.   American Public Works Assn., 1975.   522 pp. $15.00 (for AWPA members, $12.00)
This text presents the theory and practice of public works management. It focuses on the four basic elements of management-planning, organizing, directing, and controlling.

A03-0621
Stanley, C. M.   The Consulting Engineer.   John Wiley, 1961. 258 pp. $8.00

A03-0622
Stone, P. A.   Building Design Evaluation.   John Wiley, 1975. 240 pp. $15.95
Explains and illustrates the techniques of cost-in-use. This technique enables the designer to rationalize his knowledge and hunches about present costs and uses and those anticipated for the future in order to obtain the best returns from the resources used. Shows how to compute the costs and benefits of different solutions, both for complete buildings and developments and for separate building components.

A03-0623
Swanson, L. A. and Pazer, H. L.   Pertsim: Text and Simulation.   Intext, 1969.   241 pp. $4.95
A decision-making simulation or game which includes text, this book is computerized and is written in FORTRAN IV. It demonstrates the basic fundamentals of PERT and CPM and points out the variety of uses for these techniques by using a construction project for demonstration. The inclusion of the hand simulator is featured.

A03-0624
Teets, R.   Construction Management for the Subcontractor. McGraw-Hill, 1976.   298 pp. $16.50
Intended for the individual contractor who lacks experience in formal management techniques, this book presents methods of planning, scheduling, monitoring, and control. Included are such typical day-to-day problems as making collections and payments, the uses of bank borrowing, the effective utilization of working capital, a practical system for developing budgets and forecasting sales, techniques for negotiating contracts and other requirements, and the effective use of people within the company.

A03-0625
Tomson, B. and Coplan, N.   Architectural and Engineering Law. Van Nostrand Reinhold, 1967.   392 pp. $17.50
Legal information covering all aspects of a construction enterprise, supported by citations to related cases in building law. Deals with legal problems inherent in execution of a construction project.

Management

A03-0626
Turner, D. F. Building Contracts: A Practical Guide. George Godwin Ltd. 230 pp. £3.40

A practical guide for contractors, architects and surveyors, which concentrates on the operating and commercial aspects of the contract forms. Attention is given to wording likely to lead to unforeseen difficulties. Each clause and sub-clause of the standard RIBA form of contract is discussed, followed by consideration of other related forms including Government contracts and the ICE conditions of contract.

A03-0627
Vancil, R. Leasing of Industrial Equipment. McGraw-Hill, 1963. 383 pp. $21.50

Addressed to the executive who must make leasing decisions and the analyst (accountant) charged with making comparative evaluations of leasing plans, this book demonstrates how sound lease-or-buy decisions can be reached. It describes types of leasing arrangements available.

A03-0628
Volpe, S. P. Construction Management Practices. John Wiley, 1972. 181 pp. $12.25

A guide to methods of operation for the construction industry. Shows how a contractor organizes and controls his work. The book will be useful not only to new or growing firms, but also to anyone associated with construction such as bankers, insurance men, architects, engineers or owners.

A03-0629
Walker, N., et al. Legal Pitfalls in Architecture, Engineering, and Building Construction. McGraw-Hill, 1968. 270 pp. $17.50

Written in non-legal language, this book covers difficulties in law, in owner-architect and owner-contractor relationships, liens and bonds, arbitration, special contract provisions, and more.

A03-0630
Walker-Smith, D., et al. The Standard Forms of Building Contract. Charles Knight and Co., 1971. 306 pp. £8.50

A practical treatise on the law relating to building contracts carried out in Britain under the standard conditions of contract. Each clause of the contract is annotated and the legal and practical significance of the wording explained. The loose-leaf format permits the publication of replacement pages to account for future amendments of the contracts.

A03-0631
Wass, A. Construction Management and Contracting. Prentice-Hall, 1972. 298 pp. $11.50

Presents an overview of construction management and contracting, and uses current specimen documents in each chapter where they are defined. Theme of the book is people, their needs and ambitions; and the assembly of men, machines, and materials, to safely and profitably erect a satisfying and durable structure. Instructs how to get started in, and profitably stay in, the building and contracting industry.

A03-0632
Watson, D. Specifications Writing for Architects and Engineers. McGraw-Hill, 1964. 228 pp. $11.95

The author tells how and where to obtain all the information needed to prepare different types of specifications. He discusses office organization and preparation of construction documents, construction management, agreements and contracts, materials and standards, reference and source materials and working with government agencies. The book includes a sample set of specifications for a small commercial building.

A03-0633
Willis, A. J. and George, W. N. Architects in Practice. Beekman Publishers, 1974. 325 pp. $16.50

Presents to the reader in simple language some of the elementary duties that the architect owes to client and contractor alike.

A03-0634
Wilson, W. Concepts of Engineering System Design. McGraw-Hill, 1965. 255 pp. $10.25

This book introduces the practice of engineering as it is today. Discusses system design, computers, mathematical models and methods, tools of optimization, feedback control, etc.

A03-0635
Zehner, J. R. Builder's Guide to Contracting. McGraw-Hill, 1975. 270 pp. $15.00

A practical, step-by-step guide to contracting and purchasing function, this volume discusses in detail how best to solicit and evaluate bids, select subcontractors and suppliers, and more. Whether in a specialized department in a large firm or in a small general contractor's office, this manual will be useful to all involved with the purchasing and supply aspects of the building and contracting industry.

Section   A04

DESIGN: ARCHITECTURAL

      Acoustics
      Architecture--by Building Types
      Architecture--by Region
      Architecture--Historical Periods
      Design Details

A04-0650
Abraben, E. Resort Hotels: Planning and Management. Van Nostrand Reinhold, 1965. 304 pp. $25.00
   This book covers the complete range of resort hotel planning, design, operation, administration, and management, taking into account the combination of natural resources and man-made components. It provides information on land utilization, designing, planning, equipment selection, construction and operating costs.

A04-0651
Acton, H. Great Houses of Italy: The Tuscan Villas. Viking Press, 1973. 288 pp. $28.50
   The book spans the centuries from the days when Tuscan villas were inhabited by Sicilian noblemen to today's present owners. Interior and garden photos.

A04-0652
AIA. Award Winning Architecture/USA. American Institute of Architects, 1973. $35.00 (for AIA members, less 10%)
   A pictorial of the best in architecture in 1972 with over 250 award-winning entries.

A04-0653
AIA. Creating the Human Environment. American Institute of Architects, 1970. $4.95 (for AIA members, less 10%)
   This report of the AIA Committee on the Future of the Profession examines what architects will face in creating the environment of the future. Outlines social influences on the physical environment, effects of technological improvements and imperative changes for the professional.

A04-0654
AIA Membership Directory. American Institute of Architects, $25.00 (for AIA members, $5.00)
   Membership in the American Institute of Architects' various chapters is given, along with chapter officers and headquarters staff. Affiliated trade and professional organizations, foreign societies and architectural schools are also listed.

A04-0655
AJ Editors. Church Buildings. Architectural Press, 1970. 160 pp. £1.75
   This book was prepared by a team of clergy of various denominations, working together with architects. Beginning with

Design--Architectural

some historical background and general design data for places of Christian worship the book goes on to consider the special requirements of the main church groups.

A04-0656
AJ Editors. Principles of Hotel Design. Architectural Press, 1970. 80 pp. £3.00

This book aims at providing a basis for fruitful collaboration between architect, hotelier and developer. It begins by looking at market feasibility study techniques, profitability and investment, developments in catering and changing concepts in service. In the light of these factors it then sets out some general design principles which introduce more specific data on the layout of guest bedrooms, reception, dining and other public areas, and services.

A04-0657
Alander, K., ed. Viljo Revell: Works and Projects. Praeger, 1967. 120 pp. $10.00

Revell is considered one of the most important figures in Finnish architecture. Before his untimely death in 1964, Revell worked closely with the editor on the choice of works to be included in this book.

A04-0658
Allatios, L. The Newer Temples of the Greeks. Pennsylvania State Univ. Press, 80 pp. $6.50
Translated, annotated.

A04-0659
Allen, E. Stone Shelters. MIT Press, 1969. 224 pp. $13.50

Using text, photographs, maps and drawings, this book documents the development of vernacular architecture in southern Italy.

A04-0660
Allsopp, B. The Study of Architectural History. Praeger, 1971. 128 pp. $7.50 (paperback $3.95)

This book shows how the present state of architectural historical studies has evolved and establishes a disciplinary framework for future developments. It provides information on the growth of architectural styles and forms, and on the architects who have determined these trends. Also useful as source material for the social historian, it shows architectural history to be germane to considerations of the evolving environment and to an international cultural continuity.

A04-0661
Altherr, A., ed. Three Japanese Architects. Architectural Book Publishing Co., 1968. 196 pp. $19.50

Buildings from 1950 to 1967 are presented with photographs, drawings, and plans. 400 illustrations.

A04-0662
Ambasz, E. Italy: The New Domestic Landscape: Achievements

and Problems of Italian Design.   N.Y. Graphic Society, 1972.   432 pp. $15.00
A survey of recent design developments in Italy covering modes of contemporary living from permanent home to trailer.

A04-0663
Amery, C., ed.   Domestic Historic Buildings and their Details.
Architectural Press, 1974.   192 pp.   £6.00
Between 1920 and 1930 the Architectural Press published, under the title A Practical Exemplar of Architecture, a series of loose-leaf folders illustrating historic buildings in Britain and giving measured drawings of facades and noteworthy interior and exterior details.   Long out of print, these folders have been sought after, for in many cases they form the only record of buildings that have since disappeared; and the measured drawings of such details, as stonework, woodwork, fireplaces, doorways, etc. are of value to architects, builders and craftsmen engaged in restoration work.   This book is a selection of the best domestic buildings from the series.

A04-0664
Andrews, W.   Architecture in America.   Atheneum Publishers, 1960. $15.00
A photographic excursion into the history of American architecture.   Introduction by Russell Lynes.

A04-0665
Andrews, W.   Architecture in Chicago and Mid-America: A Photographic History.   Harper & Row, 1968.   212 pp. $4.95
In 260 photographs, here are the important buildings of the Middle West from the days of the Greek and Gothic revivals to the present.   Mansions in the Chicago area and Cleveland and Grosse Point are shown, along with the masterpieces of Louis Sullivan, Frank Lloyd Wright and the Saarinens.

A04-0666
Andrews, W.   Architecture in New York: A Photographic History.
Harper & Row, 1969.   212 pp. $4.95
The pictures range from the manor houses of the eighteenth century to the palatial residences of the Victorian era, the Brooklyn tenements of Alfred White and finally modern New York, and the achievements of Louis Sullivan, Frank Lloyd Wright, Mies van der Rohe, Eero Saarinen and others.

A04-0667
Andrews, W.   Architecture in New England: A Photographic History.   Stephen Greene Press, $16.95
Ranging from colonial to ultra-modern, these 230 photographs, each with historical overview and descriptive caption, offer a grand tour of New England's architectural landmarks.

A04-0668
Angus, M.   The Old Stones of Kingston.   Univ. of Toronto Press,

Design--Architectural 119

1966. 120 pp. $10.00
Although Kingston is rich in buildings with both historic and aesthetic appeal, few people are able to identify them. This book will make such identification possible.

A04-0669
Apple, J. M. Plant Layout and Materials Handling. Ronald Press, 1963. 447 pp. $10.00
A presentation of modern plant layout and materials handling practices, stressing the important interrelationships with management planning, product and process engineering, methods engineering, and production control.

A04-0670
Architects' Emergency Committee. Great Georgian Houses of America. Dover. 2 vols., 1970. 518 pp. $10.00
A set of books containing close to 500 illustrations of facades, floor-plans, interiors and decorative details from 77 of America's most beautiful homes of the Georgian era (1714-1830). All characteristics of major styles in New England, New York and New Jersey, Pennsylvania, and the South are shown.

A04-0671
Architect's Journal, eds. Architect's Working Details. Vol. 1. Crane, Russak, 1953. 160 pp. $8.50
Drawn mainly from the work of leading British architects. Covers windows, doors, staircases, walls, roofs, balconies, heating, and furniture and fittings.

A04-0672
Architect's Journal, eds. Architect's Working Details. Vol. 2. Crane, Russak, 1954. 160 pp. $8.50
Drawn mainly from the work of leading British architects. Covers windows, doors, staircases, walls, roofs and ceilings, balconies, heating, lighting, furniture and fittings.

A04-0673
Architect's Journal, eds. Architect's Working Details. Vol. 3. Crane, Russak, 1955. 160 pp. $8.50
Drawn mainly from the work of leading British architects. Covers windows, doors, walls, roofs and ceilings, balconies, heating, lighting, furniture and fittings.

A94-0674
Architect's Journal, eds. Architect's Working Details. Vol. 4. Crane, Russak, 1957. 160 pp. $8.50
Drawn mainly from the work of leading British architects. Covers windows, doors, staircases, walls, roofs and ceilings, balconies, furniture and fittings, and miscellaneous items.

A04-0675
Architect's Journal, eds. Architect's Working Details. Vol. 5. Crane, Russak, 1958. 160 pp. $8.50
Drawn mainly from the work of leading British architects.

Covers windows, doors, staircases, walls, roofs and ceilings, heating, lighting, water supply and sanitation, furniture and fittings, and miscellaneous items.

A04-0676
Architect's Journal, eds.   Architect's Working Details.   Vol. 6.
    Crane, Russak, 1959.   160 pp. $8.50
    This volume shows the work of architects in nine European countries, the U.S. and Canada. Covers windows, doors, staircases, walls, roofs and ceilings, balconies, heating and lighting.

A04-0677
Architect's Journal, eds.   Architect's Working Details.   Vol. 7.
    Crane, Russak, 1960.   160 pp. $8.50
    This volume returns to the work of English architects, with emphasis on functionalism. Covers windows, doors, staircases, walls, roofs, water supply and sanitation, furniture and fittings, and miscellaneous items.

A04-0678
Architect's Journal, eds.   Architect's Working Details.   Vol. 8.
    Crane, Russak, 1961.   160 pp. $8.50
    Work includes that of European and American architects. Covers windows, staircases, walls, roofs and ceilings, lighting, furniture and fittings, and miscellaneous items.

A04-0679
Architect's Journal, eds.   Architect's Working Details.   Vol. 9.
    Crane, Russak, 1962.   160 pp. $8.50
    Covers windows, walls, roofs and ceilings, balconies, heating, lighting, furniture and fittings, and miscellaneous items.

A04-0680
Architect's Journal, eds.   Architect's Working Details.   Vol. 10.
    Crane, Russak, 1964.   160 pp. $8.50
    Work includes that of European and U.S. architects. Covers windows, doors, staircases, walls, roofs and ceilings, balconies, heating, furniture and fittings, and miscellaneous items.

A04-0681
Architect's Journal, eds.   Architect's Working Details.   Vol. 11.
    Crane, Russak, 1965.   160 pp. $8.50
    Covers windows, doors, staircases, walls, roofs and ceilings, lighting, water supply and sanitation, furniture and fittings, and miscellaneous items.

A04-0682
Architect's Journal, eds.   Architect's Working Details.   Vol. 12.
    Crane, Russak, 1968.   160 pp. $8.50
    Work includes that of European and U.S. architects. Covers windows, doors, staircases, walls, roofs, balconies, canopies, heating, furniture and fittings.

Design--Architectural 121

A04-0683
Architect's Journal, eds. Architect's Working Details. Vol. 13.
    Crane, Russak, 1969. 160 pp. $8.50
    Covers windows, doors, staircases, walls, roofs, balconies, canopies, furniture and fittings, and miscellaneous items.

A04-0684
Architect's Journal, eds. Architect's Working Details. Vol. 14.
    Crane, Russak, 1971. 160 pp. $8.50
    The 14th volume contains an index of the entire 14-volume series and a Glossary of terms in English, French, German, and Spanish. Highlights of the volume are the updating by the Finns of the log-burning fireplace, a sandpit in a children's play area in Sweden, stairs lit from the underside of handrails (Finland), office lights and air diffusers incorporated into a single unit, and in Germany clip-on aluminum tiles being used to clad the front of a reinforced concrete office building.

A04-0685
Architect's Journal, eds. Hotels: Principles of Hotel Design.
    Wittenborn, 1968. 87 pp. $7.50
    Reprinted from a series in the Architect's Journal in which the principles of hotel design were given.

A04-0686
Architectural Journal. Church Buildings. Herman Publishing Co.,
    1970. 160 pp. $6.95
    Prepared by a team of clergy of various denominations, working with architects. Also covers pastoral centres, music in worship, organ design. Illus. Intro.

A04-0687
Architectural Record. Apartments and Dormitories. McGraw-Hill,
    1958. 238 pp. $14.00
    48 examples of apartment houses, college residence halls, and other multiple dwellings, designed by leading architects.

A04-0688
Architectural Record. The Architectural Record Book of Vacation
    Houses. McGraw-Hill, 1971. 248 pp. $9.95
    A two-room vacation house for $5,000, or a rural place for $100,000? This book provides floor plans, site plans, elevations, photographs, and structural details for vacation houses at both prices--and for 58 others in between. All plans are for houses that have actually been built in America or Europe, and they represent the best work of today's most distinguished architects.

A04-0689
Architectural Record. Campus Planning and Design. McGraw-Hill,
    1972. 288 pp. $25.00
    This book contains a collection of articles and illustrations published originally by the Architectural Record.

A04-0690
Architectural Record. Houses Architects Design for Themselves.
McGraw-Hill, 1974. 240 pp. $16.95
 Some of the most creative work done by architects is to be found in their own houses. In this tear-sheet book, 57 residences designed by and for architects are presented. This book should be of interest to architects as well as anyone interested in architecture.

A04-0691
Architectural Record. Interior Spaces Designed by Architects.
McGraw-Hill, 1974. 230 pp. $25.95
 This new volume on the design of interior spaces covers the considerations that must be made in the planning of interiors for various types of buildings from community and civic to drama and dance edifices. Also included is coverage on teaching, worship, business, dining, selling and living spaces and their purposes and designs.

A04-0692
Architectural Record. Motels, Hotels, Restaurants and Bars.
McGraw-Hill, 1960. 327 pp. $18.50
 63 successful establishments are presented as case studies which demonstrate the important relationship between good design and good business.

A04-0693
Architectural Record. Office Buildings. McGraw-Hill, 1961.
256 pp. $14.50
 Photographs, diagrams, and floor plans of over 40 outstanding office buildings, complete with details of their architectural, structural, and mechanical features.

A04-0694
Architectural Record. Record Houses of 1970. McGraw-Hill, 1970. 480 pp. $3.50
 Presents the 20 trend-setting houses that have won Architectural Record's "Award of Excellence for Design." Features a 16-page article on low-rise garden apartments.

A04-0695
Architectural Record. Record Houses of 1973. McGraw-Hill, 1974. 124 pp. $4.50
 This annual presents the winners of Architectural Record's Award of Excellence in Residential Design for 1973. Those selected, twenty houses and eight apartments, are examples of the best in creative design throughout the United States, and reflect today's casual life style. A description of each is given, along with interior and exterior photographs and floor plan.

A04-0696
Architectural Record. Record Houses of 1976. McGraw-Hill, 1976. 150 pp. $5.00

Design--Architectural

A growing concern for the preservation of the environment and the conservation of energy is reflected in the twenty houses and eight apartments shown in this year's edition of Record Houses. These award winners embrace a wide range of cost, style, and location. They also illustrate unusual solutions to the problem of building custom housing compatible with specific sites. A description of each house and apartment is given, along with interior and exterior photographs and floor plans.

A04-0697
The Architecture of England. Architectural Press, 1965. 48 pp. £0.75
This book presents in text and pictures the evolution of English architecture, and explains its relation to the social background.

A04-0698
Argan, G. C. The Renaissance City. Braziller, 1969. 128 pp. $5.95 (paperback $2.95)
The author provides an analysis of architectural theories, ideal military cities, political utopias, the work of famous architects, and actual cities in Italy, England, France, and the Netherlands. A descriptive appendix lists cities and provides short biographies.

A04-0699
Arregger, H. and Glaus, O. Highrise Building and Urban Design. Praeger, 1967. 204 pp. $20.00
The advantages and disadvantages of highrise buildings are analyzed in this book.

A04-0700
Artistic Houses, Being a Series of Interior Views of a Number of the Most Beautiful and Celebrated Homes in the United States. Benjamin Blom, 1969. 1883, 400 pp. $38.50
A pictorial study of American domestic architecture of the late 19th century; descriptions of interiors based on conversations with architects, designers, and owners.

A04-0701
ASCE. Analytical Treatment of Problems of Berthing and Mooring Ships. American Society of Civil Engineers, $6.00 (for ASCE members, $3.00)

A04-0702
ASCE. Small Craft Harbors. American Society of Civil Engineers, 1969. $8.00 (for ASCE members, $4.00)

A04-0703
Ashihara, Y. Exterior Design in Architecture. Van Nostrand Reinhold, 1973. 143 pp. $12.50
A handbook that covers the theory and practice of designing

exterior space--the exterior of buildings, the setting for groups of buildings, city plazas, gardens, etc.

A04-0704
Aurenhammer, H. J. B. Fischer von Erlach. Harvard Univ. Press, 1973. 188 pp. $15.95
Johann Bernhard Fischer von Erlach was one of the greatest Baroque architects and the virtual creator of the Austrian Baroque style. In this book the author sets Fischer in the context of his time. He takes into account the surviving buildings and examines the symbolical meaning of decorative elements.

A04-0705
Bachmann, J. and Von Moos, S. New Directions in Swiss Architecture. Braziller, 1969. 128 pp. $7.95 (paperback $3.95)
An examination of present Swiss architecture with a view to its historical basis and future trends.

A04-0706
Bacon, E. N. Design of Cities. Viking Press, 1973. 336 pp. $15.00
History of development of the city. Photographs.

A04-0707
Badawy, A. Architecture in Ancient Egypt and the Near East. MIT Press, 1966. 232 pp. $10.00
The evolution of architecture and building in Egypt, Mesopotamia, Anatolia, The Levant, and Cyprus from earliest times to the Hellenistic Period is the subject matter of this study. This volume treats the architectural achievements of each country separately, subdividing the buildings into groups: domestic, religious, funerary, and military.

A04-0708
Bailey, R. F. Pre-Revolutionary Dutch Houses and Families in Northern New Jersey and Southern New York. Dover, 1968. 612 pp. $5.00
A study of early Dutch houses and families of Brooklyn, Queens, Staten Island, nearby New York and New Jersey and western New Jersey. Land titles, types of houses, etc. are described accompanied by a collection of plates. 179 illustrations.

A04-0709
Baldwin, C. C. Stanford White. DaCapo Press, 1976. 399 pp. $20.00 (paperback $4.95)
Tracing the development and growth of White's aesthetic contributions, Baldwin explores the early years with architect H. H. Richardson, the travels through New England and Southern France, and the establishment of the firm of McKim, Mead, and White. His account includes White's plans for imposing churches and apartment buildings, and luxurious Long Island and Newport homes, as well as sketches for the Gorham, the first Tiffany

Design--Architectural 125

Building, the original Madison Square Garden, the Farragut Monument, the Washington Square Arch.

A04-0710
Banham, R. Guide to Modern Architecture. Van Nostrand Reinhold, 1963. 159 pp. $7.95
This book illuminates the main preoccupations of modern architects--function, form, construction and space--and comments on a world-wide selection of modern buildings.

A04-0711
Banham, R. Modern Architecture: The Age of the Masters. Architectural Press, 1974. 176 pp. £3.50 (paperback £1.50)
This is a revised version of a book written over a decade ago. The masters whose work was discussed have died and the whole concept of the master-architect has come under increasing criticism. Banham re-examines the period in this new light in his introduction and has made numerous changes in the text as well as adding some new buildings to the survey of key works.

A04-0712
Banham, R. Theory and Design in the First Machine Age. Praeger, 1970. 338 pp. $12.50 (paperback $3.95)
The masters of modern architecture--Gropius, Mies van der Rohe, Le Corbusier, and others--used their writings to justify their buildings. First published in 1960 and unavailable for several years, this book includes the significant theoretical writings and the buildings, projects, industrial designs, paintings, and sculptures that they represent.

A04-0713
Barley, M. W. The House and Home: A Review of 900 Years of House Planning and Furnishing in Britain. N.Y. Graphic Society, 1971. 208 pp. $12.50
A pictorial presentation of the British home over a period of nearly a thousand years.

A04-0714
Barnard, J. The Decorative Tradition in Nineteenth Century Architecture. Architectural Press, 1974. £5.00
Examines the mass-produced decorative element in Victorian architecture, concentrating on the period between 1860 and 1910. This was the great age of suburban building in Britain. This book draws attention to an element in exterior building design that is fast disappearing.

A04-0715
Barry, A. The Life and Works of Sir Charles Barry. Benjamin Blom, 1867. 420 pp. $22.00
Focuses on his most celebrated work, the complex of buildings erected for the Houses of Parliament (begun 1835), and

discusses the controversy with A. W. Pugin and his son over the attribution of the design itself.

A04-0716
Bassi, E. The Convento della Carita. Pennsylvania State Univ. Press, 1974. 252 pp. $32.50

Only part of the original design of the Convento della Carita in Venice, Palladio's attempt to re-create a grandiose Roman house, is preserved. The present volume is a well-documented study of the building's conception, execution, and subsequent vicissitudes.

A04-0717
Bastlund, K. Jose Louis Sert: Architecture, City Planning, Urban Design. Praeger, 1967. 244 pp. $20.00

Sert has been concerned with the relation of architecture to the shaping of the city throughout his career. His involvement with the relationship between all the arts is reflected in his work.

A04-0718
Baumgart, F. A History of Architectural Styles. Praeger, 1971. 300 pp. $12.00 (paperback $4.95)

This book traces changes in architectural tastes and designs from the temples of the Hittites to twentieth-century factories. The author summarizes the historical events that led to the development of each new style. He also provides examples of changing trends, describing the most significant as well as the most typical buildings from each period.

A04-0719
Bayes, K. and Franklin, S. Designing for the Handicapped. George Godwin Ltd., 1971. 87 pp. £1.50

After a general survey of designing for the handicapped and their needs, contributions from those with specialized experience in the field draw out the fundamental issues in designing for a particular handicap.

A04-0720
Beacham, H. The Architecture of Mexico: Yesterday and Today. Architectural Book Publishing Co. 256 pp. $12.95

Pre-Hispanic Mexico as revealed through its architecture. Shows influence in today's structures. 320 photographs.

A04-0721
Becher, B. and Becher, H. Anonyme Sculpturen. Wittenborn, 1970. 220 pp. $15.00

Industrial constructions differing basically in function and building methods from domestic or factory buildings: lime kilns, cooling towers, blast furnaces, winding towers, water towers, gas tanks, silos. These constructions develop without any regard for esthetics and their shape derives from calculation. Most of the objects shown are found in Eastern European and English industrial areas.

Design--Architectural

A04-0722
Benevolo, L. History of Modern Architecture. MIT Press, 1971.
Vol. 1, 374 pp. $12.50; Vol. 2, 470 pp. $17.50
The first volume describes the component parts of modern architectural thought, discovers their origins, and follows their convergence from 1760 to 1914. It traces the physical events that gave birth to the modern European city through 1890 and discusses the influences that led to the thought of Owen, Ruskin, and Morris. The second volume is concerned with the modern movement proper, from 1914 to 1966. The author emphasizes the unity of the movement, rejecting the usual treatment that allots to the individual architects separate and unconnected biographical accounts.

A04-0723
Beranek, L. L. Music, Acoustics and Architecture. John Wiley, 1962. 586 pp. $21.50

A04-0724
Berriman, S. G. and Harrison, K. C. British Public Library Buildings. Academic Press, 1971. $27.00

A04-0725
Besset, M. New French Architecture. Praeger, 1967. 236 pp. $17.50
The period after World War II presented a building boom and shifting emphasis to state-backed housing schemes requiring teams of architects, engineers, and sociologists. Since 1955, a new generation of architects has shown a common tendency to consider architectural problems in the light of a total humanism. In the manipulation of technical resources, they are chiefly influenced by the masters of reinforced concrete--Freyssinet, Coyne, Lafaille--and by Jean Prouve's experiments with prefabrication.

A04-0726
Besterman, T. Art and Architecture. Rowman and Littlefield, 1971. 216 pp. $10.00

A04-0727
Bibiena, F. G. L'Architettura Civile. Benjamin Blom, 1970. 295 pp. $37.50
Treatise on the art of baroque stage design whose influence can be traced throughout the 18th century.

A04-0728
Birks, T. Building the New Universities. David and Charles, 1971. 128 pp. $9.95
A survey, with photographs, of problems faced and solved in planning new British universities. Includes sites and plans; discusses residences and libraries.

A04-0729
Blake, V. B. Rural Ontario. Univ. of Toronto Press, 1969. 174 pp. $15.00

The text consists of two essays on the social and architectural history of southern Ontario which explain how the countryside came to assume its characteristic appearance.

A04-0730
Blaser, W. Mies van der Rohe: The Art of Structure. Praeger, 1965. 228 pp. $25.00

The author had the advice of van der Rohe himself in this book's preparation. The book surveys van der Rohe's work from his earliest days in Germany to his most recent buildings in the Americas.

A04-0731
Blunt, A. Neopolitan Baroque and Rococo Architecture. Abner Schram, 1974.

The author covers the period from the rise of a local Baroque style with its own characteristics in the early 17th century to the full flowering of the Rococo in the 1730s and 1740s.

A04-0732
Boaga, G. and Boni, B. The Concrete Architecture of Riccardo Morandi. Praeger, 1966. 234 pp. $20.00

In working drawings, models, and photographs, the authors present Morandi's most mature designs--from the Storms River Bridge in South Africa to the nuclear power station at Garigliano--and discuss the features that characterize each piece.

A04-0733
Boesiger, W. Le Corbusier. Praeger, 1972. 260 pp. $7.50 (paperback $3.95)

The text of this edition, designed especially for students, includes summaries in Le Corbusier's own words of his theories, and extracts from his diaries, conversations, and letters. The photographs, plans, and sketches include illustrations of his principle works.

A04-0734
Boesiger, W., ed. Le Corbusier: Last Works. Praeger, 1970. 204 pp. $25.00

Although Le Corbusier's death in 1965 prevented him from realizing some of his greatest projects, his associates have in many cases been able to complete works that were in the planning or construction stage. This volume contains detailed accounts of projects completed up to 1969; included are previously unpublished descriptions of the Museum of Knowledge and the Tower of Shadows in Chandigarh. Other works include a hospital in Venice, the Museum of the 20th Century at Nanterre, the House of Youth and Culture at Firminy, France, and the Sluice at Kembs-Niffer.

A04-0735
Boesiger, W., ed. Richard Neutra: Buildings and Projects. Praeger. Vol. 1, Richard Neutra 1923-50, 1966, 223 pp. $18.50; Vol. 2, Richard Neutra 1950-60, 1959, 240 pp.

Design--Architectural   129

$18.50; Vol. 3, Richard Neutra 1961-66, 1966, 256 pp. $20.00
The first book of this three-volume series encompasses Neutra's work from his arrival in America until 1950. It shows the development of his private houses, as well as many health and education buildings and large housing developments. The second volume covers the decade 1950-60, and is devoted primarily to buildings serving community needs and city planning, with examples from Sacramento, Elysian Park Heights in Los Angeles, and Guam. The final volume follows his work almost to the present.

A04-0736
Boesiger, W. and Girsberger, H. Le Corbusier 1910-65. Praeger, 1967. 352 pp. $25.00
This illustrated book traces the career of Le Corbusier from the conception of his "machine for living" in 1923 to his death in 1965. Included are studies of finished and planned urban projects, such as the "radiant city" theory and the modular system of proportionate design.

A04-0737
Borras, M. L. Contemporary Japanese Architecture. Wittenborn, 1971. 202 pp. $12.50
Some of the foremost Japanese architecture of this decade is shown, including the sports stadium by Kenzo Tange and a library by Arata Isozaki.

A04-0738
Boston Society of Architects. Architecture Boston and Cambridge. Crown Publishers, 1976. 192 pp. $12.95 (paperback $5.95)
Beginning with a brief sketch of the history of Boston as seen through its architecture, this volume includes essays written by prominent architects on the various aspects of the city. The essays discuss both specific architects and trends in the field from the early planning of Boston by Bullfinch to the recent work of Gropius, Saarinen, Breuer, and Pei.

A04-0739
Boston Society of Architects. Boston/Architecture. MIT Press, 128 pp. $8.95 (paperback $2.95)
Originally prepared to introduce Boston to members of the American Institute of Architects meeting there in June 1970, this book now serves a wider purpose of presenting America's most architecturally interesting city to both architects and non-architects. Boston's architecture is marked by diversity and juxtaposition of styles, periods and purposes.

A04-0740
Boudon, P. Lived-in Architecture: Le Corbusier's Pessac Revisited. MIT Press, 1972. $7.95
In the mid-nineteen-twenties--at Pessac near Bordeaux--Le Corbusier built his first large-scale project, the Quartiers

Modernes Fruges, consisting of some 70 housing units. Acting simultaneously as architect and town planner, he wished to provide people with low-cost, homogeneous, cubist structures--empty containers that their presence alone would activate and fulfill. This book describes what happened.

A04-0741
Bouwcentrum, ed. General Hospitals: Functional Studies on the Main Departments. North-Holland, 1961. 301 pp. $18.20
The part which hospital design plays in inducing a state of well-being in both staff and patients has only recently been recognized. This book suggests the specifications to which the various sections and departments of modern hospitals should conform. A detailed analysis has been carried out for a number of hospitals built in various countries during the post-war period. For valid comparisons, a uniform analytical procedure was applied and a number of ground plans are included to illustrate points in the text. The work was done by a group of experts from the Dutch Hospital Study Organization.

A04-0742
Bowker, R. R., Co. American Architects Directory. R. R. Bowker, $35.00
More than 20,000 American architects are listed by professional affiliation, educational credentials and other biographical information.

A04-0743
Bowyer, J. History of Building. Beekman Publishers, 1973. 275 pp. $8.95
The author, a lecturer at the Croydon Technical College, presents the influences and technical skills that have resulted in the evolution of architectural styles. The book is of special interest to building, surveying, and architectural students. The five main headings of the book are: Evolution of Structure, Historical Development of Building Materials and Components, Building in the Middle Ages, Growth, Urbanization, Industrialization, and Legislation, and the Rise of Professionalism.

A04-0744
Boyd, J. S. Practical Farm Buildings: A Text and Handbook. Garden Way Publishing, 1973. 265 pp. $7.95
The book includes varied designs, materials and tools, computing loads, insulation, storage data, actual construction steps, tables, diagrams and drawings.

A04-0745
Boyd, R. New Directions in Japanese Architecture. Braziller, 1968. 128 pp. $7.95 (paperback $3.95)
An examination of present Japanese architecture with a view to its historical basis and future trends.

Design--Architectural 131

A04-0746
Bradley, A. D. The Geometry of Repeating Design and Geometry of Design for High Schools. AMS Press, 1933. $10.00
Part I is a study of repeating geometrical designs and the development of such designs. Part II provides specific applications of the geometry of repeating design, together with some related topics, to courses in high school geometry.

A04-0747
Braham, A. and Smith, P. Francois Mansart. Abner Schram, 1974. 280 pp. 2 vols. $95.00
This monograph is the first full-length study of Mansart and his architecture. The first volume discusses all his major works and his social and architectural background. The catalogue which follows contains many hitherto unpublished documents. The second volume contains 563 plates, illustrating all Mansart's surviving buildings and drawings, and related material, including many reconstructions of lost or altered buildings.

A04-0748
Braithwaite, D. Fairground Architecture: The World of Amusement Parks, Carnivals, and Fairs. Praeger, 1968. 196 pp. $12.50
The author traces the history of fairs from ancient Greek festivals to the Industrial Revolution's introduction of steam-driven diversions. He analyzes the architecture and engineering of joy rides, as well as the layout of the fairground and the transportation and assembly of equipment. Anecdotes, photographs and drawings, scale plans, and templates used for the traditional carved work are included.

A04-0749
Branner, R. Gothic Architecture. Braziller, 1961. $5.95 (paperback $2.95)

A04-0750
Brawne, M. Libraries: Architecture and Equipment. Praeger, 1970. 200 pp. $21.50
Today's libraries, challenged by the growth of visual media and technical storage of data, have had to devise new ways of carrying out their basic functions. Thirty international examples that have successfully met this challenge are presented in this study. Community, research and university libraries, and national libraries are dealt with, and a section on the work of Alvar Aalto is included.

A04-0751
Brawne, M. The New Museum: Architecture and Display. Praeger, 1965. 208 pp. $20.00
The author surveys fifty important examples of museum architecture. He analyzes the most significant developments of recent years, including the boom in museum construction in the United States, the new Italian design theories, and the concept of

total environment, which characterizes many museums in the Scandinavian countries.

A04-0752
Brett, L. Architecture in a Crowded World. Schocken Books, 1970. 181 pp. $2.75
 The author examines how architecture affects the lives of people and how architectural beliefs have not served people's needs. Mr. Brett, a leading British architect and town planner, shows how we have failed and points the way to a new beginning.

A04-0753
Brierley, J. Parking of Motor Vehicles. Applied Science, 1972. 347 pp. £9.00
 The book includes parking in European countries, the United States of America and Canada. Contains information on town planning, design of multi-level car parks, roof car parks, mechanical parking devices, underground parking and parking of commercial vehicles.

A04-0754
Broadbent, G. Design in Architecture. John Wiley, 1973. 504 pp. $22.50
 A review of systematic design methods and their application in architectural design. Relates systematic design methods to the evolution of architectural design concepts, and to the behavioral and environmental aspects of architectural design.

A04-0755
Brodshaug, M. Buildings and Equipment for Home Economics in Secondary Schools. AMS Press, 1931. $10.00
 This study attempts a determination of the characteristics of well designed home economics plants. Topics include the evolution of facilities, the nature of home economics education, space allotment, building problems, and equipment.

A04-0756
Brooks, H. A. The Prairie School. Univ. of Toronto Press, 1972. 373 pp. $25.00
 Inspired by Louis Sullivan and given guidance and prominence by Frank Lloyd Wright, the members of the Prairie School sought to achieve an original architectural expression. Their designs were characterized by angular forms and sophisticated interior arrangements which gave a sense of space that belied the actual size. This book discusses the masters and the work of their contemporaries. It traces the course of the movement, how and why it came into existence, what it achieved, and what caused its abrupt end.

A04-0757
Brown, E. M. New Haven: A Guide to Architecture and Urban Design. Yale Univ. Press, 1976. 225 pp. $15.00 (paperback $3.95)

Design--Architectural

This guidebook views New Haven as the product of an urban community. The author concentrates not only on the magnificent architecture of the Yale Campus and the spectacular works of modern architects, but also on the vernacular production of the anonymous builders whose competent designs have provided the scenery of urban life. The book is divided into fifteen tours, accessible by foot, bicycle, or car. Capsule descriptions and a map accompany each tour.

A04-0758
Brown, L. C., ed. From Madina to Metropolis: Heritage and Change in the Near Eastern City. Darwin Press, 343 pp. $16.95
In this study of modern Near Eastern cities, architects, demographers, urban planners and other specialists discuss preservation of historical sites, spatial organization of cities, and traditional versus contemporary architecture and urban planning.

A04-0759
Brown, P. Indian Architecture. Vol. 1. Buddhist and Hindu Periods. International Publications, 1965. 216 pp. $9.50
Covers the Indus civilization and the Vedic culture, the early Mauryan dynasty. Asoka and early Buddhism, rock-cut architecture, the Gupta period, Chalukyan architecture, rock architecture, the Dravidian style, the Indo-Aryan style, the Jain temples, the Hoysala style, Brahmanical buildings, and the styles in Nepal, Ceylon, Burma, Cambodia, Siam, Java and Bali.

A04-0760
Brown, P. Indian Architecture. Vol. 2. Islamic Period. International Publications, 1968. 161 pp. $12.50
Sources of Islamic architecture; the Delhi of Imperial style; provincial style, including the Punjab, Bengal, Jaunpur, Gujarat, Malwa, the Deccan, Bijapur and Kashmir; the Mughal period, including Babur, Mumayun, Akbar, and Shah Jahan; the medieval palaces and civic buildings; and the modern position.

A04-0761
Brown, T. M. The Work of G. Rietveld, Architect. MIT Press, 1970. 198 pp.
This is a monograph on the work through 1958 of the late Dutch architect. He is known primarily for his significant role in the De Stijl movement, Holland's contribution to 20th-century design. The architect's total production, ranging from small lamps and typography to large industrial buildings and housing is studied.

A04-0762
Bruce, C. and Aidala, T. The Great Houses of San Francisco. Alfred A. Knopf, 192 pp. $15.00
In the nineteenth century, San Francisco produced some remarkable houses and public buildings. This book celebrates those that survive to this day.

A04-0763
Bruce, C. and Grossman, J. Revelations of New England Architecture: People and Their Buildings. Viking Press, 1975. 192 pp. $15.00

The human and timeless architectural environment of New England are shown through the homes, the meetinghouses, the churches, and the mills--from the first rough-hewn Puritan structures through the Georgian mansions of the Yankee traders to the final flowering of the native New England style in the mid-1800s.

A04-0764
Bruggink, D. J. and Droppers, C. H. Christ and Architecture. Eerdmans, 1964. $20.00

Discusses the theological aspects of traditional and modern architecture, and delineates practical procedures for transmitting these elements into good contemporary architecture. A tool to help building committees and their architects effect an intelligent church building program.

A04-0765
Bruggink, D. J. and Droppers, C. H. When Faith Takes Form. Eerdmans, 1971. $3.95

An introductory volume on the basic problems of church architecture. Intended for all denominations.

A04-0766
Brumbaugh, T. B., ed. Architecture of Middle Tennessee Historic American Buildings Survey. Vanderbilt Univ. Press, 1974. 160 pp. $18.50

Photographs, drawings, and details of 35 examples of Middle Tennessee architecture selected for inclusion in the Historic American Buildings Survey. Such important buildings as the Hermitage, Cragfont, Fairvue, and Oaklands are included; less well known but just as significant architecturally are such buildings as the Grange Warehouse in Clarksville, Castalian Springs in Sumner County, and Bear Springs Furnace in Stewart County. The survey includes commercial buildings as well as residences.

A04-0767
Brunskill, R. W. Illustrated Handbook of Vernacular Architecture. Universe Books, 1971. 230 pp. $9.00

A book that delves into British architectural traditions.

A04-0768
Bryan, J. and Sauer, R. Via: Structures Implicit and Explicit. Wittenborn, 1972. 220 pp. $6.00

This volume is a collection of texts centering on the question of what structures and structuring mean to man. It would be valuable for those who have become interested in the notions of structuralism and would like to see them applied to environmental design and built structures, and for those who have started from the other end--as architects, engineers, designers, even users-- and became curious about structures generally. This is a student

Design--Architectural

publication of the Graduate School of Fine Arts, University of Pennsylvania.

A04-0769
Building Construction and Design. Howard W. Sams. 400 pp. $5.95

Covers heat transmission and insulation, acoustics, noise insulation, water, fire protection, and more. Examples of good and bad building design and construction procedures are presented throughout.

A04-0770
Bullock, O. M. The Restoration Manual. Herman Publishing Co., 1966. 196 pp. $15.50

Contents: The Architect; The Development of Programs; Selecting the Period to Be Restored; Historical Research: Sources, Techniques; Archaeological Research: Purpose, Excavation Procedure, Publication; Architectural Research: Purpose, Superficial Examination, Detailed Examination, Dating Sections; Execution of a Restoration; Specifications; After Completion.

A04-0771
Bullock, O. M., Jr. The Restoration Manual. American Institute of Architects, 1966. 181 pp. $9.95 (for AIA members, less 10%)

For architects and builders specializing in the authentic restoration of old buildings. Shows how to "read" older structures to preserve and reconstruct in a manner compatible with original design and construction.

A04-0772
Bullrich, R. New Directions in Latin American Architecture. Braziller, 1969. 128 pp. $7.95 (paperback $3.95)

An examination of present Latin American architecture with a view to its historical basis and future trends.

A04-0773
Burchard, J. E. Bernini Is Dead?: Architecture and the Social Purpose. McGraw-Hill, 1976. 700 pp. $30.00

Giovanni Bernini, the architectural genius of the Baroque period, is invoked as a critical touchstone in a provocative narrative about the state of contemporary architecture. Burchard, disturbed by the lack of beauty and human dimension in what passes for civil planning today, maintains that architects have forgotten or compromised the social dimension of their art by abandoning the humane and the beautiful, and have hence accelerated the decay of our urban, social aesthetic. The author explores the human environment, past and present, and its role in fostering the fullest development and expression of the human spirit.

A04-0774
Burchard, J. E. The Voice of the Phoenix: Postwar Architecture in Germany. MIT Press, 1966. 208 pp. $12.50 (paper-

back $3.95)
Germany's postwar physical reconstruction and architectural traditions and education are placed in perspective and compared to American architecture.

A04-0775
Burden, E. E. Architectural Delineation: A Photographic Approach to Presentation. McGraw-Hill, 1971. 288 pp. $18.50
Treats the camera and perspective, layout preparation, technique of drawing, and various types of presentations.

A04-0776
Burke, A., et al. Architectural and Building Trades Dictionary. American Technical Society, 377 pp. $7.15
Thousands of terms and hundreds of illustrations were selected to clarify definitions.

A04-0777
Burris-Meyer, H. and Cole, E. C. Theatres and Auditoriums. Van Nostrand Reinhold, 1964. 384 pp. $25.00
New problems of theatre construction are discussed with particular reference to the use of new materials, new construction, and operational techniques.

A04-0778
Bussagli, M. Oriental Architecture. Harry N. Abrams, 1974. 436 pp. $35.00
This book presents the architectural glories of the East: Peking's Forbidden City and the Great Wall of China; the sanctuary of Borobudur in Java; India's Great Stupa at Sanchi and the painted caves of Ajanta; the temples and shrines at Nara and Kyoto in Japan; the temple complexes of Angkor in Cambodia; the imperial city of Hue in Vietnam; the royal necropolis of Seoul in Korea; and the fortress-cities of Toprak-kala and Ajaz-kala in Central Asia. More than 500 photographs, plans, and diagrams.

A04-0779
Butler, G. D. Recreation Areas, Their Design and Equipment. Ronald Press, 1958. 174 pp. $8.00
This book contains information on the design and equipment of recreational areas, especially the neighborhood playground, playfield, and athletic field. Though the book is not a construction manual, it contains information useful for those who prepare construction plans.

A04-0780
Byrne, L. Check List Materials for Public School Building Specifications, Covering the General Specifications. AMS Press, 1931. $10.00
The primary purpose of the investigation is to produce a body of Check List material from which a school administrator may select a Check List of the size he desires to use, and to

Design--Architectural 137

provide a ready-made franework for future investigations in the field of building specifications.

A04-0781
Callender, J. Time-Saver Standards. McGraw-Hill and R. S. Means, 1974. 1042 pp. $32.50
This workbook of architectural design standards and construction reference data includes detailed studies of space frames, suspension structures, steel domes, curtain walls, skylights, plastics, useful curves, motel design, homes for the aged, glass, plaster, and other important areas. Covers structural design, reinforced and prestressed concrete, schools, heating and air conditioning. Contains proven solutions to hundreds of problems.

A04-0782
Cambridge Historical Commission. Survey of Architectural History in Cambridge. MIT Press, 1967-74. 4 vols. $7.95 each vol.
The report is a history of the development of the Old Cambridge area and its architecture. It includes an account of the growth of Harvard Square, the formation and subsequent subdivision of the Tory estates along Brattle Street, a study of the architecture of Harvard University, and an analysis of architectural styles, using the most distinctive buildings in Cambridge as examples.

A04-0783
Campion, D. Computers in Architectural Design. Applied Science, 1968. 320 pp. £6.50
The book begins with a general introduction to computer techniques, currently available computer peripheral equipment and methods of programming computers, and deals briefly with several specific computer applications with references to other published work. It concludes with more detailed examples of some architectural applications including perspective drawing, network analysis, design brief assessment and analysis, information scheduling, and design simulation.

A04-0784
Candilis, G. Planning and Design for Leisure. International Publications, 1972. 144 pp. $27.50
This is a survey of leisure time projects ranging from weekend houses, to row and atrium houses, apartment blocks, hotels, motels, etc. to regional planning.

A04-0785
Cantacuzino, S. New Uses for Old Buildings. Watson-Guptill, 1975. 264 pp. $29.95
Today's increasingly sophisticated demands for standards and services, as well as tighter codes for fire and safety, make the conversion of an old building to a new one a formidable task. But as land, materials, and labor costs soar, renovation is becoming an increasingly attractive alternative to new construction. This lavishly illustrated book--based on a special issue of The Architectural Review devoted to the subject--describes and illustrates 73

examples of significant recent conversion projects in Germany,
Ireland, France, Yugoslavia, Italy, Canada, England and the United
States. The buildings range from churches and schools through
town houses, barns, and breweries to mills, warehouses, and
pumping stations.

A04-0786
Canter, D.   Psychology for Architects.   John Wiley, 1974.   127 pp.
$12.50
Provides an introduction to psychology to enable environmental decision-makers to understand its relevance, basic concepts, and methods and thus to deal more effectively with environmentally oriented studies.

A04-0787
Canter, D. and Lee, T.   Psychology and the Built Environment.
Architectural Press, 1974.   300 pp.   £7.50
Environmental psychology deals with the interaction between man and his environment. It covers topics ranging from the effect on behaviour of such physical phenomena as light, heat and noise to aesthetic responses to buildings. It is the fastest growing area of psychological studies and this book brings together some 30 papers describing recent research.

A04-0788
Carter, P.   Mies van der Rohe at Work.   Praeger, 1973.   240 pp.
$25.00
This book documents the thought processes and the technical means by which Mies van der Rohe achieved his mastery of design and structure. The text discusses how Mies conceived buildings such as the Barcelona Pavilion, the Lake Shore Drive Apartments in Chicago, and the Seagram Building and the proportions, details, and construction materials used for the buildings.

A04-0789
Casey, M., et al.   Early Melbourne Architecture 1840 to 1888.
Oxford University Press.
Apart from its value as a permanent record of the more architecturally interesting of Melbourne's early buildings, this book will be enjoyed by all who are interested in the development of the Australian city.

A04-0790
Casson, H., ed.   Inscape: The Design of Interiors.   Architectural
Press, 1968.   208 pp.   £3.15
Over 230 illustrations depict the most imaginative interiors designed in England. There are also eight articles written by Specialists: The Consciousness of Space, Furniture Art, Graphics, The Education of the Interior Designer, and four short articles which discuss the effects upon interior design of recent developments in air-conditioning, lighting, acoustics and component assembly.

Design--Architectural 139

A04-0791
Castedo, L. A History of Latin American Art and Architecture.
 Praeger, 1969. 320 pp. $8.95 (paperback $4.95)
 This volume traces 3,000 years of Latin American art and architecture. The author discusses both well-known and unrecognized achievements, emphasizing the extent to which Latin American art reveals a fusion of Indian, European and African elements.

A04-0792
Catalogue of the Royal Institute of British Architects Library.
 Rowman & Littlefield. Vol. 1, 1,138 pp.; Vol. 2, 514 pp.; 2-vol. set, $80.00
 The entries form a conspectus of architectural literature--from farm buildings to Acts of Parliament, from pattern books to the Modern Movement, the Acropolis to the Crystal Palace, town planning in India to baroque theatre design. The Author Catalogue (volume one) is supplemented by the Classified Catalogue (volume two), set out under the Universal Decimal Classification (UDC), and the Classified Catalogue is supplied with an alphabetical subject index.

A04-0793
Caudill, W. Toward Better School Design. McGraw-Hill, 1954.
 271 pp. $13.50
 An approach to the design of schools of all kinds, from elementary through college. Contains 91 case studies.

A04-0794
Chamberlain, S. A Tour of Old Sturbridge Village. Hastings House, 1973. 72 pp. $1.25
 Photographs and captions portray this recreated old New England community, consisting of original buildings and their authentic furnishings moved onto a single site in southern Massachusetts.

A04-0795
Chamberlain, S. and Flynt, H. N. Historic Deerfield: Houses and Interiors. Hastings House, 1973. 196 pp. $10.00
 A collection of photographs of the eighteenth-century architecture and interior decoration still extant in Deerfield, Massachusetts. Detailed captions accompany the pictures.

A04-0796
Chambers, W. Designs of Chinese Buildings, Furniture, Dresses, Machines, and Utensils. Benjamin Blom, 1757. 23 pp. 21 illustrations. $29.50
 Reprint of the first English book to give an authentic picture of Chinese design.

A04-0797
Chambers, W. A Treatise on Civil Architecture. Benjamin Blom, 1791, 50 plates. $28.50

Designed as the Englishman's Palladio and Vignola; one of the seminal books on English architecture history.

A04-0798
Chermayeff, I. Observations on American Architecture. American Institute of Architects, 1973. $16.45 (for AIA members, less 10%)
An interesting mix of photographs representing the work of world famous architects as well as buildings where the designer is unknown, many in color, with author comments.

A04-0799
Chermayeff, I. and Erwitt, E. Observations on American Architecture. Viking Press, 1972. 143 pp. $16.95
Many of the photographs in this collection were a part of the "Architecture: U.S.A." exhibit at the World Exposition in Osaka, Japan. Commentary by the authors accompanies the photographs.

A04-0800
Chiarelli, R. San Lorenzo and the Medici Chapels. International Publications, 1971. 72 pp. $4.00
Three of the greatest artists of all time, Brunelleschi, Donatello and Michelangelo, took part in the construction and decoration of San Lorenzo church, the Medici Chapels and the Laurentian Library. Their contribution is illuminated here in 75 color plates.

A04-0801
Choay, F. The Modern City: Planning in the 19th Century. Braziller, 1969. 128 pp. $5.95 (paperback $2.95)
During the 19th century European and American cities were critically examined by men like Haussmann and Olmstead. The author considers developments from urban parks to linear cities in the context of far-reaching planning theories and contemporary socioeconomic conditions.

A04-0802
Christovich, M. L., et al. New Orleans Architecture. Volume II. The American Sector. Pelican, 1972. 244 pp. $17.50
The work focuses on the area presently known as the Central Business District and is designed to serve as a handbook for modern-day renovation and restoration. The primary source for material was the New Orleans Notarial Archives, which contains thousands of architectural drawings and original specifications of many buildings included in the book. More than 500 photographs, including 22 original watercolor drawings in full color.

A04-0803
Clapham, A. W. Romanesque Architecture. Oxford University Press, 1936. 224 pp. $9.25
Mr. Clapham presents a short survey of the development of

Design--Architectural   141

Romanesque Architecture throughout Europe. Each of the great areas of Romanesque Art is considered in a separate chapter, beginning with Italy and going on to France, Spain, England, Germany, and Scandinavia.

A04-0804
Clarke, B. F. L. Parish Churches of London. Architectural Book Publishing Co. 372 pp. 197 illustrations. $30.00
This volume gives complete, often difficult to find information on London's parish churches.

A04-0805
Clasen, W. Expositions, Exhibits, Industrial and Trade Fairs. Praeger, 1968. 208 pp. $17.50
The author has assembled 100 examples of the most recent approaches to industrial and trade fair design. Includes photographs and plans.

A04-0806
Clifton-Taylor, A. The Pattern of English Building. Watson-Guptill. 466 pp. $29.50
This work describes the traditional buildings of England with their distinctive characteristics which change from region to region.

A04-0807
Cobblestone, T., ed. World Architecture: An Illustrated History. International Publications, 1966. 348 pp. $15.00
This survey provides an up-to-date text, a glossary of architectural terms, and a detailed index.

A04-0808
Coffin, L. A., Jr. and Holden, A. C. Brick Architecture of the Colonial Period in Maryland and Virginia. Dover, 149 pp. $3.50
More than 200 photos show 58 important buildings, both domestic and official, many of which no longer survive. Text gives house-history, architectural features, general background.

A04-0809
Cole, D. From Tipi to Skyscraper: History of Women in Architecture. Braziller, 1974. 150 pp. $8.95 (paperback $4.95)
Historically, women have been major collaborators, if not prime instigators, in the field of architecture in this country. Yet today, only two per cent of the practicing architects in this country are women and less than half of the women who have earned degrees are registered architects. Doris Cole analyzes the underlying social and economic reasons for the present situation and suggests ways to improve it by attracting more women to the profession of architecture.

A04-0810
Colman, S. Nature's Harmonic Unity: A Treatise on Its Relation to Proportional Form. Benjamin Blom, 1912. 336 pp. $18.75

The laws governing proportional form are applied to the works of the architect, the painter, and the sculptor.

A04-0811
Colvin, B. Land and Landscape: Evolution, Design and Control. International Publications, 1970. 412 pp. $21.00

An introduction to aesthetic awareness and control of the environment: the history, principles, problems, and solutions: a broad history of landscape design, an analysis of types of landscape and a study of topography and vegetation; practical questions of siting and design; interplay of agriculture and urban development; the landscape of towns; provision for recreation.

A04-0812
Colvin, H. M. A Biographical Dictionary of English Architects, 1660-1840. International Publications, 1954. 835 pp. $17.50

The dictionary proper (p. 27-748) concentrates on professional work of architects concerned (over 1,000 entries). Chronological list of buildings designed by the more important architects. Bibliographical notes. Appendix: "Public offices held by architects, 1660-1840." Indexes of persons and places.

A04-0813
Condit, C. W. American Building. Univ. of Chicago Press, 1968. 329 pp. $10.00 (paperback $3.95)

Covers the whole range of structural activities. The materials of building, their physical properties and appropriate structural forms, technical innovations, and engineering works that met the needs of industry are all discussed.

A04-0814
Condit, C. W. Chicago, 1930-1970. Univ. of Chicago Press, 1974. $12.50

In this work Condit concludes the biography of an American city which he began in Chicago, 1910-1930. In both works the author is concerned with cities as living places, and in this regard he believes the modern city has failed.

A04-0815
Condit, C. W. The Chicago School of Architecture. Univ. of Chicago Press, 1964. 238 pp. $10.75 (paperback $4.50)

This survey places the Chicago School in its historical setting and assesses its achievements.

A04-0816
Condit, C. W. Chicago Since 1910. Univ. of Chicago Press, 1973. $12.50

In this illustrated study, Condit introduces a new kind of

Design--Architectural

urban history: a technical biography of the American urban environment. Focusing on Chicago, a city that has manifested the best and the worst of urban technology, this volume chronicles the building activity that took place from 1910 until the depression brought building to a halt. Condit has described the continuing evolution of the city as well as trying to determine the extent to which the building achievements answered the needs of those who lived and worked in Chicago.

A04-0817
Congdon, H. W. Old Vermont Houses, 1763-1850. William L. Bauhan, 1968. 224 pp. $3.95

Beautiful old Vermont houses are illustrated in over 140 photographs. Mr. Congdon traces the history of Vermont architecture from its beginnings to the 1850's. He describes the details of these buildings and the story of their builders--the early craftsmen-farmers of Vermont.

A04-0818
Connely, W. Louis Sullivan. Horizon Press, 1971. $8.95

Sullivan's life is set against the background of the rise of modern architecture.

A04-0819
Conrads, U., ed. Programs and Manifestoes on 20th Century Architecture. MIT Press, 1971. 190 pp. $10.00

Many of the master builders of this century have held passionate convictions regarding the philosophic and social basis of their art. The most influential of these are collected here in chronological order from 1903 to 1963. Taken together, they constitute a subjective history of modern architecture; compared one with another, their great diversity of style reveals in many cases the basic differences of attitude and temperament that produced a corresponding divergence in architectural style.

A04-0820
Conrads, U. and Sperlich, H. G. The Architecture of Fantasy Utopian Building and Planning in Modern Times. Praeger, 1962. 188 pp. $16.00

An illustrated study of the fantastic, the unusual, and the visionary in twentieth-century architecture.

A04-0821
Cook, J. W. and Klotz, H., eds. Conversations with Architects: Philip Johnson, Kevin Roche, Paul Rudolph, Bertrand Goldberg, Morris Lapidus, Louis Kahn, Charles Moore, Robert Venturi. Praeger, 1973. 256 pp. $12.50

A collection of interviews with eight of America's foremost architects, who discuss themselves, their attitudes toward their work, and their observations about the work of fellow architects. The architects were selected because they are informed spokesmen for some of the major movements determining the direction of architecture today.

A04-0822
Cook, P. Architecture: Action and Plan. Van Nostrand Reinhold, 1967. 96 pp. $2.45
The buildings and experimental projects illustrated are taken largely from the twentieth century. The author attempts to read into their design a series of ideas based on fundamental human motives and aspirations. The judgment of what is normally thought of as "great architecture" is discussed against this background.

A04-0823
Cook, P. Experimental Architecture. Universe Books, 1970. 160 pp. $7.95
The author surveys the work of experimenters in architecture and shows their relationship to new materials, methods and devices, as well as contemporary ideas on planning.

A04-0824
Cook, P., ed. Archigram. Praeger, 1973. 144 pp. $12.50 (paperback $6.95)
A study of the Archigram group and its contributions over the past 10 years.

A04-0825
Coombs, R., ed. Perspecta: The Yale (University) Architectural Journal, 1971, nos. 13/14. Wittenborn, 1972. 351 pp. $25.00
More than 800 illustrations, architectural drawings, plans and models are shown.

A04-0826
Coope, R. Salomon de Brosse and the Development of the Classical Style in French Architecture from 1565 to 1630. Pennsylvania State Univ. Press, 1972. 456 pp. $32.50
Although de Brasse was the designer of the chateaux of Coulommiers and the architect of the Luxembourg Palace, he is not as well known today as many less significant architects. In this book, all of de Brosse's buildings are analyzed in detail.

A04-0827
Costonis, J. J. Space Adrift: Landmark Preservation and the Marketplace. Univ. of Illinois Press, 1973. $7.50
Over half the 12,000 buildings listed in the Historic American Buildings Survey are gone. The author presents his "Chicago Plan" for saving structures, together with an analysis of the economic causes behind the plight of America's urban landmarks.

A04-0828
Couperie, P. Paris Through the Ages. Braziller, 1971. $12.50

A04-0829
Cowan, H. J. Dictionary of Architectural Science. Applied Science or John Wiley, 1973. 354 pp. $10.95
This dictionary contains about 4500 entries ranging in length

Design--Architectural 145

from one line to 1-1/2 pages. The author has concentrated on the scientific aspects of architecture, but has also included the most frequently encountered terms from neighbouring fields, such as fine art, the history of architecture, the craft traditions of the building industry, structural, mechanical and electrical engineering, materials science, physics and chemistry.

A04-0830
Cowan, H. J. An Historical Outline of Architectural Science. North-Holland, 1966. 175 pp. $7.30
 The contrasting approaches between the methods of teaching science and engineering (taught as a logical development from experimental data) and architecture (primarily taught through study of the great masters), reflects the distinction between the scientific mind and that of the creative artist. This volume attempts to steer a middle course in dealing with those aspects of science and engineering which have influenced current architectural design.

A04-0831
Cram, R. A. Impressions of Japanese Architecture and the Allied Arts. Dover, 1966. 242 pp. $2.00
 American architect surveys highpoints in Japanese architecture, commenting on aesthetics, technical features, cultural factors, etc.

A04-0832
Crandall, C. They Chose to be Different: Unusual California Homes. Chronicle Books, 1972. 150 pp. $9.95
 The ingenuity of California architects and designers is evident in this study of innovative California homes. Photographs.

A04-0833
Crane, T. Architectural Construction: The Choice of Structural Design. John Wiley, 1956. 433 pp. $14.95

A04-0834
Creese, W. L., ed. The Legacy of Raymond Unwin: A Human Pattern for Planning. MIT Press, 1967. 234 pp. $17.50
 Sir Raymond Unwin (1863-1940) was the best-known town and regional planner of the last quarter of the nineteenth and first third of the twentieth century. Largely as a result of his efforts, whole towns were built, comprehensive and farsighted building laws enacted, and several professional societies founded. This selection of Unwin's writings includes early unpublished essays and excerpts from all of his major books and articles.

A04-0835
Creighton, T. H. Contemporary Houses: Evaluated by Their Owners. Van Nostrand Reinhold, 1961. 224 pp. $15.00
 The owners of 36 custom-designed houses discuss the advantages of their new homes, describing how site plans, room arrangements, flexible use of space, natural materials, etc. have worked out for them. The author adds to the owners' evaluations. Illustrated with 209 photographs and floor plans.

A04-0836
Creswell, K. A. C. A Bibliography of the Architecture, Arts, and Crafts of Islam to 1st January, 1960. Oxford University Press, 1961. 1382 pp. $53.50

A04-0837
Crosby, T. The Necessary Monument: Its Future in the Civilized City. N.Y. Graphic Society, 1970. 128 pp. $8.50
  A practicing architect advocates the continued use and involvement of great 19th century buildings in the economic and social life of cities and uses the histories of the Paris Opera, London's Tower Bridge and New York's Pennsylvania Station to illustrate his theories.

A04-0838
Dahinden, J. New City Structures. Praeger, 1971. 320 pp. $20.00
  These essays present radical new concepts of the way we can and will live--in three-dimensional grid cities, bridge cities, Cosmos, cell agglomerates, floating cities, biotectures, even massive clip-on and plug-in structures. The author evaluates these plans according to standards of flexibility, controlled economic growth, social environment, and careful attention to human needs in matters of measurements and dimensions. Designs are drawn from the United States, Japan, and Europe.

A04-0839
Damaz, P. Art in Latin American Architecture. Van Nostrand Reinhold, 1963. 232 pp. $15.00
  A critical analysis of the integration of the artist and the architect in Latin America.

A04-0840
Dannatt, T., ed. The Architects' Year Book. Elek Books. Vol. 8, 256 pp.; Vol. 9, 368 pp.

A04-0841
Davern, J., ed. Lewis Mumford: Architecture as a Home for Man. McGraw-Hill, 1975. 224 pp. $15.00
  How architecture shapes our lives--the human issues--is the theme of this book. Organized into five "mini-books," this volume includes a 1928 series of essays on "Modern Architecture Today"; a 1958 series on Matthew Nowicki, a brilliant architect who died tragically in a 1950 plane crash; the 1962-63 series on "the Future of the City"; and a collection of ten essays (1937-1968) on a variety of subjects from "The Case Against 'Modern Architecture' " to "The Highway and the City" and "For Older People--Not Segregation but Integration."

A04-0842
Davern, J. M. and eds. of Architectural Record. Architectural Record Book of Places for People: Motels, Hotels, Restaurants, and Bars. McGraw-Hill, 1976. 320 pp. $16.50

Design--Architectural

Intended for architects and contractors of commercial buildings, this new edition has been updated to include the latest in innovative, imaginative design concepts. With the emphasis on buildings and facilities for leisure-time use, a wide variety of hotels, resorts, lodges, restaurants, clubs, bars, camps, parks, plazas, and playgrounds are included. Several hundred black and white illustrations.

A04-0843
Davies, L. and Petty, D. J. Building Elements. Architectural Press, 1956. 386 pp. £2.75
A textbook dealing with the function and design of structural elements.

A04-0844
Day, B. F., et al. Building Acoustics. Applied Science, 1969. 120 pp. £1.50
Presents to the building and architecture professions a concise account of the basic principles involved in, for example, the techniques of sound insulation, the propagation of noise from motorways, and sound absorption.

A04-0845
Deane, P. Constantinos Doxiadis, Master Builder for Free Men. Oceana, 1965. 160 pp. $5.00
This biography introduces the man, his work and ideas. His approach to the outstanding challenge of our times, he calls "Ekistics." This is to house the world's exploding population and to shape its cities.

A04-0846
Debaights, J. Shop Fronts. Hastings House. 196 pp. $36.00
Shows the most inviting, eye-catching, interesting and attractive shop fronts of Europe. 16 color illustrations, 195 black & white photographs, 50 plans and technical drawings.

A04-0847
De Chiara, Joseph and Callender, John, eds. Time Saver Standards of Building Types. American Institute of Architects or R. S. Means, 1973. $32.50 (for AIA members, less 10%)
This collection of architectural planning data and design elements for each of the major building types is designed to serve as a companion volume for the forthcoming fifth edition of Callender's Time-Saver Standards for Architectural Design Data. The handbook covers a wide variety of buildings, including residential, governmental, commercial, and more.

A04-0848
Defoe, D. A Tour Thro' London About the Year 1725. Benjamin Blom, 1929. 144 pp. $18.75
An intimate glimpse of London life and architecture, with new notes on old buildings and places.

A04-0849
DeMare, E. The Nautical Style. Architectural Press, 1973. 100 pp. £3.95

The sea and objects associated with it--ships, piers, lighthouses and the clutter of equipment and machinery that is to be found around any harbour--have long been a powerful influence on British design. This book on the nautical tradition is a rich source of ideas for designers and decorators.

A04-0850
Dent, R. Principles of Pneumatic Architecture. John Wiley, 1972. 236 pp. $18.75

Providing an explanation of the engineering principles involved in pneumatic architecture, the author illustrates the various constructional uses to which these principles have been put.

A04-0851
Dept. of Landscape Arch., U. of Ga., and N.E. Ga. Area Planning and Development Comm. Madison--A Community Civic Design Study. Univ. of Georgia Press, 78 pp. $3.00

This book is an analysis of the physical appearance of the business district and environs of a northeast Georgia community with a population of about 3,000. It has 24 full-page illustrations including maps and "before" photographs with "after" sketches. Problems and opportunities are discussed with emphasis on the importance of creating a competitive shopping district while preserving the city's legacy of rare historical homes.

A04-0852
Dercsenyi, D. Historical Monuments in Hungary, Restoration and Preservation. International Publications, 1969. 180 pp. $16.00

An analysis of the history, methods and results achieved in the restoring and preserving of ruins, castles, country houses and manor houses, public buildings, ecclesiastical buildings, dwelling houses, rural and industrial buildings and statues and mural paintings.

A04-0853
Devereau, M. Architects' Working Details Revisited. Architectural Press, 1964. 198 pp. £2.00

This companion volume to the A.W.D. series reports on how 200 of the details selected performed in practice.

A04-0854
Diamant, R. M. E. The Chemistry of Building Materials. Cahners, 1970. 258 pp. $14.95

An up-to-date and comprehensive treatment of building materials from a chemical/scientific point of view. Covers materials used in industrialized building methods such as plastics and gas concrete as well as standard building materials.

Design--Architectural 149

A04-0855
Dixon, J. Architectural Design Preview, U.S.A. Van Nostrand Reinhold, 1962. 224 pp. $15.00
More than 500 fully captioned drawings, sketches, and photographs of models combine to make a source of ideas and applicable information on a wide range of structures.

A04-0856
Doane, D. A Book of Cape Cod Houses. Chatham Press, 1970. 91 pp. $7.95
Illustrated examples of the various styles, basic floor plans and authentic interiors.

A04-0857
Dober, R. P. Campus Planning. Van Nostrand Reinhold. 320 pp. $25.00
Provides design programs for existing facilities and institutional expansion. The coordinated programs and over-all plans outlined include complexes of buildings, housing, research laboratories, and individual structures.

A04-0858
Doelle, L. Environmental Acoustics. McGraw-Hill, 1972. 246 pp. $21.75
This book has been written primarily for the reader who is not an expert in sound or noise control. Divided into four parts, the book has an introduction (basic fundamentals), a section on room or space acoustics, noise control methods, and a final section on the execution, supervision, and checking of acoustical works. In addition, there are important reference tables contained in the appendices.

A04-0859
Downes, K. Hawksmoor. Praeger, 1970. 216 pp. $8.50
This volume is a condensed version of the author's standard study of Hawksmoor. The book is illustrated with photographs of all surviving buildings by Hawksmoor, along with examples of his draftsmanship.

A04-0860
Downing, A. F. and Scully, V. J., Jr. Architectural Heritage of Newport, Rhode Island, 1640-1915. Crown Publishers, 1967. $22.50
A survey in words and pictures. 500 illustrations.

A04-0861
Downing, A. J. The Architecture of Country Houses. Dover, 1850. 560 pp. $4.00
This is the basic book for Hudson River Gothic architecture of the middle Victorian period. Contains discussions of general aspects of housing, architecture, style, decoration, furnishing, together with scores of detailed house plans, illustrations of specific buildings, accompanied by full text. Perhaps the most influential

single American architectural book. Reprint of 1850 edition.

A04-0862
Doxiadis, C. A. Architecture in Transition. Oxford University Press, 1963. 200 pp. $7.50 (paperback $1.75)
An analysis of the modern architectural dilemma is offered in this book by an architect-engineer, world-renowned as a city and area planner. The author reflects upon the confused state of contemporary architecture, and attacks the problem of defining the architect's role, obligations, and methods in a scientific age.

A04-0863
Doxiadis, C. A. Ekistics: An Introduction to the Science of Human Settlements. Oxford University Press, 1968. 558 pp. $35.00
The author defines the scope and methodology of ekistics and describes the characteristics of human settlements, rural and urban, static and dynamic, and their interrelations within "ekistic networks." From a discussion of ekistic evolution and a consideration of ekistic pathology, he moves toward a solution based on the needs of individuals and groups and the forces these exert on growth patterns.

A04-0864
Drexler, A. Architecture of Skidmore, Owings & Merrill, 1963-1973. Architectural Book Publishers, 1973. 240 pp. $27.55
Architects, designers, engineers and city planners will welcome this book on the renowned architectural firm that has been a unique force in the building of the modern city. This work contains photographs, site and building plans, elevation drawings, and accompanying explanatory text showing the pioneering work done the last decade by this firm.

A04-0865
Dulaney, P. S. The Architecture of Historic Richmond. Univ. Press of Virginia, 1968. 208 pp. $2.45
Pictures over two hundred buildings. Construction date, distinctive features, and historical significance are provided for each illustration. The various architectural designs are discussed in detail against the background of a historical account of the city. The author also considers the problems involved in preserving these buildings and recommends steps the city might take to incorporate Richmond's architectural heritage into the modern city.

A04-0866
Dunbar, J. G. The Historic Architecture of Scotland. Architectural Book Publishers. 264 pp. 199 illustrations. $25.00
A guide to the famed castles, towers and abbeys and less familiar types of Scottish architecture.

A04-0867
Dunlop, I. The Chateaux of the Loire. Taplinger, 1969. 215 pp.

Design--Architectural 151

$12.00
Architectural history woven with stories of the men and women who peopled the chateaux. Architectural details, photographs.

A04-0868
Dunlop, I. Palaces and Progresses of Elizabeth I. Taplinger, 1970. 222 pp. $10.00
An account of architectural history--the great houses of the Elizabethan era where the Queen and her retinue periodically resided.

A04-0869
Dunlop, I. Versailles. Taplinger, 1970. 228 pp. $12.00
A historical architectural study of the French palace from its hunting lodge days to its present state of renovation.

A04-0870
Duprey, K. Old Houses on Nantucket. Architectural Book Publishing. 256 pp. $13.95
Details of decoration and furnishings as well as the unique architecture of this community. 336 illustrations.

A04-0871
Dutton, M. E. Students Guide to Model Making. Pergamon, 1971. 112 pp. $3.50
Describes the basic steps in making architectural models and interior models appropriately mounted in landscape or display settings.

A04-0872
Eastman, C. M., ed. Spatial Synthesis in Computer-Aided Building Design. John Wiley, 1975. 333 pp. $37.50
Collection of papers provides a detailed survey of the current state of the art in having the computer contribute to spatial synthesis problems. The concerns in space planning include the representation of spatial synthesis problems--both the representation of the physical alternatives that may be evaluated as potential solutions, and the representation of the criteria used in evaluating them--and the processes used for transforming a problem situation into a solution.

A04-0873
Eaton, L. K. American Architecture Comes of Age: European Reaction to H. H. Richardson and Louis Sullivan. MIT Press, 1972. 288 pp. $14.95
Eaton argues that both H. H. Richardson and Louis Sullivan profoundly affected architectural practice in the last decade of the nineteenth century and the first decade of the twentieth throughout Europe--except, interestingly, in the Romance countries.

A04-0874
Eaton, L. K. Two Chicago Architects and Their Clients: Frank Lloyd Wright and Howard Van Doren Shaw. MIT Press,

1969. 272 pp. $10.00

The architect is dependent on the immediate availability of patrons and clients and is constrained by their needs, funds, and wishes. A study of these patrons and clients is a vital if neglected part of architectural history, and this book marks an end to this neglect by revealing the backgrounds, personalities, and attitudes of two groups of clients involved in the dramatic confrontation between Frank Lloyd Wright and the established eclecticism around the turn of the century.

A04-0875
Eckbo, G. Urban Landscape Design. McGraw-Hill, 1964. 280 pp. $19.95

Concerned with the general physical landscape and with the improvement of the urban landscape.

A04-0876
Ede, C. M., ed. Canadian Architecture 1960/70. Universe Books, 1972. 264 pp. $19.50

The contemporary works of 18 architects are shown in categories of educational, public, commercial, industrial, religious and residential buildings. 400 photographs.

A04-0877
Edgell, G. H. The American Architecture of Today. AMS Press, 1928. $17.50

A04-0878
Egan, M. D. Concepts in Architectural Acoustics. McGraw-Hill, 1972. 224 pp. $18.75

Designed to give maximum help with basic acoustics problems without involving elaborate theories or highly detailed technical explanation. The book presents most of its information through quick-reference concept sketches, explanatory figures, and charts.

A04-0879
Eisenman, P., et al. Five Architects: Eisenman, Graves, Gwathmey, Hejduk, Meier. Wittenborn, 1972. 160 pp. $17.50

If these architects hold something in common it is their joint concern for the intrinsic quality and integrity of their ideas and forms. This book is the outcome of the Conference of Architects for the Study of the Environment (1969, Mus. of Modern Art). This book not only attempts to present these five architects' work of 1969 and of more recent date, but also to establish the ideological context of their work.

A04-0880
Ellsworth, R. E. Academic Library Buildings: A Guide to Architectural Issues and Solutions. Colorado Assoc. University Press, 1973. 530 pp. $10.00

This is an annotated photographic record--1500 pictures taken of 130 academic libraries in all parts of the country built since the end of World War II.

Design--Architectural 153

A04-0881
Elmes, J. Lectures on Architecture. Benjamin Blom, 1821. 440 pp. $20.00
Elmes (1782-1862) was a historian of art as well as of architecture and his lectures reveal his broad understanding of aesthetics and of the relationship between architecture and other arts.

A04-0882
End, H. Interiors Book of Hotels and Motor Hotels. Watson-Guptill. 264 pp. 256 illustrations. $16.50
Intended for designers, architects, and hotel men, this book examines the design principles of in-town, resort, and international hotels and motels.

A04-0883
ERDA. General Design Criteria Engineering Handbook-Appendix 6301. Energy Research and Development Admin., 1972. 150 pp.
Contains general design criteria to be used by the AEC and its contractors in the design of new buildings and facilities and of modifications to existing buildings. Basic design section covers architectural, structural and interior mechanical, electrical, and communications systems. Types of buildings and facilities are described in another section. Outside facilities and services, and protective construction are covered in final sections.

A04-0884
Ericsson, H. Sixty Years a Builder: The Autobiography of Henry Ericsson. Arno Press, 1942. $19.00
Ericsson was a Chicago builder who specialized in public and industrial structures. In this book he details his pioneering use of reinforced concrete, his innovations in architecture for heavy industry, and his designs for more efficient hotels.

A04-0885
Etienne-Louis Boullee: Theoretician of Revolutionary Architecture. Braziller, 1974. $6.95
Boullee was the principal theorist of neo-classicism. This work explains how the aesthetic of a return to antiquity was transformed into ideas relevant to those of the 20th century.

A04-0886
Evans, B. H. and Wheeler, C. H., Jr. Architectural Programming /Emerging Techniques. American Institute of Architects, 1968. $5.00 (for AIA members, less 20%)
Describes the means through which data about the needs of the ultimate building user are determined and expressed for the instruction of the architect to the development of design solution. Diagrams and charts illustrate techniques and methods of architectural programming and the development of the program instrument from which designs are produced.

A04-0887
Evenson, N. Le Corbusier: The Machine and The Grand Design. Braziller, 1969. 128 pp. $5.95 (paperback $2.95)
 A critical evaluation of the planner who established one of the most pervasive urban images of our time. Le Corbusier's urban concepts are here analyzed from their formation in the twenties to the culmination realized in Chandigarh.

A04-0888
Faber, T. Arne Jacobsen. Praeger, 1964. 200 pp. $20.00
 In this presentation of his achievements throughout a career of nearly forty years, the most important examples of the distinctive style of Arne Jacobsen, one of Denmark's leading architects, are illustrated with 300 photographs, sketches, and plans.

A04-0889
Faber, T. New Danish Architecture. Praeger, 1968. 220 pp. $20.00
 Disciplined form, choice of effective materials, and quality of craftsmanship are part of the Danish tradition. This is the background against which the author evaluates postwar Danish architecture, from single-family houses to city planning. He shows various developments and their most outstanding exponents--Kay Fisker, Arne Jacobsen and Jorn Utzon.

A04-0890
Fabian, D. Aquatic Buildings. Vol. II, Reference Book on Aquatic Buildings, Establishments and Facilities: Construction-Plant-Equipment-Operation-Economy. International Publications, 1970. 560 pp. $32.50
 The standard work. All texts in English and German. Contains the latest international guiding principles, Mexico 1968, and the details of systematic research into the structural and technical requirements of bath buildings. Each object is analyzed and described with plans and pictures. Vol. 1 (in prep.) will be concerned with the fundamentals and details of construction and equipment.

A04-0891
Fairweather, L., et al. Prison Architecture. Architectural Press, 1974. 350 pp.
 Ideas about prison design have changed considerably in recent years. A new generation of prisons is now being built which reflects these ideas, but the information on what is being done in various countries tends to be scattered and hard to obtain. This book brings together data on prison design and construction procedures from some twenty countries.

A04-0892
Farrar, E. F. and Hines, E. Old Virginia Houses Along the Fall Line. Hastings House. 256 pp. 175 illustrations. $15.00
 Photographs depict the mansions, architectural features and interiors.

Design--Architectural 155

A04-0893
Farrar, E. F. and Hines, E. Old Virginia Houses: The Northern Peninsulas. Hastings House, 1973. 256 pp. $15.00
A rendition of what life was like in Colonial Virginia with a presentation of photographs of houses and interiors.

A04-0894
Faulkner, W. Architecture and Color. John Wiley, 1972. 146 pp. $17.50
Shows how colors are applied to building materials; how they may be selected and specified in order to harmonize in a complete building. Describes the uses of color in architecture; symbolic, functional and esthetic, both past and present. Outlines the scientific concepts of color, beginning with the phenomena of vision and light. The basic systems by which color can be measured and specified are discussed.

A04-0895
Favero, G. B. The Villa Emo and Fanzolo. Pennsylvania State Univ. Press, 1973. 234 pp. $29.50
This volume discusses the building's design in relation to other Palladian villas and provides an extensive treatment of the pictorial decoration.

A04-0896
Fengler, M. Restaurant Architecture and Design. Wittenborn, 1972. 176 pp. $18.50
Over 350 photos, architectural plans, details in this international survey of design for dining.

A04-0897
Fengler, M. Restaurant Architecture and Design: An International Survey of Eating Places. Universe Books, 1972. 200 pp. $18.50
Hundreds of illustrations show innovative accomplishments in the planning and design of contemporary restaurants, coffee shops, cafeterias, and canteens.

A04-0898
Fernandez, J. A. Architecture in Puerto Rico. Architectural Book Publishing. 256 pp. $15.00
The finest examples of historical, traditional, and contemporary architecture in Puerto Rico.

A04-0899
Ferrey, B. Recollections of A. W. N. Pugin and His Father, Augustus Pugin. Benjamin Blom, 1861. 474 pp. $18.50
One of the first contemporary accounts and still one of the chief sources for information on the architects.

A04-0900
Ferriss, H. Power in Buildings: An Artist's View of Contemporary Architecture. Columbia Univ. Press, 1953. 102 pp.

$8.50

Through sixty drawings and a nontechnical text, Hugh Ferriss has set out to illustrate the nation's power in buildings. His drawings portray outstanding structures designed since 1929.

A04-0901
Feuerstein, G. New Directions in German Architecture. Braziller. 128 pp. $7.95 (paperback $3.95)

An examination of present German architecture with a view to its historical basis and future trends.

A04-0902
Fink, A. Adobes in the Sun: Portraits of a Tranquil Era. Chronicle Books. 160 pp. $14.95

These homes are truly part of California--constructed of its soil and grounded in its history. Whether the simple shelters of the earliest settlers, or the gracious townhouses of Colonial Monterey, they exemplify the beauty inherent in man's most basic interaction with his environment.

A04-0903
Fitch, J. M. American Building: The Environmental Forces that Shape It. Schocken Books, 1975. $5.95

The approach to architecture that gives the relationship between what man builds and the natural environment.

A04-0904
Fitch, J. M. American Building: The Historical Forces that Shaped It. Schocken Books, 1973. 364 pp. $4.95

A comprehensive history of architecture in the United States, and a cultural history of the country from colonial days to the present. Takes into account the materials, needs, technical equipment, esthetic theory and creative genius that shaped saltbox and skyscraper.

A04-0905
Fitch, J. M. Architecture and the Esthetics of Plenty. Columbia Univ. Press, 1961. 304 pp. $9.00

Professor Fitch discusses the conflict between quantity and quality in American design. He deals with the American esthetic standards in architecture, and what they have produced, both good and bad, examines them in the light of work being done elsewhere, and concludes that a scientific analysis of concrete human requirements is necessary if Americans are to begin to construct economical, comfortable, and working buildings.

A04-0906
Fletcher, B. A History of Architecture. Charles Scribner's Sons, 1961. 652 pp. $18.95

Comprehensive history from ancient monuments to present day structures. In this 17th edition, the chapters on Renaissance architecture have been considerably expanded as have the accounts of architecture since 1830. Includes three chapters on the 19th

Design--Architectural 157

and 20th century architecture in Britain, Europe and North and South America.

A04-0907
Flynn, John E. and Segil, Arthur W. <u>Architectural Interior Systems: Lighting, Air-Conditioning, Acoustics.</u> Van Nostrand Reinhold, 1970. $14.95
   Presents a study of the purposes and functions of building design in terms of the sensory and behavioral needs of the occupant and the mechanical technology of our time. Attempts to bridge gap between the technology of environmental control and the art of architecture; suggests that modern buildings require mechanically-coordinated architectural forms.

A04-0908
Ford, K. M. and Creighton, T. H. <u>Quality Budget Houses.</u> Van Nostrand Reinhold, 1954. 224 pp. $8.00
   100 architect-designed houses in the low and medium-priced category. Each house included is described from the viewpoint of what today's home buyer wants to know: cost, material, design, etc., to be able to obtain the best value from a limited budget. 350 photographs, plans and drawings.

A04-0909
Forman, H. C. <u>Maryland Architecture: A Short History from 1634 through the Civil War.</u> Cornell Maritime, 1968. 128 pp. $6.00
   The author has presented a concise history of the subject, stressing Early Georgian, Late Georgian, Maryland-German and Classical Revival, and giving some account of their development and sources. Buildings and floor plans, gardens, early colonists' artifacts, portraits, pew diagrams and molding profiles are illustrated.

A04-0910
Forman, H. C. <u>Old Buildings, Gardens and Furniture in Tidewater Maryland.</u> Cornell Maritime, 1967. 342 pp. $12.50
   The historical text is supplemented by over 580 illustrations. A vast range of objects is pictured, from chairs, secretaries, panelling, tools, and handmills to forts, mansions, house floor plans, churches, pew plans, sorry houses, outbuildings, moldings, fences, gateways, and garden plans.

A04-0911
Forman, H. C. <u>Virginia Architecture in the Seventeenth Century.</u> Univ. Press of Virginia, 1957. 79 pp. $1.25
   The author traces the origins and development of the principal styles of architecture which flourished in Virginia between 1600 and 1700. The essay is supplemented with photographs, drawings, and diagrams by the author.

A04-0912
Forrester, J. W. <u>Urban Dynamics.</u> MIT Press, 1969. 256 pp.

$17.50

Two types of simulation models are developed in the book. One, a growth model, generates the life cycle of a city from its founding to its reaching a state of equilibrium, a period of 250 years. The other model, a variation on the first, begins with the equilibrium conditions and is used to examine how various policies would cause the condition of the city to be altered over the succeeding 50 years. The variables include different levels and kinds of employment, housing, and industry.

A04-0913
Francois Mansart. Abner Schram, 1974. 280 pp. 567 illus. 2 vols. $95.00

This monograph is the first full-length study of Mansart and his architecture. The first volume contains a discussion of his social and architectural background, together with an analysis of all his major works. The second volume contains 563 plates, illustrating all Mansart's surviving buildings and drawings, and more related material, including many reconstructions of lost or altered buildings.

A04-0914
Frankl, P. Principles of Architectural History. MIT Press, 1968. 240 pp. $12.50

Frankl employs two interdependent methodological systems, one critical, the other historical. The critical system describes architectural forms in terms of four categories: spatial composition; mass and surface; light, shadow, and color; and social function. The historical system divides post-medieval architecture into four phases: 1420-1550, (the Renaissance), 1550-1700 (the Baroque), the eighteenth century (Rococo), and the nineteenth century (Neoclassicism).

A04-0915
Frary, I. T. Early Homes of Ohio. Dover, 1970. 334 pp. $3.50

A survey of early buildings (18th and 19th centuries) and their features in a region with an interesting architectural history. Homes, public buildings, old bridges, churches, taverns and four historic towns are pictured and discussed. 203 illustrations.

A04-0916
Fraser, D., et al., eds. Essays in the History of Architecture. Praeger, 1969. 382 pp. $17.50

These twenty-six essays discuss the pioneering work of Rudolf Wittkower in such fields as the Italian Baroque, and explore new aspects of Wittkower's work. Among the topics are the emigration and interpretation of symbols, problems of proportion and perspective, the iconographic interpretation of art, Italian Renaissance sculpture and architecture, Baroque art in all its manifestations, Palladio and Palladianism, and English architecture.

Design--Architectural

A04-0917
Freeland, J. M. Architecture in Australia. International Publications, 1968. 328 pp. $12.50
The history of architecture as it has evolved in Australia from 1788 to 1967.

A04-0918
Friedmann, A., et al. Interior Design: An Introduction to Architectural Interiors. American Elsevier, 1970. 303 pp. $14.00
A brief architectural history and discussion of professional training and practice of interior design are followed by a survey of interior construction, the mechanical systems, their fabrication and assemblage.

A04-0919
Frigand, S. J. The New City: Architecture and Urban Renewal. N.Y. Graphic Society, 1967. 48 pp. $1.95
Four projects commissioned by The Museum of Modern Art explore possible solutions to such problems as housing and renewal, the developments of misused land, and modifications of an existing layout of streets and parks. The sites are all on Manhattan Island, but the ideas put forward are relevant to many other American cities.

A04-0920
Fry, M. and Drew, J. Tropical Architecture. Van Nostrand Reinhold, 1964. 261 pp. $18.50

A04-0921
Furneaux, J. R. A Concise History of Western Architecture. Harcourt Brace Jovanovich, 1970. 359 pp. $9.95 (paperback $5.95)
Illustrated survey of architectural styles from classical Greece to the present.

A04-0922
Fyodorov, B. and Bartenev, I. North Russian Architecture. Beekman Publishers, 1972. 332 pp. $17.50
In this illustrated book, we travel throughout North Russia-- through the White Sea Coast, the Transonega, and along many rivers--and witness the skillful blending of churches and domestic buildings. A detailed narrative discusses each building architecturally and historically.

A04-0923
Galardi, A. New Italian Architecture. Praeger, 1967. 204 pp. $20.00
Alberto Galardi sketches the development of modern Italian architecture since the nineteenth century, but concentrates on outstanding buildings of the last ten years.

A04-0924
Garlake, P. S. The Early Islamic Architecture of the East African Coast. Oxford University Press, 1966. 218 pp. $29.00

An exhaustive study of the distinctive architecture of the Muslim people of the East African Coast, giving an account of the buildings of this civilization when it was at its height before the sixteenth century, and tracing the later development of the architecture down to the middle of the nineteenth century.

A04-0925
Gatz, K., ed. Detail; Contemporary Architectural Design. Vol. 2. Butterworth, 1965. 284 pp. £5.00

A selection of outstanding examples with structural details of steel-framed and reinforced concrete buildings. Details of interiors include sculptured brick murals, suspended and folded ceilings, balustrades, circular stairs and galley stairs. Fireplaces include examples from houses in Zurich, Tokyo and California. Unity of detail deals with five complete building projects notable for their excellent design.

A04-0926
Gatz, K., ed. Detail; Contemporary Architectural Design. Vol. 3. Butterworth, 1967. 284 pp. £5.00

A pictorial survey discussing details of structures and external works; unity of detail; and details of interiors and fittings. Examples of the structural use of natural stone walls and composite construction in Europe are given. Concrete and precast concrete structures include a comprehensive examination of a preparatory school in the USA, the Chichester Festival Theatre, a Swiss Zoo and the Pan-American skyscraper, New York. Unity of detail deals with seven differing building projects notable for their excellent design.

A04-0927
Gatz, K., ed. Detail; Contemporary Architectural Design. Vol. 4. Butterworth, 1969. 272 pp. £6.50

This fourth volume maintains a high standard of line drawings and photographs, and gives the solutions to various problems of detailing worked out by leading architects when designing some of the most interesting buildings recently completed. Examples are selected from a wide range of buildings and countries.

A04-0928
Gatz, K., ed. Detail; Contemporary Architectural Design. Vol. 5. Butterworth, 1972. 268 pp. £7.50

Detail 5 presents a pictorial study of contemporary architecture. It contains a selection of outstanding examples taken from the German periodical Detail-Zeitschrift fur Architektur und Baudetail, which is itself devoted entirely to the subject of detailing.

A04-0929
Gatz, K. and Achterberg, G. Color and Architecture. Architectural Book Publishers, 1966. 304 pp. 598 illustrations.

Design--Architectural   161

$22.50
The color properties of the range of external building materials on the market and how best to utilize them.

A04-0930
Gauldie, S. Architecture. Oxford University Press, 1969. 204 pp.
$10.00 (paperback $5.95)
Through the use of a variety of buildings from many places and societies, the author demonstrates that buildings may be comprehended as the culmination of a process in which functional, structural, and aesthetic intentions combine.

A04-0931
Gebhard, D. Rudolph M. Schindler. Wittenborn, 1971. 216 pp.
$7.95 (paperback $3.95)
This book, part of the series, Pioneers of Modern Architecture, is about two architectural geniuses: Pugin and Schindler.

A04-0932
Geerlings, G. K. Metal Crafts in Architecture. Charles Scribner's Sons, 1972. $10.00
This volume presents an analysis of the characteristics of metal embellishing buildings, a description of the manner in which these materials are worked and a historical background of metals from various countries.

A04-0933
Geerlings, G. K. Wrought Iron in Architecture. Charles Scribner's Sons, 1972. $10.00
This book points out the best examples of wrought iron architecture from each country and considers design possibilities and limitations for each metal. This information gives the architect some insight into the craftsman's work giving him a sounder design concept.

A04-0934
Geretsegger, H., et al. Otto Wagner 1841-1918. Praeger, 1970. 276 pp. $25.00
This illustrated volume shows the range and diversity of Wagner's designs; its highlights include new photographs of all extant Wagner buildings (accompanied by a wide variety of details and a street map showing their location in Vienna), analyses of Wagner's most significant writings, and an outline of the main events in the life of this innovative designer.

A04-0935
Germann, G. Gothic Revival in Europe and Britain Sources, Influences, and Ideas. MIT Press, 306 pp. $25.00
Most analyses of the nineteenth-century Gothic revival in architecture have treated it as an essentially English phenomenon. This work is one of the few to consider the important parallel developments in Europe and thus set the movement in proper international perspective.

A04-0936
Gibbs, J. A Book of Architecture. Benjamin Blom, 1728. 28 pp. 150 plates. $35.00
Among the central designers in 18th century English and American architectural history whose influence can be seen in countless buildings in both countries.

A04-0937
Gieselmann, R. New Churches. Architectural Book Publishing, 1973. 176 pp. $22.50
A new wave of church architectural design has evolved throughout the world. Specifically, the main trends have been to move the altar from a high position in front of the congregation and place it where worshippers can group around it. The building is now less imposing than surrounding buildings. And as the Church has become a center for secular activities, areas for worship and secular activities are combined. These new trends have an important bearing on church architecture.

A04-0938
Gilford, C. L. S. Acoustics for Radio and Television Studios. Int'l. Scholarly Book Services (Peter Peregrinus Ltd.), 1972. 305 pp. $17.00
A review of studio acoustic design presented so that the logical, scientific and practical basis of design is covered at every stage. The emphasis is on achieving the most economical solutions to satisfy agreed subjective criteria.

A04-0939
Gillon, E. V., Jr. Early Illustrations and Views of American Architecture. Dover, 1971. 296 pp. $6.95
Includes more than 700 line illustrations of Colonial through Victorian churches, residences, prisons, courthouses, stores, banks, theatres, depots, windmills, lighthouses, hospitals, hotels, etc.

A04-0940
Gillon, E. V., Jr. and Lancaster, C. Victorian Houses: A Treasury of Lesser Known Examples. Dover, 1973. 128 pp. $3.50
116 full page photographs of East coast Victorian houses. Sequenced to show stylistic development of eclectic Victorian architecture.

A04-0941
Girbau, L. D. Contemporary Spanish Architecture. Praeger, 1971. 240 pp. $20.00
This book describes every aspect of architecture in contemporary Spain. There is a complete account of the work of Jose Antonio Coderech and his contemporaries. There are also extensive descriptions of new resorts, the integration of new buildings into historical settings, the factories being set up in newly created

Design--Architectural 163

areas of development, and the luxurious private homes of the wealthy.

A04-0942
Girouard, M. The Victorian Country House. Oxford University Press, 1971. 240 pp. $41.00
The combination of a prosperous upper-class with huge numbers of industrial-revolution new rich setting up as country landowners made the Victorian age a boom period for country housebuilding. The results are fascinating because of the houses' social context, their often enormous size, their heavyweight technological equipment, their complex and stratified planning, and the variety of their architecture.

A04-0943
Girouard, M. The Victorian Public House. Architectural Press, 1974. £10.00
This study of public houses of the period is a contribution to social and architectural history in the 19th century.

A04-0944
Givoni, B. Man, Climate and Architecture. Applied Science, 1969. 364 pp. £5.50
Describes the climatic factors affecting human comfort and building design, the biophysical relationship between man and thermal environment, physiological and sensory effects of heat and cold stress in relation to work. Includes a description and evaluation of various thermal indices used to express the combined effect of various environmental factors.

A04-0945
Glaeser, L. The Works of Frei Otto. New York Graphic Society, 1971. 128 pp. $10.00
A pictorial survey of the work of one of the most imaginative of contemporary architects.

A04-0946
Gleis, F. Kleinkirchenbau. Wittenborn, 1968. 140 pp. $5.00 (paperback $3.50)
Architectural details of German-Lutheran county churches from Small Church Conference, 1967, Plon/Holstein.

A04-0947
Goldfinger, M. Villages in the Sun: Mediterranean Vernacular Architecture. Praeger, 1970. 228 pp. $18.50
Through photographic and textual analysis of the village architecture of Greece, Italy, Spain, Morocco, and Tunisia, the author shows what community architecture teaches at a time when neighborhood units are losing their identities and the land around cities is being distorted by featureless suburbs.

A04-0948
Goldsmith, S. Designing for the Disabled. McGraw-Hill, 1968.

207 pp. $19.50

A guide dealing with the needs of the disabled, and how architects and interior designers can provide accommodation of the handicapped.

A04-0949
Goodman, C. and Von Eckardt, W. Life for Dead Spaces: The Development of the Lavanburg Commons. Harcourt Brace Jovanovich. 127 pp. $12.50

Published for the Fred L. Lavanburg Foundation. Practical ways of turning the open spaces in high-rising housing developments into neighborhood centers.

A04-0950
Goodwin, G. A History of Ottoman Architecture. Johns Hopkins Univ. Press, 1971. 511 pp. $30.00

This is the first comprehensive survey of Ottoman architecture in English and the first to deal seriously with the buildings of the last two centuries. The author places architectural developments in historical perspective and describes not only the mosques but also the layout and function of the buildings grouped around them.

A04-0951
Goody, J. E. New Architecture in Boston. MIT Press, 1965. 70 pp. $5.00 (paperback $2.95)

Fifty of the most significant works of architecture to appear in metropolitan Boston and environs in the last twenty-five years are presented in this book. Many of these contemporary buildings have been designed by such noted architects as Walter Gropius, Alvar Aalto, and Le Corbusier.

A04-0952
Gotch, J. A. Inigo Jones. Benjamin Blom, 1928. 271 pp. $12.50

All Jones's great achievements in architecture and stage design, including his royal masques and entertainments.

A04-0953
Gowans, A. Church Architecture in New France. Univ. of Toronto Press, 1955. 218 pp. $10.00

Behind the spired churches of Quebec lies a tradition which began in tiny cabanes of branches and bark and culminated in the great twin-towered churches of the mid-eighteenth century. The author examines the relationship of this tradition to architecture in France and in the American colonies. Catalogue of chapels and churches erected from 1615 to 1760 in New France.

A04-0954
Grady, J. Architecture of Neel Reid in Georgia. Univ. of Georgia Press, 1973. 226 pp. $29.75

A survey of Reid's contribution to the architecture of Atlanta,

Design--Architectural   165

Macon, LaGrange, Athens and Columbus. Illustrations convey his
mastery of scale.

A04-0955
Graf, D.   Don Graf's Data Sheets.   Reinhold Books, 1975.   816 pp.
    $11.95
    Thousands of facts about building materials, planning and
construction. Covers stock shapes of materials, stress statistic,
termite control, methods, leg room needs in a theater, and more--
all indexed.

A04-0956
Granger, A. H.   Charles Follen McKim: A Study of His Life and
    Work.   Benjamin Blom, 1913.   159 pp. $13.75
    This illustrated book includes a brief biography, followed by
a discussion of McKim's most important works: The Boston Public Library, the World's Fair at Chicago, Pennsylvania Terminal,
and the American Academy at Rome.

A04-0957
Gregotti, V.   New Directions in Italian Architecture.   Braziller.
    128 pp. $7.95 (paperback $3.95)
    An examination of present Italian architecture with a view
to its historical basis and future trends.

A04-0958
Greiff, C. M., ed.   Great Houses from the Pages of Antiques.
    Charles Scribner's Sons, 1973.   160 pp. $15.00 (paperback
    $8.95)
    America's finest homes and interiors are discussed by ex-
perts of Antiques magazine. From the early articles on "Dating of
New England Houses" to those later photographic essays on "History in Houses," here are photographs and important information
on furnishings and the structures which house them.

A04-0959
Greiff, C. M., et al.   Princeton Architecture: A Pictorial History
    of Town and Campus.   Princeton Univ. Press, 1967.
    $12.50
    Because few important buildings in Princeton have been de-
stroyed, the town is a museum of American architectural history
with examples of every important style from the Colonial period to
the present. The authors discuss the architecture of both town
and campus in relation to its history and to American architecture.

A04-0960
Gropius, W.   Apollo in the Democracy.   McGraw-Hill, 1968.
    224 pp. $12.50
    This collection of papers and addresses represents a com-
mentary on the architectural situation of our time. 69 illustrations.

A04-0961
Gropius, W.   The New Architecture and the Bauhaus.   MIT Press,

1965. 112 pp. $2.45

A reissue of one of the most important books on the modern movement in architecture, Professor Gropius' book poses some of the fundamental problems presented by the relations of art and industry and considers their possible, practical solution.

A04-0962
Gropius, W. Town Plan for the Development of Selb. MIT Press, 1970. 84 pp. $15.00

The development of a master plan for Selb, a town of 20,000 in northeast Bavaria, was the last extended work of Walter Gropius, undertaken in collaboration with other members of his firm. This book presents that plan, together with the concurrently developed traffic control system.

A04-0963
Grube, O. W. Industrial Buildings and Factories. Praeger, 1971. 180 pp. $18.00

This book presents 40 outstanding examples of recent industrial buildings from all over the world. In addition to good design, the principal criteria for their selection were facilities for smooth coordination between supplies and production, and the flexibility of layout to allow for changes in production methods and expansion of production machinery. Other factors analyzed are speed of construction, use of prefabricated elements, and problems of lighting, ventilation, air conditioning, power, and internal transportation.

A04-0964
Guinness, D. and Ryan, W. Irish Houses and Castles. Viking Press, 1971. 352 pp. $35.00

Development of Irish architecture from castle stronghold to Gothic revival.

A04-0965
Guinness, D. and Sadler, J. T., Jr. Mr. Jefferson, Architect. Viking Press, 1973. 200 pp. $14.95

Thomas Jefferson's architectural accomplishments, illustrated with plans, drawings, engravings, old and new photographs.

A04-0966
Gutheim, F., ed. In the Cause of Architecture: Frank Lloyd Wright for the Record. McGraw-Hill, 1975. 247 pp. $16.50

This is the only complete edition of Frank Lloyd Wright's famous articles written for Architectural Record, in which he articulated his architectural philosophy. In addition to Wright's own articles there is a symposium of essays by those who knew and worked with him and have become leaders in current architectural thought.

A04-0967
Habraken, N. J. Supports: An Alternative to Mass Housing. Praeger, 1972. 105 pp. $7.95 (paperback $3.95)

Design--Architectural

Summarizes the ideas and theories of one of the most influential planner-architects of our time and puts forward a revolutionary alternative called "supports"--steel frameworks into which individual dwelling units can be slotted and connected, providing the purchaser with a wider choice of designs and fittings.

A04-0968
Halfpenny, W. and Halfpenny, J. The Art of Sound Building. Benjamin Blom, 1725. 56 pp. $12.50
18th century English carpenter-architects prepared these pattern books for the working builder, complete with accurate measurements and notes on construction, all in the best styles of the period.

A04-0969
Halfpenny, W. and Halfpenny, J. Chinese and Gothic Architecture. Benjamin Blom, 1752. 12 plates. $12.50

A04-0970
Halfpenny, W. and Halfpenny, J. Practical Architecture. Benjamin Blom, 1730. 56 pp. 24 plates. $12.50

A04-0971
Halfpenny, W. and Halfpenny, J. Rural Architecture in the Chinese Taste. Benjamin Blom, 1755. 64 plates. $12.50

A04-0972
Halfpenny, W. and Halfpenny, J. Useful Architecture. Benjamin Blom. 21 plates. $12.50

A04-0973
Halprin, H. Cities. MIT Press, 1973. 240 pp. $15.95 (paperback $4.95)
The author examines the basic elements of the cityscape: the open spaces that give a city its character and the spaces within which its life takes place (streets, plazas, parks, the private living places and small gardens); street furniture (kiosks, benches, light); the city floor (asphalt, brick, concrete); water; trees; roofviews; and what Halprin calls the "choreography" of the city.

A04-0974
Halprin, L. The RSVP Cycles: Creative Processes in the Human Environment. Braziller, 1970. $15.00 (paperback $5.95)
Discusses the processes of environmental creation and suggests many innovations for the future.

A04-0975
Halse, A. The Use of Color in Interiors. McGraw-Hill, 1968. 164 pp. (19.50
An illustrated guide to understanding the use of color.

A04-0976
Ham, F., ed. Theatre Planning. Univ. of Toronto Press, 1972.

292 pp. $27.50

Basic design principles are re-stated in present-day terms in this book. Includes information on the design of performance and auditorium spaces and their ancillaries, and such matters as acoustics, sight lines, heating and ventilating and the comparative economics of various design solutions.

A04-0977
Hamlin, T. F. Greek Revival Architecture in America. Dover, 1944. 439 pp. $4.50

The basic study of important form, 1820-1860; late Colonial architecture, Jeffersonian classicism, origin and spread, local modifications, decline. 221 photos, 101 figures and drawings show buildings, special features.

A04-0978
Hamlin, T. F., ed. Forms and Functions of Twentieth-Century Architecture. Columbia Univ. Press, 1952. $25.00 ea. Vol. 1, The Elements of Building. Vol. 2, The Principles of Composition. Vol. 3, Building Types: Residence, Gatherings, Education, Government. Vol. 4, Building Types: Commerce and Industry, Public Health, Transportation, Social Welfare and Recreation.

Volumes 1 and 2, written almost entirely by Professor Hamlin, deal with the structural and aesthetic aspects of architecture. Volumes 3 and 4 contain more than fifty chapters, each on one of the various building types required by our civilization. 3,745 illustrations.

A04-0979
Hammond, P. Liturgy and Architecture. Columbia Univ. Press, 1961. 191 pp. $6.00

Peter Hammond expresses the hope that this book "may go some way towards meeting the needs of the many thoughtful people who share my concern at the present state of church architecture on this side of the Channel."

A04-0980
Hancocks, D. Animals and Architecture. Praeger, 1971. 224 pp. $12.50

This book focuses on the structures that men have built for animals since their first domestication and also looks at the feats achieved by animals themselves as builders. Among the buildings created by humans that Mr. Hancocks describes are stables, menageries, aquariums, reptiliaries, aviaries, and children's zoos.

A04-0981
Handler, A. B. Systems Approach to Architecture. American Elsevier, 1970. 184 pp. $16.40

Examines architecture in terms of technical, environmental, human, symbolic, and economic performance, and delineates the role of each in the architectural system. The author also presents an analysis of economic performance in relation to the building industry, building operations, and the economy. A book intended for those directly concerned with planning and building.

A04-0982
Hansen, H. J., ed. Architecture in Wood: A History of Wood Building and Its Techniques in Europe and North America. Viking Press, 1971. 280 pp. $40.00
Color plates, photographs, drawings.

A04-0983
Harbron, D. The Conscious Stone: The Life of Edward William Godwin. Benjamin Blom, 1949. 208 pp. $12.50.
Godwin as architect, as pioneer stage designer, innovator in costume design and lighting, as a champion of Japanese design in architecture and furniture, and as intimate of theatre, literary, and artistic figures of the 1880s and 1890s.

A04-0984
Harker, J. Studio and Stage. Benjamin Blom, 1924. 283 pp. $12.50
A personal account by one of the outstanding stage designers of our era, describing productions over a 50-year span, particularly the early 20th century. Chapters on Irving's Lyceum, Beerbohm Tree, the circus, American manager, and changes in stage scenery.

A04-0985
Harper, G. Computer Applications in Architecture and Engineering. McGraw-Hill, 1968. 256 pp. $18.00
Discusses the practical applications of the computer in architectural and engineering firms. Provides an instructional guide to applications in such areas as specifications, estimating, accounting, networking, building information systems, etc.

A04-0986
Harris, C. M. Dictionary of Architecture and Construction. McGraw-Hill, 1975. 512 pp. $29.50
This dictionary provides definitions of terms encountered in the everyday practice of architecture and construction, as well as the more specialized vocabularies found in such areas as the history of architecture, church architecture, restoration, and art history. The meanings for terms in both traditional and recently developed materials, finishes, coatings, and surfacing, various types of units, assemblies, systems, and modules, along with tools, devices, machines, and construction equipment for their assemblage and construction are covered.

A04-0987
Harris, J. and Lever, J. Illustrated Glossary of Architecture. Crown Publishers. 316 pp. $12.50
A new departure in glossary writing whereby an actual structure is shown in a photograph and each feature is labeled right on the photograph.

A04-0988
Hatch, A. Buckminster Fuller: At Home in the Universe. Crown Publishers, 1974. 320 pp. $7.95
An intimate biography of one of the most remarkable men of our time by a lifelong friend. Provides personal detail never before revealed.

A04-0989
Hawkins, H. L. Appraisal Guide for School Facilities. Pendell Publishing Co., 1972. 96 pp. $5.00
The reader can use this guide to formulate a permanent record; highlight specific needs; examine the need for new facilities; evaluate the need for renovation; and, serve as an instructional tool. The author selects the criteria for appraisal as follows: site, structural, mechanical, building environment, school safety, space utilization, and maintainability.

A04-0990
Hayashi, M. House Design in Today's Japan. Wittenborn, 1969. 182 pp. $18.00
Japanese, English text. 42 postwar Japanese houses are shown in ground-plans and site views.

A04-0991
Hegemann, W. and Peets, E. The American Vitruvius: An Architects' Handbook of Civic Art. Benjamin Blom, 1922. 298 pp. $37.50
This book graphically traces the achievements and failures of urban planning in the United States and western Europe. The text gives the specific concepts underlying the authors' work as architects and city planners. 1200 illustrations.

A04-0992
Helick, R. M. Merchant Built Houses in Western Pennsylvania. Regent Graphic Services, 1966/1972. 108 pp. $15.00
This book is a compilation of 50 selected builders' houses of conservative design, all of which have been built and sold in western Pennsylvania during the past decade. Selection is based on quality of floor plan, general adaptability and good sales history. Each model is in the form of a photo-reduced architectural drawing in sufficient detail to secure estimates.

A04-0993
Helick, R. M. The Regent System of Townhouse Variations. Regent Graphic Services, 1969/1971. 108 pp. $13.95
A folio of 44 basic townhouse floorplans, drawn and dimensioned at quarter-inch scale. Various spatial relationships are categorized to the end that practically all room combinations are represented either by the basic plans or by variants of the basic plans.

A04-0994
Helick, R. M. Varieties of Human Habitation. Regent Graphic

Design--Architectural 171

Services, 1970. 110 pp. $17.50
A reference atlas, opening to 34 inches and containing 177 specimen floorplans of a variety of dwellings: primitive, historical, modern. It compares in a graphic way different modes of domestic life from the beginnings of Egyptian civilization to the typical American tract house. Contains information and bibliographic material unavailable in even the most comprehensive architectural histories.

A04-0995
Herrmann, W. The Theory of Claude Perrault. Abner Schram, 1973. $30.00
The book is the first full-size study that deals in detail with Perrault's theoretical writing and gives an interpretation of his views on architecture.

A04-0996
Hersey, G. L. High Victorian Gothic: A Study in Associationism. Johns Hopkins Univ. Press, 1972. 254 pp. $15.00
High Victorian Gothic was one of the earliest attempts to create a new style, rather than simply purifying an existing one or reviving an old one. It was a style characterized by polychrome, shiny materials, eclectic sources and a strong ecclesiastical flavor.

A04-0997
Hesselgren, S. The Language of Architecture. Applied Science. 663 pp. £6.00
It is estimated that in a short time 80% of the world's populations will live in towns. The houses and buildings constructed in these new communities must not only provide shelter against the elements, they must also "speak" to their inhabitants. This book analyzes this "language of architecture" in hope that such an analysis may make a contribution to a more effective approach to architectural problems.

A04-0998
High Rise Storage: Modern Materials Handling Guidebook. Cahners, 1971. 48 pp. $4.50
In-depth view of a new concept in storage with enormous potential. Illustrated.

A04-0999
Hildebrand, G. Designing for Industry: The Architecture of Albert Kahn. MIT Press, 1974. 224 pp. $25.00
This book documents Kahn's career, including the unique team practice that he originated. Of the more than two thousand factories designed by his firm, all representative examples are discussed in detail. Major nonindustrial buildings produced by the firm are also included.

A04-1000
Historic American Buildings Survey. Lenox Hill Publishers, 1941. 470 pp. $25.00

Catalog of the measured drawings and photographs of the survey in the Library of Congress, 2d edition.

A04-1001
Hitchcock, H. R. The Architecture of H. H. Richardson and His Times. MIT Press, 1966. 343 pp. $3.95
A study of his architecture and of the setting in which he worked. It includes detailed discussions of the social, economic, and technical history of the era and examines the architecture of Richardson's contemporaries and of those countries in which he traveled and lived.

A04-1002
Hitchcock, H. R. Early Victorian Architecture in Britain. DaCapo, 1976. 522 pp. $45.00 (paperback $7.95)
The leading authority on 19th-century architecture vividly recreates the iconoclastic era of British Victorian architecture from 1837 to 1851 in this pictorial study. Over five hundred illustrations depict the three characteristic styles of the period: Pugin's 15th-century Gothic formula for churches; Barry's revival of the Italian Renaissance palazzo for estates and public buildings; and the "Jacobethan" style for country houses, derived from British civil architecture of the late 16th and early 17th centuries.

A04-1003
Hitchcock, H. R. Rhode Island Architecture. MIT Press, 1969. $4.95
The author devotes over half the book to photographs, arranged in chronological and topical order, with a corresponding critical commentary which includes sketches and plans of many of the buildings photographed.

A04-1004
Hitchcock, H. R. Rococo Architecture in Southern Germany. Praeger, 1969. 276 pp. $20.00
Here is a survey of eighteenth-century churches, monasteries, and various secular buildings in South Germany, Switzerland, and Austria. The effect of light and space achieved was outstanding--the result of elaborate wall painting enhanced by concealed lighting. Architects who are discussed in detail range from the Asam brothers to Johann Michael Fischer and J. B. Neumann.

A04-1005
Hoag, J. D. Western Islamic Architecture. Braziller, 1963. $5.95 (paperback $2.95)

A04-1006
Hoffmann, D. The Architecture of John Wellborn Root. Johns Hopkins Univ. Press, 1973. 288 pp. $13.50
The office of Burnham & Root was pre-eminent during the early years of the Chicago School of commercial architecture. As the designing partner, John Root pioneered in developing the distinctively American type of building, the commercial skyscraper.

Design--Architectural 173

In the first intensive study of Root's career, the author compares his contributions to those of Adler & Sullivan and argues that Root's achievement at least equals theirs.

A04-1007
Hoffmann, H. Row Houses and Cluster Houses: An International Survey. Praeger, 1967. 176 pp. $18.50
The author investigates the problems of occupancy and space planning and shows how they have been solved with low-rise housing in European and Mediterranean countries, and the United States. He shows that row houses can be handsome, spacious, and private-- and that these qualities can be achieved at a relatively low cost.

A04-1008
Hofmann, W. and Kultermann, U. Modern Architecture in Color. Viking Press, 1971. 528 pp. $35.00
112 color plates, 100 plans. Works since 1850 from many countries.

A04-1009
Hofstatter, H. Living Architecture: Gothic. Grosset and Dunlap, 1970. $7.95

A04-1010
Hogg, G. A Guide to English Country Houses. International Publications, 1969. 160 pp. $6.25
Illustrated account of 75 great houses in 33 English counties, each with a brief description, a discussion of its outstanding features, when open for viewing and how to reach them.

A04-1011
Hohauser, S. Architectural and Interior Models: Design and Construction. Van Nostrand Reinhold, 1970. 211 pp. $20.00
A guide to model-making, and handbook of planning, estimating, building, and photographing models. The book combines an account of what has been done with a catalog of tools and materials available to the model builder.

A04-1012
Hohl, R. Office Buildings: An International Survey. Praeger, 1969. 176 pp. $18.50
Many new office buildings are being constructed all over the world to accommodate the increasing administrative and managerial work force. In this book, Reinhold Hohl shows forty recent examples from the United States, Europe, and Australia. Functional considerations and decorative aspects are treated. The selection includes low-rise buildings, office towers, and examples of isolated office buildings in rural environments as well as those in urban areas.

A04-1013
Holmes, M. Elizabethan London. Praeger, 1969. 138 pp. $5.95
A portrait of Tudor London, with descriptions of the people

A04-1014
Hornbeck, J. Stores and Shopping Centers. McGraw-Hill, 1962. 178 pp. $12.75
A collection, selected from Architectural Record, of the best articles and design presentations on retail stores and shopping centers. 426 illustrations.

A04-1015
Hornbostel, C. Materials for Architecture: An Encyclopedic Guide. Van Nostrand Reinhold, 1961. 624 pp. $22.50
This reference enables comparison of the basic physical and chemical properties of structural products, and shows the uses of building products, describing their advantages and limitations. Numerous tables, charts, diagrams, and photographs.

A04-1016
Horrobin, P. J., ed. Housing the Elderly. Herman Publishing Co., 1974. 240 pp. $18.50
Elderly have special housing requirements with comfort, convenience and safety vital. For those with reduced physical abilities and mobility, every aspect of planning and design has to be subjected to careful scrutiny. Social needs also have to be considered to ensure companionship while retaining privacy. This book sets down standards and gives appraisals of housing schemes in use. Edited from material prepared for the UK's Department of the Environment by the Building Research Establishment.

A04-1017
Housing the Family. Cahners, 1974. 272 pp. $17.50
Written for designers and planners, this book discusses the intelligent use of space and materials; describing in detail the living patterns of the modern family and their implications for residential design. Based on extensive design experience and surveys of family housing by England's Department of Environment, the book shows how even limited space, thoughtfully planned, can be adaptable to all stages of the family life cycle.

A04-1018
Howard, H. Structure--An Architect's Approach. McGraw-Hill, 1966. 288 pp. $14.50
Shows how to analyze and correlate knowledge of structures with the overall architectural design of the building. Seven buildings are described in detail, providing examples of construction in wood, steel, and reinforced concrete. 199 illustrations.

A04-1019
Howells, J. M. Lost Examples of Colonial Architecture. Dover, 1963. 248 pp. $3.50
Full-page photographs of buildings that have disappeared or

Design--Architectural

been so altered as to be denatured, including many designed by major early American architects. 245 plates.

A04-1020
Huber, B. and Steinegger, J. C. Jean Prouve: Prefabrication, Structures and Elements. Wittenborn, 1971. 208 pp. $25.00
Illustrated study of the life and work of the "father of industrialization" in architecture, who seeks to combine original creative design with prefabrication.

A04-1021
Hubert, J. The Carolingian Renaissance. Braziller, 1970. $30.00

A04-1022
Hudenberg, L. Planning the Community Hospital. McGraw-Hill, 1967. 751 pp. $22.00
Fully analyzing the functions of hospitals and their relationship to hospital planning and design, this book is particularly useful to hospital administrators, trustees, and architects.

A04-1023
Hunt, W. D., ed. Comprehensive Architectural Services. American Institute of Architects, 1965. $10.50 (for AIA members, less 10%)
This book covers all facets of the professional requirements, responsibilities, and involvement of architects. An authoritative source on matters of direct concern to the architect in his professional capacity and the conduct of his business.

A04-1024
Hunt, W. D., Jr. Total Design: Architecture of Welton Becket and Associates. Wittenborn or McGraw-Hill, 1972. 244 pp. $24.95
Topics discussed include the concept, framework, business, practice, and future of total design. 300 photographs of plans, models and architecture.

A04-1025
Hussain, F. Living Underwater. Praeger, 1971. 128 pp. $7.50 (paperback $3.95)
This book describes projects aimed at creating controlled aquatic environments--underwater complexes where men can live, work, can chart their surroundings to exploit the ocean's full potential.

A04-1026
Huxtable, A. L. Four Walking Tours of Modern Architecture in New York City. N.Y. Graphic Society, 1961. 76 pp. $.95
A pocket guide to Manhattan's modern architecture, this

book covers the most significant modern office buildings, residences, stores and public structures of the midtown area, with maps.

A04-1027
Huxtable, A. L. Have You Kicked a Building Lately? Harper & Row, 1976. 228 pp. $10.00
The author reviews the designs of pioneering modern architects such as Louis Kahn, Venturi and Rauch, and Alvar Aalto, and writes about the major controversies over important buildings. Mrs. Huxtable has been in the forefront of the preservation movement and has also explored the nuts and bolts of industrialized building and futuristic constructions of visionary architects. She concludes that the environmental concern of recent years has resulted in significant gains for both new design and preservation.

A04-1028
INBEX '70 Digest of Seminars--Industrialized Building Exposition and Congress. Cahners, 1970. 85 pp. $4.95
Devoted to factory-built housing and new building systems. More than 100 papers presented on all phases of industrialized building, technology, design, production and marketing. Complete speakers index. Illustrated.

A04-1029
INBEX '71 Digest of Seminars--Industrialized Building Exposition and Congress. Cahners, 1972. 180 pp. $10.95
Experts, opinion-makers, and movers in the fastest growing area of home building gathered at the Second Annual INBEX at Louisville, Kentucky. Over 150 talks are presented.

A04-1030
Insall, D. The Care of Old Buildings Today: A Practical Guide. Crane, Russak, 1973. 197 pp. $12.00
This work provides information on all aspects of the care of old buildings, from acquisition to specific restoration techniques.

A04-1031
Isaac, A. R. G. Approach to Architectural Design. Univ. of Toronto Press, 1971. 112 pp. $15.00
This study uses a practical rather than a theoretical approach and emphasizes the importance of human perception in environmental design. It applies the principles and processes of design to the aspects of planning, space, spectators' route, colour, lighting, unity, and scale.

A04-1032
Isham, N. M. and Brown, A. F. Early Connecticut Houses: An Historical and Architectural Study. Dover. 303 pp. $3.00
29 representative dwellings were studied for this survey of design and construction in colonial America. 118 figures.

Design--Architectural

A04-1033
Ishimoto, T. and Ishimoto, K. Japanese House, Its Interior and Exterior. Crown Publishers, 1963. 128 pp. $5.00
Contains more than 200 photographs and text describing every element of the Japanese home and how it can add new dimensions to American design.

A04-1034
Itoh, T. The Classic Tradition in Japanese Architecture: Modern Versions of the Sukiya Style. Weatherhill, 1972. 280 pp. $35.00
An illustrated volume on modern Japanese architecture in the sukiya style, this book features the work of five of Japan's outstanding architects in nineteen examples of buildings erected within the past two decades.

A04-1035
Itoh, T. The Elegant Japanese House: Traditional Sukiya Architecture. Weatherhill, 1969. 220 pp. $30.00
The sukiya style of architecture, with foundations in the ceremonial tea-house, is one of the most sophisticated types of Japanese residential architecture.

A04-1036
Itoh, T. Kura, Design of Traditional Japanese Barns and Storehouses. Harper & Row, 1973. 252 pp. $50.00
From rural storehouse to the still-used imperial treasure house of the 8th century, the Shoso-in, storehouses reflect an awareness of and respect for nature and have had a major influence on the development of Japanese architecture. Mr. Itoh considers all aspects of kura, the historical and cultural characteristics; the development of architectural styles attuned to regional and functional requirements; the relation of the structures to the environment and examples of different types of kura.

A04-1037
Itoh, T. Traditional Domestic Architecture of Japan. Weatherhill, 1972. 176 pp. $12.50

A04-1038
Izenour, G. C. Theater Design. McGraw-Hill, 1976. 480 pp. $29.50
The author introduces the general terms and types of theater design, detailing and defining auditoriums, stages, sight and hearing lines, and the basics necessary to an appreciation of theater design. From this he moves into a discussion of theater design from ancient times down to the present. Concert halls, with their specific acoustical characteristics, are also fully covered. Client relationships, budgeting, and building codes--all major, modern-day design considerations are also treated.

A04-1039
Jackson, A. The Politics of Architecture. Univ. of Toronto Press,

1970. 219 pp. $8.50

The history of British architecture since 1930 has been one of controversy between the old idiom and the new and between various social and technological viewpoints. This book spells out the issues and describes how they arose.

A04-1040
Jackson, T. G. Byzantine and Romanesque Architecture. Hacker Art Books, 1971. $50.00
Reproduction of 1913 edition.

A04-1041
Jackson, T. G. Renaissance of Roman Architecture. Hacker Art Books, 1971. $75.00
Reproduction of 1921 edition.

A04-1042
Janke, R. Architectural Models. Praeger, 1968. 140 pp. $8.95

The practice of model-building is represented here and is illustrated with a selection of international examples--including works of Le Corbusier, Revell, Saarinen, Nervi, Fuller, Mies van der Rohe, and Skidmore, Owings & Merrill. The book treats the organization of the workshop, equipment, tools and materials, building techniques, execution, and coloring. There is also a chapter on photographing the model.

A04-1043
Jencks, C. Architecture 2000, Predictions and Methods. Praeger, 1971. 128 pp. $7.50 (paperback $3.95)

This book applies methods of forecasting and prediction used in other disciplines to the evolution of architecture through the year 2000. After a critique of predictive models and the ideologies that produce them, the author describes the six major traditions of architecture. He discusses actual predictions, taken from forecasters in architecture and other fields, such as technology, sociology, and biology, and emphasizes the importance of separating interrelated planning problems into individual elements that can be solved independently.

A04-1044
Jencks, C. Le Corbusier and the Tragic View of Architecture. Harvard Univ. Press, 1973. $13.95

Charles Jencks discusses Le Corbusier's career from its early beginnings in Switzerland to the final international years, setting his private life and public image, his theories and his buildings, his paintings and his propaganda in relationship with one another.

A04-1045
Joedicke, J. Architecture Since 1945, Sources and Directions. Praeger, 1969. 180 pp. $18.50

This illustrated survey of post-World War II architecture places today's buildings in an historical perspective of movements

Design--Architectural

that can be traced back to the turn of the century. After analyses of works by Wright, Le Corbusier, Gropius, Neutra, Aalto, Niemeyer, and other pioneers, the author discusses their successors, giving special attention to such third-generation architects as Saarinen and Tange. He evaluates specific post-war architectural movements from the Brauhaus through the International style to the New Brutalism, Formalism, and Mannerist trends.

A04-1046
Jones, G. R., et al. Teach Yourself Acoustics. Dover. 179 pp. $2.50
The fundamentals of acoustics for the layman: the ear and hearing, psychoacoustics, environmental acoustics, electroacoustics, sonics. Accurate, yet non-technical explanations of sonic boom, methods of recording sound, etc.

A04-1047
Jordy, W. H. Progressive and Academic Ideals at the Turn of 20th Century. Wittenborn or Doubleday, 1972. 441 pp. $15.00
From the series, American Buildings and Their Architects.

A04-1048
Joseph, S., ed. Actor and Architect. Univ. of Toronto Press, 1964. 118 pp. $4.25
This book is not only for architects, those who commission theatres and those who work in them, but it is also for the ordinary public. It aims at increasing informed public opinion about the relationship between actor, architect and audience, and how important each is to the other.

A04-1049
Kaspar, K. Shops and Showrooms: An International Survey. Praeger, 1967. 168 pp. $15.00
This selection of fifty shops in the United States, Australia, and nine European countries presents successful solutions to problems of modern shop design. Included are specialty shops for shoes, jewelry, men's and women's fashions, furniture, electrical appliances, photographic equipment, and gourmet foods; bookstores; travel bureaus and beauty shops. Both remodeled and new buildings are considered. Attention is given to interesting ways of handling light, storage, and traffic; and the use of design motifs, color, and decorative elements.

A04-1050
Kaspar, K. Vacation Houses: An International Survey. Praeger, 1967. 168 pp. $18.50
This selection of fifty houses demonstrates the variety available in vacation homes. The styles range from weekend cottages to villas, the architects from the Finnish Saarinen to the Australian Seidler. The author shows how the economical basic rectangular bungalow can be changed to an L-shape or cruciform,

or adjusted to a T-form; there is a sampling of homes for almost every type of terrain or climate.

A04-1051
Kaufmann, E. Architecture in the Age of Reason: Baroque and Post-Baroque in England, Italy, France. Dover, 1968. 293 pp. $4.00
Often considered the most important work on 18th and early 19th century architecture; examines the new ideals that caused revolution in architectural style. Work of every major architect analyzed, dozens of buildings.

A04-1052
Kaufmann, E., Jr., ed. The Rise of an American Architecture. Praeger, 1970. 240 pp. $10.00
This study of the growth of American architecture during the nineteenth century appeared in conjunction with an exhibition of the same name held at The Metropolitan Museum of Art. The essays that constitute the book were written by Henry-Russel Hitchcock, Albert Fein, Winston Weisman, and Vincent Scully.

A04-1053
Kaufmann, E. and Raeburn, B., eds. Frank Lloyd Wright: Writings and Buildings. Horizon Press. $5.95
This book presents the achievement of "the greatest of all architects"--in his own words and works. Text and pictures recreate the art and insights of Frank Lloyd Wright. Over 150 illustrations, and a list of executed Wright buildings, keyed to a map of the United States.

A04-1054
Kelemen, P. Baroque and Rococo in Latin America. 2 vols. Dover, 1967. 488 pp. $6.00
17th and 18th century Latin American art presents a highly imaginative fusion of European styles with native American elements. Covering architecture, religious sculpture, church furnishings, this is the basic account, Mexico to Peru.

A04-1055
Keller, H. The Renaissance in Italy. Harry N. Abrams, 1974. 382 pp. $25.00
The book encompasses the new themes and inventions during the century-long development of Renaissance art and architecture. The inspiration drawn from ancient art, the great cities of the time, the solutions to artistic problems such as color and perspective, isolated masterpieces and vast projects of decoration are all explored. 432 illustrations.

A04-1056
Kelly, J. F. Early Domestic Architecture of Connecticut. Dover, 1924. 210 pp. $4.50
Based on personal observation of surviving examples and research into colonial records, this book includes 242 measured

Design--Architectural

diagrams (windows, door frames, construction details, etc.) and 192 photographs of more than 150 homes, 1650 to 1800.

A04-1057
Kemper, Alfred M. Drawings by American Architects. American Institute of Architects or John Wiley, 1973. 613 pp. $36.25 (for AIA members, less 10%)
 A collection of renderings from over 100 architectural offices illustrating a wide range of techniques and viewpoints.

A04-1058
Kennedy, D. and Kennedy, M. I., eds. Architects Year Book 14: The Inner City. John Wiley, 1974. 228 pp. $22.95
 Partial list of contributors include: D. Kennedy, M. I. Kennedy, L. Plotnikov, R. A. Fireden, B. M. Mann, Greater London Council, Van Ginkel Associates, R. Ridley, A. Prakash, S. Manda.

A04-1059
Kennedy, R. W. The House and the Art of Its Design. Van Nostrand Reinhold, 1959. 560 pp. $9.95
 One hundred and seventy-five photographs, plans, and diagrams illustrate this book. Subjects covered include style, environment, site, structure, livability and activities.

A04-1060
Ketchum, M., Jr. Shops and Stores. Van Nostrand Reinhold, 1957. 264 pp. $15.00
 Photographs, plans, sketches and text are used to express ideas on equipment, lighting, materials, structural design, store fronts, and a variety of stores such as department stores, drive-in shops and shopping centers.

A04-1061
Keyes, M. N. Nineteenth Century Home Architecture of Iowa City. Univ. of Iowa Press, 1971. 126 pp. $4.50 (paperback $3.25)
 This specialized book gives a brief history of Iowa City (Iowa) and a description of the architectural styles, starting with the earliest houses and ending with a description of composite or eclectic style houses. Illustrated.

A04-1062
Kicklighter, C. E. and Baird, R. S. Architecture: Residential Drawing and Design. Goodheart-Wilcox, 1973. 492 pp. $9.96
 Provides instruction in preparing architectural working drawings. Helps the student develop the necessary technical skills which will enable him to communicate his architectural ideas in an efficient and accurate manner. Covers industrialized housing, estimating, financing, workmanship specifications, career opportunities. Discusses all aspects of residential design, construction and graphic presentations.

A04-1063
Kidney, W. C. The Architecture of Choice: Eclecticism in America 1880-1930. Braziller, 1974. $15.00 (paperback $5.95)
A discussion of the movement in American architecture that produced Colonial houses, Gothic churches, Byzantine synagogues, and Roman banks.

A04-1064
Kidson, P. and Murray, P. A History of English Architecture. Arco, 1962. 256 pp. $5.95
Covers English architecture from Anglo-Saxon times to the present day. Included are nearly 120 photographs of famous churches, cathedrals, and country houses.

A04-1065
Kilham, W. H., Jr. Raymond Hood, Architect of Ideas. Architectural Book Publishing, 1973. 208 pp. $10.00
Raymond Hood (1881-1934) was a powerful figure in the transitional period in American architecture. Mr. Kilham, an architect who worked with Hood, tells how the ideas and designs developed for many modern landmarks. He reveals Hood's method of working on actual projects, with the personalities and politics involved as well as structural details, financial and aesthetic considerations, materials and use of space. Hood's functionalism and aesthetic sense, and his psychological and technical tactics as an innovator are spotlighted by the author who was present during much of the action.

A04-1066
Kimball, F. American Architecture. AMS Press, 1928. $20.00

A04-1067
Kimball, F. Domestic Architecture of the American Colonies and of the Early Republic. Dover, 1966. 314 pp. $4.00
Foremost architect and restorer of Williamsburg and Monticello covers nearly 200 homes between 1620 and 1825. Architectural details, construction, style features, special fixtures, floor plans, etc. 219 illustrations of houses, doorways, windows, capital mantels.

A04-1068
Kinzey, B. Y. and Sharp, H. W. Environmental Technologies in Architecture. Prentice-Hall, 1963. 788 pp. $17.20
Treatise on architectural design and building equipment as they relate to human needs for heat, light, sound, sanitation, and electrical power distribution.

A04-1069
Kira, A. The Bathroom. Viking Press, 1975. 320 pp. $15.00
According to the author the bathroom has a long and embarrassing history as the victim of poor design, the sum of ill-fitting and inefficient component parts. Without mincing words, he analyzes each function of the bathroom, the history of our attitudes,

Design--Architectural

and the possible ways in which we can achieve the most effective design. He accounts for the special demands of public facilities and those for the aged and disabled.

A04-1070
Klose, D. Metropolitan Parking Structures: A Survey of Architectural Problems and Solutions. Praeger, 1965. 248 pp. $18.50
This book reviews planning concepts, data on specific requirements, operations and economics, and some outstanding parking structures in Europe, Japan and the U.S.

A04-1071
Knox, B. The Architecture of Poland. Praeger, 1971. 240 pp. $18.50
The author has recorded the development of Polish architectural ideas from the twelfth century to the present. He describes the brick churches of the Baltic, Renaissance churches and palaces, and the stylistic currents from Italy and the Low Countries influential in buildings of the Polish Baroque and Rococo. There is also a treatment of Polish architecture in the twentieth century up to 1969.

A04-1072
Knudsen, V. O. and Harris, C. M. Acoustical Designing in Architecture. John Wiley, 1950. 457 pp. $18.25

A04-1073
Koeper, F. Illinois Architecture. Univ. of Chicago Press, 1968. 304 pp. $10.00 (paperback $1.95)
This is a pictorial survey of significant Illinois architecture from the early settlement buildings to the work of Sullivan, Wright, Mies van der Rohe, and Moholy-Nagy. Representative examples of the early French influence, Greek and Gothic revivals, post-war Victorian styles, and Romanesque and Classic-Renaissance revivals are included. The accompanying text highlights the architectural and historical importance of these structures.

A04-1074
Konya, A. and Burger, A. The International Handbook of Finnish Sauna. John Wiley, 1973. 176 pp. $15.95
Deals with the Finnish sauna, what it is, what it does, how it is constructed, and how it is used. Beautifully illustrated throughout.

A04-1075
Korn, A. Glass in Modern Architecture of the Bauhaus Period. Braziller, 1968. $15.00

A04-1076
Kornwolf, J. D. M. H. Baillie Scott and the Arts and Crafts Movement: Pioneers of Modern Design. Johns Hopkins Univ. Press, 1972. 618 pp. $27.50

In this first biography of Baillie Scott, the author outlines the influence of English and American domestic architecture of the 1880s on the young Englishman. He relates the work of Baillie Scott to that of such contemporaries as Adolf Loos, J. M. Olbrich, Josef Hoffmann, Peter Behrens, and Frank Lloyd Wright and establishes a strong connection between Baillie Scott's work and that of the next generation of architects--Walter Gropius, Mies van der Rohe, and Le Corbusier.

A04-1077
Kraemer, K. One-Family Houses in Groups/Einfamilienhauser in der gruppe: A Collection of Examples. International Publications, 1966. 144 pp. $20.00
Gathers European examples of "houses in groups" or "clusters" which display the various qualitative and quantitative advantages of the traditional detached house but require only a minimal site. Numerous photos, floor plans, site layouts, graphs, etc.

A04-1078
Kubler, G. The Religious Architecture of New Mexico: In the Colonial Period and Since the American Occupation. Univ. of New Mexico Press, 1972. 259 pp. $15.00

A04-1079
Kuenzlen, M. Playing Urban Games: The Systems Approach to Planning. Braziller. $6.95 (paperback $3.95)
This young architect-urbanist investigates planning in highly-developed capitalist societies, concluding that great changes must occur in the planning profession and in society before either can be truly rational and human.

A04-1080
Kuhne, G. New Restaurants: An International Survey. Architectural Book Publishing, 1973. 156 pp. $29.95
The problems which arise from the development of the initial concept through to the completion of a restaurant are presented. The primary considerations were the various needs for restaurants in urban areas: in businesses, city hotels, in theatres, in meeting halls, airports; and in vacation villages and hotels, canteens and eating places for students. Shown are particular kinds of restaurants, the most popular of each type and how solutions were arrived at.

A04-1081
Kuhnel, E. Islamic Art & Architecture. Cornell Univ. Press, 1966. 200 pp. $9.50
The elements characterizing the art and architecture of the Islamic world are described and illustrated in this survey. From its beginnings into the modern period, Ernst Kuhnel analyzes the distinctive qualities of this art: unity transcending differences of place and time; emphasis on the decorative over the naturalistic, on the horizontal rather than the vertical; and the intermingling of secular and sacred.

Design--Architectural 185

A04-1082
Kulski, J. Architecture in a Revolutionary Era. Aurora Publishers, 1971. $20.00

A04-1083
Kultermann, U. New Directions in African Architecture. Braziller, 1969. $7.95 (paperback $3.95)
An examination of present African architecture with a view to its historical basis and future trends.

A04-1084
Kurtz, S. A. Wasteland, Building the American Dream. Praeger, 1973. 128 pp. $7.95 (paperback $3.95)
A book that translates the values of the counterculture into architectural criticism. The author sees modern architecture's major failure as its lack of involvement with human needs and desires. He suggests that, if individuals are given a choice in the shaping of their environment, and if new architectural forms that respect both the land and the needs of people are developed, America need not become one endless Levittown.

A04-1085
Kuttruff, H. Room Acoustics. Applied Science, 1973. 298 pp. £9.00
The science of room acoustics has been enriched during the past two decades by many new ideas concerning the physical understanding of sound propagation in enclosures. In particular, the following achievements should be underlined: the statistical treatment of sound fields in enclosures based on the theory of room modes; the reexamination and further development of the classical reverberation theory; new understanding of the significance of sound field diffusion and its subjective correlate, pertaining to the impression of "spaciousness"; improvements in the procedures for measuring reverberation; and the application of correlation techniques to problems of room acoustics.

A04-1086
Lafever, M. Beauties of Modern Architecture. Da Capo, 1968. $17.50
Reproduction of 1835 edition.

A04-1087
Lamb, H. The Dynamical Theory of Sound. Dover. 307 pp. $3.50
All differential equations are constructed from their physical bases and solved in this mathematical treatment of theory of vibrations, general theory of sound, equations of motion of strings, pipes, membranes, etc. Mathematical details are explained throughout.

A04-1088
Landau, R. New Directions in British Architecture. Braziller. $7.95 (paperback $3.95)

An examination of present British architecture with a view to its historical basis and future trends.

A04-1089
Landy, J. The Architecture of Minard Lafever. Columbia Univ. Press, 1970. 313 pp. $17.50

Minard Lafever (1798-1854), best known for his popular The Modern Builder's Guide (1833), is examined in this study in the light of his own architecture as it matured in New York City between 1825 and 1855. It was in the Gothic Revival mode that he made his most significant contribution to the developing architecture of New York. Each of his New York buildings is surveyed with emphasis on history and style.

A04-1090
Lang, A., ed. 101 Select Dream Houses. Hammond, 256 pp. $9.95 (paperback $5.95)

101 most popular designs selected from the newspaper column, "The House of the Week." The book offers 5 basic styles arranged by total square footage of living space. An architect's rendering, basic floor plans, a list of necessary materials are furnished with each illustration. Information on financing, planning, contract and trade practices will aid potential home builders.

A04-1091
Lang, J., et al., eds. Designing for Human Behavior: Architecture and the Behavioral Sciences. Dowden, Hutchinson & Ross, or John Wiley, 1974. 432 pp. $20.00

In 1971, the Philadelphia Chapter of the American Institute of Architects organized a conference and exhibit on "Architecture for Human Behavior" at the Franklin Institute. The conference was designed to help bridge the gap separating research and practice. The topics covered included the process of environmental behavior, cognition, spatial behavior, and obtaining and using behavioral information.

A04-1092
Langley, B. The Builder's Director. Benjamin Blom, 1751. 184 pp. $12.50

Together with the Halfpennys, Langley kept England's (and America's) carpenter-builders supplied with pattern books, giving accurate models of buildings and their decoration.

A04-1093
Langley, B. The Builder's Jewel. Benjamin Blom, 1746. 100 plates. $12.50

A04-1094
Langley, B. The City and Country Builder's and Workman's Treasury of Designs. Benjamin Blom, 1750. 240 pp. $28.50

Design--Architectural 187

A04-1095
Langley, B. and Langley, T. Gothic Architecture. Benjamin
   Blom, 1747. 67 plates. $25.00
   Batty Langley (1696-1751) is best known for his system of
Gothic orders, fully described in this volume, which contains exact
measurements and several different views of everything from windows to chimney pieces and pavilions.

A04-1096
Lawrence, A. Architectural Acoustics. Applied Science, 1970.
   235 pp. £5.00
   The rapid growth of architectural acoustics in the last few
years, including the increasing provision of noise control requirements in building and planning regulations, points to the necessity
of having current information available in a readily accessible
form. Practicing acousticians should find the book useful.

A04-1097
Lawson, F. Designing Commercial Food Service Facilities. Watson-Guptill, 1974. 148 pp. $24.95
   A guide to the physical design of such commercial food service facilities as cafeterias (public and institutional), kitchens,
company dining rooms, and storage areas. Provides technical information needed in the industry: management efficiency, production line techniques, design of facilities, selection of equipment,
etc.

A04-1098
Lawson, F. Hotel Planning and Design. Architectural Press,
   1974. £16.00
   Taking account of changes in catering technology and new
patterns of customer demand, this is a practical handbook covering everything from siting to the selection of equipment.

A04-1099
Leacroft, R. The Development of the English Playhouse. Cornell
   Univ. Press, 1973. 368 pp. $29.50
   In this illustrated history of English playhouses, the author
records the economic and social factors that contributed to their
development, and reveals their structural evolution through a wealth
of three-dimensional open-sided models of the buildings.

A04-1100
Le Corbusier. Creation Is a Patient Search. Praeger, 1966.
   312 pp. $17.50
   The figure of the genius emerges from this cornucopia of
his sketches, text, drawings, exclamations, renderings, plans,
prose poems, clippings, reproductions of his paintings and tapestries, aphorisms, and photographs of his architecture and sculpture.

A04-1101
Le Corbusier. Towards a New Architecture. Praeger, 1970.
   268 pp. $7.50 (paperback $3.95)

This work has greatly influenced contemporary architectural thought. Appearing in English in 1927, it was the first popular exposition of the modern movement in architecture that gradually established itself in Europe during the early part of this century. This is a new facsimile of the original English edition.

A04-1102
Lindermann, G. A History of German Art: Painting, Sculpture, and Architecture. Praeger, 1971. 232 pp. $9.95
The author discusses every aspect of German architecture, sculpture, painting, and the graphic arts. Landmarks discussed include architecture as diverse as the buildings of the Romanesque and Gothic periods and the buildings conceived at the Bauhaus.

A04-1103
Lindley, K. Appreciation of Architecture--Landscape and Buildings. Pergamon, 1971. 68 pp. $5.50
The first of a new series aimed at providing the basis for an examination of environment, free from the stylistic approach to the study of architecture.

A04-1104
Lindsay, R. B., ed. Acoustics: Historical and Philosophical Development. Dowden, Hutchinson & Ross, 1973. 480 pp. $24.00
The book begins with an essay on the science, technology, and art of acoustics, followed by the editor's "The Story of Acoustics." All other selections, including materials by Vitruvius, Galileo Galilei, Robert Boyle, Sir Isaac Newton, John Tyndall, and Lord Rayleigh are presented in chronological order. This is a compendium of the most important papers which have influenced the development of all branches of acoustics up to 1900.

A04-1105
Link, D. E., ed. Residential Designs: How to Get the Most for Your Housing Dollar. Cahners, 1974. 163 pp. $14.50
Provides builders, architects, and designers with good, marketable design ideas for houses and apartments. The ideas are taken from the popular Design Lab feature that appears regularly in Professional Builder magazine. Included are floor plans and sketches on kitchens, bathrooms, land plans, full floor plans on low-cost housing, condominiums, apartments, duplexes and fourplexes.

A04-1106
Linley, J. Architecture of Middle Georgia: The Oconee Area. Univ. of Georgia Press, 1972. 208 pp. $17.50
From 1807 to 1869 Milledgeville was the capital of Georgia, and many superb examples of antebellum architecture are still to be found there in Baldwin County and in the surrounding counties of the Oconee area. The author draws on these antebellum examples and other structures ranging from prehistoric Indian monuments to contemporary buildings. Included in this survey are

Design--Architectural

indigenous types from the late eighteenth and early nineteenth centuries, examples of Federal and Greek Revival architecture, and a variety of stylistic Victorian and post-Victorian buildings.

A04-1107
Linstrum, D. Sir Jeffry Wyatville: Architect to the King. Oxford University Press, 1972. 296 pp. $32.00
Sir Jeffry Wyatville was the architect who remodelled Windsor Castle for George IV; but before this he had been a successful country-house architect. In this monograph, his work is set against the contemporary background of patronage and the development of the architectural profession.

A04-1108
Lissitzky, E. Russia: An Architecture for a World Revolution. MIT Press, 1970. 208 pp. $10.00
Between the October Revolution of 1917 and the imposition of "Socialist Realism" under Stalin in the early thirties, the arts, architecture among them, were developing with a great freedom of invention and expression in the Soviet Union. The book reveals how the Russian practitioners took advantage of this opportunity by experimenting with structural innovations and by simplification of form. Lissitzky presents drawings and photographs of the work--both projected and constructed--of the principal Soviet architects of the modern school.

A04-1109
Little, B. Birmingham Buildings. David and Charles. 128 pp. $9.95
The architectural story of a midland city. An illustrated survey.

A04-1110
Little, B. English Historic Architecture. Architectural Book Publishers, 1974. $6.95
England's building achievement from Saxon times down to 1914 and how styles were influenced by political, social and economic factors.

A04-1111
Lloyd, N. A History of English Brickwork. Benjamin Blom, 1925. 449 pp. $28.50
Examples and notes of the architectural use and manipulation of brick from medieval times to the end of the Georgian Period.

A04-1112
Lloyd, N. A History of the English Houses. Architectural Press, 1974. £12.00
This long out-of-print classic is still the most authoritative book on the subject and is being re-issued for European Architectural Heritage Year.

A04-1113
Lockwood, Crosby. Bricks and Brownstone: The New York Row House, 1783-1929: An Architectural and Social History. American Institute of Architects, 1973. $17.95 (for AIA members, less 10%)
   A detailed examination of the architectural styles of the New York row house.

A04-1114
Lundberg, D. E. The Hotel and Restaurant Business. Cahners, 1970. 301 pp. $12.95 (paperback $9.95)
   Covers the economics, psychology, management, food technology, food chemistry, engineering, architecture, accounting, marketing and law necessary to a successful operation. Famous hotels and restaurants cited as examples.

A04-1115
Lynch, K. The Image of the City. MIT Press, 1960. 194 pp. $12.00 (paperback $2.95)
   Mr. Lynch, supported by studies of Los Angeles, Boston, and Jersey City, formulates a new criterion--imageability--and shows its potential value as a guide for the building and rebuilding of cities.

A04-1116
Lynch, K. Site Planning. MIT Press, 1962. 248 pp. $10.00
   The art of arranging the external physical environment in all its detail: analysis of site and purpose, land use and circulation design, visual form, climate, site engineering, landscaping, plus such special topics as housing, shopping centers, industrial estates, and planning for institutions. 41 figures and 221 marginal drawings.

A04-1117
Lynes, J. A. Principles of Natural Lighting. Applied Science, 1968. 212 pp. £3.50
   This book contains an account of recent research on natural lighting and an introduction to experimental techniques for research workers. The latest mathematical and graphical techniques for daylight design are developed from first principles starting from a discussion of the visual process. The book contains charts and tables to simplify the prediction of daylight levels.

A04-1118
Maass, J. The Victorian Home in America. Hawthorn, 1972. $19.95

A04-1119
McC. Groff, S. New Jersey's Historic Houses: A Guide to Homes Open to the Public. A. S. Barnes, 1971. 247 pp. $6.95 (paperback $1.95)
   Surveys nearly 100 houses in 44 cities, giving a historical and an architectural account of each house, and describing its

Design--Architectural 191

decor and furnishings. There are also photographs of interiors, showing Colonial bedrooms, Federal dining rooms, and Victorian kitchens.

A04-1120
McCall, E. B. Old Philadelphia Houses on Society Hill, 1750-1840. Architectural Book Publishing, 1966. 184 pp. 150 illustrations. $12.50
 In text and photographs, the outstanding colonial and early American mansions and historical buildings of the area.

A04-1121
McCoy, E. Five California Architects. Van Nostrand Reinhold, 1960. 208 pp. $10.95
 The history of the rise of the California school of architecture as seen through the work of five men who gave it impetus and direction: Maybeck, Gill, Greene and Greene, Schindler.

A04-1122
Macleod, R. Style and Society: Architectural Ideology in Britain, 1835-1914. Wittenborn, 1971. 144 pp. $12.00
 Victorian and Edwardian architectural ideology is expressed in the Gothic revival, Queen Anne, William Morris and Philip Webb, and the Moderns.

A04-1123
Malo, P. Landmarks of Rochester and Monroe County: A Guide to Neighborhoods and Villages. Syracuse Univ. Press, 1973.
 This book is designed as a guide for viewing neighborhoods and downtown areas in Rochester and villages in Monroe County in a progression that does not isolate the "notable buildings" but relates them to the area that surrounds them. Architectural and historical information. Over 100 photographs.

A04-1124
Malt, H. Furnishing the City. McGraw-Hill, 1970. 256 pp. $12.50
 Discusses the subject of street furniture (traffic signs, mailboxes, light fixtures, directional signals, and similar items). The author examines the urban product environment with its problems, goals, and system design approach--and explores man's relationship to this synthetic setting.

A04-1125
Mankovsky, V. S. Acoustics of Studios and Auditoria. Hastings House, 1972. 416 pp. $16.50
 A book on an increasingly important subject. The whole of acoustics, sound insulation and noise control is covered with special reference to studios and places of public performance. There is sufficient basic theory to give the student a thorough understanding of the subject and a mass of design information which enables the book also to serve as a work of reference.

A04-1126
March, L., ed. The Architecture of Form. Cambridge Univ. Press, 1976. 522 pp. $29.50
The editor's introduction sets out the relationship between mathematics, science and design, and provides a framework for the 14 papers that follow. Divided into three main topics, the papers describe architectural form through the different models used, the analysis and prediction of architectural form using computer simulation, and the problems of evaluation and application of value theory in design situations.

A04-1127
March, L. and Steadman, P. The Geometry of Environment. Wittenborn, 1971. 360 pp. $18.50
An introduction to spatial organization in design and architecture. Includes mappings and transformations, translations, rotations and reflections, symmetry, stacking, nesting and fitting, and spatial allocation procedures.

A04-1128
Markus, T. A. Building Performance. Applied Science, 1972. 281 pp. £8.00
This book describes the theory and instruments developed by a research unit comprised of an architect, an operational research scientist, a psychologist, a quantity surveyor, a systems analyst and a physicist. Starting with a theoretical description of the relationships between buildings and the people who use them, the book goes on to develop this description in the design context and to indicate the cost implications which derive from this approach. The results of a series of appraisals of nearly 50 comprehensive schools in Scotland is described as a way of illustrating the techniques developed by the Research Unit.

A04-1129
Mason, B. S. and Koch, F. H. Cabins, Cottages and Summer Homes. A. S. Barnes, 1972. 168 pp. $4.98
Frame, stone, adobe, as well as timber and log construction are explained in detail. The book teaches how to figure the cost, how to select a site and how to select and use materials. There are twenty complete plans.

A04-1130
Maxwell, R. New British Architecture. Praeger, 1973. 200 pp. $25.00
This illustrated book surveys the most recent accomplishments of young British architects. It covers the period when the world looked to Britain for innovations in building. The author discusses and shows private homes, universities, and civic and commercial buildings. Plans, elevations, and photographs.

A04-1131
Mazzotti, G. Palladian and Other Venetian Villas. International Publications, 1966. 510 pp. $40.00

Design--Architectural   193

A survey of the many villas scattered throughout the countryside of Venetia. 700 illustrations.

A04-1132
Merrilees, D. and Loveday, E. Pole Building Construction. Garden Way Publishing, 1974. 48 pp. $3.00
This book will save money, labor, time and materials in building a tool shed, animal shelter, barn or home. Pole construction is economically and ecologically sound, involving only limited grading, no excavation, flexibility in site, good wind resistance, few materials. A "do-it-yourself" book, illustrated with more than 40 plans and drawings.

A04-1133
Meyer-Bohe, W. Bauten fur die Jugend/Structures for Children. Wittenborn, 1972. 180 pp. $26.50
German and English text, 370 photographs. Recent German architecture of kindergartens, youth hostels and country summer schools with many design details.

A04-1134
Michaelides, C. E. Hydra: A Greek Island Town, Its Growth and Form. Univ. of Chicago Press, 1967. 94 pp. $10.50
Architectural drawings and photographs of the entire town, and a brief history of Hydra's development as a maritime community (particularly between 1774 and 1815), with a discussion of the form of the town, its development, and its components (streets, houses, and clusters of houses).

A04-1135
Mielziner, J. The Shapes of Our Theatres. Clarkson Potter, 1970. $6.95
Ideas and conclusions form a collection of essays on the theatre's shape, future costs, and the impact of new technology. Non-technical, with suggestions for sight lines and stage configurations for ideal theatres.

A04-1136
Miller, H. R. Great Houses of Washington, D.C. Crown Publishers, 1969. 216 pp. $25.00
Photographs by Charles Baptie. A tour of the outstanding historical houses of the capital city. 144 illustrations.

A04-1137
Millon, H. A. Baroque and Rococo Architecture. Braziller, 1961. $5.95 (paperback $2.95)

A04-1138
Millon, H. A. Key Monuments of the History of Architecture. Prentice-Hall, 1964. 535 pp. $10.95
761 illustrations present a record of man's architectural achievement from prehistoric times to the present. Over 500 black and white plates of airview, interior and exterior views, and

closeups. 200 diagrams of ground plans, city plans, sections and elevations of buildings, models and reconstructions.

A04-1139
Mills, E. D. The Changing Workplace: Modern Technology and the Working Environment. George Godwin Ltd., 1972. 140 pp. £3.75

The author discusses the ways in which technological and social changes are affecting the planning and design of the working environment. Those concerned with the commissioning and design of new buildings will find this book of interest.

A04-1140
Mills, E. D., ed. Planning: Volume 1: Architects' Technical Reference. Butterworth, 1972. 148 pp. £4.00

In keeping with the rapid innovations in the building field, a new approach has been made in this edition of Planning--which now appears in five volumes. The first of these volumes--Architects' Technical Reference--includes basic planning data for all types of building: car-parking, circulation, sanitary and storage requirements, legislations, British Standards and materials.

A04-1141
Mills, E. D. and Kaylor, H. The Design of Polytechnic Institute Buildings. UNESCO Publication Center, 1972. 92 pp. $6.50

Provides guidelines for the choice of site, layout, design and construction of polytechnic institutes for educating and training technicians at post-secondary level. Indicates and discusses factors affecting the planning of polytechnics.

A04-1142
Milstein, J. and Walker, L. R. Designing Houses: An Illustrated Guide. Viking Press, 1975. 148 pp. $8.95

An illustrated guide for people who want to be their own architect. The authors stress their own personal method--a tactile mini-project, a step-by-step process that takes you from the history of architectural shapes to choosing a contractor. Along the way you will learn how to build a model, draw a set of plans, choose a site, and how to do many of the tasks required to design a house. Useful to both layman and student of architecture.

A04-1143
Moller, C. B. Architectural Environment and Our Mental Health. Horizon Press, 1968. $5.95

This book is a contribution to the literature of architectural humanism. It demonstrates how psychological well-being is profoundly dependent upon architectural environment.

A04-1144
Moore, G. T., ed. Emerging Methods in Environmental Design and Planning. MIT Press, 1970. 304 pp. $25.00 (paperback $5.95)

Design--Architectural

The research reports in this book are concerned with new methods for solving the problems of the physical environment. Four major papers have been brought together in one section. Each is a theory of the process of design and planning and, in addition, includes descriptions of techniques. The majority of the volume is devoted to presentations and evaluations of specific new methods. Some areas covered are building layout models, problem structuring, and computer-aided design.

A04-1145
Moore, G. T., ed. Environmental Design and Planning. MIT Press. 304 pp. $3.95

The book is comprised of papers and proceedings of a conference on design methodology in 1968, sponsored by the Design Methods Group; some thirty papers together with nine commentaries on them, by Americans and Britons. A major part of the volume is devoted to presentations and evaluations of specific new methods. Some areas covered are building layout models, problems structuring, and computer-aided design.

A04-1146
Moore, J. E. Design for Good Acoustics. Architectural Press, 1967. 92 pp. £1.25

In the design of buildings, size, shape and volume are factors which have a great bearing on acoustics. This book contains all the information needed by architects during the early stages and ensures that acoustics are treated as an active component and not as a matter for adjustment later on by the specialist when he is called in.

A04-1147
Morgan, E. J. and Gilbert, S. H. Early Adelaide Architecture 1836 to 1886. Oxford University Press, 1969. 182 pp. $9.95

Contains 191 photographs of buildings erected before 1887 in the City of Adelaide. These photographs are supported by full captions giving wherever possible details of the design, construction, ownership and tenancy of the buildings illustrated.

A04-1148
Morisseau, J., ed. The New Schools. Reinhold Books, 1972. 128 pp. $7.95

Plans and data for 65 new schools in the U.S. stressing open-plan, flexible design. The schools discussed range from preschool through high school. Many plans also provide for adult education programs and community uses, and some for special students. Project and construction costs are included.

A04-1149
Morrison, H. Early American Architecture. Oxford University Press, 1952. 633 pp. $15.00

A study of architecture in the colonies from St. Augustine

in 1565 to San Francisco in 1848. Covers colonial architecture in all its manifestations.

A04-1150
Morse, E. S. Japanese Homes and Their Surroundings. Dover, 1961. 372 pp. $3.50
An authentic picture of the traditional Japanese home (mats, lamps, furniture construction and design, ornamentation, floor plans, etc.) and gardens.

A04-1151
Moynet, M. J. L'Envers du Théâtre: Machines et Décorations. Benjamin Blom, 1874. 290 pp. $13.95
Behind the scenes in the Paris theatres of the 1870s--practically a handbook on how to create the stage effects, machinery, and lighting that were the most spectacular in Europe.

A04-1152
Mullin, D. C. The Development of the Playhouse: A Survey of Theatre Architecture from the Renaissance to the Present. Univ. of California Press, 1970. $15.00

A04-1153
Mumford, L. The Brown Decades: A Study of the Arts in America 1865-1895. Dover, 128 pp. $2.00
This book discovered a "buried Renaissance" in the post-Civil War architecture of Root, Sullivan, Roebling; the paintings of Homer, Eakins, Ryder; and in other areas.

A04-1154
Mumford, L. From the Ground Up: Observations on Contemporary Architecture, Housing, Highway Building, and Civic Design. Harcourt Brace Jovanovich, 1956. 243 pp. $1.65
26 essays from the "Sky Line" department of The New Yorker.

A04-1155
Mumford, L. Sticks and Stones: A Study of American Architecture and Civilization. Dover, 1955. 113 pp. $2.00
American architecture from the medieval-inspired earliest forms to the early 20th century; evolution of structure and style, and reciprocal inflences on environment.

A04-1156
Munce, J. Industrial Architecture. McGraw-Hill, 1960. 232 pp. $16.50
A comparative survey of modern industrial building design in the United States, Great Britain, and Germany covering all aspects of industrial architecture.

A04-1157
Munz, L. and Kunstler, G. Adolf Loos: Pioneer of Modern Architecture. Praeger, 1966. 234 pp. $20.00

Design--Architectural

This account of Loos's work explores, through illustrations and text, the theories and design that have been a source of controversy for over fifty years. Three essays from Loos's own writings, including "Ornament and Crime," provide additional insight into the architect's personality and art.

A04-1158
Murray, P. The Architecture of the Italian Renaissance. Schocken Books, 1972. 268 pp. $2.95

A book with illustrations of the works of Brunelleschi, Alberti, Leonardo, Bramante, Raphael, Giulio Romano, Peruzzi, Michelangelo, Serlio and Palladio. How the times shaped the architects. The buildings range from palaces and churches to the villas of Vignola and Palladio.

A04-1159
Murray, P. History of Architecture Series. Vol. 1 Renaissance. American Institute of Architects, 1971. $27.00 (for AIA members, less 10%)

Illustrated volume concentrating mainly on Renaissance Italy.

A04-1160
Museum of Modern Art. Ludwig Mies van der Rohe. MIT Press, 1974. 70 pp. $20.00

Each of the drawings in this portfolio-sized book is an act of architectural creation. The sketches represent projects that were never constructed, and so for those interested in the full range of Mies's architectural concepts, the book serves as a supplement to those that are based largely on his built projects.

A04-1161
Muther, R. and Haganas, K. Systematic Handling Analysis. Cahners, 1969. $15.00

Systematic approach to analyzing materials handling problems. 150 projects from America and Western Europe drawn upon to develop the methods. Procedural techniques which eliminate analysis time. Contains information on classification of materials, layout, handling plans, installation.

A04-1162
Nairn, I. Britain's Changing Towns. International Publications, 1967. 168 pp. $8.75

Architectural developments affecting the individual character of sixteen British towns, both good modern building and discriminating preservation on one hand and thoughtless destruction and soulless construction on the other.

A04-1163
Natl. Aero. and Space Admin. Facilities Engineering Handbook. U.S. Gov't. Printing Office, 1974. 336 pp. $2.50. Order No. NAS 1.18:F 11/2/974 S/N 3300-00562

Recently revised, this edition presents the uniform policies and criteria to be used in carrying out the design, construction, operations, and maintenance of NASA facilities. It includes such criteria as: architectural, structural engineering, site and civil engineering.

A04-1164
Natl. Aero. and Space Admin. NASA Facilities Engineering Handbook. U.S. Gov't. Printing Office, 1974. 308 pp. $3.20. Order No. NAS 1.2:F 11 S/N 033-000-00592-0
Provides information on the standards NASA seeks to maintain in the design of its building projects. Includes recent revisions for environmental and energy-related reasons.

A04-1165
Navy Department. Design Manual: Community Facilities. U.S. Gov't. Printing Office, 1975. 176 pp. $3.05. Order No. D 209.14/2:37/3 S/N 008-050-00115-1
This manual was prepared for prospective defense contractors. It gives the criteria the Navy requires for the design of such buildings as theatres, schools, firehouses, bakeries, and for golf courses, pools, and playing courts.

A04-1166
Neff, W. Architecture of Southern California. Rand McNally, 1965. $14.95

A04-1167
Negroponte, N. The Architecture Machine. Wittenborn, 1972. 110 pp. $5.95 (paperback $2.95)
Using the computer to aid the architectural designer. Illustrated.

A04-1168
Negroponte, N. Soft Architecture Machines. MIT Press, 1974. 140 pp. $7.95
The general assumption of this book is that the architect is an unnecessary middleman between individual, continuously changing needs and the continuous incorporation of those needs into the built environment. The book proposes a new kind of architecture without architects, and even without surrogate architects.

A04-1169
Nelson, G. Problems of Design. Watson-Guptill, 1957. 206 pp. $8.95
26 essays offer factual information, illustration, and analysis of the world of modern design. Included are chapters on: problems of design; art; architecture; planning; and interiors.

A04-1170
Nervi, P. Structures. McGraw-Hill, 1956. 118 pp. $12.50
Subjects discussed include architect-client relations, training of designers and builders, theory of structures, building in reinforced concrete, and the future of reinforced concrete. Includes 88 photographs and 25 drawings of the author's most famous works.

Design--Architectural

A04-1171
Neutra, R. Building with Nature. Universe Books, 1971. 223 pp.
   $18.50
   This book, made up of a number of designs by the late author, illustrates his feeling that residential architecture should be linked with nature. Floor plans and 382 photographs.

A04-1172
Neutra, R. Survival Through Design. Oxford University Press, 1969. 384 pp. $2.95
   This book advocates design "for life" and health, especially mental health. The author contends that designing a vital setting cannot be derived from building techniques, "utilitarian" function, or fast-changing stylistic cliches. He makes suggestions for the application of the biological and behavioral sciences to architecture and community planning.

A04-1173
Nilsson, S. European Architecture in India, 1750-1850. Taplinger, 1969. 214 pp. $20.00
   An architectural history showing the mark of 100 years of European contact on Indian architecture. Discusses colonial history and administration, techniques of materials and construction, stylistic problems, climatic considerations, and more.

A04-1174
Nissen, H. Industrialized Building and Modular Design. Cement and Concrete Assn., 1972. 446 pp. £6.00
   Translated from the Danish.

A04-1175
Norberg-Schulz, C. Existence, Space, and Architecture. Praeger, 1971. 128 pp. $7.50 (paperback $3.95)
   This book embodies a new approach to the concept of space. On the basis of a theory of "existential space," the author develops the idea that architectural space may be understood as a concretization of environmental, schemata or images that form a necessary part of man's general orientation. His study is drawn from philosophical and psychological sources, as well as architectural ones.

A04-1176
Norberg-Schulz, C. History of Architecture Series. Vol. 2, Baroque. American Institute of Architects, 1971. $27.00 (for AIA members, less 10%)
   A volume with examples of the Baroque period.

A04-1177
Norberg-Schulz, C. Intentions in Architecture. MIT Press, 1966. 300 pp. $10.00 (paperback $4.95)
   "The importance of the book is that it presents architecture with its first modern, fully structured and ornamented theory, comparable to the film theory of Elsenstein, Arnheim, and Kracauer.... By exploring the nature of the building task, physical form, and the

relationships between them, plus the techniques of building, a theoretical model can be constructed to shape logical thought, whether analytical or intuitive."--Nathan Silver, Progressive Architecture.

A04-1178
Nuffield Foundation. Children in Hospital: Studies in Planning. Oxford University Press, 1963. 128 pp. $8.50
This book reports the final researches of the Nuffield Foundation's Division for Architectural Studies. Observations and recording during five years covered aspects of present management of children in hospital. The Nuffield design is for children's wards forming sizable departments attached to general hospitals. The arrangements in these wards are intended to give accommodation which can be used with considerable freedom.

A04-1179
Oakley, D. The Phenomenon of Architecture in Cultures in Change. Pergamon, 1970. 390 pp. $12.00
Introduces the intending architect, and his fellow members of physical planning and building design teams, to overall structure and concerns of architectural thought.

A04-1180
Oliver, P., ed. Shelter and Society: Studies in Vernacular Architecture. Praeger, 1970. 144 pp. $12.50
Thirteen experts on vernacular architecture show how structures were created by individual communities and ethnic groups to suit their particular needs. Pictures accompany discussions of several facets of vernacular architecture, showing how differences of climate, geography, and available materials affect the image of an ideal life expressed in nonprofessional architecture.

A04-1181
Olmstead, R., et al. Here Today: San Francisco's Architectural Heritage. Chronicle Books. 334 pp. $14.95
Area by area, in some cases block by block or even building by building, this reference work catalogs and illustrates the best of San Francisco architecture. 343 photographs.

A04-1182
Olmstead, F. and Kimball, T. Frederick Law Olmsted, Landscape Architect: 1822-1903. Benjamin Blom, 1928. 706 pp. $25.00
The fullest biographical sketch of America's first great urban designer, with all the papers connected with one of his masterpieces, New York's Central Park.

A04-1183
O'Neal, W. B., ed. The American Association of Architectural Bibliographers. Vol. 1. Univ. Press of Virginia, 1965. 128 pp. $7.50
Henry-Russell Hitchcock, by James H. Grady. Walter Gropius, by Caroline Shillaber. Philip C. Johnson, by William B.

Design--Architectural

O'Neal. Early Architecture of Virginia, by Frederick Doveton Nichols.

A04-1184
O'Neal, W. B., ed. The American Association of Architectural Bibliographers. Vol. 2. Univ. Press of Virginia, 1966. 113 pp. $7.50
Sibyl Moholy-Nagy. Philip C. Johnson, by William B. O'Neal. Holabird and Roche, by J. William Rudd. Early Architecture of Virginia, by Frederick Doveton Nichols.

A04-1185
O'Neal, W. B., ed. The American Association of Architectural Bibliographers. Vol. 3. Univ. Press of Virginia, 1966. 138 pp. $7.50
Walter Gropius.

A04-1186
O'Neal, W. B., ed. The American Association of Architectural Bibliographers. Vol. 4. Univ. Press of Virginia, 1967. 130 pp. $7.50
Carroll L. V. Meeks, by William B. O'Neal and Frederick Doveton Nichols. Charles-Louis Clerisseau, by Thomas J. McCormick. Library at Biltmore, by Stapleton Dabney Gooch IV. International Expositions, 1851-1900, by Julia Finette Davis.

A04-1187
O'Neal, W. B., ed. The American Association of Architectural Bibliographers. Vol. 5. Univ. Press of Virginia, 1968. 106 pp. $7.50
Henry-Russell Hitchcock, by James H. Grady. Architectural Comment in American Magazines, 1783-1815, by J. Meredith Neil. The Adam Style in America, 1770-1820, by Sterling M. Boyd. Calvert Vaux, by John David Sigle. Alvar Aalto, by Peter W. Beal.

A04-1188
O'Neal, W. B., ed. The American Association of Architectural Bibliographers. Vol. 6, Jefferson as an Architect. Univ. Press of Virginia, 1969. 150 pp. $7.50

A04-1189
O'Neal, W. B., ed. The American Association of Architectural Bibliographers. Vol. 7. Univ. Press of Virginia, 1970. 124 pp. $7.50
Sir Nikolaus Pevsner.

A04-1190
O'Neal, W. B., ed. The American Association of Architectural Bibliographers. Vol. 8. Univ. Press of Virginia, 1972. 208 pp. $7.50
Paradise Improved: Environmental Design in Hawaii, by J. Meredith Neil.

A04-1191
O'Neal, W. B., ed. The American Association of Architectural Bibliographers. Vol. 9. Univ. Press of Virginia, 1972. 132 pp. $7.50

A supplement to the bibliography of Walter Gropius, compiled by Ise Gropius and William B. O'Neal. A bibliography of works about Sir Christopher Wren, compiled by Gail Griffin Stringer. Benjamin Henry Latrobe, compiled by Paul F. Norton. Frank Lloyd Wright in Print, 1959-1970, compiled by James Muggenberg.

A04-1192
O'Neal, W. B., ed. The American Association of Architectural Bibliographers. Vol. 10. Univ. Press of Virginia, 1973.

A bibliography of Antonio Gaudi and the Catalan Movement, 1870-1930, compiled by George R. Collins with the assistance of Maurice E. Farinas.

A04-1193
Orr, J. M. Designing Library Buildings for Activity. Academic Press, 1972. 146 pp. $7.50

This volume presents the latest information on the design of libraries.

A04-1194
Otto, F. Tensile Structures. Wittenborn, 1972. $7.95

A one-volume edition of Pneumatic Structures and Cables, Nets and Membranes, unabridged with all the drawings of original.

A04-1195
Overdyke, W. D. Louisiana Plantation Homes. Architectural Book Publishing, 1965. 224 pp. $12.50

A pictorial album of famous colonial and ante bellum homes, with text.

A04-1196
Owen, R. D. Owen, Robert Dale, 1801-1877. Lenox Hill, 1849. 119 pp. $25.00

Hints on public architecture, containing among other illustrations, views and plans of the Smithsonian Institution.

A04-1197
Palladio, A. The Four Books of Architecture. Dover. 110 pp. $10.00

Translated into every major Western European language in the two centuries following its publication in 1570, this has been one of the most influential books in the history of architecture. Complete reprint of the 1738 Isaac Ware edition.

A04-1198
Pallister's Model Homes 1878. Glenwood Publishers, 1972. 98 pp. $4.95

First published in 1878 by Pallister, Pallister and Co. of Bridgeport, Conn., here is a reprint of residential and farm

Design--Architectural

buildings, schools, churches and public buildings of the era. The intent of the original author was to demonstrate good architectural taste to people away from cities who could not utilize the services of an architect. Each building is described and has a floor plan and exterior view.

A04-1199
Palmes, J. Le Corbusier: My Work. Watson-Guptill, 1974. 312 pp. $25.00

Le Corbusier surveys his development from his early days as a student to the completion of his last building, the Priory at La Tourette. This is a graphic self-portrait of the man and his work--describing the total creative process. Included are excerpts from his notes and sketchbooks; a wide range of photographs of his buildings, models, plans and paintings.

A04-1200
Papachristou, T. Marcel Breuer: New Buildings and Projects. Praeger, 1971. 240 pp. $22.50

This illustrated book is a record of the buildings designed by Marcel Breuer up to 1970. The author describes every facet of Breuer's career, from his bold concepts for the IBM Research Center in LaGaude, France, to the Whitney Museum of American Art in New York City. Also included are descriptions of such projects as the resort town of Flaine in France, the proposed Franklin D. Roosevelt Memorial in Washington, D.C., and the office tower that Breuer designed to rise fifty-three stories directly over Grand Central Terminal in New York City.

A04-1201
Parsons, K. C. The Cornell Campus: A History of Its Planning and Development. Cornell Univ. Press, 1968. 352 pp. $15.00

Partly an examination of the University's development and partly a critique of its architectural and campus setting. All the major buildings on the Cornell campus are considered, both in relation to prevalent styles of American university architecture and in the ways they reveal the individual talents of their architects. The efforts made to preserve and enhance the natural beauty of the site are an important part of the narrative.

A04-1202
Paul, S. Apartments: Their Design and Development. Van Nostrand Reinhold, 1967. 320 pp. $25.00

This book illustrates how good architectural design can be achieved despite all limitations and controls placed upon the undertaking, and at the same time emphasizes the vital importance of proper financing and the need for cost control throughout every phase of design.

A04-1203
Pawley, M. Architecture Versus Housing. Praeger, 1971. 128 pp. $7.50 (paperback $3.95)

An examination of public housing--its history, present, problems, and potential breakthroughs, based on the British experience and on experiments in Europe and America. The author discusses the growth of public housing in England from its birth after World War I to the mid-1950s. He shows how improved industrial techniques, the rise of a consumer society, and the rejection of outmoded architectural ideas led to the demise of traditional functionalist theories.

A04-1204
Pehnt, W. Expressionist Architecture. Praeger, 1973. 240 pp. $27.50

Examines the works of Adolf Loos, Bruno Taut, Theo van Doesburg, and Erich Mendelsohn, among others, who sought to apply the subjectivity of Expressionism to architecture. Such important structures as Mendelsohn's Einstein Tower in Potsdam and Bernhard Hoetger's Chile House in Hamburg are discussed. The author places Expressionist architecture in its primarily German, post-World War I social context and analyzes its later influence on the architecture of Nazism.

A04-1205
Pehnt, W. German Architecture 1960-1970. Praeger, 1970. 232 pp. $20.00

Over 70 examples show a variety of architectural expression which, in the last two or three years, has shown signs of moving in new directions. Among the examples are city-planning experiments in Frankfurt and innovative university complexes in Marburg, Bochum, and West Berlin.

A04-1206
Pelican Guide to Plantation Homes of Louisiana. Pelican, 1971. 128 pp. $2.50

Revised and enlarged edition of this handbook. Illustrated with thumbnail sketches of more than 240 architecturally and historically significant houses. Easy-to-follow map included. Lists all houses open to the public, hours and salient features. Describes many houses not open to the public.

A04-1207
Pentagram. Pentagram: The Work of Five Designers. Watson-Guptill, 1973. 216 pp. 186 illustrations. $7.50

This book is a representation of the architectural, graphic, and industrial designs of five professionals: Theo Crosby, architect; Alan Fletcher, Colin Forbes, and Mervyn Kurlansky, graphic designers; and Kenneth Grange, industrial designer. This book includes examples of their designs, along with the strategy and methods they use in keeping up with the latest innovations in the field.

A04-1208
Perin, C. With Man in Mind. MIT Press, 1972. 168 pp. $7.50 (paperback $2.95)

The book outlines a new organizing principle for the conduct

Design--Architectural

of work in environmental design, intended to overcome the conceptual gap between what people do to make and change the environment and what people require from it.

A04-1209
Peters, P., ed. Centers for Storage and Distribution. Reinhold, 1972. 136 pp. $12.00 (paperback $7.95)
Plans and data for 48 warehouses, midpoint storage centers, and similar facilities, with emphasis on modern high-ceiling design. Functions include storage, transshipment, distribution to outlets, processing and distribution and cold storage. Plans show flow of goods. Data include handling and storage capacity.

A04-1210
Peters, P., ed. Factories. Reinhold, 1972. 136 pp. $12.00 (paperback $7.95)
Plans and data for 71 new factories in the U.S. and abroad, small to moderate in size. Production flow for each is explained, and data include production capacity, employee and area figures. Manufactured products include chemicals, synthetics, foodstuffs, furniture, heavy machinery, etc.

A04-1211
Peters, P., ed. Libraries for Schools and Universities. Reinhold, 1972. 136 pp. $12.00 (paperback $7.95)
Plans and data for 38 new libraries in the U.S., Great Britain, Japan, Israel and Europe, with emphasis on open-plan libraries in a large range of sizes from small to very large. Types include faculty and institute, university, research, municipal, state, and national.

A04-1212
Peterson, C. E., ed. The Carpenters' Company 1786 Rule Book. Charles Scribner's Sons. $4.50
A volume on early American architecture, with 34 illustrations showing detail typical of Philadelphia buildings of the period.

A04-1213
Petzsch, H. Architecture in Scotland. International Publications, 1971. 155 pp. $10.00
An introductory guide to the building and architecture of Scotland from prehistoric times to the present day. Sections, corresponding to centuries are introduced by a short historical account and a note of the architectural character of the period. Selected buildings are discussed in detail and illustrated with photos and plans.

A04-1214
Pevsner, N. Some Architectural Writers of the Nineteenth Century. Oxford University Press, 1972. 354 pp. $29.00
The writers treated in this book range from Horace Walpole to William Morris and include not only as a matter of course Pugin

Ruskin, Viollet le Duc, Fergusson and Semper, but also such little known men as Leonce Reynaud and Edward Lacy Garbett. Robert Kerr's "English Architecture Thirty Years Hence" and William Morris's "The Revival of Architecture" are reprinted in full as appendices.

A04-1215
Pevsner, N. The Sources of Modern Architecture and Design. Praeger, 1968. 216 pp. $8.50 (paperback $3.95)
In reaction against historical imitation, original ideas and techniques in architecture and design appeared in the latter part of the nineteenth century. Architectural developments include advances in the use of iron in construction; the Chicago school and the first skyscrapers; the Domestic Revival in England; the extravagances of Art Nouveau, Behrens, Loos, and Gropius; the beginnings of modern city planning; and the garden-city experiments in England. These are discussed as germinal developments in the growth of modern architecture.

A04-1216
Pierson, W. H. The Colonial and Neoclassical Styles. Wittenborn, 1970. 329 pp. $12.50
From the series entitled American Buildings and Their Architects.

A04-1217
Platt, F. America's Gilded Age: Its Architecture and Decoration. A. S. Barnes, 1974. 352 pp. $25.00
Architects Louis Sullivan; Henry Hobson Richardson; McKim, Mead and White; Daniel H. Burnham were all architects of America's Gilded Age, 1880-1915. Mansions, resorts, hotels, train stations, libraries, and office buildings all began to reflect the new wealth architecturally with an ornate eclecticism never to be duplicated at a later time. This is a cross-country guided tour of the best examples.

A04-1218
Podd, G. O. and Lesure, J. D. Planning and Operating Motels and Motor Hotels. Hayden, 1964. 400 pp. $12.00
Information and ideas for anyone working in this area. Discusses the motel business from site selection to financial statements.

A04-1219
Poor, A. E. Colonial Architecture of Cape Cod, Nantucket & Martha's Vineyard. Dover, 1970. 135 pp. $3.50
195 photographs and 22 drawings show some of the finest examples from this rich architectural area. Shingled and clapboard one-story houses, "salt box" and "half-house" type structures, two- and three-story houses are all represented, as are interior features such as doorways, moldings, fireplace bricking, staircases, painted floor patterns, etc.

Design--Architectural

A04-1220
Portland Cement Assn. Modern Plans for Modern Farms. Portland Cement Assn., 1965. 120 pp. $1.80
Contains plans for modern concrete masonry farm buildings and concrete improvements. Construction plans were selected from the plan services of many state agricultural colleges and the U.S. Department of Agriculture. Construction details and tables that apply to all of the plans are included.

A04-1221
Portoghesi, P. Roma Barocca: The History of an Architectonic Culture. MIT Press, 1970. 490 pp. $75.00
A history of building activity in Rome from about 1600 to 1750. Includes a broad summary of the urban policies and building programs of the popes and an analytic history of the artistic culture of Rome as documented by the theoretical writings and architectural publications of the period.

A04-1222
Pothorn, H. Architectural Styles. Viking Press, 1971. 192 pp. $7.95
Chronological history covers important building styles throughout the world.

A04-1223
Pratt, R. and Gunther, R. T. The Architecture of Sir Roger Pratt. Benjamin Blom, 1928. 324 pp. $15.75
The first edition of the notebooks of Charles II's Commissioner for the rebuilding of London after the great fire. Pratt (1620-84) was a link between the architectural achievements of Inigo Jones and Wren.

A04-1224
Preiser, W. F. E., ed. Environmental Design Research. Vol. 1. Dowden, Hutchinson & Ross or John Wiley, 1973. 572 pp. $20.00
The goal of environmental design research is to further the understanding of man-environment interaction, and to devise methodologies for the analysis and synthesis of the man-constructed environment, as well as for the management of resources and influences governing human and environmental systems. This is the first of two volumes of the proceedings of the Fourth International Environmental Design Research Association Conference held April 1973.

A04-1225
Preiser, W. F. E., ed. Environmental Design Research. Vol. 2. Dowden, Hutchinson & Ross, or John Wiley, 1973. 508 pp. $20.00
The second volume of the proceedings of the Fourth International Environmental Design Research Association Conference will record the other two types of events at the conference. Symposia, with invited papers, are theory oriented. They are intended to

synthesize and assess existing knowledge and future prospects in the respective fields contributing to environmental design.

A04-1226
Priestley, H. E.   The English Home.   International Publications, 1971.   216 pp. $17.50
The development of the English home from the primitive roof and four walls to the present-day skyscraper, with a discussion of the influences of society, and why certain designs have been adopted to the present terraces, semi-attached houses, bungalows and blocks of apartments with an increasing number of modern conveniences.

A04-1227
Proksch, V.   Houses in the Alps/Maisons dans les Alpes.   International Publications, 1972.   118 pp. $11.50
Presents country houses (permanent dwellings and vacation homes) built in recent years by well-known architects in varied Alpine settings, mainly in Germany, Austria and Switzerland. With ground-plans and over 150 exterior and interior photographs.

A04-1228
Proshansky, H. M., et al., eds.   Environmental Psychology: Man and His Physical Setting.   Holt, Rinehart & Winston, 1970. 668 pp. $15.75
Essays in environmental psychology by psychologists, psychiatrists, sociologists, urban planners, geographers and novelists range over theory and methodology. The studies reflect concern of individual needs, aims of environmental planning and problems of environmental design and social institutions.

A04-1229
Pyne Press.   American Historical Catalog Collection.   Vol. 1, Architectural Elements, 1850-1880.   Charles Scribner's Sons. $4.95
This is the first of 15 volumes of exact facsimile illustrations from manufacturer's catalogs of the day.

A04-1230
Ragette, F., ed.   Engineering and Architecture and the Future Environment of Man.   Syracuse Univ. Press, 1968.   305 pp. $5.00
Papers on architecture, hydrology, semiconductors, and transportation.

A04-1231
Ragon, M.   The Aesthetics of Contemporary Architecture.   Wittenborn, 1968.   155 pp. $21.50
Eulogy of the City. The epic of the skyscrapers. Beauty of the material. Beauty of structures. Skeletons. Stairways. Piles. Beauty of the facades. Cubist beauty. Beauty of transparency. Fortresses. Architecture-sculpture. Places of worship.

Design--Architectural 209

Shells. Cables, tubes, space structures. Individual houses. Technological beauty.

A04-1232
Ramsey, C. G. and Sleeper, H. R. Architectural Graphic Standards. American Institute of Architects or R. S. Means, 1970. 659 pp. $44.95 (for AIA members, less 10%)
 A source book and standards guide for sizes, dimensions and details for all building materials and equipment for every type of building. Entirely a book of scale drawings, it will save manhours for architects, draftsmen, specification writers, and engineers, as well as serving as an authoritative reference guide book. Its 16 chapters generally correspond to the 16 divisions of the Uniform System for Construction Specifications.

A04-1233
Ramsey, S. C. and Harvey, J. D. M. Small Georgian Houses and Their Details: 1750-1820. Wittenborn, 1972. 230 pp. $17.95
 An invaluable sourcebook of Georgian houses. About 160 Georgian interiors, exteriors and details, plus 40 measured drawings.

A04-1234
Ransom, H. S., ed. The People's Architects. Univ. of Chicago Press, 1964. 126 pp. $6.95
 The essays included in this volume were written by a group of leading architects. Discussed are the social role of the architect and his responsibility to meet the needs of an expanding populous while retaining its necessary cultural and esthetic values.

A04-1235
Rapoport, A. House Form and Culture. Prentice-Hall, 1969. 150 pp. $6.95
 Proposes a conceptual framework for looking at the great variety of house types and forms and the forces that affect them.

A04-1236
Rasmussen, S. E. Experiencing Architecture. MIT Press, 1962. 245 pp. $10.00 (paperback $2.95)
 The enthusiasm of a world-known professor of architecture for designing excellence is conveyed in this discussion of architecture and design, which ranges from the Villa of Palladio to the Peking Winter Palace.

A04-1237
Rasmussen, S. E. Towns and Buildings. MIT Press, 1969. 224 pp. $15.00 (paperback $3.95)
 Concentrates on the town as a unity, as a whole composed of buildings and places. Most of the town plans are scaled to 1:20,000 for easy over-all comparison; several famous places are reduced to 1:2000 for closer comparison. The buildings are for the most part presented in three views: the first, en face; then

the same view, minus the facade; and finally a top view, as if the upper stories had been removed.

A04-1238
Rattenbury, J. and Van Eck, N. OSIRIS: Architecture and Design. Univ. of Michigan--Inst. for Social Research, 1973. 315 pp. $8.00
This monograph is an attempt to provide technical documentation for the benefit of those involved with the writing and modification of the OSIRIS package of computer programs.

A04-1239
Rawlings, J. S. Virginia's Colonial Churches: An Architectural Guide, together with Their Surviving Books, Silver, and Furnishings. Univ. Press of Virginia, 1963. 286 pp. $12.50
A guide to Virginia's forty-eight colonial churches. These range from the Jamestown Tower in James City County and Merchant's Hope in Prince George County, both constructed in the seventeenth century, to Little Fork Church in Culpeper County, built on the eve of the Revolution. Five churches are reproduced in color. In addition to structural description the author presents historical background and details about extant vestry books, silver, and furnishings.

A04-1240
Rayleigh, J. W. S. The Theory of Sound. Dover. 984 pp. $6.00
Still valuable classic by the Nobel Laureate. Standard compendium summing up previous research and Rayleigh's original contributions. Covers harmonic vibrations, vibrating systems, vibrations of strings, membranes, plates, curved shells, tubes, solid bodies, refraction of plane waves, general equations.

A04-1241
Raymond, E. Early Domestic Architecture of Pennsylvania. Charles Scribner's Sons. $6.95
A photographic study of Quaker, Moravian and other forms of 17th and 18th century building. 200 illustrations.

A04-1242
Redstone, L. Art in Architecture. McGraw-Hill, 1968. 256 pp. $24.75
A magnificently illustrated book which shows how art can be integrated with architecture to achieve the optimum blend of aesthetics with usefulness.

A04-1243
Regional Plan Assn. Urban Design: Manhattan. Viking Press, 1969. 132 pp. $17.50
Principles and objectives for urban areas.

A04-1244
Reilly, C. H. McKim, Mead and White. Benjamin Blom, 1924.

Design--Architectural

24 pp. $10.75

The British architect's account of the firm's achievement, the first major international recognition of their work, with full-page plates showing plans and elevations of buildings.

A04-1245
Rempel, J. I. Building with Wood. Univ. of Toronto Press, 1967. 304 pp. $17.50

This study treats construction techniques in the context of their environment: the history of varied methods of building in Upper Canada during the pioneering years. The author dissects a building to study its anatomy and traces the influences of the United Empire Loyalists, the Pennsylvania Germans, and the French on building techniques in Ontario. He discusses the practical aspects of construction and technical methods used in the erection of log and frame houses, mills, churches, bridges, schools, and jails.

A04-1246
Restaurant Planning and Design. Architectural Press, 1973. 120 pp. £8.50

The organization of restaurants has been revolutionized by changes in the packaging and preparation of food as well as by more sophisticated eating habits. This entails an entirely new look at the design and layout of access, kitchen and eating areas. Architects and restauranteurs will find this book a practical guide to designing restaurants for the seventies.

A04-1247
Rettig, R. B. Guide to Cambridge Architecture: Ten Walking Tours. MIT Press, 1969. 224 pp. $8.95 (paperback $3.95)

Covering Cambridge, Massachusetts, in a series of ten neighborhood tours, this book describes hundreds of buildings of all sorts, institutional, commercial, residential. Each building is illustrated with a small photograph; the date of construction and the name of the architect, when known, are given, together with a comment by the author on each building, clarifying architectural and historical points and offering evaluations.

A04-1248
Rettinger, M. Acoustic Design and Noise Control. Chemical Publishing, 1973. 577 pp. $22.50

Covers the physics of sound, room acoustics and design, noise and noise reduction. Contains charts, graphs and practical examples.

A04-1249
Reynolds, H. W. Dutch Houses in the Hudson Valley Before 1776. Dover, 1965. 469 pp. $5.00

The standard survey of the Dutch colonial house and outbuildings, with constructional features, decoration, and local history associated with individual homesteads. Introduction by Franklin D. Roosevelt. 150 illustrations.

A04-1250
Richards, J. Functional Tradition in Early Industrial Buildings. Architectural Press, 1958. 200 pp. £2.50
 The first and still the standard survey of industrial architecture in Britain.

A04-1251
Richards, J. M. A Guide to Finnish Architecture. Praeger, 1967. 224 pp. $10.00
 Beginning with Finland's ancient castles and churches and ending with the work of its contemporary architects, J. M. Richards shows how the architecture of Finland derives directly from its history, climate, topography, and the temperament of its people. Attention is given to the stone and wooden churches; to the Empire period, dominated by the work of C. L. Engel; and to Eliel Saarinen and the period of national Romanticism in the early years of this century. But the chief emphasis is on the contemporary architecture of Finland's great modern master Alvar Aalto.

A04-1252
Richards, J. M. and Pevsner, N., eds. The Anti-Rationalists. Univ. of Toronto Press, 1973. 216 pp. $17.50
 Although the leading exponents of Art Nouveau--Horta, Guimard, van de Velde, Tiffany and Gaudi--are now well-known and appreciated, a number of unacknowledged pioneers have been brought to light in this book and their contributions set beside those of the masters. Their work ranges from Spain to Czechoslovakia and Hungary, from Glasgow to Vienna and Berlin, from pubs to chapels.

A04-1253
Roberts, R. Your Engineered House. Garden Way Publishing, 1964. 237 pp. $8.95
 This book is filled with "house wisdom" and the human factors often neglected in housing. How to plan quiet areas and active rooms to make living sense for those who will use them. Details of sensible construction; where not to skimp, where to buy the least expensive line.

A04-1254
Robertson, H. Principles of Architectural Composition. Architectural Press, 1924. 180 pp. £0.75
 This book fills a gap in the literature on the theory of architectural design and has been adopted as a textbook in many architectural schools.

A04-1255
Rogers, W. G. What's Up in Architecture: A Look at Modern Building. Harcourt Brace Jovanovich. 192 pp. $4.25
 An account of new developments in architecture that includes studies of the great architects, by a man who for many years was Arts Editor of The Associated Press. Ages 14 up.

Design--Architectural

A04-1256
Rojas, P. Art and Architecture of Mexico: 10,000 B.C. to the Present. Tudor Publishing Co. $3.98
A visual record, with 132 illustrations.

A04-1257
Rolleston, S. E. Historic Houses and Interiors in Southern Connecticut. Hastings House. 208 pp. $15.00
Twenty-six historic museum houses are presented here. Photographs and descriptive text show the development of the American home from pioneering and colonial days to Victorian mansion.

A04-1258
Rosenfield, I. Hospital Architecture and Beyond. Van Nostrand Reinhold, 1969. 320 pp. $24.95
A discussion of the problem of meeting the medical needs of the modern community. Calls for new types of hospitals which bear little resemblance to the tradition-bound conglomerates of today. Shows how to determine and plan the best working and living conditions for a hospital's staff and patients.

A04-1259
Rosenfield, I. Hospital Architecture: Integrated Components. Van Nostrand Reinhold, 1971. 336 pp. $26.50
An explanation of how hospital departments can be designed to promote greater functional efficiency and better administration. Considers each department an integral part of the entire hospital facility. The book shows how these departments function within a given division, and how the division relates to the hospital's overall operation. Includes design information related to diagnostic and therapeutic facilities, patient quarters, ancillary services, administration, costs, hospital environment, light and color, and hospital originated infections.

A04-1260
Roth, A. The New Schoolhouse. Praeger, 1966. 304 pp. $15.00
This work on school architecture and design reflects the most recent theories and trends in the planning of school buildings. The most imaginative and practical designs for kindergarten, elementary, and high schools are represented in examples from Switzerland, Germany, Denmark, Italy, England, Israel, and the United States. A discussion of techniques, styles, and trends, and of the attitudes of various nations to the design and construction of educational buildings.

A04-1261
Roth, L. Monograph of the Work of McKim, Mead and White: 1879-1915. Benjamin Blom, 1915/1973. $50.00
The new critical introduction and the magnificent plates with captions make this the most complete record of the firm that was the pathbreaker in American renaissance architecture.

A04-1262
Rowland, K. History of the Modern Movement: Art, Architecture, Design. Van Nostrand Reinhold, 1973. 240 pp. $13.95
A study, evaluation, and history of the Modern Movement in art, architecture, and design from 1880 to today. It demonstrates how pure visual means replaced traditional representation and style, and assumed the importance of a new visual language when art, architecture, and design broke with the accustomed 19th century viewpoints. The book describes the effects of such visual influences as the Art Nouveau, the early work of Frank Lloyd Wright, and the spatial concepts of de Stijl.

A04-1263
Rubens, P. P. Palazzi Di Genova. Benjamin Blom, 1622/1969. 233 pp. 54 plates. $45.00
Rubens' only architectural work, essential in the history of Italian influence on North European styles.

A04-1264
Rubens, P. P. Pompa Introitus ... Ferdinandi Austriaci ... Cum Antiverpiam ... Adventu Suo Bearet, 15 Ka. Maii Anno 1665. Benjamin Blom, 1665. 212 pp. $49.50
Rubens at the height of his creative powers in a book that is a major document for baroque stage design.

A04-1265
Rutledge, A. Anatomy of a Park: The Essentials of Recreation Area Planning and Design. McGraw-Hill, 1971. 192 pp. $18.75
This always interesting "nontechnical technical book" presents a comprehensive system for evaluating the worth of a park design plan and a view of design criteria based upon behavioral science findings relating to people's needs--both aesthetic and functional.

A04-1266
Saalman, H. Medieval Cities. Braziller, 1968. 128 pp. $7.95 (paperback $3.95)
A critical look at the structure of the medieval city, at the buildings and spaces which were its characteristic components, and at the economic, social and political forces which gave this complex its shape.

A04-1267
Sabine, W. C. Collected Papers on Acoustics. Dover, 1964. 279 pp. $3.50
11 papers on architectural acoustics that form the scientific basis of the field: building materials and pitch, reverberation and pitch, air currents and temperature and their influence, whispering galleries, etc.

A04-1268
Safdie, M. Beyond Habitat. MIT Press, 1970. 300 pp. $12.50

Design--Architectural

(paperback $3.45)

When the apartment complex Habitat (Moshe Safdie, architect) appeared at Montreal's EXPO 67, it was greeted as a solution to the housing crisis in the urban centers of the world. Beyond Habitat is a personal statement covering the author's life and career before Habitat, his struggle to transform Habitat from a "design solution" on paper to a reality, his more recent architectural work and his general ideas on housing and other matters since its completion.

A04-1269
Safdie, M. For Everyone a Garden. MIT Press, 1974. 280 pp. $25.00

This book is a synthesis of words and pictures. Much of it is devoted to illustrations, supported by substantial captions--while the text proper puts these into perspective from four thematic points of view: the idea of the three-dimensional community; the requirements and possibilities of human habitation; the techniques of building in the factory; and the attributes of well-planned urban meeting places.

A04-1270
Salokorpi, A. Modern Architecture in Finland. Praeger, 1971. 192 pp. $4.50

This book is a guide to the development of architectural and environmental awareness in Finland during the twentieth century. It describes early rationalist and romantic trends, Finnish functionalism, and the synthesizing work of the 1950s and 1960s. It also examines post-World War II rebuilding, the new demands of industrial architecture, and the vigorous renewal of town planning in modern Finland.

A04-1271
Salvadori, M. Mathematics in Architecture. Prentice-Hall, 1968. 173 pp. $8.95

Covers all the mathematics needed by the architect doing preliminary design evaluations in structures, air conditioning, estimation or comparison of systems. Introduced through essentially intuitive, elementary means to facilitate a more intelligent dialogue between the architect and his technical consultants. Each topic is presented through the geometrical or physical interpretation of the problem to be solved and is illustrated by means of problems of practical significance.

A04-1272
Santini, P. C. and Marini, G. L., eds. Catalogo Bolaffi dell' Architettura Italiana, 1963-1966. Wittenborn, 1966. 648 pp. $45.00

An alphabetical listing of major works by outstanding Italian architects, with notes and technical details revealing the progressive achievements of these years in building, planning and domestic housing. 2231 illustrations.

A04-1273
Saphier, M. Office Planning and Design. McGraw-Hill, 1968. 193 pp. $18.95
Discusses office design problems in terms of operational aims, aesthetic goals, and financial limitations. 54 illus.

A04-1274
Saylor, H. H. Dictionary of Architecture. John Wiley, 1963. 221 pp. $6.00

A04-1275
Schmertz, M., ed. Office Building Design. McGraw-Hill, 1975. 288 pp. $18.95
This revised edition of a popular Architectural Record book offers up-to-date coverage on low- and medium-rise office buildings, high-rise office buildings, high-rise technology, and selected office building interiors.

A04-1276
Schmitt, K. W. Multistory Housing. Praeger, 1966. 216 pp. $18.50
The author traces the recent history of housing construction around the world. He discusses effective large-scale housing and shows how various architects have been successful or unsuccessful in their resolutions of the problem.

A04-1277
Schulze-Fielitz, E. Stadtsysteme I/Urban Systems. International Publications. Vol. 1, 1971. 60 pp. $8.50; Vol. 2, 1973. $8.50
Only some 20% of the building construction of the year 2000 exists today. Proposals and suggestions are given for space structures, occupation and density, communication, distribution, sociology and psychology.

A04-1278
Schuyler, M. American Architecture and Other Writings. Atheneum Publishers, 1964. $2.45
A selection of essays on developing American architecture from 1876 to 1914. Abridged by W. H. Jordy from the book originally published by Harvard University Press.

A04-1279
Schwab, G. The Architecture of Paul Rudolph. Praeger, 1970. 240 pp. $20.00
This book is a study of the work of an American master architect of the post-World War II period. Over 150 illustrations trace the development of Paul Rudolph's buildings as an innovative force in the American architectural context.

A04-1280
Scully, V. American Architecture and Urbanism. Praeger, 1969. 272 pp. $18.50

Design--Architectural

This history of architecture and city living in America shown as inseparable, in both immediate effect and ultimate significance. This book traces a circle of development from Pueblo Indian cliff dwellings to proposed urban megastructures of the future, describing the qualities shared in common by architecture as diverse as the mission churches of the Southwest and seventeenth- and eighteenth-century New England houses. It explores the effects of such movements as the New Bauhaus and the International Style on American architecture.

A04-1281
Scully, V. Modern Architecture. Braziller, 1974. 160 pp. $9.95 (paperback $4.95)
This revised edition focuses on the major new developments in architecture in the last twelve years. Unlike the earlier edition, it devotes considerable attention to city planning, contrasting the towers of Sullivan, Mies, and Le Corbusier with Moshe Safdie's Habitat and the Mobile Home.

A04-1282
Seaborne, M. The English School: Its Architecture and Organization. Univ. of Toronto Press, 1970. 427 pp. $25.00
Britain has a rich heritage of school buildings dating from the later Middle Ages to the present. While some of these schools have attracted the attention of architectural historians, they have not previously been considered from the educational viewpoint. This book is a study both of architectural history and the development of educational ideas and practices from the fourteenth to the nineteenth century.

A04-1283
Senkevitch, A., Jr. Soviet Architecture, 1917-1962: A Bibliographic Guide to Source Material. Univ. Press of Virginia, 304 pp. $13.50
This annotated bibliography contains over 1,000 titles of monographs and articles on the history and ideology of Soviet architecture from 1917 to 1962. It includes Russian and Western-language publications, the Russian works forming the largest portion of the material available. Each entry contains a bibliographic description and an annotation that evaluates the nature and content of the work. Abstracts have been included when the material is of particular interest. The entries are numbered and indexed.

A04-1284
Serlio, S. The Book of Architecture. Benjamin Blom, 1969 [1611]. $40.00
This five-part treatise is one of the seminal documents on architecture and theatre design.

A04-1285
Service, A., ed. Edwardian Architecture and Its Origins. Architectural Press, 1974. 300 pp. £8.00
From its earliest issues The Architectural Review published

classic studies of Victorian and Edwardian architecture, but these are now dispersed and are frequently unavailable. This illustrated book collects over twenty of these essays together with twenty newly-commissioned articles. A source book for amateur and professional architectural historians.

A04-1286
Seto, W. W.   Acoustics.   McGraw-Hill, 1970.   224 pp. $4.95
    Covers vibrations and waves, plane acoustic waves, spherical acoustic waves, transmission of sound, architectural and underwater acoustics, and ultrasonics.

A04-1287
Severino, R.   Equipotential Space: Freedom in Architecture.
    Praeger, 1970.   224 pp. $12.50 (paperback $4.50)
    This book presents a plea for total architectural planning and proposes that technology serve as the key to a new architecture enabling man to mold his environment. The author shows how applied technology can be used by architects to create basic standardized units that can be manipulated in a variety of ways. These inexpensive and readily available components will enable individuals to control the quality of their housing, rather than their being controlled by the forces of industrialization.

A04-1288
Sharp, D.   The Picture Palace: And Other Buildings for the
    Movies.   Praeger, 1968.   224 pp. $12.50
    A history of the movie house from the nickelodeon to the drive-in, from the picture palace to the art theater, in America, England, and continental Europe. Treats problems of sight-lines, ventilation, acoustics, projection, and decoration; discusses the styles and innovations of important theater architects; and assesses the latest developments in the design of movie houses all over the world.

A04-1289
Sharp, D.   Twentieth Century Architecture.   Wittenborn, 1972.
    304 pp. $27.50
    The accomplishments of architecture during the past 70 years, from private homes to factories, from art nouveau curves and Bauhaus functionalism to glass-wall cubes. A chronology of buildings erected between 1900 and 1970. Bibliography, 1,000 photos, 50 color plates.

A04-1290
Sharp, D.   A Visual History of Twentieth Century Architecture.
    N.Y. Graphic Society, 1972.   304 pp. $27.50
    An illustrated chronology of the major architectural works of this century. All significant buildings completed during the period are examined, with photographs, drawings and plans.

A04-1291
Sheringham, G. and Laver, J.   Design in the Theater.   Benjamin

Design--Architectural

Blom, 1927. 158 pp. $17.50
A collection of 120 full-page plates showing work by scores of designers, among them Bakst, Adrian Allinson, Cecil Beaton, Reginald Brill, André Derain, Dulac, R. E. Jones, R. Boyd Morrison, L. Popova, and Sheringham.

A04-1292
Shinohara, K. 16 Houses and Architectural Theory. Wittenborn, 1971. 174 pp. $22.50
Japanese, English text. 123 photographs of outstanding Japanese houses and the theory of functional space.

A04-1293
Shipway, V. C. and Shipway, W. Decorative Design in Mexican Homes. Architectural Book Publishing, 1966. 272 pp. $14.95
The fourth volume in the Shipways' series on Mexican homes reflects the timelessness of cultures on which Mexican architecture and decorative motifs are based.

A04-1294
Shipway, V. C. and Shipway, W. Houses of Mexico, Origins and Traditions. Architectural Book Publishing. 200 pp. $14.95
The latest in the Shipways' series on Mexican design delves into the roots of the Mexican style. Many photographs from Spain and Portugal reflect the Moorish influence.

A04-1295
Shipway, V. C. and Shipway, W. Mexican Homes of Today. Architectural Book Publishing, 1964. 272 pp. $14.95
This depicts the furniture and architectural details of modern homes.

A04-1296
Shipway, V. C. and Shipway, W. The Mexican House: Old and New. Architectural Book Publishing, 1960. 208 pp. $13.95
A record of a way of life in which architecture, climate and tradition combine to produce the charm of the past.

A04-1297
Shulman, J. Photographing Architecture and Interiors. Watson-Guptill. 160 pp. $14.95
This is a "how to" book for amateur and professionals. The author explains equipment, composition in general, and architectural photography in particular. The liaison between photographer and architect or designer is explored as well as rates and rights for photographic work. Designers of buildings and interiors can learn how to make the most of photography in recording and promoting their work.

A04-1298
Siegel, A., ed. Chicago's Famous Buildings: A Photographic

Guide to the City's Architectural Landmarks and Other Notable Buildings. Univ. of Chicago Press, 1969. 272 pp. $2.45

The guide is organized into five categories: buildings in historic styles; buildings of the Chicago School (including those officially designated as Landmark buildings); Prairie School houses; buildings of general interest; and recent buildings. Architectural plans of the Landmark buildings are supplied. There is a list of other notable buildings in the general metropolitan area, not illustrated by photographs, but identified by the name of the architect, date of completion, and address.

A04-1299
Siegel, C. Structure and Form in Modern Architecture. Van Nostrand Reinhold, 1962. 309 pp. $21.50

A book that covers the theoretical and practical aspects of all meaningful structural forms used in contemporary design and building.

A04-1300
Silverman, M. and Bowman, N. A. Theatre Architecture: An Illustrated Survey and a Checklist of Publications, 1946-1964. N.Y. Public Library, 1965. $10.00

A04-1301
Skudrzyk, E. J. Foundations of Acoustics, Basic Mathematics and Basic Acoustics. Springer-Verlag, 1972. $73.80

A04-1302
Sleeper, H. R. Architectural Specifications. John Wiley, 1940. 822 pp. $31.75

A04-1303
Sleeper, H. R. Building Planning and Design Standards for Architects, Engineers, Designers, Consultants, Building Committees, Draftsmen, and Students. John Wiley, 1955. 331 pp. $16.00

A04-1304
Smith, C. B. Builders in the Sun: Five Mexican Architects. Architectural Book Publishing. 224 pp. $12.95

The work of Juan O'Gorman, Felix Candela, Mathias Goeritz, Luis Barragan and Mario Pani.

A04-1305
Smith, C. R. The American Endless Weekend. American Institute of Architects, 1973. $7.00 (for AIA members, less 20%) Recreation problems and planning in America.

A04-1306
Smith, D. L. Amenity and Urban Planning: Origin and Role of the Aesthetic Element in Modern Practice. Beekman Publishers, 1974. 247 pp. $15.00

Design--Architectural 221

The author's approach will appeal to those who wish to take a new look at the traditional values of urban planning. Uses the findings of research to discuss popular valuations, future trends, and lessons for local government.

A04-1307
Smith, E. B. Dome: A Study in the History of Ideas. Wittenborn, 1972. 240 pp. $3.95

A04-1308
Smith, J. L. and Hoppe, T. Building to Scale: A Manual for Model Home Construction. Prentice-Hall, 1971. 158 pp. $9.50

A guide for transforming two-dimensional plans into three-dimensional models. Complete details of actual model construction, from procedure to various tools and materials employed in building a scale model. Offers suggestions to aid development of furniture and fixtures for models and how to properly mount models and prepare landscaping.

A04-1309
Smith, P. C. The Design and Construction of Stables and Ancillary Buildings. British Book Centre, 1974. $8.95

The author, a practicing architect and an enthusiastic horseman, has written this book for those interested in the construction of new stables or in the extension, alteration or adaptation of existing buildings. For the architect, whether experienced or not in equestrian matter, this book will prove of particular use. Layouts have been designed to assist him at all stages of a project.

A04-1310
Smith, R. and Owens, J. The Majesty of Natchez. Pelican, 1969. 96 pp. $4.95

In color, featuring interior and exterior views of the outstanding old homes of antebellum Natchez.

A04-1311
Society of Mexican Architects, ed. Four Thousand Years of Mexican Architecture. William Heinman, 1956. $40.00

A04-1312
Soleri, P. Arcology: The City in the Image of Man. MIT Press, 1969. 256 pp. $25.00 (paperback $8.95)

The sprawling cities and suburbs that are eating up the surface of the earth are absurd and unworkable. Each new building exposes more of a built-in flaw--the distances and travel times that separate individuals from their institutions (or from nature) do not conform to the human scale, and the obstacles between the two exceed human tolerance limits. Soleri proclaims an alternative; an ideal against which to measure the direction and extent of future change.

A04-1313
Soleri, P. Sketch Books of Paolo Soleri. MIT Press, 1971.
432 pp. $27.50 (paperback $12.95)
These Sketchbooks contain the seed of the giant Arcologies, those self-contained city-buildings of comparable extent in all three dimensions. Most of the material presented here is taken from sketchbooks of the early 1960s and pertains to the development of a hypothetical city of 2,000,000 on a plateau and all its satellites and servicing agents. This is the "Mesa City" concept, a pre-Arcology that contains in potential form what was later realized in Arcology: The City in the Image of Man.

A04-1314
Southern, R. Proscenium and Sight Lines. Theatre Arts. $8.25

A04-1315
Sowden, H. Towards an Australian Architecture. International Publications, 1970. 273 pp. $20.00
A survey of modern Australian architecture in urban and rural settings. Shows the work of nineteen architects or architectural firms, including public buildings, homes and apartment compounds, government offices, theatres, banks, churches, etc. Illustrated.

A04-1316
Spreiregen, P. D., ed. On the Art of Designing Cities: Selected Essays of Elbert Peets. MIT Press, 1968. 243 pp. $22.50
The civic design of Washington, D.C. is the concern of the first sections of this book. Other places and other planners--London, Rome, Sitte, Haussman--are also discussed.

A04-1317
Spreiregen, P. D. Urban Design: The Architecture of Towns and Cities. American Institute of Architects or McGraw-Hill, 1965. 243 pp. $17.50 (for AIA members, less 10%)
An exploration of the design and layout of communities. Interrelates commercial and industrial development with residential and institutional areas, tremendously enhancing overall attractiveness and livability.

A04-1318
Spyer, G. Architect and Community: Environmental Design in an Urban Society. International Publications, 1971. 168 pp. $17.50
The author examines the ideas and achievements of past and present townplanners and discusses their influence. He considers also the impact of industrial technology on architecture. Finally, he stresses the need for greater coordination between and teamwork among professionals and government authorities.

A04-1319
Stanton, P. B. The Gothic Revival and American Church Architecture: An Episode in Taste, 1840-1856. Johns Hopkins

Design--Architectural 223

Univ. Press, 1968. 350 pp. $15.00
Focusing on the mid-nineteenth century, the author explores the influence of the English ecclesiological Gothic revival on American church architecture. This fundamental conservative movement provided a foundation for a new aesthetic in American architecture. The book surveys the architectural principles and models most influential in America. Traces the development of American style and theory in terms of the work and writings of individual architects and of their major organization, the New York Ecclesiological Society.

A04-1320
Stanton, P. Pugin. Viking Press, 1972. 216 pp. $3.95
A reassessment of the most important exponent of Gothic Revival in English architecture.

A04-1321
Stern, R. A. M. New Directions in American Architecture. Braziller, 1969. 128 pp. $7.95 (paperback $3.95)
An examination of present American architecture with a view to its historical basis and future trends.

A04-1322
Stevens, G. P. Restorations of Classical Buildings. American School of Classical Studies, 1958. 49 pp. $1.10
This collection includes twenty restored drawings of Greek, Roman and Byzantine buildings in Athens, Corinth, Tivoli and Leptis Magna as well as a plan of the Acropolis at Athens and a photograph of Mr. Stevens' model of the Acropolis. A brief explanatory text and a bibliographical reference accompany each plate.

A04-1323
Steyert, R. D. Economics of High-Rise Apartment Buildings of Alternate Design Configuration. American Society of Civil Engineers, 1972. 171 pp. $6.00 (for ASCE members, $3.00)
This book is a study of the economic problems of conceptual design of high-rise apartment buildings. Definitions and techniques useful in total cost analysis are developed. The economics of buildings of alternate height, floor size, shape, and room size are explored quantitatively. Implications of these quantitative results are considered relative to zoning and to apartment design.

A04-1324
Storrer, W. A. The Architecture of Frank Lloyd Wright: A Complete Catalog. MIT Press, 1974. 289 pp. $15.00
This work is the only publication that documents all of the buildings designed by Frank Lloyd Wright that were actually constructed. It includes a significant number of authentic buildings not found in previous listings. The book also offers a short commentary on each building and a picture of each extant structure.

A04-1325
Street, A. E. Memoir of George Edmund Street, R. A. 1824-1881.
    Benjamin Blom, 1881. 441 pp. $18.50
    The primary source for biographical and critical information on the architect who helped shape the English gothic architecture revival in the nineteenth century. Street's academy lectures on the art, style, and practice of architecture are included.

A04-1326
Structure, Space, Mankind; Expo '70. International Publications, 1970. 326 pp. $15.00
    A photographic interpretation of the architecture and all pavilions, display plans, construction, areas on the site, total floor space, height, structure, display themes, numbers of persons admitted, construction costs, etc. Illustrations and floor plans.

A04-1327
Studio Dictionary of Designs and Decoration. Viking Press, 1973. 544 pp. $28.50
    This illustrated dictionary of the decorative arts covers all the main figures, movements, technical developments, and periods in Europe and America. Over two thousand alphabetized entries on decorators and their designs, architects and their buildings, industrial designers and their products; information is provided on furniture, china, glass, silver, and other crafts; on home structure and building materials; on fabrics, wood, synthetics; and on the terminology used in all these fields. In addition, biographical entries are included on designers and architects.

A04-1328
Sturgis, R. A Dictionary of Architecture and Building. Gale Research Co., 1902/1966. 3 vols. 1856 pp. $45.00 set.
    More than 65 experts in American antiquities, ecclesiology, decorative art, perspective and projection, structural engineering, and acoustics contributed to the dictionary. The 3-volume set covers the development and nomenclature of the architecture of all cultures and ages. The work of hundreds of architects is described.

A04-1329
Sullivan, L. H. The Autobiography of an Idea. Dover, 1924. 333 pp. $3.50
    The early creative years of the architect who has perhaps been America's most influential architectural theorist.

A04-1330
Summerson, J. The Classical Language of Architecture. MIT Press, 1966. 80 pp. $5.95 (paperback $2.95)
    The author explains the origin and significance of the Orders, and the influence of classical disciplines on Renaissance and Baroque architects and others nearer our own time.

A04-1331
Sunset Eds. Cabins and Vacation Homes. Lane Magazine and Book Co., 1967. $1.95

Design--Architectural

A04-1332
Sunset Eds. Planning and Landscaping Hillside Homes. Lane Magazine and Book Co., 1965. $1.95

A04-1333
Sunset Eds. Sunset Ideas for Planning Your New Home. Lane Magazine and Book Co., 1967. $1.95

A04-1334
Suri, R. L. Acoustics: Design and Practice. Asia Publishing House, 1966. 539 pp. $14.50
Volume 1 deals with general acoustics, sound absorbing materials and their applications, methods of vibrations and shock control in machinery, design and construction of sound-test chambers, and more.

A04-1335
Swan, A. A Collection of Design in Architecture. Gregg International Publishers, 1972. $27.95
Reprint of 1757 edition.

A04-1336
Swan, A. Upwards of One Hundred Fifty New Designs for Chimney Pieces. Gregg International Publishers, 1972. $20.15
Reprint of 1768 edition.

A04-1337
Sweeney, J. J. and Sert, J. L. Antoni Gaudi. Praeger, 1972. 192 pp. $18.50
This book incorporates the latest research and criticism into an evaluation of Gaudi's work. The authors describe how his structural innovations and unusual designs resulted in revolutionary buildings. Nearly 200 plates show how Gaudi's buildings seem to grow out of their surroundings, carrying the organic philosophy of Art Nouveau to its ultimate expression.

A04-1338
Tange, K. and Kawazoe, N. Ise: Prototype of Japanese Architecture. MIT Press, 1965. 212 pp. $17.50
The Ise shrine is both old and new. The buildings date to at least 685 A.D., but every twenty years the shrine is completely rebuilt. Thus far Ise has been rebuilt 59 times. This book is of importance as a photographic document, for the inner areas of the Ise Shrine, which are ringed by four fences and contain the key buildings, are barred to visitors.

A04-1339
Tange, K. and Kultermann, U. Kenzo Tange. Praeger, 1971. 304 pp. $29.50
Tange has been strongly influenced by Le Corbusier; he acknowledges this in his treatment of rough concrete and structuring of space (building on stilts, creating roof gardens, and planning freely divisible interior space). Among the projects discussed are Tange's

plan to extend Tokyo into Tokyo Bay, the Olympic Arenas in Tokyo, St. Mary's Cathedral in Tokyo and the Tsukiji urban development program.

A04-1340
Taylor, J.   Model Building for Architects and Engineers.   American Institute of Architecture, 1971.   150 pp. $15.75 (for AIA members, less 10%)

Four basic types of models are discussed in detail, plus special model categories: structural models, interior models, planning models, landscape models. What materials to use and where to get them, short cuts, precise steps in preparing a sturdy base, constructing contours, building structural shape. Nearly 200 photographs, drawings, diagrams, and plans.

A04-1341
Taylor, N.   Looking at Cathedrals.   International Publications, 1968.   48 pp. $2.25

The cathedrals of Bristol, Canterbury, Durham, Birmingham (St. Chad's), Gloucester, Lichfield, Lincoln, Liverpool, Salisbury, St. Paul's, Wells and York.   Illustrations and plans.

A04-1342
Tempel, E.   New Finnish Architecture.   Praeger, 1968.   192 pp. $20.00

The author begins with a historical survey; he treats the medieval castles and early wooden churches, the age of Empire architecture in nineteenth-century Helsinki, and the historical romanticism of the early 20th century, with its outstanding exponents, Eliel Saarinen and Lars Sonck. He shows how the work of Aalto and Bryggman engendered an architectural expression intimately related to the Finnish landscape.

A04-1343
Tempel, E.   New Japanese Architecture.   Praeger, 1970.   220 pp. $18.50

Modern Japanese architecture has synthesized Japan's ancient indigenous architecture and the modern design techniques of the West. Faced with a burgeoning population, Japanese architects explore all the possibilities of prefabrication and use entirely new criteria of space planning and architectural shaping. This volume illustrates these developments with examples ranging from single-family houses to municipal buildings to utopian city plans.

A04-1344
Thompson, J. D. and Goldin, G.   The Hospital: A Social and Architectural History.   Yale Univ. Press, 1976.   377 pp. $25.00

A study combining modern thinking about hospital design with a deep understanding of the architectural, medical and cultural history of the hospital ward.   More than 270 illustrations.

Design--Architectural

A04-1345
Tibbs, R. King's College Chapel, Cambridge (England). Wittenborn, 1970. 95 pp. $6.00
The story and renovation. 54 photographs.

A04-1346
Tidworth, S. Theatres: An Architectural and Cultural History. Praeger, 1973. 224 pp. $18.50
A survey of theatres from ancient to contemporary times. Theatre architects such as Palladio, Wren, Inigo Jones, Cuvillies, Sullivan, and Gropius are represented by nearly 200 illustrations, including many plans and drawings.

A04-1347
Torroja, E. The Structures of Eduardo Torroja. McGraw-Hill, 1958. 198 pp. $9.00
An autobiography which explains in text, photographs, and drawings, 30 of the author's most significant structures, including bridges, dams, sports arenas, factories, and churches.

A04-1348
Trill, J. Teamwork in Building Construction. Architectural Press, 1974. 150 pp. £1.50
A paper on "industrial studies" is now mandatory in all courses on building at craft level, and also under the heading of "structure of the industry" and "building administration" in the Ordinary National Certificate and the Construction Technicians Course respectively. This book fills the requirements of the syllabus in providing a concise description of the roles of the various members of the building team and explaining their relationship to each other at every stage of the building project in England.

A04-1349
T'Serstevens, M. Living Architecture: Chinese. Grosset and Dunlap, 1972. $7.95

A04-1350
University Press of Africa. Homes for Kenya. International Publications, 1969. 108 pp. $3.00
A description of the National Housing Corporation and what the Councils are doing; basic designs and plans of different houses; how to get a mortgage; the lending companies and their services; professional associations in Kenya building; and the building industry in Kenya today.

A04-1351
Unwin, R. Town Planning in Practice. Benjamin Blom, 1934. 438 pp. $18.75
Principles of planning in specific European cities from the early Middle Ages, the streets, squares, buildings, and environment they create. A practical guide to the creation of cities.

A04-1352
Vandenberg, M. and Elder, A. J., eds. Handbook of Building Enclosure. Architectural Press, 1973. 250 pp. £7.00
This handbook emphasizes building performance. Each of the enclosing elements of the building (floors, walls, roofs, etc.) is examined and construction methods are discussed in the context of functional performance. This handbook contains technical studies, information sheets and design guides; and includes data on the thermal and acoustic performance of building enclosures.

A04-1353
Van Rensselaer, M. G. Henry Hobson Richardson and His Works. Dover, 1969. 152 pp. $4.00
A study of one of the great American architects. Contains a full biography, plans, photographs, and discussions of all of his major buildings.

A04-1354
Vaux, C. Villas and Cottages. Dover, 1970 [1864]. 348 pp. $3.00
Vaux is considered to have founded the Hudson River School of Architecture with this book. It contains 39 designs for well-styled, efficient and low-priced houses, actual examples of which may still be seen. Each design is supplemented with detailed floor plans, and perspective views.

A04-1355
Vazquez, P. R. The National Museum of Anthropology, Mexico. Harry N. Abrams, 1968. 258 pp. 283 illustrations. $25.00
The entire concept of the museum building, completed four years ago at a cost of 20 million dollars, is presented by the author who is the museum's architect, and by the museum's director, curator and others. The building, its displays and designs, incorporates the most recent advances in museography.

A04-1356
Venture, R. Complexity and Contradiction in Architecture. New York Graphic Society, 1967. 144 pp. $5.95 (paperback $3.95)
The views of a contemporary American architect on architecture today. Presents architectural criticism and an explanation of the formation of his own ideas and work.

A04-1357
Venturi, R., et al. Learning From Las Vegas. MIT Press, 1972. 208 pp. $25.00
The book is divided into three parts. The first is an illustrated study of the iconography and symbolism of Las Vegas, with special attention to the Las Vegas "Strip." In the middle part the authors prescribe a needed antidote: a new modesty, an architectural populism. The last part of the book illustrates how the theory is translated into reality: it presents the projects undertaken over

Design--Architectural

the past several years by the firm of which the authors are members.

A04-1358
Vogt-Goknil, U. Living Architecture: Ottoman. Grosset and Dunlap, 1966. $7.95

A04-1359
Volwahsen, A. Living Architecture: India. Grosset and Dunlap, 1969. $7.95

A04-1360
Vohwahsen, A. Living Architecture: Islamic Indian. Grosset and Dunlap, 1970. $7.95

A04-1361
Vostell, W. and Higgins, D. Fantastic Architecture. Wittenborn, 1971. 192 pp. $6.95
An anthology of architectural and spatial projects, some realizable, others less so.

A04-1362
Wagner, W., ed. Great Houses for ... View Sites, Beach Sites, Sites in the Woods, Meadow Sites, Small Sites, Sloping Sites, Steep Sites, Flat Sites. McGraw-Hill, 1976. 248 pp. $16.95
68 houses selected from the pages of Architectural Record with suggestions and comment on planning a house suited to almost any site make up this volume. Over 300 photographs and many simple line drawings outline the principles of design that create an interesting, workable, and convenient house that takes advantage of any kind of site. This book will be useful to prospective home-buyers, custom house clients, architects, builders and developers.

A04-1363
Walker, J. Glossary of Art, Architecture and Design Since 1945. Shoe String Press, 1973. 240 pp. $10.50
Arranged in alphabetical order, the glossary provides definitions of the great variety of art movements, groups and styles which have evolved internationally since World War II. The terms are derived from the published literature on art and concentration is on the main conceptual and theoretical notations.

A04-1364
Wall, D. R. Visionary Cities: The Arcology of Paolo Soleri. Praeger, 1970. 152 pp. $20.00 (paperback $5.95)
This book presents Soleri's concepts for replacing urban sprawl with cities conceived as single buildings--"objects" of immense size that can support a million people per cubic mile. He calls his city plans "arcologies"--a synthesis of architecture and ecology.

A04-1365
Walsh, J. W. T.   The Science of Daylight.   Dover.   285 pp.
$6.50
Assembles data on daylight from physical, engineering, architectural, etc., sources: measurement of daylight, sky luminance, color and spectral distribution of daylight, window design, daylight calculation, sunlight control, building requirements, town planning, artificial adjuncts, etc.

A04-1366
Waterman, T. T. and Barrows, J.   Domestic Colonial Architecture of Tidewater, Virginia.   Dover.   191 pp. $4.00
The Adam Thoroughgood house, Greenspring, Bacon's Castle, Fairfield, Ampthill, The President's House, Stratford, Westover, Rosewell, Carter's Grove--in all 15 great houses. Elevations, photographs, prints, construction details, groupings of domestic buildings, historical background, famous occupants.

A04-1367
Watterson, J.   Architecture: A Short History.   Norton and Co., 1968. $10.00

A04-1368
Weisskamp, H.   Hotels: An International Survey.   Praeger, 1968.   212 pp. $20.00
This text probes the relationships between hotel architects and decorators and between architects and clients, particularly in large luxury chains. The author analyzes the special requirements of all types of hotels, from the Hiltons and Ritzes to tourist camps and motels. Presents examples from outstanding designs throughout the world, including works of architects such as Neutra and Yamasaki, as well as those of lesser-known designers.

A04-1369
Wesdert, W.   Private Homes: An International Survey.   Praeger, 1967.   168 pp. $16.00
This book presents a selection of homes in which the needs of the particular families and the potentials of specific sites were the main factors in design. Many of the houses illustrated were designed by the architects for their own families.

A04-1370
West, T.   The Timber-Frame House In England.   Hastings House.   212 pp. $7.95
Shows how this house construction has endured. Much practical advice for anyone wishing to unravel the history of an old house to preserve one.

A04-1371
Wheeler, E.   Hospital Design and Function.   McGraw-Hill, 1964.   296 pp. $14.50
A description of a general hospital's functional program requirements, and the specific manner in which each hospital department may be designed to meet these requirements.

Design--Architectural

A04-1372
Whiffen, M. American Architecture Since 1780: A Guide to the Styles. MIT Press, 1969. 320 pp. $7.95
The author has designed this guide for the layman as an aid to his experiencing architecture. It is concerned entirely with the visual characteristics of the various styles as they can be seen from the outside. The book's aim is to serve as a guide to architectural styles, not as a history or criticism of them. The periods covered begin with the colonial and extend to the modern.

A04-1373
Whiting, P. New Houses. Architectural Press, 1964. 168 pp. £1.50
An architect-housewife's selection of thirty houses.

A04-1374
Whiting, P. New Single-Storey Houses. Architectural Press, 1966. 176 pp. £1.50
Thirty-one selected architect-built bungalows discussed and illustrated.

A04-1375
Whittick, A. Symbols for Designers. Gale Research Co., 1935. 1968 pp. $14.00
A handbook on the application of symbols and symbolism to design. For the use of architects, sculptors, ecclesiastical and memorial designers, commercial artists and students of symbolism. The first part treats the meaning of symbolism, types of symbols used in pictorial design, and symbols for the modern designer. The main part provides the meanings and history of several hundred symbols with emphasis on their suitability for design.

A04-1376
Wilkinson, J. Storage Specifics. Cahners, 1969. 104 pp. $3.95
A guide to proper planning and designing of receiving and storage areas. Gives essential control measures for storage of food.

A04-1377
Williams, H. L. and Williams, O. K. A Guide to Old American Houses 1700-1900. A. S. Barnes, 1962. 168 pp. $2.98
This volume will enable antique collectors, historians, architects, students and all lovers of fine houses to identify the main styles of houses built between 1700 and 1900 and to detect earlier or later features grafted on to them. The reader will know where, when and how the architectural details orginated, and he will be able to detect the identifying features of each style and what they represent, as well as recognize where styles have overlapped.

A04-1378
Wills, R. B. Better Houses for Budgeteers. Hastings House. 110 pp. $6.50
Sketches and plans of low-cost houses. Scarce copies

discovered in warehouse, improperly bound, without dust jackets. Architects and collectors keep asking for this rare book.

A04-1379
Wills, R. B. Houses for Good Living. Architectural Book Publishing. 112 pp. $6.50
How to personalize a style, budget costs, more.

A04-1380
Wills, R. B. More Houses for Good Living. Architectural Book Publishing. 128 pp. $10.00
A guide for those who want a comfortable home of the traditional kind. Includes elevations and plans.

A04-1381
Wilson, F. Conversations with Lev Zetlin. Cahners, 1975. 224 pp. $19.95
A chronicle of the ideas of a great and original mind, this book grew out of a series of interviews, casual conversations, and question-and-answer sessions conducted by the author with the widely known structural engineer Lev Zetlin. The purpose of the book is to describe the frame of reference within which an engineer solves problems. The book is intended to bridge the gap between the layman's curiosity and the engineer's specialized knowledge and to serve as an inspiration to architects, engineers, builders, construction managers, teachers, and students.

A04-1382
Wilson, F. Structure: The Essence of Architecture. Van Nostrand Reinhold, 1972. 96 pp. $5.95 (paperback $2.95)
This book begins with the art of building, which man has mastered and ends with the art of understanding our built environment.

A04-1383
Wilson, J. and Leaman, A. Color in Decoration. Wittenborn, 1971. 160 pp. $24.95
129 color photographs of interior decoration to the exterior finishing of houses.

A04-1384
Wilson, S., Jr. and Lemann, B. New Orleans Architecture. Volume 1, The Lower Garden District. Pelican, 1971. 172 pp. $12.50
The first of a five-volume series on the architectural treasures of this city. Volume 1 focuses on the Lower Garden District area of mid-19th century New Orleans, constructed primarily during the Golden Period and recognized today as one of the most comprehensive 19th century Greek Revival communities remaining in America. Mansions, country cottages, row houses and sophisticated urban dwellings are depicted, with detailed illustrations of significant features.

Design--Architectural

A04-1385
Winckel, F.   Music, Sound and Sensation: A Modern Exposition.
   Dover, 1966.   189 pp. $2.50
      Psychoacoustics: loudness, dissolution power of ear, influence of hall properties, function of time variation, sound spectrum, musical space, electroacoustic sound structure, and similar important modern topics. Non-technical.

A04-1386
Wingler, H. M.   The Bauhaus: Weimar Dessau Berlin Chicago.
   MIT Press, 1969.   696 pp. $55.00
      This is a documentary on the German School of design, from its beginnings in Weimar to its rebirth in Chicago. The material is organized chronologically and includes architectural plans and realizations, craft and industrial model designs, stained glass, etc. of such masters as Gropius, Hannes Meyer and Mies van der Rohe.

A04-1387
Witney, D.   The Barn: A Vanishing Landmark in North America.
   N.Y. Graphic Society, 1972.   256 pp. $25.00
      A pictorial history of the American barn, its architecture, construction methods and materials from the 17th century onward. 413 illustrations.

A04-1388
Wittkower, R.   Gothic vs. Classic: Architectural Projects in Seventeenth Century Italy.   Braziller, 1974. $12.50 (paperback $4.95)
      A study of the Gothic during the three centuries from Alberti to the Enlightenment by a famous architectural historian.

A04-1389
Wolgensinger, B.   Vacation Homes in the Sun/Ferienhaeuser in der Sonne.   International Publications, 1971.   160 pp. $20.00
      Presents 34 homes--mainly by Italian, Greek, Spanish and French architects--in photos showing the exterior and interior design. Floorplans and descriptive text with every house.

A04-1390
Wolgensinger, B.   Vacation Houses of Europe.   C. E. Tuttle, 1960. $15.00

A04-1391
Wood, A. B.   Acoustics.   Dover.   594 pp. $5.00
      Advanced text offers broad coverage of sound: wave motion, forced vibrations, resonators, reflection, refraction, the ear, pitch, intensity, and similar topics. Basic work for physicists.

A04-1392
Woods, S.   Candilis-Josic-Woods: Building for People.   Praeger, 1968.   226 pp. $20.00
      These three architects have been working together since

1951 in the area of collective and low-cost housing. Aware of the social implications of their buildings, they work within a framework of town planning. This book demonstrates some of the views and working methods they have developed. Each project, whether building or city, is treated in terms of function, architectural elements, and over-all pattern and structure.

A04-1393
Wright, F. L.   An American Architecture.   Horizon Press.
     $17.50
     Ranging over a lifetime of building and writing, every chapter in this volume is devoted to a major aspect of the architect's work. Illustrated with 250 plans, photos, and drawings.

A04-1394
Wright, F. L.   The Drawings of Frank Lloyd Wright.   Horizon Press. $17.50
     This collection privileges us to watch Frank Lloyd Wright at work through all the renowned structures that have shaped the course of architecture.

A04-1395
Wright, F. L.   The Early Work.   Horizon Press. $17.50
     This is a new edition of one of the rarest of all the books ever published by and about Frank Lloyd Wright. The first book (coming after the portfolio sheets in 1910) to exhibit the architect's revolutionary work.

A04-1396
Wright, F. L.   The Future of Architecture.   Horizon Press.
     $12.50
     This survey contains, complete, several rare works long out of print: The Princeton Lectures, The Chicago Lectures, The London Lectures, and several later works, including the "Conversations." Illustrated.

A04-1397
Wright, F. L.   Genius and the Mobocracy.   Horizon Press, 1971.
     $20.00
     This book is an account of Wright's relationship with Louis Sullivan, with whom he began his career.

A04-1398
Wright, F. L.   The Industrial Revolution Runs Away.   Horizon Press. $40.00
     First published in 1932 and almost immediately thereafter out of print, the writing of this book was an act of prophecy. Regarded as the classic indictment of the City, it is a protest against the inhumanity that surrounds us in rampant urbanism. The book reproduced here is the copy in which his handwritten revisions were made, and each facsimile page is gaced by the newly set text.

Design--Architectural 235

A04-1399
Wright, F. L. The Living City. Horizon Press. $12.50
    In this book, Frank Lloyd Wright embodied all his ideas on the ideal city, and illustrated them with many new drawings especially made for this volume.

A04-1400
Wright, F. L. The Natural House. Horizon Press. $8.95
    This book brings together for the first time all of Mr. Wright's writing on low-cost and moderate-cost houses. It also gives a step-by-step account of the method which enables the potential home owner to build his own Frank Lloyd Wright house.

A04-1401
Wright, F. L. An Organic Architecture: The Architecture of Democracy. MIT Press, 1970. 88 pp. $10.00
    This book is a verbatim text of four talks given by Frank Lloyd Wright to some young British architects in May, 1939. His emphasis in these lectures was upon a universalized concept of organic architecture.

A04-1402
Wright, F. L. A Testament. Horizon Press. $17.50
    The architect tells his own story. 210 photographs, plans and original drawings, from 1888 through the 70 years of his work. Included, in a foldout four-pages high, is Mr. Wright's detailed presentation of the 20th-century miracle of building: The Mile-High Skyscraper.

A04-1403
Wright, F. L. The Work of Frank Lloyd Wright (The Wendingen Edition). Horizon Press. $25.00
    Identical with the original but now printed at lower cost on both sides of the paper, this work becomes available to a larger audience. 200 photographs, drawings and plans.

A04-1404
Wurman, R. S. Various Dwellings Described in a Comparative Manner. MIT Press, 1972. 64 pp. $4.95
    The drawings on display here span the variety of human habitations over many periods, in many places. They range in time from an Egyptian house of 2000 B.C. to a Louis Kahn dwelling of the 1960s, in place from India to the Yucatan, and in elaboration from a simple Norwegian farmhouse to the Chateau de Chambord. The pictured buildings are all drawn to the same scale, 1:384 or 1 inch to 32 feet, so that the relative sizes of the various dwellings can be directly apprehended.

A04-1405
Wurman, R. S. and Chermayeff, I. The Design Necessity. MIT Press, 1973. 80 pp. $6.95
    The book documents these assertions: there are proven criteria for judging design effectiveness; design is a requirement, not

a cosmetic addition; design can save money and time; it enhances communication; it simplifies use, manufacture and maintenance; the design necessity is present in projects ranging in scale and complexity from a postage stamp to a highway system; the absence of design is a hazardous kind of design on any given project, designers and government officials have the same basic goal--performance.

A04-1406
Wurman, R. S. and Feldman, E. The Notebooks and Drawings of Louis I. Kahn. 2d ed. MIT Press, 1974. 92 pp. $17.50
This is a collection of statements and drawings by a master architect. The book consists of two sections: the first is a group of sketches produced during his European travels, reproduced actual size; the second incorporates rough sketches and finished renderings of some of Kahn's buildings and visions. The text is largely based on reworked versions of addresses and talks delivered by Kahn.

A04-1407
Yamamoto, T., et al., eds. Japan's Contemporary Houses. Wittenborn, 1970. 176 pp. $18.00
English text, 80 houses, each documented, ground-plans with 3 photographs each. Showing recent Japanese houses and interiors.

A04-1408
Yarwood, D. The Architecture of Europe. Architectural Book Publishing, 1973. 632 pp. $18.95
This architectural reference work gives an account of European architecture from its earliest development to the 1970's, and is illustrated with photographs and 1093 drawings by the author.
All the major historical styles are examined, from the ancient civilizations, through early Christian, Romanesque and Gothic building, on to the Renaissance, Mannerist, Baroque, Rococo and Neoclassical, the study being rounded off with a survey of nineteenth-century Eclecticism and twentieth-century Modernism.

A04-1409
Yoshida, T. The Japanese House and Garden. Praeger, 1969. 224 pp. $15.00
Aided by nearly 200 photographs, this volume shows how the traditional Japanese house is an integrated organism--in which the house fits naturally into its garden and the garden, a miniature natural landscape, enters the house. The author describes how materials are chosen and used to provide a natural, harmonious environment; he also deals with construction details, emphasizing the ancient Japanese modular technique derived from the standard-sized "mat" that is the basis for room measurement.

A04-1410
Zevi, B. Architecture as Space: How to Look at Architecture. Horizon Press, $15.00 (paperback $8.95)

Design--Architectural

This is a newly revised and enlarged edition of Bruno Zevi's work in the esthetics of architecture. Many illustrations.

A04-1411
Zucker, P. Town and Square from the Agora to the Village Green. 1959. MIT Press, 1970. 416 pp. $4.95
"This book tries to develop the history and aesthetics of the artistically shaped void, which finds its most characteristic form in the square, in the plaza, the focal point in the organization of the town."

A04-1412
Zuk, W. and Clark, R. H. Kinetic Architecture. Van Nostrand Reinhold, 1970. 160 pp. $14.95
This volume offers a working knowledge of emerging, evolutionary concepts and ideas in architecture which are flexible to the speed, scale, and nature of change today.

Section   A05

DESIGN: CIVIL AND STRUCTURAL

        Computer-Aided Design
        Hydraulics
        Pollution, Environmental Impact
        Seismic and Windstorm Design
        Soil and Rock Mechanics

A05-1450
Aalami, B. and Williams, D. G. Thin Plate Design for Transverse Loading. John Wiley, 1975. 176 pp. $19.50
 Comprehensive presentation of design data, enabling designers to check deflection and stresses in greater detail. Describes the phenomena, derives the basic equations, gives examples of the applications and specific design studies.

A05-1451
Abbett, R. W. American Civil Engineering Practice. John Wiley. Vol. 2, 1956. 917 pp. $30.25; Vol. 3, 1957. 1282 pp. $36.75

A05-1452
Abeles, P. W. An Introduction to Prestressed Concrete. Cement and Concrete Assn. Vol. 1, 1964. 384 pp. £3.00; Vol. 2, 1966. 358 pp. £3.60

A05-1453
ACI. Analysis of Structural Systems for Torsion. American Concrete Institute, 1973. 438 pp. $12.75 (for ACI members, $9.75)
 Evaluation and summarization of current information about analysis of torsional forces. Included is a "State of the Art" report and several practical papers introducing simplified analytical and numerical procedures for torsion analysis of commonly used structural systems.

A05-1454
ACI. Computer Applications in Concrete Design and Technology. American Concrete Institute, 1967. 144 pp. $10.00 (available from University Microfilm)
 The eight papers of this symposium volume deal with the use of computers in design, analysis, and detailing of concrete structures, and in proportioning and quality control of concrete mixes.

A05-1455
ACI. Concrete Thin Shells. American Concrete Institute, 1971. 442 pp. $15.75 (for ACI members, $11.75)
 The intent of this symposium volume is to aid the architect, engineer, and contractor in achieving that high level of competence and interdisciplinary collaboration which, due to the inherent intricacy of shell construction, is essential. Specific papers cover:

Cylindrical Shells and Folded Plates; Hyperbolic Paraboloids; Cable-Supported Concrete Shells; Specialized Computer Programs and many more ... Conceptual Design and Construction; Historical View of Shell Construction; Outstanding Recently Completed Examples; Problems of Analysis and Design.

A05-1456
ACI. Cracking, Deflection, and Ultimate Load of Concrete Slab Systems. American Concrete Institute, 1971. 388 pp. $15.75 (for ACI members, $11.75)

Sixteen papers are contained in this volume, the majority of which were delivered at the International Symposium on the Cracking, Deflection, and Ultimate Load of Concrete Slab Systems held at the Annual ACI Convention in Denver, 1971. The theme of this determined by crack control in one-way and two-way slab systems, sheer strength and moment transfer in column regions and concentrated load zones, and limit capacity of concrete slabs and plates, including membrane and arch action effects.

A05-1457
ACI. Deflections of Concrete Structures. American Concrete Institute, 1974. 637 pp. $13.00 (for ACI members, $10.00)

Covers current information and the state-of-the-art of deflection calculations. Includes an overview of the treatment of deflection evaluation in a number of codes, a review of one-way floor systems, two-way slab systems, prestressed flexural members, simple calculation methods, and complete analysis procedures.

A05-1458
ACI. Designing for Effects of Creep, Shrinkage and Temperature in Concrete Structures. American Concrete Institute, 445 pp. $15.75 (for ACI members, $11.75)

Much has been written on the damaging failures resulting from designs without adequate provision for the effects of creep, shrinkage and temperature. The object of this symposium volume is to combine the laboratory and actual field experience into a working procedure for estimating the effects of volumetric changes. By comparing the experiences of researchers and design engineers this publication aims for an eventual mathematical procedure that could be as well agreed upon as moment distribution. 21 pages.

A05-1459
ACI. Fatigue of Concrete. American Concrete Institute, 1974. 350 pp. $11.00 (for ACI members, $8.25)

Covers the effects of age, moisture, condition, and the rate of loading on the fatigue of strength of plain concrete. Includes research reports on fatigue properties of rebar, welded wire fabric, and reinforced concrete.

A05-1460
ACI. Impact of Computers on the Practice of Structural Engineering in Concrete. American Concrete Institute 1972. 320 pp. $15.75 (for ACI members, $11.75)

Design--Civil and Structural

This volume examines computer systems, typical design applications and computer graphics applications. The papers provide an overview of the contribution computers are making and could be making to the practice of structural engineering in concrete. This volume includes 17 papers covering: proportioning of concrete mixes; structural analysis of slabs, mass concrete, and machine foundations; design optimization; design of buildings, bridges, and beams; and an appendix listing computer program information available through ACI.

A05-1461
ACI. Models for Concrete Structures. American Concrete Institute, 1970. 448 pp. $15.00 (for ACI members, $12.00)

Seventeen papers dealing with all aspects of modeling concrete structures. Emphasis is placed on modeling the true, inclastic behavior of concrete structures. Includes annotated bibliography.

A05-1462
ACI. Probabilistic Design of Reinforced Concrete Buildings. American Concrete Institute, 1971. 260 pp. $12.50 (for ACI members, $9.50)

This volume contains seven papers and one committee report on applications of probabilistic principles to design of reinforced concrete structures. The first five papers were presented in a Symposium on Probabilistic Design as a part of the Annual Convention of the American Concrete Institute in Denver, Colorado, March 12, 1971.

A05-1463
ACI. Realism in the Application of ACI Standard 214-65. American Concrete Institute, 1973. 224 pp. $9.00 (for ACI members, $6.50)

The statistical concepts of ACI Standard 214-65 are explored. The first section deals with the use and understanding of ACI 214-65. Additional papers report on practical examples and the final discusses various types of computer equipment that can be used to aid in the analysis of strength test data.

A05-1464
ACI. Reinforced Concrete Design Handbook. American Concrete Institute, 1965. 271 pp. $4.50 (for ACI members, $3.00)

Explains methods for mastering the design of flexural members, stirrups, columns, square spread footings, and pile footings. Tables cover a wide range of working stresses encountered in general practice. It reduces the design of members under combined bending and axial load to the same simple form as that used in common flexural problems.

A05-1465
ACI. Response of Multistory Concrete Structures to Lateral Forces. American Concrete Institute, 1972. 320 pp. $12.00 (for ACI members, $9.00)

Selected papers dealing with the structural analysis and design of tall reinforced concrete buildings. Subjects covered include: frame-shear wall systems; framed tube structures; optimization, load distribution analysis and symmetric shear-wall analysis. Four papers deal with various aspects of the problems of constructing tall buildings in seismic areas.

A05-1466
ACI. Shear in Reinforced Concrete. American Concrete Institute, 1974. 2 vols. 965 pp. $19.50 (for ACI members, $14.50)
Includes the latest data and concepts on shear strength. 39 papers by authors from 10 countries covering the basic mechanics of shear transfer, and shear in medium-slender to slender beams. Also covered is shear in deep beams, brackets, walls, columns, and slabs. Descriptions of experimental and analytical studies of mechanisms by which shear is transferred in beams are presented.

A05-1467
ACI. Symposium on Reinforced Concrete Columns. American Concrete Institute, 1966. 378 pp. $20.00 (available from University Microfilms)
Selected papers dealing with experimental and analytical studies of reinforced and prestressed concrete columns, are contained in this volume. Compilation of the papers has been carried out under the sponsorship of the joint ACI-ASCE Committee 441, in accordance with the committee's three main objectives: (1) to review and correlate information on the analysis and design of concrete columns, including the methods required by various building codes; (2) to encourage experimental research on column behavior; (3) to propose design criteria for concrete columns.

A05-1468
ACI. Ultimate Strength Design Handbook. Vol. 1. American Concrete Institute, 1973. 432 pp. $14.00 (for ACI members, $10.50)
Prepared to save time and effort for those who design reinforced concrete structural elements using ultimate strength theory in accordance with the ACI 318-71 Building Code. There are design examples showing specific applications of ultimate strength design based on aids presented in this book; design aids-tables, graphs, and drawings which simplify shear, reinforcement, flexure, deflection and other calculations; and commentary presenting the assumptions, equations, derivations, and explanations for all the design aids.

A05-1469
ACI. Ultimate Strength Design Handbook. Vol. 2, Columns. American Concrete Institute, 1970. 226 pp. $9.00 (for ACI members, $6.00)
This volume replaces SP-7, Ultimate Strength Design of Reinforced Concrete Columns, and is based on the ultimate strength design provisions in Building Code Requirements for Reinforced Concrete (ACI 318-63). The tables and charts presented are for

Design--Civil and Structural

use in design of eccentrically loaded columns by the ultimate strength method, in accordance with Part IV-B of ACI 318-63. Spiral bound to lie flat.

A05-1470
AISC. Plastic Design in Steel. American Institute of Steel Construction. 104 pp. $6.00

Plastic design, an improved method for designing steel structures, is based upon the behavior of continuous steel structures loaded beyond the yield point. Studies show that savings of as much as 15 to 20 percent can be realized in the steel requirements for structures designed by this method as compared with those based on elastic analysis. This new book includes a discussion of methods of analysis and modifying factors, design examples based upon the rules for plastic design adopted by AISC, and numerous charts and formulas.

A05-1471
AISC. Plastic Design of Braced Multistory Steel Frames. American Institute of Steel Construction, $7.00

This book is basically a design manual. Enough theory is presented for the engineer to understand the structural behaviour involved, but the focus is on practical design aspects. A large portion of the book is devoted to the design of a 24-story, three-bay, braced steel apartment house frame. This design example will serve as a guide to the practicing engineer as he applies the principles of plastic design to his own work. Design aids are included to reduce design time and simplify calculations.

A05-1472
Allen, A. H. Safe-load Tables for Solid Slabs. Cement and Concrete Assn., 1973. 56 pp. £1.50

A05-1473
Allen, H. G. Analysis and Design of Structural Sandwich Panels. Pergamon, 1969. 300 pp. $7.00

Deals with the analysis, design and testing of structural sandwich beams, struts and flat panels. Analyses are given for bending and buckling problems in the elastic range, including a thorough study of wrinkling.

A05-1474
Allgood, J. R. and Swihart, G. R. Design of Flexural Members Static and Blast Loading. American Concrete Institute, 1970. 155 pp. $10.00 (available from University Microfilms)

Monograph examines blast-resistant structures and offers up-to-date methods to cope with them. It provides a direct design method and an explanation of the behavior of flexural elements, not available elsewhere in a single source.

A05-1475
Almen, J. and Black, P. Residual Stresses and Fatigue in Metals.

McGraw-Hill, 1963. 226 pp. $14.50

55 case histories and other experimental data support theories and principles discussed in the study of residual stresses and fatigue in metal.

A05-1476
Ambartsumyan, S. A. Theory of Anisotropic Plates: Strength, Stability, Vibration. Technomic, 1970. 248 pp. $7.00

A05-1477
American Water Works Assn. Water Quality and Treatment: A Handbook of Public Water Supplies. McGraw-Hill, 1971. 640 pp. $24.50

Leaders in all branches of sanitary engineering, water chemistry, and purification and treatment methods have pooled their knowledge and experience to make this handbook the acknowledged authority in its field. Treatment methods such as aeration, coagulation and flocculation, mixing and sedimentation, chlorination, tastes and odors, filtration, corrosion, softening, and fluoridation are discussed.

A05-1478
Andersen, P. Statically Indeterminate Structures--Their Analysis and Design. Ronald Press, 1953. 318 pp. $8.95

In contrast to most volumes on the subject of statically indeterminate structures, which are confined to the theorems of analysis, this book directly associates structural analysis with structural design. The majority of the design problems deal with steel structures because the separate courses in reinforced concrete usually include the application of indeterminate structures in that material. Throughout, great emphasis is placed on deflections and their importance in understanding the behavior of a structure under load.

A05-1479
Andersen, P. Substructure Analysis and Design. Ronald Press, 1956. 336 pp. $9.95

This book presents a comprehensive treatment of the analysis and design of those portions of a structure that are usually below the surface of the ground, of the water, or both. The subject is approached from the viewpoint of the designing engineer rather than that of the soil technician or the construction engineer.
Theory is developed largely by application to practical design problems, and methods and procedures are presented by which the engineer can develop safe and economical designs notwithstanding the uncertainties that exist in this kind of engineering.

A05-1480
Andersen, P. and Norby, G. M. Introduction to Structural Mechanics. Ronald Press, 1960. 340 pp. $10.95

A modern textbook for a civil engineering first course in structural analysis. In adopting the viewpoint that the "art" of structural design is built upon a firm scientific foundation, the

Design--Civil and Structural

authors have provided a sound basis for solution of the problems encountered in engineering practice. The book deals both with basic theory and with applications to traditional structures such as roof trusses, railway bridges, and highway bridges.

A05-1481
Anthrop, D. E. Noise Pollution. Lexington Books, 1973. 176 pp. $12.50

This study provides the information necessary for understanding the physics of noise and its measurement. Explanations of technical terminology as well as tables illustrating concepts such as the comparative levels of various types of noise: noise in dwellings, construction noise, motor-vehicle noise, and aircraft noise.

A05-1482
APWA. Control of Infiltration and Inflow into Sewer Systems. American Public Works Assn., 1970. 121 pp. $1.25 (available from Gov't. Printing Office, Order No. 11022EFF 12/70)

A survey by APWA of 212 public jurisdictions in the United States and Canada produced 20 recommendations for corrective action other than relief sewers or enlarged treatment plants.

A05-1483
APWA. Practices in Detention of Urban Stormwater Runoff. American Public Works Assn., 1974. 231 pp. $12.50 (for APWA members, $10.00)

Based on a study conducted for the Office of Water Research, U.S. Dept. of the Interior, this report covers such areas of stormwater management as local flooding, pollution from combined sewer overflows, and problems associated with the beneficial use of stormwater. For public works officials facing problems that could be alleviated through better stormwater management.

A05-1484
APWA. Problems of Combined Sewer Facilities and Overflows. American Public Works Assn., 1967. 189 pp. $1.00 (available from Gov't. Printing Office, Order No. WP-20-11)

The effects and means of correcting combined sewer overflows and separate storm and sanitary sewer discharges were inventoried nationally by APWA in 1967 and compiled in this report.

A05-1485
APWA. Water Pollution Aspects of Urban Runoff. American Public Works Assn., 1969. 272 pp. $1.50 (available from Gov't. Printing Office, Order No. WP-20-15)

A study of the causes and remedies of water pollution from surface drainage. Street refuse--litter--and catch basins were found to be potential major sources of pollution. The report evaluates the efficiency of street cleaning methods and the limitations of commonly used equipment.

A05-1486
Architectural Institute of Japan. Design Essentials in Earthquake Resistant Buildings. North-Holland, 1970. 294 pp. $22.70

Since Japan is seismically active all buildings in that country must be earthquake resistant. This book presents information obtained through experience gained from a long series of destructive earthquakes, including the unusual Niigata earthquake of 1964. It will contribute to the proper interpretation of structural design and calculation standards for earthquake resistant buildings and for a wide variety of construction operations in the building industry.

A05-1487
Architectural Institute of Japan, ed. Design Essentials in Earthquake Resistant Buildings. American Elsevier, 1970. 304 pp. $22.00

The present book, originally published in Japanese in 1966 by the Architectural Institute of Japan presents information obtained through experience gained from a long series of destructive earthquakes, including the unusual Niigata earthquake of 1964. It will contribute to the proper interpretation of structural design and calculation standards for earthquake resistant buildings and for a wide variety of construction operations in the building industry.

A05-1488
Argyris, J. H. and Kelsey, S. Energy Theorems and Structural Analysis. Butterworth, 1960. 88 pp. £3.00

A contribution to the subject of stress analysis, set out with diagrams and explanations of procedure.

A05-1489
ASCE. Bibliography on Bolted and Riveted Joints. American Society of Civil Engineers, 1967. $8.00 (for ASCE members, $4.00)

A05-1490
ASCE. Design of Cylindrical Concrete Shell Roofs. American Society of Civil Engineers, 1952. $5.00 (for ASCE members, $2.50)

A05-1491
ASCE. Fourth American Conference on Soil Mechanics and Foundation Engineering. American Society of Civil Engineers, 1971. 301 pp. $8.00

Includes keynote address, discussions, list of registrants and index.

A05-1492
ASCE. Fourth American Conference on Soil Mechanics and Foundation Engineering. Vol. 1. American Society of Civil Engineers, 1971. 366 pp. $8.00

Covers such subjects as standard penetration test, slope stability in residual soils, allowable settlement of structures,

Design--Civil and Structural

effects of foundation construction on nearby structures, and field deformation measurements.

A05-1493
ASCE. Fourth American Conference on Soil Mechanics and Foundation Engineering. Vol. 2. American Society of Civil Engineers, 1971. 402 pp. $8.00

Contains the 33 selected papers from the Conference and deals with subjects covered in Vol. 1, as well as the business and practice of foundation engineering.

A05-1494
ASCE. Guide for Design of Steel Transmission Towers. American Society of Civil Engineers, 1971. $4.00 (for ASCE members, $2.00)

This manual has been compiled to provide the engineer with specific loading recommendations for self-supporting steel transmission towers. It contains design formulas for use in structural analysis, fabrication and testing recommendations, foundation nomenclature, survey information, and construction methods. Also included are sections covering overload factors, slenderness ratios, detailing, fasteners, right-of-way preparation, erection, and full-scale testing.

A05-1495
ASCE. Outstanding Papers of Thomas R. Camp. American Society of Civil Engineers, 410 pp. $12.00 (for ASCE members, $6.00)

Twenty-four outstanding technical papers and discussions from various sources are arranged by subject in chronological order.

A05-1496
ASCE. Plastic Design in Steel-A Guide and Commentary. American Society of Civil Engineers, 1971. $10.00 (for ASCE members, $5.00)

A05-1497
ASCE. The Proceedings of the August 1972 International Conference on the Planning and Design of Tall Buildings. American Society of Civil Engineers, 1972. 5200 pp. $50.00 set.
Vol. 1, Introduction, Conference Proceedings, Miscellaneous Discussion, Project Descriptions, Abstracts, Indexes. Vol. 1a, Systems and Concepts. Vol. 1b, Criteria and Loading. Vol. 2, Structural Design: Steel. Vol. 3, Structural Design: Concrete and Masonry.

The 5200-page set of five hard-bound volumes is a treatise on the current state of the art for high-rise planning and design. The broad spectrum of design problems and methodology unique to modern skyscrapers is covered. Urban problems related to tall buildings are treated in the context of the future city as a more civilized place for realization of human potential.

A05-1498
ASCE. Proceedings of the Conference on Performance of Earth and Earth-Supported Structures. American Society of Civil Engineers, Vol. 1, 74 papers, 1555 pp. $15.00; Vol. 2, Session Leaders' Reports. 154 pp. $2.00; Vol. 3, Summaries, $5.00

The 1972 Soil Mechanics and Foundations Division Specialty Conference focused on the observed behavior of full-scale structures constructed of, or supported on, earth. Analytical tools and design procedures were examined and predictions of performance compared with field measurements. To this end, papers reporting the results of field measurements were solicited; the Conference Proceedings described the current state-of-the-art.

A05-1499
ASCE. Safety and Reliability of Metal Structures. American Society of Civil Engineers, 1972. 449 pp. $5.00

Contains the papers from the November 1972 Specialty Conference. Papers deal with loadings, strength of structural components, fracture and fatigue of structures, and design.

A05-1500
ASCE. Sewage Treatment Plant Design. American Society of Civil Engineers, 1959. $7.00 (for ASCE members, $3.50)

A05-1501
ASCE. Structural Failures: Modes, Causes, Responsibilities. American Society of Civil Engineers, 1972. $4.00 (for ASCE members, $2.00)

The papers in this book were presented at the National Structural Engineering Meeting of the ASCE in April 1972. To extract the maximum amount of information possible from structural failures, causes of failures, and responsibilities for structural failures. Causes of failures concern descriptions of structural reasoning based on pathological studies of failures. Responsibilities for failures concern parties involved with the structure from its inception to the time of its failure who may be professionally or legally liable for the failures.

A05-1502
ASCE. Survey of Current Structural Research. American Society of Civil Engineers, $8.00 (for ASCE members, $4.00)

A05-1503
Ashdown, A. J. The Design of Prismatic Structures. Cement and Concrete Assn., 1958. 85 pp. £0.45 Order No. 12.031

A05-1504
Ashton, J. E. and Whitney, J. M. Theory of Laminated Plates. Technomic, 1970. 153 pp. $7.00

Contents include the theoretical development and solutions for plates of anisotropic materials and a clear foundation in the theory of laminated plates, including problems of bending, stability, and free vibration.

Design--Civil and Structural 249

A05-1505
ASME. Computer Software in Structural Analysis. American Society of Mechanical Engrs., 1970. 32 pp. $6.00 (for ASME members, $4.80)
 Discusses the use of structural software in large and small installations, including special purpose programs, accessibility for the intermittant user, computer analysis by remote control and automated and graphical display of input and output.

A05-1506
ASME. General Purpose Finite Element Computer Programs. American Society of Mechanical Engrs., 1970. 152 pp. $15.00
 Explains the development and use of general purpose programs for finite element analysis. Gives the reader insight into the power of this new tool for design and analysis, and documents the current state of the art as a guide to the development of the next generation of general purpose programs.

A05-1507
ASME. 1970 National Incinerator Conference Proceedings. American Society of Mechanical Engrs., 1970. 330 pp. $25.00 (for ASME members, $20.00)
 Covers the more technical details of incinerator design, here and abroad.

A05-1508
ASME. Proceedings of 1968 National Incinerator Conference. American Society of Mechanical Engrs., 1968. 354 pp. $20.00 (for ASME members, $16.00)
 This volume contains forty-one technical papers which cover municipal and industrial incineration. Air and water pollution are given prominence. Several papers are devoted to foreign experience and practice. Also included are planning, general design factors, materials of construction and plant performance.

A05-1509
ASME. Structures and Materials. American Society of Mechanical Engrs., 1969. 471 pp. $22.00 (for ASME members, $15.00)

A05-1510
ASME. Thermal Structural Analysis Programs: A Survey and Evaluation. American Society of Mechanical Engrs., 1972. $8.50
 Outlines a format for thermostructural program evaluation. Categories include: method of analysis; scope or applicability of a particular computer program; special features and limitations; graphical adjuncts; documentation; and cost estimating. An attempt was made to make the subdivisions of the analysis thought sequential, providing the information in the order required by the qualified user.

A05-1511
ASME. Three-Dimensional Continuum Computer Programs for Structural Analysis. American Society of Mechanical Engrs., 1972. $6.00
A symposium of 19 summaries of three-dimensional continuum structural analysis computer programs.

A05-1512
Asplund, S. O. Structural Mechanics: Classical and Matrix Methods. Prentice-Hall, 1966. 474 pp. $16.95
Presents a modern course in the theory of (stable) structures, stressing matrix methods. Follows the latest techniques of structural analysis and stresses the fundamentals of mechanics. Employs matrix language throughout, starting with elementary problems. Defines structures by subdividing them into members of known flexibility.

A05-1513
ASTM. Achievement of High Fatigue Resistance in Metals and Alloys. American Society for Testing and Materials, 1970. 298 pp. $28.75 (for ASTM members, $23.00)
Covered in twelve papers are parameters important to mechanisms and processes for achieving high fatigue resistance. Useful for design engineers and metallurgists.

A05-1514
ASTM. Determination of the In Situ Modulus of Deformation of Rock. American Society for Testing and Materials, 1970. 200 pp. $19.25 (for ASTM members, $15.40)
Problems in rock engineering and construction involve either the strength of the in situ rock mass or the compressibility of the rock mass. Here in 11 papers are some thoughts concerning on site testing particularly applicable to dams, heavy bridge piers, high-rise structures and tunnels.

A05-1515
ASTM. Effects of Environment and Complex Load History on Fatigue Life. American Society for Testing and Materials, 1970. 332 pp. $22.00 (for ASTM members, $17.60)
This book contains 17 papers dealing with cumulative damage, random load effects, and effects of corrosive environments.

A05-1516
ASTM. Performance of Deep Foundations. American Society for Testing and Materials, 1969. 399 pp. $20.00 (for ASTM members, less 20%)
This book contains 16 papers discussing the analysis of the behavior of piles under both static and dynamic loads; correlation of pile-load-test data with soil data; controlled pile group tests; and measurement in the real structure of load deformation.

A05-1517
ASTM. Stress Analysis and Growth of Cracks. American Society

Design--Civil and Structural

for Testing and Materials, 1972. 307 pp. $27.50 (for ASTM members, $22.00)
Proceedings of the 1971 National Symposium on Fracture Mechanics. Part 1 contains 17 papers dealing with both stress analysis of cracks and subcritical flaw growth caused by repeated loads, environment and combinations.

A05-1518
ASTM. Vibration Effects of Earthquakes on Soils and Foundations. American Society for Testing and Materials, 1969. 267 pp. $18.50 (for ASTM members, $14.80)
A book covering the response of foundations and soils to earthquake problems. Nine papers give a cross section of current studies including foundation vibrations, design applications and dynamic soil behavior.

A05-1519
Au, T. Elementary Structural Mechanics. Prentice-Hall, 1963. 521 pp. $16.95
Basic concepts and principles of mechanics in the analysis of structures. Deals with statically determinate structures subjected to static or equivalent static loads. Treats the deformations as well as the internal forces of these structures.

A05-1520
Baker, A. L. L. Limit-State Design of Reinforced Concrete. Concrete Construction Publications. 345 pp. $16.50
This edition applies limit-state concepts, methods and criteria to the general principles and philosophy of design. Strength and stiffness of members subjected to bending, compatibility calculations for the analysis of structures, prestressed concrete, and design practice are dealt with. Slabs, columns and struts, shear and bond, secondary effects and others, as well as estimating, costing, and progress charts are included.

A05-1521
Baker, A. L. L. Limit-State Design of Reinforced Concrete. Cement and Concrete Assn., 1970. 360 pp. £4.00

A05-1522
Baker, A. L. L. Raft Foundations. Cement and Concrete Assn., 1965. 148 pp. £0.75

A05-1523
Baker, E. and Rish, F. Structural Analysis of Shells. McGraw-Hill, 1972. 384 pp. $29.95
This book provides instruction, procedures, basic solutions to facilitate the expedient static structural analysis of shells and an introduction to and reference for the practical analysis of shells. Emphasis is on the presentation of solutions and their applications.

A05-1524
Baker, J. and Heyman, J. Plastic Design of Frames. Vol. 1,

Fundamentals. Cambridge University Press, 1969. $12.50

A05-1525
Baker, J. F. Steel Skeleton. Vol. 1, Elastic Behaviour and Design, $17.50; Vol. 2, Plastic Behaviour and Design. Cambridge Univ. Press, 1954-56. $24.95

A05-1526
Baker, W. E., et al. Similarity Methods in Engineering Dynamics: Theory and Practice of Scale Modeling. Hayden, 1973.
This text and source book develops basic modeling techniques and gives an understanding of the transient response of structures, fluids, gases, and soils to mechanical and thermal impulses. It also uses experimental test data on thirty-five complex engineering problems to illustrate how and why similarity work.

A05-1527
Bakos, J. D., Jr. Structural Analysis for Engineering Technology. Charles Merrill, 1973. 336 pp.
A presentation of structural analysis for the elementary course in structural theory. Emphasizes analysis of statically determinate structures, but includes a chapter on statically indeterminate structures as basis for further study. Derivatives and sample problems give insight to both theory and practical applications.

A05-1528
Bares, R. Tables for the Analysis of Plates, Slabs and Diaphragms: Based on the Elastic Theory. Adler's Foreign Books, 1971. 626 pp. $40.00
Plates, slabs, diaphragms are terms variously used to denote a wide range of structural elements employed in most branches of engineering construction. The rapid and precise analysis of such elements, which may be subjected to loads acting transversely and/or parallel to their planes, is therefore of the greatest importance to the designer. Based on the elastic theory of structural behavior, this book contains a set of formulas and tables accompanied by examples illustrating their use.

A05-1529
Bares, R. and Massonet, C. Analysis of Beam Grids and Orthotropic Plates: By the Guyon-Massonnet-Bares Method. Frederick Ungar, 1968. 480 pp. $13.50
An explanation of the theory and practice of the design analysis of beam grids and orthotropic plates, with particular respect to applications in bridge construction. The stress distribution and the deformations of all systems of this kind are solved in essentially the same manner regardless of the structural material employed. The working procedures have been developed as an extension of the flexural theory of orthotropic plates in the case of beam grids and composite systems, and the analysis is based on the concept of lateral load distribution.

Design--Civil and Structural

A05-1530
Barkan, D. Dynamics of Bases and Foundations. McGraw-Hill, 1962. 496 pp. $19.95
   The book combines and examines the basics of soil mechanics, the elasticity theory, and the dynamics of vibratory motion to develop usable solutions for machine foundation design problems.

A05-1531
Bartlett, R. E. Pumping Stations for Water and Sewage. John Wiley, 1974. 150 pp. $12.00
   Contents--Types of Pumps, Water Supply and Land Drainage, Sewage Pumping Stations, Some Special Installations for Sewage, etc., Prime Movers, Pump Starters, Controls, and Other Accessories. Ancillary Equipment, Buildings, Pumping Mains, Specification and Testing, Operation and Maintenance.

A05-1532
Bathe, K. J. and Wilson, E. Numerical Methods in Finite Element Analysis. Prentice-Hall, 1976. 544 pp. $27.95
   This book explains the use of the numerical method in studying the behavior of structures. Geared for easy adaptation on the computer, this book will prove an invaluable reference source for civil and aeronautical engineers and those involved with the study of earthquakes. The author has divided the book into three sections. The first deals with matrix and linear algebra, the second explains the use of digital computers in implementing the numerical method and the third part presents the solutions of finite element equilibrium equations.

A05-1533
Beaufait, F. W., et al. Computer Methods of Structural Analysis. Prentice-Hall, 1970. 543 pp. $18.95
   Covers development of methods of structural analysis suited to computer procedures.

A05-1534
Beckett, D. Limit State Design of Reinforced Concrete Structures. John Wiley, 1974. 134 pp. $14.95
   Concise introduction for students and engineers to the application of limit state theory to the design of the more common forms of reinforced concrete construction, the slab, beam, and column.

A05-1535
Beckett, D. The Ultimate Load Design of Continuous Concrete Beams. Plenum Press, 1968. 116 pp. $15.00
   This work explores principal notations and metric conversions, and supplies recent information on the limitations of elastic design. Examining the behavior of reinforced concrete under load, this volume also reports on such vital topics as plastic design of concrete beams, shear and torsion, and factors of safety and serviceability.

A05-1536
Beedle, L. S.   Plastic Design of Steel Frames.   John Wiley, 1958.
    406 pp. $21.25

A05-1537
Benjamin, B. S.   Structural Design with Plastics.   Van Nostrand
    Reinhold, 1969.   288 pp. $15.00
    Examines the relatively few types of plastics that can be employed efficiently for structural purposes. Provides the necessary data for structural engineers to originate their own designs, and alerts them to the properties most desirable in plastics, so that they can evaluate the design possibilities of new plastic products appearing on the market.

A05-1538
Benjamin, J. and Cornell, C.   Probability, Statistics and Decisions for Civil Engineering.   McGraw-Hill, 1970.   640 pp. $19.50
    Realistic and common engineering examples involving economic decisions under conditions of uncertainty are stressed in this book which presents the principles of applied probability and statistics needed by civil engineers to optimize and understand the relation of analysis to practical engineering decisions.

A05-1539
Bennett, E. W.   Structural Concrete Elements.   John Wiley, 1973.
    314 pp. $16.50
    Partial contents--Notation, Mechanics of Structural Concrete Elements: Uncracked Members with Linear Deformation, Cracked Members with Linear Deformation, Members Subject to Time-Dependent Deformation under Normal Loading. Non-Linear Deformation and Failure.

A05-1540
Bickerdike, et al.   Design Failures in Buildings.   George Godwin
    Ltd., 1973.   32 pp.   £2.80
    These technical information sheets state why a detail has failed and what the correct solution should have been. Based on examples of actual failures in use, these information sheets will be of use to all concerned with providing trouble-free detailing in line with good building practice.

A05-1541
Biggs, J. M.   Introduction to Structural Dynamics.   McGraw-Hill,
    1964.   341 pp. $13.95
    Covers numerical analysis of simple systems, rigorous analysis of one-degree systems, lumped-mass multidegree systems, structures with distributed mass and load, earthquake analysis and design, and more.

A05-1542
Billington, D.   Thin Shell Concrete Structures.   McGraw-Hill,
    1965.   332 pp. $22.50

Design--Civil and Structural

Uses an understanding and analysis of structural behavior of such systems in a form suitable for the computer to develop structural design.

A05-1543
Blaszkowiak, S. and Kaczkowski, Z. Iterative Methods in Structural Analysis. Pergamon, 1965. 592 pp. $23.50
A survey of the applications of the method of successive approximations, known in structural analysis as the Cross method. Each chapter illustrated by numerical examples, and the book is supplemented with tables.

A05-1544
Bleich, F. Buckling Strength of Metal Structures. McGraw-Hill, 1952. 498 pp. $16.95
Examines all current factual material on the stability of metal structures for engineers, naval architects, and others concerned with the analysis and design of compression members.

A05-1545
Borg, S. F. Fundamentals of Engineering Elasticity. Van Nostrand Reinhold, 1962. 276 pp. $8.95

A05-1546
Borg, S. F. and Gennaro, J. J. Modern Structural Analysis. Van Nostrand Reinhold, 1969. 320 pp. $12.95
A presentation of the merging of the modern methods of analysis with classical and historic methods. Slope-deflection and stiffness matrix methods of computation as well as moment distribution method. Computer analysis is included.

A05-1547
Borrego, J. Space Grid Structures and Stressed Skin Systems. Wittenborn, 1972. 208 pp. $12.50 (paperback $4.95)
The applications both in spanning large spaces in roof construction as well as in floor construction, where services can be integrated within the structural depth.

A05-1548
Borrego, J. Space Grid Structures: Skeletal Frameworks and Stressed Skin Systems. MIT Press, 1968. 192 pp. $12.50
Part One considers models, experimental projects, and permanent structures actually built. These include two and three-way true space grids, two and three-way lattice space grids, and special projects. Part Two is concerned with stressed skin space grids, including tetrahedral, pentahedral, and hexagonal pyramids, elongated tetrahedra, and space trusses. Part Three consists of a pictorial catalogue of space grid geometries, two-way and three-way, true and lattice. The networks are shown both isometrically and from the top. In addition, the top layer, the bottom layer, and the interconnecting vertical and diagonal webbing are diagramed.

A05-1549
Bowes, W. H. and Russell, L. T. Stress Analysis by the Finite Element Method for Practicing Engineers. Lexington Books, 1975. 224 pp. $16.95

Unlike other works in this area, this book develops the theory for the solution to real stress problems and attempts to develop an understanding of the physical significance of each step in the theory. In the seven main programs presented, theory is developed, flow charts are shown, sample programs are given, and the FORTRAN statements are listed. Tables, figures, appendix, index.

A05-1550
Bowles, J. E. Analytical and Computer Methods in Foundation Engineering. McGraw-Hill, 1974. 503 pp. $14.50

This volume covers the design and/or analysis of foundation engineering structures while placing equal emphasis on both fps and SI units which are also included in examples in every chapter. Computer programs in FORTRAN IV are part of each major topic with program description, variable identification, and program limitations given for every computer program. Methods of problem checking are displayed. The finite element method is extensively utilized and current codes are used.

A05-1551
Bowles, J. E. Engineering Properties of Soils and Their Measurements. McGraw-Hill, 1970. 192 pp. $11.95

This book offers a complete series of standard soil mechanics laboratory tests with each step of each test procedure spelled out in careful detail. The theoretical treatment is brief, but pertinent.

A05-1552
Bowles, J. E. Foundation Analysis and Design. McGraw-Hill, 1968. 672 pp. $21.00

A discussion of soil mechanics and structural design--soil exploration methods, refined bearing capacity and settlement computations, and more.

A05-1553
Boyes, F. G. H. Structural and Cut-Off Diaphragm Walls. John Wiley, 1975. 181 pp. $22.50

The development of the technique of slurry stabilization method of excavation and the equipment used in construction are traced and detailed descriptions are provided of the various methods used in forming the completed wall or foundation structure. The difficulties and procedures associated with the stages of construction are explained in detail.

A05-1554
Brater, E. F. and King, H. Handbook of Hydraulics. McGraw-Hill, 1976. 608 pp. $24.50

This new edition features tables, particularly those for pipe

Design--Civil and Structural

and open channel problems, that cut solution time for these problems. Also included in this edition is a new chapter covering numerical methods and the use of computer programming to solve hydraulic problems. Hydrostatics, flow through gates, orifices and culverts, weirs, and flow in pipes and open channels are among the many subject areas discussed with extra coverage given to special hydraulics areas such as water hammer, unsteady flow, flow in high velocity, transition flow measurement, and coastal hydraulics.

A05-1555
Brebbia, C. A. and Connor, J. J. Fundamentals of Finite Element Techniques: For Structural Engineers. John Wiley or Butterworth, 1974. 259 pp. $15.50

An introductory text covering basic equations of linear elasticity; principles of virtual displacement and virtual force; links between the conventional matrix structural analysis approach and finite element methods; application to plane stress; three-dimensional bodies; plate bendings; and shell structure. Introductory subjects on matrix algebra, solid mechanics (strength of materials), and elementary structural analysis are prerequisites.

A05-1556
Breneman, J. Strength of Materials. McGraw-Hill, 1965. 200 pp. $6.95

Helps the technical student understand the fundamentals of the subject within the total engineering design. Treats the design of simple structures, beams, shafts, and columns.

A05-1557
Bresler, B., ed. Reinforced Concrete Engineering. Vol. 1, Materials, Structural Elements, Safety. John Wiley, 1974. 529 pp. $23.75

A reference work for structural engineering designers, researchers, and advanced students in which authorities summarize latest developments in the behavior of reinforced concrete structures.

A05-1558
Bresler, B., et al. Design of Steel Structures. John Wiley, 1968. 830 pp. $22.75

The emphasis in this text is on a rational approach to design based on the knowledge of the fundamental principles of structural mechanics, on the understanding of the behavior of actual and idealized structures, and on the appreciation of practical requirements including safety, feasibility, and economy.

A05-1559
British Geotechnical Society. Settlement of Structures. John Wiley, 1975. 841 pp. $65.00

Contains the proceedings of a conference organized by the British Geotechnical Society held at Cambridge University. Each session includes a state of the art paper covering existing knowledge.

A05-1560
Brown, R. J. E. Permafrost in Canada. Univ. of Toronto Press, 1970. 234 pp. $12.50
An analysis of permafrost--its origin, definition and occurrence, and the effect it has on industry and agriculture.

A05-1561
Browne, J. S. C. Basic Theory of Structures. Pergamon, 1966. 256 pp. $6.95
For students in colleges of technology this book supplies the needs of those in the initial stages of theory of structures courses.

A05-1562
Brush, D. O. and Almroth, B. O. Buckling of Bars, Plates and Shells. McGraw-Hill, 1975. 379 pp. $21.00
The methods of structural stability analysis are discussed within the context of elementary nonlinear bending analysis. Equilibrium and stability equations for structural members are systematically developed. The structural members considered are columns, beams, beam-columns, flat plates, circular rings, cylindrical shells, and general shells.

A05-1563
Bryan, E. R. The Stressed Skin Design of Steel Buildings. John Wiley, 1973. 160 pp. $13.50
This book outlines an analysis of clad buildings, describing the profound effect cladding can have on the behavior of buildings and how the effect may be simply and logically allowed for in design. This volume deals with research carried out over the last 12 years and sets down the principles of stressed skin construction as applied to single-bay, single-story portal frame buildings. It also shows how the method leads to a considerable economy of steel compared with existing design methods.

A05-1564
Bull, F. B. and Sved, G. Moment Distribution in Theory and Practice. Pergamon, 1964. 304 pp. $4.95
Conventional processes of moment distribution and several new methods are discussed in this book for university and college of technology students.

A05-1565
Byars, E. F. and Snyder, R. D. Engineering Mechanics of Deformable Bodies. Intext, 1969. 444 pp. $12.50
This text on strength of materials covers basic concepts of stress and strain mechanical properties of engineering materials, types of loads and statics, analysis of simple structure elements, and design considerations.

Design--Civil and Structural

A05-1566
Calcote, L. R. The Analysis of Laminated Composite Structures. Van Nostrand Reinhold, 1969. 208 pp. $12.00
  The implications and possibilities for the future application of composite structural techniques are investigated and the newly evolving technology of composite material is analyzed in this text.

A05-1567
Carpenter, S. T. Structural Mechanics. John Wiley, 1960. 538 pp. $15.00

A05-1568
Cassie, W. F. Fundamental Foundations. Applied Science, 1968. 226 pp. £4.00
  The book covers the problems of foundations from the architect's point of view. The fundamentals of the subject, the problems of bearing capacity and settlement are covered.

A05-1569
Castigliano, C. A. The Theory of Equilibrium of Elastic Systems and Its Applications. Dover, 1966. 399 pp. $4.50
  Major treatise (1879) by a pioneer, and still a penetrating exposition of structural mechanics. Part I: theory (general equations of elastic equilibrium of a solid, theory of straight beams, columns, curved ribs, etc.); Part II: applications (beams strengthened by trusses and tie-bars, various roof constructions, bridge problems, etc.).

A05-1570
Cernica, J. N. Strength of Materials. Holt, Rinehart and Winston, 1966. 416 pp. $12.50
  For the basic course in Strength of Materials, this book develops the fundamental relationship between the applied forces and the internal effects. Fundamental definitions, equations, and derivations.

A05-1571
Charlton, T. M. Model Analysis of Plane Structures. Pergamon, 1966. 124 pp. $3.50
  For final year students of civil and structural engineering in universities and colleges of technology as well as for practising engineers concerned with design and analysis of structures.

A05-1572
Chellis, R. Pile Foundations. McGraw-Hill, 1961. 704 pp. $21.50
  The relationship between borings, soil mechanics, and pile foundations are covered. Includes factors determining pile capacities, types, protection and repair, and more.

A05-1573
Chilver, A. H., ed. Thin-Walled Structures. John Wiley, 1967. 303 pp. $10.00

Contents include buckling of frames composed of thin-walled members, flexural buckling of thin-walled elastic bars, unsymmetrical beams and axially-loaded thin-walled struts. Buckling of plates under various loadings, including decking systems, are an important part of the book.

A05-1574
Chou, P. C. and Pagano, N. Elasticity: Tensor, Dyadic and Engineering Approaches. Van Nostrand Reinhold, 1967. 312 pp. $10.50

In this text, equations and principles are first explained in a one-dimensional or two-dimensional case, and later in the more general three-dimensional case. Then a more general treatment is provided through Cartesian tensor and dyadic notations.

A05-1575
Chow, V. Handbook of Applied Hydrology. McGraw-Hill, 1964. 1,468 pp. $45.00

Presentation of the principles and data required for the study and management of water and water resource projects. Discusses the fundamental sciences underlying hydrology and the hydrologic cycle, with due attention to applications in different areas, and flood and flow control. The handbook surveys legal, economic, and political factors and gives guidance on planning, designing, and developing water resource systems and projects.

A05-1576
Clough, R. and Penzien, J. Dynamics of Structures. McGraw-Hill, 1975. 576 pp. $15.95

This book provides a comprehensive treatment of the theory of structural dynamics and its application to the solution of practical problems arising from the action of dynamic loads on structures. General methods, applicable to any kind of structure from buildings and bridges to airplanes, spacecraft, and ships are fully covered. Extensive treatment is given to deterministic analysis theory and applications. A special section is included on aspects of seismology and earthquake engineering.

A05-1577
Conner, J. Theory of Structural Member Systems. Ronald Press, 1975. 700 pp.

This text contains a treatment of the structural mechanics of a member element and matrix formulation of member systems. Its central theme is to derive the governing equations starting with the basic concepts of equilibrium, deformation, and material behavior, and then show how solution schemes such as the "stiffness" and "flexibility" methods evolve naturally by manipulating the governing equations.

A05-1578
Connolly, W. Design of Prestressed Concrete Beams. McGraw-Hill, 1960. 264 pp. $13.50

Gives a simplified method for the design of pretensioned and posttensioned concrete members.

Design--Civil and Structural

A05-1579
Cook, R. D. Concepts and Applications of Finite Element Analysis: A Textbook for Beginning Courses in the Finite Element Method, as Used for the Analysis of Displacement, Strain and Stress. John Wiley, 1974. 402 pp. $18.75
Contents--The Stiffness Method and the Plane Truss. Potential Energy and the Rayleigh-Ritz Method. Use of Assumed Displacement Fields. The Isoparametric Formulation. Some Modifications of Elements. Solids of Revolution. Thin Shells of Revolution. Bending of Flat Plates. General Shells. Coordinate Transformation. Dynamics and Vibrations. The Initial Stress Stiffness Matrix and Linear Stability. Geometric Nonlinearity. Material Nonlinearity. Detecting and Avoiding Errors.

A05-1580
Coull, A. and Dyke, A. Fundamentals of Structural Theory. McGraw-Hill, 1972. 292 pp. $14.00
This basic book treats fundamental aspects of structural theory and illustrates it by numerous examples based on different structural forms. The virtual work concept is used throughout, and SI units are used in all numerical examples.

A05-1581
Cowan, H. J. Architectural Structures: An Introduction to Structural Mechanics. American Elsevier, 1971. 400 pp. $23.80
This textbook examines the entire range of available structures and the fundamental theory structures. Frequent references are made to the physical behaviour of structures. Included are problems and solutions, and appendices containing design aids for the student, and tables for the conversion of foot-pound units to metric units.

A05-1582
Cowan, H. J., et al. Models in Architecture. Applied Science, 1968. 228 pp. £5.00
This book describes the purpose of using models for the architectural and engineering design of buildings, the techniques for making models and the methods of testing models. Evaluation of numerical results by dimensional theory is derived, with examples, for structural models (including the effects of wind and earthquakes, of thermal stresses, etc.), for thermal, acoustical and lighting models. Special attention is paid to the approximations in theory, construction and testing which cause minor, admissible errors, and those which cannot be tolerated for adequate design.

A05-1583
Crawley, S. W. and Dillon, R. M. Steel Buildings: Analysis and Design. John Wiley, 1970. 397 pp. $18.75
Discusses and designs each element in a steel building individually and the total structure. The authors examine the elements of structural design and the theory behind design decisions. Stresses structural performance, safety, and economy.

A05-1584
Croll, J. G. A. and Walker, A. C. Elements of Structural Stability. John Wiley, 1972. 400 pp. $15.00
  Presents a systematic classification of the buckling behavior and analysis of thin structures. Emphasizes the role of simple mechanical models and avoids complex mathematics.

A05-1585
Cross, F. L., Jr. and Noble, G. Handbook on Hospital Solid Waste Management. Technomic, 1973. 114 pp. $15.00
  This is an analysis of hospital solid waste problems and the various systems, techniques, and equipment now available to deal with them. Collection, processing, and disposal systems are examined along with economic factors for each alternative. Directory of equipment manufacturers.

A05-1586
CRSI Design Handbook--Working Stress Design. Concrete Reinforcing Steel Institute, 1968. 650 pp. $5.00
  Based on 1963 Building Code. Complete with safe load tables for columns, footings, walls and floor systems. $f'_c = 3,000$ psi; 3,750 psi; 5,000 psi.

A05-1587
CRSI Handbook. Concrete Reinforcing Steel Institute, 1973. 750 pp. $15.00
  Time saver and problem solver. Based on 1971 ACI Code "Strength Design." Eliminates algebraic formulas and calculations. Read in values for load and span (or load and moment) and read off complete designs directly from tables. For design of beams, slabs, joists, footings, and retaining walls; $f'_c = 3000$ and 4000 psi; $f'_y = 60,000$ psi. For columns: $f'_c = 4000$, 5000, and 6000 psi; $f'_y = 60,000$ psi. All for normal weight concrete and Grade 60 rebars.

A05-1588
CRSI Handbook-Ultimate Strength Design. Concrete Reinforcing Steel Institute, 1970. 610 pp. $3.50
  1963 ACI Building Code $f_y = 60,000$ psi. Chapters 1-7: Columns and footings with $f'_c = 3,750$ psi and 5,000 psi. Chapters 8-15: One-way solid and joist slabs, solid and waffle flat slabs, beams. Theory and practical economics of splices.

A05-1589
Cusens, A. R. and Pama, R. P. Bridge Deck Analysis. John Wiley, 1975. 278 pp. $27.50
  A book on the analysis and design phases of bridge superstructures which will provide the designer and student with methods he can use; it is aimed at postgraduate students, research workers and practitioners in civil engineering.

A05-1590
Czerniak, E. Reinforced Concrete Columns. Vol. 1, Working Stress Design for Concrete Columns. Frederick Ungar,

$14.00

This text explains the mechanics of designing concrete columns subjected to axial loads and bending moments. Uniaxial and biaxial bandings are considered, with solutions given to eccentricity problems.

A05-1591
Czerniak, E. Reinforced Concrete Columns. Vol. 2, Working Stress Design Charts for Spiral Columns. Frederick Ungar, $8.00. Vols. 1 and 2, $20.00

Over 250 design charts reduce the drudgery of repetitive calculations of stress in spiral columns.

A05-1592
Daniels, S. R. Inelastic Steel Structures. Univ. of Tennessee Press, 1966. 190 pp. $7.50

This book is a design-oriented study of the most common applications of plastic theory. The method is recommended for steel building frames of one or two stories which are subjected to static loads. It is also suitable for continuous beams; single-story, single-bay industrial buildings; and such special structures as missile facilities, blast-resistant fallout shelters, "hardened" military complexes, and containment vessels for nuclear explosions.

A05-1593
Davies, C. Steel Concrete Composite Beams for Building. John Wiley, 1975. 125 pp. $18.50

Deals with the theory, design, and construction of steel-concrete composite beams as used in building. Gives coverage of the theoretical and practical aspects of composite beam design, considering both elastic and ultimate load behavior.

A05-1594
Davies, J. D. Structural Concrete. Pergamon, 1964. 170 pp. $3.50

Deals with the behavior of reinforced and prestressed concrete structures in relation to elastic and ultimate load theories.

A05-1595
Davies, R. M., ed. Space Structures: A Study of Methods and Developments in Three-dimensional Construction. John Wiley, 1967. 1233 pp. $46.50

Nearly all of the 107 contributions to this book were originally presented at the 1966 International Conference on Space Structures. Topics are developments in analytical theories, stability, construction, influence of the electronic computer on analysis and design.

A05-1596
Davis, C. and Sorenson, K. Handbook of Applied Hydraulics. McGraw-Hill, 1969. 1680 pp. $45.50

The handbook gives broad, definitive coverage to the principles of hydrology and basic hydraulics, water conveyance, dams and water use, control and disposal. It incorporates the most recent advances and demonstrates the practical applications of these principles by examples drawn largely from the records of recently constructed projects.

A05-1597
Den Hartog, J. P. Strength of Materials. Dover, 1949. 323 pp. $3.50
Introduction, refresher, school text with full treatment of basic material (tension, torsion, bending, compound stress, deflection of beams), plus advanced material on engineering methods.

A05-1598
Desai, C. S. and Abel, J. F. Introduction to the Finite Element Method. Van Nostrand Reinhold, 1972. $18.50
Fundamental theory, assumptions, and limitations. Computer code included in appendix.

A05-1599
Design of Tilt-Up Buildings. 2d ed. Know-How Publications, 1963. 170 pp. $12.50
This publication outlines the practical engineering and architectural design of Tilt-Up buildings that insures structural integrity with construction economy. Attention is given to details that provide reserve resistance to wind and earthquake. Topics covered include foundation design, girder and column design, prestressed panels, sandwich wall panels, and pick up analysis.

A05-1600
Disque, R. Applied Plastic Design in Steel. Van Nostrand Reinhold, 1971. $14.95
Covers all aspects of plastic design in steel, incorporating the 1969 American Institute of Steel Construction Specifications for the Design and Erection of Structural Steel for Buildings. The basic theory of plastic design is presented, but the actual design and selection of members is emphasized.

A05-1601
Dober, R. P. Environmental Design. Van Nostrand Reinhold, 1969. 288 pp. $18.50
The book lays the groundwork for an interdisciplinary approach to creating a viable human environment--an approach emphasizing methodology and opportunities.

A05-1602
Dravid, P. S. Analysis of Continuous Beams and Rigid Frames. Asia Publishing House, 1965. 95 pp. $5.00
The author presents Kani's method of rotation contribution for the structural analysis of continuous beams and multi-storied frames with the help of the well-known slope deflection equations and the free body equations of individual storeys. The subject matter has been graded to cover the usual cases in logical sequence. Worked-out examples have been given for different cases.

Design--Civil and Structural

A05-1603
Dreyer, W. The Science of Rock Mechanics. Vol. 1, Strength Properties of Rocks. Trans Tech Publications, 1972. 500 pp. $25.00

This volume covers the relationships between the state of stress, stress of rocks and their textures. Several series of biaxial and triaxial deformation tests and model studies examine stress conditions in room and pillar systems.

A05-1604
Duerden, C. Noise Abatement. Butterworth, 1970. 304 pp. £5.00

An account of noise detection and control. Provides an introduction to the science of sound waves and their propagation. Further chapters give a description of sound meters and methods of using them, the best systems for noise detection and analysis, and the legislation which has been enacted to lessen disturbance from excessive noise.

A05-1605
Dugdale, D. S. Elements of Elasticity. Pergamon, 1968. 160 pp. $4.50

Examines the physical properties of stress and strain and applies them to an elastic body. Stress distribution set up by plane deformation and by torsion are calculated. Mathematical treatments are taken in easy stages.

A05-1606
Dugdale, D. S. and Ruiz, C. Elasticity for Engineers. McGraw-Hill, 1972. 304 pp. $15.00

Explaining various methods of stress analysis used in the design of structures and machines by civil, mechanical, and structural engineers, the author's main aim is to provide a clear introduction to certain topics concerning both students and practicing engineers.

A05-1607
Dunham, C. Advanced Reinforced Concrete. McGraw-Hill, 1964. 480 pp. $16.50

Deals mainly with indeterminate structures, ultimate strength analysis, and analysis and design of rigid frames.

A05-1608
Dunham, C. Theory and Practice of Reinforced Concrete. McGraw-Hill, 1966. 629 pp. $18.50

An introductory text on reinforced concrete (prestressed concrete), and working stress and ultimate strength design.

A05-1609
Durgar'yan, S. H. Theory of Shells and Plates. Israel Program for Scientific Translations, 1964. 960 pp.

A collection of 200 papers presented to the Soviet IV All-Union Conference on the Theory of Shells and Plates. Deals with general problems of the theory of shells and plates, stability and nonlinear problems, dynamics, thermoelasticity, creep, plasticity, and structural mechanics.

A05-1610
Eagleman, J. R., et al. Thunderstorms, Tornadoes and Building Damage. Lexington Books, 1975. 352 pp. $23.00
  In this study of the effects of tornadoes on buildings the authors given an in-depth analysis of the formation and behavior of these storms as well as a thorough discussion of design and construction practices that minimize storm damage. Finding definite wind flow patterns exist inside the tornado, the authors develop for the first time in the laboratory a travelling tornado-like vortex, an experiment which contributes to an understanding of damage patterns to buildings.

A05-1611
Eaton, K. J., ed. Wind Effects: Proceedings of the Fourth International Conference on Wind Effects on Buildings and Structures, September 1975. Cambridge Univ. Press, 1976. 650 pp. $47.50
  The effects of wind upon structures, particularly skyscrapers, is a topic of major interest to engineers, architects and urban planners. The papers presented in this volume describe recent developments in the areas of meteorology, aerodynamics and structural engineering that relate to the effects of wind on buildings and structures.

A05-1612
Eckardt, O. W. Strength of Materials. Holt, Rinehart and Winston, 1969. 480 pp. $11.00
  The author presents a mathematically simple development of the basic theory; the book contains no calculus. Many sample problems.

A05-1613
Eckenfelder, W. Industrial Water Pollution Control. McGraw-Hill, 1966. 288 pp. $18.50
  This text presents the procedure needed to evaluate an industrial waste problem, collects experimental data, and then develops the process design for required treatment facilities.

A05-1614
Eckenfelder, W. W., Jr. Water Quality Engineering for Practicing Engineers. Cahners, 1970. 328 pp. $10.95 (paperback $7.95)
  Contents: General Concepts of Water Quality Management. Sewage and Industrial Waste Characterization. Analysis of Pollutional Effects in Natural Waters. Characteristics of Municipal Sewage. Industrial Wastes. Wastewater Treatment Processes. Pretreatment and Primary Treatment. Oxygen Transfer and Aeration. Biological Treatment. Tertiary Treatment. Sludge Handling

Design--Civil and Structural

Disposal. Miscellaneous Processes. Economics of Wastewater Treatment.

A05-1615
EERI. Earthquake and Blast Effects on Structures. Earthquake Engrg. Research Institute, 1952. 322 pp. $16.50 (available from University Microfilms)
Proceedings of the Symposium held in Los Angeles, California in 1952.

A05-1616
EERI. First World Conference on Earthquake Engineering. Earthquake Engrg. Research Institute, 1956. 536 pp. $12.00
Proceedings of the Conference held in Berkeley, California in 1956.

A05-1617
Eriksen, B. Theory and Practice of Structural Design Applied to Reinforced Concrete. Cement and Concrete Assn., 1953. 402 pp. £1.25 Order No. 12.011

A05-1618
Evdokimov, P. D. and Sapegin, D. D. Stability, Shear and Sliding Resistance, and Deformation of Rock Foundations. Israel Program for Scientific Translations, 1964. 152 pp.
The authors describe the research carried out by Soviet institutions to ensure stability of hydro structures. Results of laboratory tests as well as field tests carried out at the sites of famous Soviet hydro developments, on concrete blocks and rock foundations, are given.

A05-1619
Everard, N. J. and Tanner, J. L. Reinforced Concrete Design: Including 200 Solved Problems. McGraw-Hill, 1967. 330 pp. $4.50
Covers materials and parts for reinforced concrete construction, gravity loads, lateral loads, working stress design and ultimate stress design. Also bond, anchorage, splices, short and long columns, retaining walls, slabs, and torsion.

A05-1620
Fairhurst, C. Failure and Breakage of Rock: Proceedings, 8th Symposium on Rock Mechanics. Society of Mining Engrs. of AIME, 1967. 581 pp. $20.00 (for AIME members, $17.00)
Covers rock failure, underground design and excavation, surface design and construction, drilling, and blasting.

A05-1621
Fawcett, P. Timber Structures. Butterworth. (in preparation)
Covers basic structural principles, structural properties of timber materials, design of elements in bending and axial load, connections and assembly, and structural forms.

A05-1622
Feltham, P. Deformation and Strength of Materials. Plenum Press, 1966. 142 pp. $10.00

An examination of the theoretical foundations of materials science and the basic concepts of the mechanical properties of materials. The author investigates such topics as elasticity, viscoelastic behavior, damping capacity, and the strength of real crystals. Dislocation theory, fracture, fatigue, and the behavior of non-Newtonian fluids are also discussed.

A05-1623
Fenves, S. J. Computer Methods in Civil Engineering. Prentice-Hall, 1967. 242 pp. $16.15

A collection of methods, procedures, ideas, and examples for application to the many computational, data processing, and logical problems in civil engineering. Emphasizes the development of general procedures and their use in complete, general and flexible computer programs. Illustrates the methods developed with specific examples from many branches of civil engineering.

A05-1624
Fenves, S. J., et al., eds. Numerical and Computer Methods in Structural Mechanics. Academic Press, 1973.

Subjects in the conference included finite element methods, displacement models, fracture mechanics, thin shell analysis, computer-aided design, numerical analysis, elastic-plastic and creep analysis.

A05-1625
Feodosyev, V. Strength of Materials. Beekman, 1968. 570 pp. $17.50

This introduction to the theory of the strength of materials includes tension, compression, shear and torque, and the geometrical characteristics of bar sections. The theory of bending and present theories of strength are outlined and discussed, and the author describes in simplified form an advanced method of strength analysis by limit states.

A05-1626
Ferguson, P. M. Reinforced Concrete Fundamentals. John Wiley, 1973. 718 pp. $20.75

Covers concrete materials and specifications, beams and columns, flexture, shear and bond stresses and retaining wall design. Slabs, composite beams and footings are also examined. The ACI Building Code requirements for reinforced concrete is explained.

A05-1627
Fertis, D. G. Dynamics and Vibrations of Structures. John Wiley, 1973. 485 pp. $21.75

Contents--Fundamentals of Structural Vibrations. Dynamic Response of Spring-Mass Systems. Idealized Beams, Frames, and Simple Buildings. Systems with Infinite Degrees of Freedom.

Design--Civil and Structural

Modal Analysis. Methods of Vibration. Structures with Members of Variable Stiffness. Fourier and Laplace Transforms. Variational Approach. Approximate Methods for Dynamic Response. Blast and Earthquake. Stochastic Approach to Structural Dynamics.

A05-1628
Field, J. Lessons from Failures of Concrete Structures. American Concrete Institute, 1964. 179 pp. $10.00 (available from University Microfilms)
Monograph examines engineering failure with the philosophy that failure is noncomformity with design expectations which may or may not result in structural collapse. Discusses failures in categories: design deficiencies, construction problems, durability and compatibility of materials, and foundation problems.

A05-1629
Filonenko-Borodich, M. Theory of Elasticity. Beekman, 1968. 388 pp. $15.00
Contains information on the theory of stresses, the geometrical theory of strain, the torsion of prismatic bars, and the bending of plates. It sets forth the theoretical fundamentals of the solution of elasticity problems in terms of displacements and stresses and shows the general methods of solving problems in the theory of elasticity.

A05-1630
Fischer, L. Theory and Practice of Shell Structures. Adler's Foreign Books, 1968. 541 pp. $53.00
The following problems are analyzed: simple problems of translational shells, hyperbolic paraboloids, cylindrical shells, and shells of revolution. Also covered are conoid shells, the theory of membrane shells, and the bending theories for shallow shells, translational shells, cylindrical shells, axially symmetrical cylindrical shells and shells of revolution under axially symmetrical shells.

A05-1631
Fischer, R. Architectural Engineering and New Structures. McGraw-Hill, 1964. 220 pp. $12.50
Describes structural concepts, systems, and design procedures as applied to the gamut of modern materials and building requirements.

A05-1632
Fisher, J. W. and Struik, J. H. A. Guide to Design Criteria for Bolted and Riveted Joints. John Wiley, 1974. 314 pp. $22.00
Contents--General Provisions, Rivets, Bolts, Symmetric Butt Splices, Truss-Type Connections, Shingle Joints, Lap Joints, Oversize and Slotted Holes, Filler Plates between Surfaces, Alignment of Holes, Surface Coatings, Eccentrically Loaded Joints, Combination Joints, Gusset Plates, Beam and Girder Splices, Tension-type Connections, Beam-to-Column Connections.

A05-1633
Fletcher, G. A. and Smoots, V. A. Construction Guide for Soils and Foundations. John Wiley, 1974. 420 pp. $21.25

Partial subject list includes subsurface exploration and sampling, soil and rock characteristics, tests of soil samples and test reports. Also included are shoring, bracing, foundations and piles made of timber, steel, cast-in-place concrete, and precast and prestressed concrete.

A05-1634
Fluegge, W. Stresses in Shells. Springer-Verlag, 1973. 525 pp. $22.40

Deals with shell theory as it applies to engineering; attempts to bridge gap between those who apply rules of thumb and elementary formulas and those for whom shell theory is principally a mathematical exercise.

A05-1635
Forsyth, B. Unified Design of Reinforced Concrete Members. McGraw-Hill, 1971. 480 pp. $34.00

Covers the entire range of flexural problems relating to design of beams, beam-columns and columns. Problems involving flexure or flexure in combination with axial loads and even some column design procedures are all based upon concept of the Section Modulus. The work is general in character and applicable to all design work performed under the main codes that govern the design of reinforced concrete members in the USA.

A05-1636
Gallagher, R. H. and Zienkiewicz, O. C. Optimum Structural Design: Theory and Applications. John Wiley, 1973. 358 pp. $24.95

Partial contents: Fully Stressed Design, Optimality--Criterion-based Algorithms, Mathematical-Programming Methods--a Critical Review, Linear Programming in Structural Analysis and Design, Shape Optimization and Sequential Linear Programming, Feasible-direction Methods in Structural Optimization, Penalty-function Methods, Dynamic Programming and Structural Optimization. Discrete Variables in Structural Optimization, Limit Design in the absence of a Given Layout: A Finite-element, zero-one Programming Problem, Design for Reliability--Concepts and Applications, Optimum Design of Multi-story Rigid Frames. Reinforced-concrete Design, Structural-steel Design.

A05-1637
Gartner, R. Statically Indeterminate Structures. Cement and Concrete Assn., 1958. 126 pp. £0.90

A05-1638
Gaylord, E. H. and Gaylord, C. N. Design of Steel Structures. McGraw-Hill, 1972. 700 pp. $19.50

The second edition discusses the design of structural members and their connections, and applies these principles to steel

Design--Civil and Structural 271

bridges, buildings, and other structures. This revision includes new articles on fatigue and brittle fracture; a new chapter on the stability and postbuckling strength of flat plates in compression; a treatment of the buckling and postbuckling strengths of beam and plate-girder webs; an expanded discussion of composite beams; and up-to-date examples of design calculations for members as components of structures.

A05-1639
Gaylord, E. and Gaylord, C. Structural Engineering Handbook. McGraw-Hill, 1968. 1216 pp. $35.00

This handbook assembles material related to the planning, design, and construction of engineered structures. Discusses soil exploration, foundations, working-stress, and ultimate-strength design of reinforced concrete structures, industrial and high-rise buildings, concrete shells, towers, etc., with special attention to design for dynamic loads, fatigue and brittle fracture, and the solution of engineering problems by computer.

A05-1640
Gere, J. M. and Weaver, W., Jr. Analysis of Framed Structures. Van Nostrand Reinhold, 1965. 488 pp. $14.95

A05-1641
Gerstle, K. Basic Structural Design. McGraw-Hill, 1967. 670 pp. $18.00

Covers the basic methods common to stress analysis of structural members, regardless of material, loading condition, or range of stress.

A05-1642
Ghali, A. and Neville, A. M. Structural Analysis: A Unified Classical and Matrix Approach. Intext, 1972. 769 pp. $17.50

Designed for beginning or advanced courses in structural analysis, this text develops the classical and modern methods of structural analysis concurrently.

A05-1643
Gibson, J. E. The Design of Cylindrical Shell Roofs. Van Nostrand Reinhold, 1961. 285 pp. $11.50

Contents include the general theory of cylindrical shells, the design of a simple long shell without edge beams, edge beam theory, design of long shells with edge beams, shells with prestressed edge beams, reinforcement, and more.

A05-1644
Gibson, J. E. Linear Elastic Theory of Thin Shells. Pergamon, 1965. 194 pp. $4.50

Introduces shell theory at an intermediate level so that it may be understood by final year students. Starting from an elementary membrane theory for cylindrical shells, more complicated theories are developed using simple mathematical derivations.

A05-1645
Gill, S. Structures for Nuclear Power. Applied Science, 1964. 398 pp. £8.00
This book records the experience in structural planning, design and construction gained during the first decade of development of the peaceful uses of nuclear power. Present practice and future trends of development are explained.

A05-1646
Glushkov, G. S., et al. Handbook of Formulas for the Analysis of Complex Frames and Arches. Israel Program for Scientific Translations, 1966. 352 pp.
This book gives formulas for the computation of joint and support moments, and of support reactions for complex frames and arches under a great variety of loading conditions. Frames with stepped stanchions are analyzed, as well as multibay and multistory frames. One chapter deals with one, two, and three-bay arches variously connected at their supports.

A05-1647
Godden, W. G. Numerical Analysis of Beam and Column Structures. Prentice-Hall, 1965. 309 pp. $18.30
Simple numerical method for computing deflections, buckling loads, and natural frequencies of beam systems. Contains solutions suitable for slide rule computation and small desk calculator.

A05-1648
Goodfriend, L. S. Noise Pollution. CRC Press, Inc., 1974. 200 pp. $26.00
Noise Pollution provides the background for understanding the nature of noise and its effects on man and the principles and practices of noise pollution control. The physics of sound and their relation to the generation, radiation, and propagation of sound are reviewed from the aspect of real machines and processes. The response of man to noise from a physiological and experimental psychology point of view is also reviewed, but the interaction of man with noise in the community is discussed from the point of view of the latest studies carried out by various federal agencies.

A05-1649
Graf, W. Hydraulics of Sediment Transport. McGraw-Hill, 1971. 544 pp. $23.50
The author organizes and examines recent developments in the field and includes: (1) a short history of sediment transport; (2) hydrodynamics of fluid particle systems; (3) sediment transport in open channels; and (4) sediment transport in closed pipes.

A05-1650
Granet, I. Strength of Materials for Engineering Technology. Prentice-Hall, 1973. 448 pp. $13.95
Incorporates all of the advances of the latest edition of the AISC Steel Construction Manual. Features a large number of illustrative problems grouped by topic in each chapter, and an extensive

Design--Civil and Structural

table of structural shapes for use as a reference from the 7th Edition of the AISC Manual.

A05-1651
Granholm, H.   A General Flexural Theory of Reinforced Concrete.
John Wiley, 1965.   204 pp. $11.00
Covers deformation and strength properties of concrete reinforcement, fundamental equations of the new theory, prestressed concrete, tests of under and over-reinforced beams, and much more.

A05-1652
Griffel, W.   Handbook of Formulas for Stress and Strain.   Frederick Ungar, 1966.   440 pp. $18.00
This book offers a compilation of simplified formulas to reduce the laborious tasks of stress computations. A wide field of mechanics of materials is covered, with emphasis on flat plate design, pressure vessels, beams, rigid frames, contact stresses, column design, concentration factors, and many other, including ogival shapes for those concerned with missile nose cone design.

A05-1653
Griffel, W.   Plate Formulas.   Frederick Ungar, 1968.   250 pp. $9.00
Presents a series of tables containing computed data for use in the design of components of structures which can be idealized as flat, circular, rectangular, square, triangular and elliptical plates. A total of 139 tabulated cases with most common--and some not so common--loadings and supports typical of those encountered in design cover the field. Also offered is a detailed treatment of large deflections of plates and a solution of statically indeterminate plates.

A05-1654
Grosvenor, N. E. and Paulding, B. W., Jr., eds.   Status of Practical Rock Mechanics: Proceedings, 9th Symposium on Rock Mechanics.   Society of Mining Engrs. of AIME, 1968. 496 pp. $19.50 (for AIME members, $16.50)
This volume surveys the capabilities and possibilities that rock mechanics can offer mining, construction and petroleum engineers. Topics covered include surface and underground openings, measurement of in-situ stresses, measurement of mechanical properties, design and support of openings, excavation, etc.

A05-1655
Gurfinkel, G.   Wood Engineering.   Southern Forest Products Assn., 1974.   537 pp. $9.95
Describes the characteristics of wood, its strength properties, allowable stresses of structural wood, glued-laminated timber and plywood, and the durability of well-designed wood structures. There is material on fastenings, connections, behavior and design of beams and columns, design of buildings and bridges and miscellaneous applications including arches and rigid frames, trusses,

foundations and pole-frame structures. Examples of wood design, including construction details, are given.

A05-1656
Gurney, T. R. Fatigue of Welded Structures. Cambridge Univ. Press, 1968. $15.50

A05-1657
Guyon, Y. Limit-State Design of Prestressed Concrete. Vol. 1: The Design of the Section. John Wiley, 1972. 450 pp. $40.00

Based on a series of lectures given at the Centre des Hautes Etudes du Beton Arme et du Beton Precontrainte (Centre for Advanced Studies of Reinforced and Prestressed Concrete) which clarified some of the more difficult aspects of the subject. Emphasizes the extending scope of prestressing to include applications in which some degree of tensile strain is permitted. Points out the limits within which this type of construction could be applied.

A05-1658
Guyon, Y. Limit-State Design of Prestressed Concrete. Vol. 2: The Design of the Member. John Wiley, 1974. 469 pp. $38.50

Topics include the cable profile, resistances to shear, beams with deflected tendons, beams of varying depths, cantilevered bridges, and block reinforcement, and prestressed beams with pre-tensioned steel.

A05-1659
Hahn, J. Structural Analysis of Beams and Slabs. Frederick Ungar, 1967. 300 pp. $12.50

This volume contains tables and diagrams on most types of slab designs. Many of the analytical methods are of recent derivation.

A05-1660
Hall, A. S. and Woodhead, R. W. Frame Analysis. John Wiley, 1967. 329 pp. $14.00

Allows the reader to understand the physical significance of the mathematical processes, while incorporating better methods of explanation. Original data has become more specific and new, more detailed material has been added. Problems have been provided for all relevant chapters.

A05-1661
Halperin, D. A. Statics and Strength of Materials for Technology. John Wiley, 1975. 300 pp.

Presents easy to understand methods of solution to the problems of statics, strength of materials, and structural design in building construction. Leads the student progressively from the study of external forces through the internal reaction of the pieces of a structure into plastic design in steel and ultimate strength design in concrete.

Design--Civil and Structural

A05-1662
Hardenbergh, W. A. and Rodie, E. B. Water Supply and Waste Disposal. Intext, 1961. 502 pp. $12.50
Presents the fundamentals of water supply and sewage treatment. Simultaneous discussion of water and wastes simplifies the task of stressing sanitary engineering principles, and applies them to both design and operation of water and sewage processes.

A05-1663
Harr, M. Foundations of Theoretical Soil Mechanics. McGraw-Hill, 1966. 381 pp. $16.50
The basic principles of soil mechanics for those concerned with the action of structures of, on, or in the earth when subject to natural or induced loadings.

A05-1664
Harr, M. Groundwater and Seepage. McGraw-Hill, 1962. 315 pp. $18.50
The author offers an analytical approach to the solution of groundwater and seepage problems and an understanding of the design of earth structures that impound water.

A05-1665
Harris, C. Handbook of Noise Control. McGraw-Hill, 1957. 194 pp. $27.00
An exposition of noise, its nature, measurement, and techniques for its control in buildings, industry, transportation, and the community. Provides fundamentals, practical engineering techniques, and reference data in connection with topics such as the effect of noise on human efficiency, hearing, and speech; vibration problems and control; the control of industrial noise, and treatment of special types of noise sources.

A05-1666
Harris, C. O. Elementary Structural Design. American Technical Society, 163 pp. $6.50
A practical text on design for the building tradesman and the construction technician.

A05-1667
Harris, C. O. Strength of Materials. American Technical Society, 231 pp. $8.50
This 3rd edition contains review problems and summarizes information on new materials, in addition to the text.

A05-1668
Harrison, H. B. Computer Methods in Structural Analysis. Prentice-Hall, 1973. 400 pp. $19.00
Develops the matrix methods of analysis of framed structures using the concept of influence lines. Lists eleven short Fortran programs.

A05-1669
Hart, F., Henn, W. and Sonntag, H. Multi-Storey Buildings in Steel. John Wiley, 1975. 356 pp. $39.50
This book will be of special interest to firms and engineers who are engaged in the fabrication and erection of structural steel work. Includes about 2300 illustrations and about 300 photographs.

A05-1670
Heins, C. P. Bending and Torsional Design in Structural Members. Lexington Books, 1975. 369 pp. $24.95
Here in one volume are both the theory and the applied examples needed by student and engineer to understand the bending and torsional forces widely used in bridge and building construction. The two forces, heretofore considered separately, are explained in terms of past theories and developments, including appropriate design examples. The theories are then extended to producing new formulations and design information for student, design engineer, and consultant.

A05-1671
Hendry, A. W. Elements of Experimental Stress Analysis. Pergamon, 1964. 204 pp. $8.50
An introduction to the more important techniques of experimental stress analysis, indicating appropriate applications in field and laboratory investigations.

A05-1672
Hermann, G., ed. Dynamic Stability of Structures. Pergamon, 1967. 324 pp. $16.50
Stability of continuous systems. Dynamic buckling of elastic structures. Survey of problems of structural dynamic stability in vehicle design. Buckling of long slender ships due to wave-induced shipping. Stability and vibration problems of mechanical systems under harmonic excitation. Dynamic plastic buckling.

A05-1673
Heyman, J. Beams and Framed Structures. Pergamon, 1964. 148 pp. $4.50
Illustrates the use of the fundamental equations of the theory of structures, and applies these equations to elastic and plastic frames.

A05-1674
Hodge, B. Computers for Engineers. McGraw-Hill, 1969. 224 pp. $11.00
This work covers all the basic principles and techniques which any engineer can use to apply computers to his field of specialty. Emphasizes the problem solving possibilities and limitations of digital computers.

A05-1675
Hodge, P. G., Jr. Plastic Analysis of Structures. McGraw-Hill, 1959. 364 pp. $15.50

Covers bending of beams and frames and structures under combined stresses. Discusses general theory, limit analysis, and loading for both topics.

A05-1676
Hodgkinson, A., ed. AJ Handbook of Building Structure. Architectural Press, 1974. 320 pp. £7.50
The object of this handbook is to refresh the architect's knowledge of the principles of structural engineering. The technical studies are intended to give background understanding and to cover general principles. Information sheets give specific data that can be applied directly by the designer and the design guide provides advice for the architect undertaking simple structures on his own.

A05-1677
Hoff, N. J. The Analysis of Structures: Based on the Minimal Principles and the Principle of Virtual Displacements. John Wiley, 1956. 493 pp. $19.50

A05-1678
Hollis, E. P. Bibliography of Earthquake Engineering. Earthquake Engrg. Research Institute, 1971. 247 pp. $12.00

A05-1679
Hopkins, R. B. Design Analysis of Shafts and Beams. McGraw-Hill, 1970. 475 pp. $24.50
Provides background information and formulas and procedures required for the analysis of stress, deflection, and fatigue life of beams, shafts, and machine members which serve as beams.

A05-1680
Horne, M. R. and Merchant, W. The Stability of Frames. Pergamon, 1965. 188 pp. $4.95
Deals with the stability of frames in both the elastic and elasto-plastic ranges.

A05-1681
Hough, B. K. Basic Soils Engineering. Ronald Press, 1969. 590 pp. $14.95
Emphasizes basic concepts and theoretical developments and stresses practical design and construction applications. In this second edition, the chapter on spread foundations contains simple design charts developed by computer to indicate the proper footing size for specified column load and limitation of earth settlement. Other special features include a discussion of the developments of osmotic pressure in clays and its effect on soil structure and compressibility, a theory for the migration of water into zones of freezing, and a clarification of rupture theory for clays.

A05-1682
Howells, D. A., et al., eds. Dynamic Waves in Civil Engineering. John Wiley, 1971. 575 pp. $32.00

Proceedings of a conference organized in 1970 by the Society for Earthquake and Civil Engineering Dynamics. Scientists and engineers discuss behavior of waves of various types and their interaction with structures placed in their path, and investigate the origins and transmissions of waves and the dynamic properties of structures.

A05-1683
Hsieh, Y. Y. Elementary Theory of Structures. Prentice-Hall, 1970. 432 pp. $16.15

Basic concepts of structures, statically indeterminate structures including the theory of elastic deformations, slope deflection, moment distribution, and a unified treatment of structures by the matrix methods.

A05-1684
Hughes, B. P. Limit State Theory for Reinforced Concrete. Pitman and Sons, 1973. 446 pp. £4.00 (paperback £3.00)

This book presents the fundamental principles and introduces current design practice. New design requirements are summarized and the book contains in appendixes what is in effect a designer's handbook in miniature. Examples and exercises are also included. The book is in SI units throughout, and includes an explanation of the draft Unified Code of Practice for the structural use of concrete.

A05-1685
Hurty, W. C. and Rubinstein, M. F. Dynamics of Structures. Prentice-Hall, 1964. 455 pp. $17.95

Considers conservative and nonconservative systems with linear damping. Develops both the normal mode and the frequency response methods in dealing with the response to periodic and nonperiodic excitations. Extends the treatment to the response of structures to random excitations such as gusts, blasts, and earthquakes.

A05-1686
Institution of Structural Engineers. International Symposium on Computer-Aided Structural Design. Peter Peregrinus, 1972. 832 pp. $46.00

The symposium was organized by the Department of Engineering, University of Warwick, in conjunction with the Institution of Structural Engineers and held in July, 1972. The proceedings cover the application of optimization techniques, skeletal frames, component design, slab plate and shells, graphics, foundations and retaining structures, and computer-aided systems.

A05-1687
International Atomic Energy Agency. Aseismic Design and Testing of Nuclear Facilities. International Publications, 1964. $1.00

Design--Civil and Structural

A05-1688
International Atomic Energy Agency. Earthquake Guidelines for Reactor Siting. International Publications, 1972. $2.00

A05-1689
International Conference on Wind Effects on Buildings and Structures, 3rd, Tokyo, Sept. 1971. International Publications, 1972. 1294 pp. $50.00
 Proceedings contain 113 papers, together with discussions comments. Part 1: Turbulent structure and statistics of wind. Part 2: Drag and pressure distribution on structures; effects of fluctuating pressures, etc. Part 3: Design problems of structures to winds; interaction of structures and shedding vortices.

A05-1690
Ippen, A. Estuary and Coastline Hydrodynamics. McGraw-Hill, 1966. 650 pp. $29.50
 Here is a well-illustrated, organized study of coastal engineering, tidal hydraulics, and nearshore oceanography.

A05-1691
Jaeger, L. G. Cartesian Tensors in Engineering Science. Pergamon, 1966. 132 pp. $4.50
 Provides the engineer with an account of the symmetric cartesian tensor and demonstrates its applicability in several areas of engineering science.

A05-1692
Jaeger, L. G. Elementary Theory of Elastic Plates. Pergamon, 1964. 118 pp. $5.25
 An account of the small deflection theory of elastic, initially flat plates.

A05-1693
Jenkins, W. Matrix and Digital Computer Methods in Structural Analysis. McGraw-Hill, 1969. 200 pp. $14.50
 The basic concepts and applications of matrix methods in structural analysis are covered in this book which provides an introduction to the use of digital computers for structural calculations. The dual methods of flexibility and stiffness are first developed in parallel in order to emphasize basic concepts, then individually in a more formal way which develops further applications and refinements.

A05-1694
Jensen, A. and Chenoweth, H. H. Applied Strength of Materials. McGraw-Hill, 1975. 365 pp. $12.95
 It presents in easily understood language the basic knowledge required for the successful design of machine parts, structural members, and connections. Design formulas are developed through many illustrative problems actually showing the applications of these formulas in finished form. Methods of manufacture and the properties of various engineering materials are also included.

A05-1695
Jensen, J. and Chenoweth, H. Statics and Strength of Materials. McGraw-Hill, 1967. 384 pp. $10.95

This revised guide gives the theory behind the development of formulas used in the design of machine parts, structural members, and connections.

A05-1696
Jeppson, R. W. Analysis of Flow in Pipe Networks. Ann Arbor Science, 1976. 170 pp. $15.00

With widespread use of digital computers in engineering practice, improved methods have been developed for analyzing flow rates and pressures throughout large fluid distribution systems. This book is devoted exclusively to these methods with emphasis on how they are implemented in computer programs.

A05-1697
Johnson, R. Structural Concrete. McGraw-Hill, 1968. 271 pp. $15.50

The principal theories which underlie current design methods by structural concrete are discussed and related to observed behavior of concrete and steel. The book covers ultimate strength theories; recent advances in the yieldline theory; and the behavior of members under combinations of axial force, bending moment, shear force, and torsion.

A05-1698
Johnson, R. P. Composite Structures of Steel and Concrete. Vol. 1, Beams, Columns, Frames, and Applications in Building. John Wiley, 1975. 176 pp. $19.50

Reviews current knowledge of the behavior of composite structures of steel and concrete.

A05-1699
Johnson, S. and Kavanagh, T. The Design of Foundations for Buildings. McGraw-Hill, 1968. 416 pp. $21.95

Reviews current practices and standards for design of foundations, and considers potential problem areas. Based mainly on work to develop provisions of a new building code for New York City.

A05-1700
Johnston, B. G., ed. Column Research Council Guide to Design Criteria for Metal Compression Members. John Wiley, 1966. 217 pp. $14.75

Covers centrally loaded columns, compression member details, unsupported beams, plate girders, beam-columns and other steel structures.

A05-1701
Joiner, J. H. Essentials of the Theory of Structures. Hart Publishing, 1968. 277 pp. $12.50

This book concentrates on basic formulae, and methods and

Design--Civil and Structural 281

principles of analysis without getting too involved in complicated theoretical considerations and proofs. Full summary of formulae, design stresses for structural material, beam shear force, bending moment and deflection values, fixed end moments for beams, and geometric properties of bending moment diagrams. Includes material on techniques such as electron beams, ultrasonic, friction and diptrans for welding.

A05-1702
Jumikis, A. R.   Foundation Engineering.   Intext, 1971.   828 pp. $18.50
This text covers the selection and design of structural foundations. While the principles of soil mechanics are used, emphasis is placed on the practical design of buildings and bridges. It includes a brief historical survey of foundation engineering, descriptions and analyses of the preparation of excavations and coffer-dams for laying foundations in the dry, shallow foundations, deep foundations, and such topics as scouring of river beds around bridge piers and seismic effects on structures.

A05-1703
Jumikis, A. R.   Introduction to Soil Mechanics.   Van Nostrand Reinhold, 1967.   448 pp. $9.95
The text provides a fundamental knowledge of the properties of various soils and their behavior in relation to foundations, pavings, tunnels, and structures built on or beneath the ground. Examples illustrate the application of principles to problems. Definitions of various soil types and their characteristics at various temperatures, under various loads, and with and without the presence of water and ice are given.

A05-1704
Juvinall, R.   Engineering Considerations of Stress, Strain, and Strength.   McGraw-Hill, 1967.   569 pp. $16.95
Provides basic material relative to the stress, strain, and strength considerations of mechanical and structural components. Treats four areas: static and dynamic loading, environment, and experimental methods.

A05-1705
Kardestuncer, H.   Elementary Matrix Analysis of Structures. McGraw-Hill, 1974.   540 pp. $19.00
With this text, it is possible to teach matrix analysis of structures without any other structural analysis prerequisite. The author used a method-oriented organization and includes over a hundred problems and numerous illustrations to reinforce the material.

A05-1706
Khachatupian, N. and Gurfinkle, G.   Prestressed Concrete. McGraw-Hill, 1969.   384 pp. $19.50
While emphasizing the fundamentals of prestressed concrete, this book offers a balance between systematic and theoretical concepts and practical engineering design. Includes illustrative examples.

A05-1707
Kobayashi, A. S. Experimental Techniques in Fracture Mechanics. Iowa State Univ. Press, 1973. 150 pp. $6.50

Written to help solve practical problems of sudden structural failures, this volume covers theory of fracture mechanics, acoustic emission techniques, compliance measurements, testing problems and associated instrumentation, and photoelasticity techniques.

A05-1708
Komendant, A. Contemporary Concrete Structures. McGraw-Hill, 1972. 688 pp. $27.50

This book discusses practical applications of essential and applicable structural theories, material characteristics and appropriate construction methods relating to concrete. It examines such topics as design and its critical evaluation, various carrying actions, pre-fab housing, high-rises, bridges, etc.

A05-1709
Kulski, J. E. Architecture in a Revolutionary Era. Wittenborn, 1971. 320 pp. $30.00

Air, water and noise pollution of space are discussed in the framework of total environmental architecture.

A05-1710
Kurtz, M. Comprehensive Structural Design Guide. McGraw-Hill, 1969. 328 pp. $19.50

This guide presents detailed solutions to a number of structural engineering problems, reviews the theory underlying each topic, and affords an intimate knowledge of standard design codes.

A05-1711
Kuzmanovic, B. and Willems, N. Steel Design for Structural Engineers. Prentice-Hall, 1976. 544 pp. $21.95

The authors provide a complete description of design procedures based on thorough knowledge of steel as a structural material and comparison of design specifications. Especially useful are the design methods which the authors develop and the fully listed computer programs. The authors offer explanations for all important formulas of respective specifications to show importance, accuracy, and limitations. Where necessary, the authors introduce fatigue design.

A05-1712
Lancaster, P. and Mitchell, D. The Mechanics of Materials. McGraw-Hill, 1970. 308 pp. $6.95

This book treats the mechanics of materials as a study in itself rather than concentrating on the solution of specific problems. Describes the Hardy Cross relaxation method which replaces the "theorem of three moments" for dealing with continuous beams, discusses the energy methods in bending and torsion, and explains strain equations.

Design--Civil and Structural

A05-1713
Large, G. E. and Chen, T. Y. Reinforced Concrete Design. Ronald Press, 1969. 634 pp. $14.95
This third edition presents the two standard approaches to concrete design calculation: ultimate strength and working stress. Column formulas of the American Concrete Institute Building Code are developed and explained in detail. Slender column theory is derived from ideal short columns; the Morsch procedure for reinforcing rectangular columns is introduced, and this procedure is extended to a wider range of bending effects. The study of creep and of the effects of shrinkage and temperature change is designed to remove the guesswork from the prestressing of concrete.

A05-1714
Laursen, H. Structural Analysis. McGraw-Hill, 1969. 486 pp. $19.50
Elementary concepts of analysis, indeterminate analysis, matrix methods (displacement and force methods) and the basic concepts of structural stability and dynamics are discussed. Material is presented in such a way that computers can be used to solve more complex problems. Covers fundamentals of matrix algebra and computer programs.

A05-1715
Lee, D. H. An Introduction to Deep Foundations and Sheet Piling. Cement and Concrete Assn., 1961. 260 pp. £1.00

A05-1716
Lenczner, D. Elements of Loadbearing Brickwork. Pergamon, 1972. 128 pp. $7.50
An introduction to loadbearing brickwork as a structural material which can be analysed and designed on the same scientific principles as steel and concrete. An attempt has been made to include the most up-to-date advances in brickwork technology including the latest recommendations on design in accordance with the U.K. Draft Revision of Code of Practice.

A05-1717
Lenczner, D. Movement in Buildings. Pergamon, 1973. 150 pp.
The importance of movements in buildings has increased greatly in recent years. This book gives the various causes of movements in buildings and their effect on the structure with details of movements resulting from elastic, time dependent environmental deformations and settlement.

A05-1718
Leonards, G. Foundation Engineering. McGraw-Hill, 1962. 1146 pp. $34.00
Experts in the field have contributed their knowledge to this full treatment of the current status of foundation engineering, with emphasis on design and construction.

A05-1719
Leonhardt, F. Prestressed Concrete Design and Construction. 2d ed. Adler's Foreign Books. 677 pp. $42.00
Fundamental concepts: materials; anchorages; tensioning devices and prestressing; degrees of prestressing; importance of bond; longitudinal mobility and sliding resistance of tendons--loss of prestressing force due to friction--tendon extension; establishing bond with post-tensioned tendons and protecting them against corrosion; transmission of the prestressing forces to the concrete. Constructional principles; design of prestressed concrete structures; analytical treatment of the effects of shrinkage and creep of concrete; ultimate strength and safety against failure; safety against fatigue under oscillating loads; stability problems of prestressed structural members; fire resistance of prestressed concrete.

A05-1720
Levinson, I. J. Mechanics of Materials. Prentice-Hall, 1970. 338 pp. $12.95
Principles of strength of materials through basic mathematics without resorting to calculus. Contains contemporary problems taken from modern engineering situations.

A05-1721
Levinson, I. J. Statics and Strength of Materials. Prentice-Hall, 1971. 498 pp. $12.95
Combines pertinent portions of two recently published second editions of Introduction To Mechanics and Mechanics of Materials. Emphasizes basic principles. Uses contemporary rather than classic examples. Limits mathematics to algebra and trigonometry. Provides worked out example problems.

A05-1722
Lightfoot, E. Moment Distribution. Barnes and Noble, or John Wiley, 1961. 363 pp. $11.25 (published by Spon)

A05-1723
Lin, T. H. Theory of Inelastic Structures. John Wiley, 1968. 454 pp. $21.50
This book presents a unified treatment of thermal, creep, and plastic strain through a generalization of Duhamel's analogy for thermal strain. This generalized analogy reduces the analysis of structures with any combination of these three strains to the analysis of an identical elastic body with an additional set of equivalent forces.

A05-1724
Lin, Y. Probabilistic Theory of Structural Dynamics. McGraw-Hill, 1967. 716 pp. $17.50
Offers a study of the applications of stochastic processes in structural dynamics problems.

A05-1725
Linsley, R. and Franzini, J. Water Resources Engineering.

Design--Civil and Structural 285

McGraw-Hill, 1972. 704 pp. $19.50
Instead of concentrating on the traditional topics of water supply and waste-water disposal, this book offers a comprehensive treatment of all aspects of water resources development and control. Applications of basic engineering sciences to the planning and design of water projects are illustrated, and the reasons why various techniques are employed to solve specific problems are discussed.

A05-1726
Linsley, R. K., et al. Hydrology for Engineers. McGraw-Hill, 1975. 482 pp. $18.50
This book offers basic concepts in the techniques of hydrologic analysis relevant to the design, planning and operation of engineering works for the management and control of water resources. This new edition covers stochastic analysis for water yield studies and the application of hydrologic simulation in such fields as river forecasting and urban drainage design.

A05-1727
Liptak, B. G., ed. Environmental Engineers' Handbook. Chilton, 1974. 2018 pp. $39.00
The types of water pollutants and the methods of waste-water treatment are covered in 150 papers. Laws affecting pollution control are given.

A05-1728
Liptak, B. G., ed. Environmental Engineers' Handbook. Vol. 2, Air Pollution. Chilton, 1974. 1340 pp. $35.00
Collection of 160 papers covering air pollution laws, types of air pollutants and their sources. Other subjects include control of pollution and removal of contaminants from the air.

A05-1729
Liptak, B. G., ed. Environmental Engineers' Handbook. Vol. 3, Land Pollution. Chilton, 1975. 1130 pp. $25.00
Contains data on causes of land pollution and their effects. Noise, thermal and mercury pollution are discussed as well as problems from radioactive wastes and pesticides. The remedies of land pollution including processing and disposing of wastes are covered.

A05-1730
Litton, E. Automatic Computational Techniques in Civil and Structural Engineering. John Wiley, 1973. 373 pp. $19.50
Contents--Applications of Automatic Techniques Using the Traditional Methods. Structural Analysis--The Applications of Automatic Techniques Using Methods Based on Modern Computational Processes. Matrix Algebra. The Computer. Structural Analysis-- An Outline of Methods Based on Modern Computational Processes. Automatic Techniques in the Laboratory and on Site. Miscellaneous Practical Problems and Their Solution. The Approach to New

Problems. Implications Arising From the Development and Use of
Automatic Computational Techniques.

A05-1731
Long, J. E.   Bearings in Structural Engineering.   Butterworth or
    John Wiley, 1974.   225 pp. $11.75
    Bearing technology is more closely allied to mechanical than
structural engineering. Many of the materials used are novel, at
least to the average structural engineer, and so the relevant properties of these, particularly elastomers and polytetrafluorethylene,
are discussed. Current British design rules and practices are
presented, although in such a developing subject formal rules are
few and tentative, so that many additional recommendations are
made by the author.

A05-1732
Lothers, J. E.   Advanced Design in Structural Steel.   Prentice-
    Hall, 1960.   583 pp. $18.95
    Practical treatise on the design of statically indeterminate
structures, with emphasis on both analysis and design.

A05-1733
Lothers, J. E.   Design in Structural Steel.   Prentice-Hall, 1972.
    561 pp. $20.00
    The new 1969 AISC and AREA specifications are presented
by means of examples throughout the book. Elastic and plastic
analysis and design of steel members are covered. Two-hinged
and hingeless bents are included in the chapter on plastic analysis
and design. Full cognizance is taken of the rapid replacement of
rivets by high-strength steel bolts.

A05-1734
Lund, H.   Industrial Pollution Control Handbook.   McGraw-Hill,
    1971.   864 pp. $37.50
    A multi-pronged approach tackles legal aspects, quality standards, community relations, various control systems, and problems
specific to the major industries. Much useful information is provided on equipment, performance testing and guarantees, department
organization, and the training of personnel. The coverage is practical, up-to-date and field oriented.

A05-1735
McCormac, J. C.   Structural Analysis.   Intext, 1967.   494 pp.
    $12.50
    Provides a background in statically determined structures
with information on statically indeterminate structures. Focus is
on analysis, with only incidental reference to design. Problems
are provided to establish an understanding of structural theory,
stressing the basics which apply to all cases.

A05-1736
McCormac, J. C.   Structural Steel Design.   Intext, 1971.   604 pp.
    $14.50

Design--Civil and Structural

Covering all the material necessary for the beginning course, the author discusses the fundamental theories and principles used in the design of simple steel structures, including the design of individual structural steel designs for entire structures.

A05-1737
Macdonald, A. J. Wind Loading on Buildings. John Wiley, 1975. 219 pp. $23.50
A handbook for architects and engineers as well as a textbook for students. Examines the history of wind loading problems, relevant meteorological information, and fundamental concepts of fluid flow and drag on immersed bodies. Wind bracing systems for all common types of buildings are discussed. Specialized problems which arise out of the dynamic response of light, slender structures to the wind are examined. Techniques for calculation are described.

A05-1738
McFarland, D. E., et al. Analysis of Plates. Hayden, 1972. 296 pp. $18.50
This text will serve well in a senior or graduate level course in the analysis of thin plates for stresses, strains, and lateral deflections. It is divided into three parts: basic relationships, classical solutions, and computer oriented solutions.

A05-1739
MacGinley, T. J. Structural Steelwork Calculations and Detailing. Butterworth, 1974. 314 pp. £3.50
The book provides an introduction to the design and detailing of structural elements and is intended to form the bases from which the complete design of buildings, including rigid frame analysis and design, can be developed. The structural frames chosen are primarily of simple pin jointed design and elements; joints, etc., are chosen from these for various problems. In most problems suitable representative loading is assumed for the element so that the design principles, detail and relevant code requirements for that element are brought out.

A05-1740
McGuire, W. Steel Structures. Prentice-Hall, 1968. 1112 pp. $22.65
An explanation of the performance of structural steel members, connections, and frames, considering their behavior in relation to current design procedures.

A05-1741
McKaig, T. Applied Structural Design of Buildings. McGraw-Hill, 1965. 512 pp. $22.50
This reliable data-packed manual of design contains shortcuts, tables, methods and formulas, etc., for architects, designers, plant and structural engineers.

A05-1742
McKaig, T. Building Failures. McGraw-Hill, 1962. 255 pp.
$16.50
Case studies of over 200 building failures in recent years are presented--failures due to defective building materials, over loading, wind action, etc.

A05-1743
McMinn, S. J. Matrices for Structural Analysis. Barnes and Noble or John Wiley, 1966. 240 pp. $9.25 (published by Spon)

A05-1744
Magill, P., et al. Air Pollution Handbook. McGraw-Hill, 645 pp. $32.75
Here is a definitive report on air pollution, how to control it, and areas requiring further research. It covers many topics including city planning and plant location in relation to pollution control, the chemistry of contaminated atmospheres, harmful pollutants, sampling procedures, test methods, and legal considerations.

A05-1745
Majid, K. I. Non-Linear Structures: Matrix Methods of Analysis and Design by Computers. Butterworth or John Wiley, 1972. 360 pp. $16.95
In the field of structural engineering, the combined effects of stability and plasticity are of paramount importance, especially in the case of modern, tall and slender structures in which non-linear and secondary effects are preponderant. The effect of non-linearity becomes more apparent with larger structures where recent advances in matrix methods of structural analysis can be applied. In using matrix methods, the problems of non-linearity follow naturally from the linear ones and are much easier to understand. The main aspects of recent work carried out on this subject are covered in this book, which outlines the use of modern methods of structural analysis suitable for computers.

A05-1746
Majid, K. I. Optimum Design of Structures. John Wiley, 1974. 264 pp. $22.50
Introduction to existing design methods, mathematical optimization, matrix force and matrix displacement methods, design of structures with variable topology.

A05-1747
Manson, S. S., ed. Metal Fatigue Damage--Mechanism, Detection, Avoidance, and Repair. American Soc. for Testing and Materials, 1971. 351 pp. $21.00 (for ASTM members, $16.80)
Five papers deal with fatigue in both the sub and normal creep range. Primarily concerned with gas turbine applications,

Design--Civil and Structural

this book contributes information to many applications of fatigue theory and practice.

A05-1748
Mantell, M. I. and Marron, J. F. Structural Analysis. Ronald Press, 1962. 423 pp. $11.95
 A unified treatment of statically determinate and indeterminate structures, with attention to both elastic and plastic conditions. This textbook presents a fresh approach to several structural concepts, including three-dimensional displacement geometry, compatibility, generalized stiffness, strain-energy methods, "frenum moments," three and four-moment equations for frames with nonprismatic numbers and sidesway, collapse or plastic methods, the generalized column analogy equation and multiple rings, and the symbolic generalizations of influence functions.

A05-1749
Manual of Steel Construction. American Institute of Steel Construction. 1000 pp. $16.00
 A complete revision, based on the February 1969 AISC Specification for the Design, Fabrication and Erection of Structural Steel for Buildings. New standard shape nomenclature, new standard properties and dimensions, new beam and column load tables, new design examples, and more.

A05-1750
Marshall, R. D. and Thom, H. C. S. Wind Loads on Buildings and Structures: Building Science Series 30. National Technical Information Service, 1971. 169 pp. $3.00. Order No. COM 71 00141

A05-1751
Martin, H. Introduction to Matrix Methods of Structural Analysis. McGraw-Hill, 1966. 331 pp. $19.50
 Structural analysis is presented from the finite element point of view in this treatment of matrix structural theory and its application in analyzing structure.

A05-1752
Martin, H. C. and Carey, G. F. Introduction to Finite Element Analysis. McGraw-Hill, 1973. 480 pp. $17.50
 The scope of the book encompasses intuitive structural concepts, through elementary problems of elasticity, heat transfer, fluid mechanics, to linear and nonlinear elasticity problems within an abstract finite element formulation, and arbitrary boundary value problems and initial value problems. Numerical mathematics and computer analysis are included and treatment of errors in finite element analysis and appropriate diagnostic procedures.

A05-1753
Matheson, J. A. L., et al. Hyperstatic Structures: An Introduction to the Theory of Statically Indeterminate Structures. Butterworth or Davey, 1971. 520 pp. $19.75

A05-1754
Maugh, I. C.   Statically Indeterminate Structures.   John Wiley, 1964.   443 pp. $15.00

A book which treats the subject of hyperstatic structures as an organic whole. An appendix has been added to this edition which explains an alternative approach to energy methods, and brings out a number of ideas.

A05-1755
Meehan, J. F., et al.   Managua, Nicaragua Earthquake of Dec. 23, 1972, Reconnaissance Report.   Earthquake Engrg. Research Institute, 1973.   214 pp. $5.00

The report describes mainly through photographs, the geology and faulting, the strong motion instrumental data, the behavior of selected buildings and the effects on utilities. The Managua earthquake is of special interest to engineers because it affected many structures that had been designed to possess various degrees of earthquake resistance. Conclusions and recommendations are included in the geology and engineering sections.

A05-1756
Meek, J. L.   Matrix Structural Analysis.   McGraw-Hill, 1971.   450 pp. $19.50

This book attempts to outline the fundamental principles of modern structural analysis, and to provide program sketches that allow the theory to be converted into usable computer programs. The author first examines basic matrix transformations and structural principles, and then proceeds to use them as the foundation for more advanced study.

A05-1757
Merritt, F.   Standard Handbook for Civil Engineers.   McGraw-Hill, 1968.   1326 pp. $36.00

This practice-oriented, data-packed handbook--prepared by a staff of specialists--covers design management, specifications, construction management, the use of computers, and municipal and regional planning in addition to the standard civil engineering subjects.

A05-1758
Merritt, F. S., ed.   Structural Steel Designers' Handbook.   McGraw-Hill, 1972.   866 pp. $27.50

This volume covers expert practice in structural steel design, from design criteria to structural performance analysis. The editor, in collaboration with a board of consultants and 14 authors from the structural engineering field, has assembled studies of 373 structural steel topics.

A05-1759
Metcalf and Eddy.   Wastewater Engineering: Collection, Treatment, Disposal.   McGraw-Hill, 1972.   782 pp. $22.50

A completely new version of the classic Metcalf-Eddy (circa

Design--Civil and Structural

1930), this work covers sewage, sewage systems, parameters of pollution, unit operations of treatment, chemical and biological unit processes. Chapters have been added on treating primary and secondary treatment processes, design of sludge systems, water pollution control, industrial waters, economics and operation of sewage systems.

A05-1760
Michalos, J. and Wilson, E. N. Structural Mechanics and Analysis. Macmillan, 1965. $12.95

A05-1761
Miroloyubov, L. Aid to Solving Problems in Strength of Materials. Beekman, 1976. 480 pp. $16.50
A study aid for engineering students, this contains a short exposition of the main principles of the theory of strength of materials. Tables of data included for solving the 1028 problems.

A05-1762
Morgan, T. L. Forces in Framed Structures. Barnes and Noble, 1952. $3.00 (published by Spon)

A05-1763
Morgan, V. A. The Analysis of Single-Bay Frames. Cement and Concrete Assn., 1968. 48 pp. £0.75

A05-1764
Morris, H. M. and Wiggert, J. M. Applied Hydraulics in Engineering. Ronald Press, 1972. 629 pp. $16.50
This book covers the principles of engineering hydraulics and hydrology, as applied in the design of structures and systems for hydraulic developments. Basic engineering principles are stressed throughout a wide selection of topics concerned with water planning, control, and utilization. The fundamentals of fluid statics and dynamics are reviewed and extensive treatment is given to both flow in closed conduits and in open channels.

A05-1765
Morris, O. E. Handbook of Structural Design. Van Nostrand Reinhold, 1963. 822 pp. $29.95
Here is information on all types of structural design for the practicing professional.

A05-1766
Munse, W. H. and Grover, L. Fatigue of Welded Steel Structures. American Welding Society, 1964. $6.00

A05-1767
Murashev, V. I., et al. Design of Reinforced Concrete Structures. Beekman, 1971. 596 pp. $22.50
The physical and mechanical properties of concrete, steel reinforcement, and reinforced concrete; the experimental fundamentals underlying the theory of the strength of reinforced concrete;

the features of axially and eccentrically compressed and tensioned elements, and of flexural and prestressed elements--these are some of the topics dealt with in the first part of this book. The second part is devoted to the general principles of designing buildings, with special attention to joint connections, flat-slab floors, roofs, beams, girders, trusses, frames, and the skin construction of roofs.

A05-1768
Nash, W. A. Strength of Materials: Including 345 Solved Problems. McGraw-Hill, 1972. 416 pp. $4.95

Covers tension and compression, direct shear stresses, torsion, shearing force and bending, centroids, elastic deflection of beams, double integration method, columns, theories of failure, and more.

A05-1769
National Bureau of Standards. Building Performance in the 1972 Managua Earthquake. U.S. Gov't. Printing Office, 1973. 150 pp. $1.60, Order No. C13.46:807 S/N 0303-01208

This report summarizes the field investigation conducted by a National Bureau of Standards team following the Managua, Nicaragua, earthquake of December 23, 1972. The major emphases in the investigation were to assist in evaluating conditions of public buildings, and to observe causes of building failure for future design and construction of earthquake resistant buildings.

A05-1770
National Bureau of Standards. Design, Siting, and Construction of Low-Cost Housing and Community Buildings to Better Withstand Earthquakes and Windstorms. U.S. Gov't. Printing Office, 1974. 132 pp. $4.85, Order No. C13.29/2:48 S/N 0303-01212

A05-1771
National Research Council of Canada. Wind Effects on Buildings and Structures. Univ. of Toronto Press, 1967. 1233 pp. $15.00

This book, containing thirty-seven papers based on the latest research and findings in the subject by authors from fifteen countries, covers such topics as climatic factors, boundary layer characteristics, wind tunnel modelling, full-scale measurements on buildings, the performance and response of particular structures including towers and bridges, and code criteria for design against wind. Proceedings of the International Research Seminar, Ottawa, 1967.

A05-1772
National Weather Service. Earthquake History of the United States. U.S. Gov't. Printing Office, 1973. 208 pp. $2.80. Order No. C 55.202:Ea 7 S/N 0319-00019

This history of prominent earthquakes in the United States, from historical times through 1970, is documented with earthquake

Design--Civil and Structural

data. Explanations of terms and tabular data have been included to assist the reader in understanding this study of earthquakes, many of which have been devastatingly destructive to life and property. Earthquakes are listed and described by region and state, and maps and photographs accompany the text.

A05-1773
National Weather Service. United States Earthquakes, 1971. U.S. Gov't. Printing Office, 1973. 176 pp. $2.10. Order No. C 55.226:971 S/N 0319-00024

Prepared annually since 1928, this volume of United States earthquakes lists epicenters of all earthquakes and associated phenomena recorded in the United States and nearby territories during 1971. Also featured is a list of principal earthquakes of the world during the year, along with brief accounts of their effects.

A05-1774
Navy Department. Design Manual, Cold Regions: Engineering. U.S. Gov't. Printing Office, 1975. 240 pp. $3.90. Order No. D 209.14/2:9/2 S/N 008-050-00114-3

Gives criteria that builders need in order to construct facilities in a cold or polar region.

A05-1775
Neal, B. G. Structural Theorems and their Applications. Pergamon, 1964. 208 pp. $7.00 (paperback $4.95)

A presentation of the theories of structural analysis showing how these stem from the principle of virtual work, with emphasis placed upon two types of transformation which can be achieved by this principle.

A05-1776
Newmark, N. M. and Rosenblueth, E. Fundamentals of Earthquake Engineering. Prentice-Hall, 1971. 640 pp. $25.95

Presents practical techniques used for earthquake-proofing every kind of structure. Features reliable methods in seismology for measuring intensity, maximum ground acceleration, velocities, displacements and the duration of ground motion. Includes methods for every operation involved in damage prevention, from testing the vulnerability of a possible construction site to choosing bracing arrangement to anticipate varying degrees of stress.

A05-1777
Noll, K. E. and Duncan, J. Industrial and Air Pollution Control. Ann Arbor Science, 1973. 354 pp. $20.00

Contents include monitoring-feasibility of reducing sulferoxide, air pollution standards for plant site selection, power generation and pollution, control for stone, concrete batch, hot mix asphalt plants, and other industrial applications.

A05-1778
Nonhebel, G., ed. Processes for Air Pollution Control. CRC Press, 1971. 736 pp. $37.50

Gas purification can be divided into two basic classes which cover a wide variety of processes: the removal of gaseous impurities and the removal of particulate impurities. Information is included on preliminary purification of crude gases; gas absorption including centrifugal gas absorbers and regenerative processes; solid chemical absorbents; purification of cryogenic gases; adsorption; catalytic destruction or conversion of impurities; removal of grit, dust, sprays and mists; and air purification.

A05-1779
Norrie, D. H. and deVries, G. The Finite Element Method: Fundamentals and Applications. Academic Press, 1973. 336 pp. $23.00
 Covers the formulation of physical problems, field problems and their approximate solutions, the variational calculus, the laplace or potential field, the Hemholtz and wave equations, the diffusion equation, and more.

A05-1780
Norris, C. H., et al. Elementary Structural Analysis. McGraw-Hill, 1976. 680 pp. $19.50
 Retaining its detailed coverage of basic and advanced structural theory, this extensive revision features important new information on the use of digital computers for solving structural analysis problems. This includes the development and use of computer codes and new procedures for the solution of analysis problems, as well as a detailed description of the force and displacement methods of transforming structural analysis data into data defining the structural response.

A05-1781
Norris, C. H., et al. Structural Design for Dynamic Loads. McGraw-Hill, 1959. 651 pp. $19.00
 A volume compiled from lecture notes which were prepared for a special program surveying this field. Topics include: behavior of materials under dynamic loading; calculation of response of structural systems to dynamic loading; modern computational systems to dynamic loading; and the application of structural design and analysis to cases involving dynamic loads.

A05-1782
Novozhilov, V. V. Thin Shell Theory. Wolters-Noordhoff, 1970. 417 pp. $18.00
 Translated from the 2nd Russian edition. Covers general theory of thin elastic shells, the membrane theory of shells, the analysis of cylindrical shells, and shells of revolution.

A05-1783
Obert, L. and Duvall, W. I. Rock Mechanics and the Design of Structures in Rock. John Wiley, 1967. 650 pp. $33.25
 The first book in English that covers the theoretical and experimental techniques for designing or evaluating the stability of

Design--Civil and Structural

underground structures in rock, such as mines, power stations, military installations, and storage chambers.

A05-1784
Oden, J. Mechanics of Elastic Structures. McGraw-Hill, 1967. 426 pp. $18.50
Geared for advanced undergraduate courses in structural engineering, this book develops basic relations of the linear theory of elastics, discusses basic field equations, and demonstrates their use.

A05-1785
Okamoto, S. Introduction to Earthquake Engineering. John Wiley, 1973. 561 pp. $36.00
Contents include: Earthquakes, Earthquake Intensity, Seismicity of Japan, Great Earthquakes in Japan. Design Earthquake Motion Procedures of Anti-Seismic Design. Aseimic Provisions. Earth Pressure During Earthquake. Roads, Railways and Rivers. Earthquake Resistance of Port and Harbor Facilities. Earthquake Resistance of Bridges, Dams. Arch Dams. Embankment Dams. Earthquake Resistance of Water Works. Earthquake Resistance of Underground Structures. Earthquake Resistance of Buildings.

A05-1786
Olsen, G. A. Strength of Materials. Prentice-Hall, 1956. 456 pp. $10.80
Covers the main topics of the usual introductory course in the strength of materials with a review to the appropriate mechanics included where necessary.

A05-1787
Olszak, W. and Sawczuk, A. Inelastic Behavior in Shells. Wolters-Noordhoff, 1967. 122 pp. $7.00
This book surveys the main trends in the theory and research of inelastic shells. Covers linear viscoelasticity, steady creep, limit analysis, elastic-plastic deformations, and carrying capacity of reinforced concrete shells.

A05-1788
Osgood, C. C. Fatigue Design. John Wiley, 1970. 523 pp. $27.00
Evaluates fatigue life as to degree of conservatism. Obtains solutions by applying the most adequate methods. The effects of shape, environment, materials, properties, and loading modes are thoroughly delineated. Contains references for data not readily available.

A05-1789
Osipova, L. N. and Tumarkin, S. A. Tables for the Computation of Toroidal Shells. Wolters-Noordhoff, 1965. 128 pp. $7.00

A05-1790
Pannell, F. N. Design Charts for Members Subjected to Biaxial Bending and Thrust. Cement and Concrete Assn., 1966. 52 pp. £1.20

A05-1791
Parcel, J. I. and Moorman, R. B. Analysis of Statically Indeterminate Structures. John Wiley, 1955. 571 pp. $19.50

A05-1792
Park, R. and Paulay, T. Reinforced Concrete Structures. John Wiley, 1975. 769 pp. $27.50
Emphasizes the basic behavior of reinforced concrete, rather than just presenting standard solutions for common problems. Examines aspects of earthquake engineering as they affect reinforced concrete structures, in particular the effects of high intensity reversed cyclic loading and the requirements of ductility.

A05-1793
Parker, H. Simplified Design of Reinforced Concrete. John Wiley, 1968. 310 pp. $14.00
Presents changes in allowable stresses allotted to concrete and steel in agreement with the Building Code Requirements for Reinforced Concrete.

A05-1794
Parker, H. Simplified Design of Structural Timber. John Wiley, 1963. 265 pp. $10.75

A05-1795
Parker, H. Simplified Engineering for Architects and Builders. John Wiley, 1975. 411 pp. $14.95
Provides a working knowledge of structural design and structural design procedures; updated to conform with current practice and industry-recommended specifications and design standards for steel, wood and reinforced concrete. Contains numerous illustrative design examples worked out step by step.

A05-1796
Parker, H. Simplified Mechanics and Strength of Materials. John Wiley, 1961. 285 pp. $9.50

A05-1797
Parker, H. and Hauf, H. D. Simplified Design of Structural Steel. John Wiley, 1974. 326 pp. $13.75
Topics covered include structural steel shapes and properties. Beam design, floor framing, joists, and connections are included as is the theory of plastic design.

A05-1798
Parker, H., et al. Simplified Design of Roof Trusses for Architects and Builders. John Wiley, 1953. 278 pp. $12.00

Design--Civil and Structural

A05-1799
Parkes, E. W. Braced Frameworks: An Introduction to the Theory of Structures. Pergamon, 1965. 208 pp. $6.25
A textbook emphasizing basic ideas of the theory of structures, which provides a clear physical picture of how structures behave and guidance in analyzing and designing them.

A05-1800
Peck, R. B., et al. Foundation Engineering. John Wiley, 1974. 514 pp. $17.50
Contains chapters on soil and rock classification and hydraulic properties, their stress-deformation-strength characteristics and methods of exploration. Foundation types and methods of construction are covered in chapters on drainage, stabilization and pile and pier foundations. Other chapters cover the basis of design for foundations and earth retaining structures.

A05-1801
Perloff, W. H. and Baron, W. Soil Mechanics--Principles and Applications. Ronald Press, 1975. 700 pp.
This book for advanced undergraduate students presents a modern and fundamental treatment of soil mechanics and many of its applications. Subject matter is introduced in a framework of published case histories, a device that illustrates the relevance of the text material to practical design situations while using the principles developed in the text to analyze the case histories presented.

A05-1802
Pisani, T. J. Essentials of Strength of Materials. Van Nostrand Reinhold, 1964. 283 pp. $6.95
This elementary book covers simple stresses, properties of materials, theory and design of beams and columns and essentials of plain and reinforced concrete.

A05-1803
Polakowski, N. H. and Ripling, E. J. Strength and Structure of Engineering Materials. Prentice-Hall, 1966. 535 pp. $15.95
Introduces three-dimensional concepts at the beginning in describing the principles of deformation and fracture. Relates the bulk material behaviors to their underlying atomic mechanisms. Develops the basic mechanical properties and numerical analysis by using the above as a foundation.

A05-1804
Pollution Control Technology. Research and Education Association, 1974. 608 pp. $17.75
The book reviews the state of the art and the technical and economic feasibility of processes, equipment and plants. There are 43 main topics, some of which include: meteorology and atmospheric dispersion, air pollution from carbon monoxide, nitrogen oxides, sulfur oxides, etc., tall stacks, incinerator design,

wastewater treatment, sewer overflows, sanitary landfills, and waste disposal for apartment houses.

A05-1805
Popko, E. Geodesics. Wittenborn, 1968. 104 illustrations. $4.00
This book is a primer of geodesic structures, part of a course supplement of the School of Architecture, University of Detroit.

A05-1806
Portland Cement Assn. Design of Multistory Reinforced Concrete Buildings for Earthquake Motions. Portland Cement Assn., 1961. 330 pp. $9.00
Furnishes information about the phenomena of ground motion, the characteristics of structures, and their behavior in earthquakes that is relevant to the earthquake-resistant design of multistory reinforced concrete buildings. Includes design considerations, code requirements, and an illustrated design example.

A05-1807
Poulos, H. G. and Davis, E. H. Elastic Solutions for Soil and Rock Mechanics. John Wiley, 1974. 411 pp. $19.25
Partial Contents--Fundamental Definitions and Relationships. Basic Solutions for Concentrated Loading. Distributed Loads on the Surface of a Semi-infinite Mass. Distributed Loading beneath the Surface of a Semi-infinite Mass. Surface Loading of a Finite Layer Underlain by a Rigid Base. Surface Laoding of Multi-Layer Systems. Rigid Loaded Areas. Stresses and Displacements in Cross-Anisotropic Media.

A05-1808
Powell, S. Water Conditioning for Industry. McGraw-Hill, 1954. 548 pp. $16.50
A guide to water conditioning describing the various types of apparatus in detail, including design features, control required for efficient performance, cost of installation and operation, and giving the technical material and engineering data needed for selecting and applying best processes for particular industrial water-treatment needs.

A05-1809
Prager, W. and Hodge, P. G. Theory of Perfectly Plastic Solids. Dover. 264 pp. $3.00
An introduction for research workers and students in engineering or applied mathematics. Contents: basic concepts; trusses and beams; torsion of cylindrical or prismatic bars; plane strain; problems with axial symmetry; general theory; specific problems; contained plastic deformation; extremum principles.

A05-1810
Preece, B. W. and Davies, J. D. Models for Structural Concrete. Applied Science, 1964. 252 pp. £4.00

Design--Civil and Structural

After an introductory chapter on the uses and application of models, the book presents the theories of direct and indirect model analysis which apply to any structural material. The special problems associated with models for structural concrete are discussed at some length.

A05-1811
Proceedings, Sixth International Conference on Soil Mechanics and Foundation Engineering. Univ. of Toronto Press, 1965. Vol. 1, 422 pp.; Vol. 2, 2587 pp.; Vol. 3, 611 pp.; $125.00 per set

An international review of the state of the art and the practical application of soil and rock mechanics in association with foundation engineering.

A05-1812
Przemieniecki, J. Theory of Matrix Structural Analysis. McGraw-Hill, 1968. 480 pp. $19.50

Discusses aspects of design analysis, and provides fundamental theorems of matrix structural analysis which are developed into practical methods suitable for the analysis of complex structures. Covers principles of virtual displacements and forces. Castigliano's theorems, minimum strain energy theorems, etc.

A05-1813
Pucher, A. Influence Surfaces of Elastic Plates. 4th ed. Springer-Verlag, 1973. 135 pp. $42.20

Introduction summarizes fundamental theories leading to the concepts of influence surfaces and demonstrates their application with seven examples; main part consists of 93 design charts, covering a variety of plate problems, varying from infinite and semi-infinite strips to rectangular plates and circular discs, all with various support conditions.

A05-1814
Ramaswamy, G. Design and Construction of Concrete Shell Roofs. McGraw-Hill, 1968. 641 pp. $32.50

Here is the first book to discuss the theory, design, and construction of both double-curvature and cylindrical or single-curvature shells.

A05-1815
Rao, H. K. Reinforced Cement Concrete--A Textbook for Polytechnic Students. Asia Publishing House, 1968. 242 pp. $2.25

Partial contents include design of simple beams, shear and bond stresses, slabs, reinforced brickwork, tee-beams and columns, retaining walls, prestressed concrete.

A05-1816
Raz, S. A. Analytical Methods in Structural Engineering. John Wiley, 1974. 393 pp. $9.95

Contents--Basic Principles in Structural Analysis. Method

of Consistent Deformations. The Slope-Deflection Method. The Moment-Distribution Method. Kani's Method. The Distribution of Deformation Method. The Method of Column Analogy. Influence Lines for Indeterminate Structures.

A05-1817
Reynolds, C. E. Basic Reinforced Concrete Design. Cement and Concrete Assn. Vol. 1, 1968. 264 pp. £1.20; Vol. 2, 1966. 224 pp. £1.20

A05-1818
Rice, P. F. and Hoffman, E. S. Structural Design Guide to the ACI Building Code. Concrete Construction Publications, 1972. 437 pp. $19.95

A guide for the ACI 1971 building code for more profitable using routine designs involving common reinforced concrete structural elements. The book covers situations involving the combination of torsion and shear design; the application of 2-way slab design to flat plates and slabs, waffle flat slabs, and slabs with beams; the use of new methods for crack control in one-way slab and joint systems, beams and girders; and two-way slabs.

A05-1819
Richart, F. E., Jr., et al. Vibrations of Soils and Foundations. Prentice-Hall, 1970. 414 pp. $17.95

Examines procedures for analysis, design, and measurement of the response of foundations to dynamic loadings; and the transmission of vibrations through soils. Although primary emphasis is directed towards vibrations of the magnitudes generated by machinery, these principles can be applied to the dynamic conditions resulting from earthquakes or blast loadings.

A05-1820
Roark, R. J. and Young, W. C. Formulas for Stress and Strain. McGraw-Hill, 1976. 512 pp. $19.50

This major revision of a highly respected, best-selling work contains approximately 30% entirely new material. There has been a tremendous increase in the amount of tabulated data included, much of which has been restructured in a style most easily used on a computer. A new chapter on curved beams has been added. Most of the equations in the book have been updated to include the most recent approaches to analysis and design of a variety of members and structures. The number of different loading cases and combinations of boundary conditions has been increased from 435 to 1,460. Engineers, designers, architects and anyone else involved in the analysis and understanding of stress and strain should find this thoroughly up-to-date revision an indispensable aid in his work.

A05-1821
Robinson, J. Integrated Theory of Finite Element Methods. John Wiley, 1973. 428 pp. $33.00

Contents--Element Characteristic Matrices and Interchangeability. Characteristic Matrices of Some Strain and Some Stress

Design--Civil and Structural

Elements. Inconsistent Elements and Their Characteristic Matrices. Isoparametric Strain Elements. Isoparametric Stress Elements. Family of Axials and Shear Panels. A Curved Beam Element in Three Dimensions. Warped Semi-Monocoque Quadrilateral Membrane Element. Family of Cracked Finite Elements. Analysis of Large Practical Structures Using Substructure Coupling Procedures.

A05-1822
Robinson, J. Structural Matrix Analysis for the Engineer. John Wiley, 1966. 344 pp. $14.00
Contents include matrix algebra, linear equations, element energy and flexibility matrices, the matrix force approach, the rank technique, and the matrix displacement approach.

A05-1823
Robinson, J. L. Mechanics of Materials. John Wiley, 1969. 345 pp. $10.95
A text that covers the entire subject from an introduction to fairly advanced applications. Numerous worked examples throughout. Subjects covered include pressure vessels, beams and columns, steel shapes, and methods of calculation by means of moment-area, double integration and other methods.

A05-1824
Rockey, K. C. The Finite Element Method: A Basic Introduction. John Wiley, 1975. 239 pp. $22.00
Designed as an introduction for practicing engineers, this book combines the theoretical basis and practical aspects of the finite method, which is gaining widespread use among engineers.

A05-1825
Rogers, G. L. and Causey, M. L. Mechanics of Engineering Structures. John Wiley, 1962. 428 pp. $16.75

A05-1826
Rogers, P. Reinforced Concrete Design for Buildings. Van Nostrand Reinhold, 1973. 288 pp. $13.95
The book includes work sheets for designing beams and columns. The procedures permit longhand computation. On the other hand, computerization of these procedures is not difficult, and the book provides many examples of computerized solutions.

A05-1827
Rogers, P. Tables and Formulas for Fixed End Moments of Members of Constant Moment of Inertia and for Simply Supported Beams. Frederick Ungar, 1965. 104 pp. $5.50
A manual that gives solutions of fixed end moments for rigid frames and continuous beams, and simply supported beam moments. Each loading condition is separately indicated with a loading diagram, mathematical formulas, tables of fixed end moment coefficients, and charts.

A05-1828
Rogers, P., et al. Steel Columns Eccentrically Loaded. Frederick Ungar, 204 pp. $7.50

A manual for the design of steel columns. The charts encompass virtually all the possible variations of axial load and moment combinations and include all the commonly used wide flange column shapes, over a range of unbraced lengths from 10 feet to 30 feet, with varying length increments. Separate pages are used for moments applied about the "strong" axis (x-x) and the "weak" axis (y-y).

A05-1829
Roland, C. Frei Otto: Tension Structures, Ideas and Experiments in Lightweight Construction. Praeger, 1971. 180 pp. $22.50

The designs of Frei Otto are being hailed as ideal solutions to the problems of enclosing large open spaces. This study of his work provides extensive examples and technical information on the field of tension structures. New structural possibilities envisioned by Otto include mutable roofs that can be adapted to changing environmental conditions and giant shelters that span whole sections of a city, enabling the inhabitants to control their own weather.

A05-1830
Rolfe, S. and Barsom, J. Fracture and Fatigue Control in Structures: Applications of Fracture Mechanics. Prentice-Hall, 1976. 608 pp. $27.50

The authors illustrate the importance of engineering experience and economics in combination with theory. In a simplified manner, the authors present the reader with examples of recently developed fracture criteria to show how those criteria are developed. Including actual case studies, the authors give fracture control plans for such structures as bridges, nuclear pressure vessels, aircraft, and floating nuclear plants.

A05-1831
Rosman, R. Tables for the Internal Forces of Pierced Shear-Walls Subject to Lateral Loads. Adler's Foreign Books. 93 pp. $6.25

This book contains tables which allow a quick determination of the internal forces of shear-walls intersected with one arbitrarily, or two symmetrically arranged vertical rows of openings. Not only the influence of an uniformly distributed load, but also trapezoidal and triangular loads as well as a concentrated load on top of the system are dealt with. These methods allow for a more precise investigation of the influence of the wind load and seismic disturbing forces.

A05-1832
Rubinstein, M. F. Matrix Computer Analysis of Structures. Prentice-Hall, 1966. 402 pp. $17.20

Provides a unified development of matrix computer methods

Design--Civil and Structural

of structural analysis applicable to civil, structural, mechanical, aeronautical and aerospace structures.

A05-1833
Rubinstein, M. F. Structural Systems: Statics, Dynamics and Stability. Prentice-Hall, 1970. 306 pp. $16.95

Theme is the synthesis of systems from constituent elements in finite element method of structural analysis and computation for stresses and strains in frames, plates, and shells. Uses coordinates to characterize differential elements, finite elements and total systems and compute stresses and strains.

A05-1834
Ruskin, A. M. Materials Considerations in Design. Prentice-Hall, 1967. 103 pp. $4.95

Application of knowledge of materials to engineering problems. Single phase materials are studied in light of the functions all materials perform; i.e., interacting with energy and reacting to forces. Emphasizes that, unlike energy which can be thermal, mechanical, electrical, magnetic, or chemical, materials can react in only a limited number of ways, and resist or yield to forces in a similarly limited number of ways.

A05-1835
Rygol, J. Nomograms for the Analysis of Frames. Cement and Concrete Assn., 1957. 60 pp. 26 nomographs. £0.90

A05-1836
Sachs, P. Wind Forces in Engineering. Pergamon, 1972. 480 pp. $36.00

The book collates wind speed and forces information to provide a fundamental background for engineers. The subject matter gives basic data for calculating wind speeds and forces, and for obtaining empirical information through wind-tunnel work. It applies these methods to structures commonly susceptible to wind forces such as building bridges, masts and chimneys, cables, cooling towers and radar aerials.

A05-1837
Salencon, J. Applications of the Theory of Plasticity in Soil Mechanics. John Wiley, 1975.

Brings together and presents all the aspects of plasticity theory which can be used in soil mechanics.

A05-1838
Salmon, C. G. and Johnson, J. E. Steel Structures: Design and Behavior. Intext, 1971. 946 pp. $18.50

Presents the theoretical background for understanding the behavior of structural steel components as related to the 1969 AISC specification requirements, with examples illustrating the procedure for selecting such components. Emphasizes elastic and inelastic stability to aid the designer in applying the AISC specification provisions which relate to buckling problems.

A05-1839
Salvadori, M. and Heller, R. Structure in Architecture. Prentice-Hall, 1963. 384 pp. $13.95
Non-mathematical presentation of basic principles of architectural structures and structural systems.

A05-1840
Salvadori, M. and Levy, M. Structural Design in Architecture. Prentice-Hall, 1967. 457 pp. $15.00
Covers methods of analysis required by preliminary design with a minimum of mathematics, and without resorting to "cook book" formulas.

A05-1841
Sanks, R. L. Land Treatment and Disposal. Ann Arbor Science, 1976. 310 pp. $20.00
Land disposal is becoming a very attractive alternative for the ultimate disposal of municipal and industrial wastewater. Covering engineering methods and design, feasibility, consequences, economics and environmental policies, cost data are given to assist the engineer to make preliminary cost estimates for an entire waste treatment project.

A05-1842
Sarkar, R. K. Structural Analysis-Influence Coefficient Method. Muko (Munich), 1971. 418 pp. £3.50 (available from Cement and Concrete Assn.)

A05-1843
Save, M. A. and Massonet, C. C. Plastic Analysis and Design of Plates, Shells and Disks. American Elsevier, 1970. 490 pp. $31.00
In this text solved problems of limit analysis and design of plates, shells and disks, at present widely dispersed in the literature, are brought together. Both metal and reinforced concrete structures are covered. The chapter on reinforced concrete plates, that clearly places the yield line method within the frame of the general theory of limit analysis, is especially developed in view of its practical importance.

A05-1844
Savin, G. N. and Fleishman, N. P. Rib-Reinforced Plates and Shells. Israel Program for Scientific Translations, 1964. 344 pp.
This book is the result of theoretical studies and investigations on the behavior of thin-walled structural elements (plates, shells, etc.) and their reinforcing ribs and braces. The solutions to most of the problems dealt with in the book are presented in tables and graphs.

A05-1845
Schroeder, W. L. Soils in Construction. John Wiley, 1975. 236 pp. $13.95

Design--Civil and Structural

Contains introductory level information on soil behavior and testing, and relates this information to construction specifications.

A05-1846
Schwartz, M. Civil Engineering for the Plant Engineer. McGraw-Hill, 1972. 290 pp. $17.50

Here is the first book written for those industrial or plant engineers who are expected to have some familiarity with subject areas traditionally thought of as belonging to the field of civil engineering. These areas include property acquisition, industrial surveying, soils and foundations, construction codes, materials, construction techniques, and others.

A05-1847
Scott, C. R. An Introduction to Soil Mechanics and Foundations. Applied Science, 1969. 316 pp. £4.00

Provides those engaged in the design of foundations and earth structures with a description of soil behavior--and of the effects of the clay minerals and of the soil water on such behavior. Encourages such persons to look critically at the traditional methods of analysis and design.

A05-1848
Scott, R. and Schoustra, J. Soil Mechanics and Engineering. McGraw-Hill, 1968. 320 pp. $11.95

The fundamental physics and mechanics of soils, and their practical applications to soil engineering problems are developed in this systematic theory-to-practice approach. It treats a wide range of situations and presents material in a form suited to computer calculations.

A05-1849
Sechler, E. E. Elasticity in Engineering. Dover. 419 pp. $3.50

Written primarily for engineers working in structural analysis, this book is of value in undergraduate and graduate courses in applied elasticity and advanced strength of materials. It brings together in a single volume the material of most interest to the engineer who needs a broad foundation in stress and deformation of elastic bodies under load.

A05-1850
Seelye, E. E. Data Book For Civil Engineers. John Wiley. Vol. 1, Design. 1960. 670 pp. $29.50; Vol. 2, Specifications and Costs. 1957. 566 pp. $26.75

A05-1851
Seelye, E. E. Foundations: Design and Practice. John Wiley, 1956. 466 pp. $20.50

A05-1852
Semerdjiev, S. Metal-to-Metal Adhesive Bonding. Cahners, 1970. 204 pp. $15.00

Examines the use of metal-to-metal bonding in various branches of industry and engineering. Covers properties, design and inspection of adhesives and adhesive bonded joints.

A05-1853
Shaw, F. S. Virtual Displacements and Analysis of Structures. Prentice-Hall, 1972. 561 pp. $18.95

Develops the analysis of structures using the virtual displacement principle only. Has a parallel treatment of some classical methods of analysis with current matrix methods. Proceeds from the linear treatment to a non-linear analysis in which both non-linear stress-strain laws and equilibrium in the deformed position are discussed.

A05-1854
Shepley, E. Continuous Beam Structures--A Degree of Fixity Method and the Moment of Distribution. Cement and Concrete Assn., 1962. 128 pp. £0.60. Order No. 12.018

A05-1855
Shermer, C. L. Design in Structural Steel. Ronald Press, 1972. 311 pp. $10.95

A concise treatment of the fundamentals of structural steel design for engineering and architectural students. Emphasis is placed on the effects of loads on structures, irrespective of the nature or source of the loads. Both elastic and plastic design methods are treated together rather than as completely separate entities. Frequent reference is made to the latest AISC design specification as well as other applicable codes and specifications. The aim throughout the book is to prepare the student to cope with practical design problems at the professional level.

A05-1856
Simitses, G. An Introduction to the Elastic Stability of Structures. Prentice-Hall, 1975. 288 pp. $18.95

This book introduces the basic concepts of elastic stability, illustrates the approaches used in solving stability problems, and discusses the different buckling phenomena observed in nature and how they relate to the mathematical models analyzed. From the relation between buckling loads and load carrying capacity to high and shallow arch problems subject to instability, this introductory book offers a complete coverage of the elastic stability of columns and a clear and comprehensive treatment of the energy criterion for stability.

A05-1857
Singer, F. L. Strength of Materials. Harper and Row, 1962. 590 pp. $12.95

An introductory text on the principles of strength of materials, reflecting modern concepts. Axial, torsional, and flexural loading are thoroughly discussed for both stresses and deformations before their combined effects are considered. Major emphasis is

Design--Civil and Structural

on elastic analysis, but the field of inelastic analysis or plasticity is also extensively covered.

A05-1858
Smith, A. and Nicholson, A., eds. Advances in Creep Design. John Wiley, 1972. 485 pp. $42.50

A05-1859
Smith, G. N. Elements of Soil Mechanics for Civil and Mining Engineers. Beekman, 1974. 370 pp. $15.00
Presents in concrete form, the applications of soil mechanics that are now used in civil engineering. Soil suction and partial saturation are also considered.

A05-1860
Smith, G. N. Introduction to Matrix and Finite Elements in Civil Engineering. Applied Science, 1971. 222 pp. £3.00
The finite element methods described in the text deal with civil, rather than structural, engineering problems, the emphasis being on continuum mechanics and not on the study of the discrete elements of a structure. Matrix and finite element methods are so intimately connected with the solution of stress and analysis problems that the author has included two chapters giving a summary of this important subject. The worked examples have been largely taken from the field of soil mechanics but most are applicable to other fields of civil engineering.

A05-1861
Smolira, M. Analysis of Structures. Cement and Concrete Assn., 1955. 183 pp. £0.90

A05-1862
Smolira, M. Analysis of Tall Buildings by the Force Displacement Method. John Wiley, 1975. 299 pp. $39.50
Presents a new approach to the subject of structural analysis as applied mainly to tall building frames and shear walls. A wide range of practical problems is considered. These are suitable for application in the office as well as for laboratory work.

A05-1863
Spangler, M. G. and Handy, R. L. Soil Engineering. Intext, 1973. 748 pp. $16.00
Examines the origin, properties, and identification of soils, focusing on the practical engineering aspects of the subject. There is a chapter on air-photo interpretation of soils and one on weather and soil profiles.

A05-1864
Spillers, W. R. Automated Structural Analysis: An Introduction. Pergamon, 1972. 182 pp. $11.00
A computer-oriented text which presents an introduction to structural analysis for students and engineers who solve structures by computer. A basic textbook for undergraduate students and a

reference source for postgraduate students and engineering research workers.

A05-1865
Spindel, P. D. Computer Applications in Civil Engineering. Van Nostrand Reinhold, 1971. $9.95

The author discusses basic computer hardware, FORTRAN as a programming language, various aspects of computer use, the problem of communication with the computer, and the use of information systems within an engineering context.

A05-1866
Spunt, L. Optimum Structural Design. Prentice-Hall, 1971. 168 pp. $16.95

A generalized approach employing mathematical criteria and constraints, with solutions obtained in parametric index form. Emphasizes the comparison between alternative shapes and material selections for fundamental classes of load transmission. Proceeds from single element optimizations to systems of elements. Provides an important complement to computer-aided design.

A05-1867
Stanek, F. J. Stress Analysis of Circular Plates and Cylindrical Shells. Dorrance and Co., 1970. 102 pp. $7.50

This book offers methods of stress analysis for flat circular plates, cylindrical shells and tapered circular plates subjected to axially symmetric loads based upon the classical small deflection theory.

A05-1868
Stark, R. and Nicholas, R. Civil Engineering Systems: Mathematical Techniques for Design. McGraw-Hill, 1971. 256 pp. $18.50

Designed to aid civil engineers interested in mathematical techniques commonly associated with operations research and systems engineering. The emphasis is on application for design. Treats methodology, optimization, and probability.

A05-1869
Stephenson, R. E. Computer Simulation for Engineers. Harcourt Brace Jovanovich, 1971. 504 pp. $15.95

A textbook for an upper-level introductory course in computer simulation. Teaches students to simulate continuous systems and solve engineering problems using analog and digital computers.

A05-1870
Stevens, L. K., et al. Finite Elements and Matrix Structural Analysis. Butterworth (in preparation)

Contents include elementary concepts in matrix algebra, basic structural concepts, one, two, and three-dimensional problems, bending of plates, and assemblages.

A05-1871
Straub, C. P.   Low-Level Radioactive Wastes: Their Handling, Treatment, and Disposal. Energy Research and Development Administration, 1964.   430 pp. $22.70 (available from University Microfilms)
A book that brings together material that was scattered throughout many project reports and published literature relating to continuous operations. It was written for those interested in low-level waste disposal: the health physicist; the water and sewage-works personnel concerned with the efficiency of water and sewage-treatment processes for the removal of radioactive materials; the personnel engaged in the design, construction, licensing, and operation of treatment facilities; and the student of nuclear technology.

A05-1872
Stuttgart Universität, Institute of Lightweight Structures. Biology and Building I: Colloquium Report. International Publications, 1971. 79 pp. $6.50
The IL is the leading research center in regard to widespanning and adaptable buildings and membrane constructions, domes, etc. All texts are in German/English.

A05-1873
Stuttgart Universität, Institute of Lightweight Structures. Biology and Building II: Soap Film Models. International Publications, 1973. 80 pp. $7.50

A05-1874
Stuttgart Universität, Institute of Lightweight Structures. Biology and Building III. International Publications, 1973. 80 pp. $12.50

A05-1875
Stuttgart Universität, Institute of Lightweight Structures. Convertible Roofs. International Publications, 1973. 400 pp. $35.00

A05-1876
Stuttgart Universität, Institute of Lightweight Structures. The Experimental Determination of Minimal Nets. International Publications, 1969. 56 pp. $6.50

A05-1877
Stuttgart Universität, Institute of Lightweight Structures. Project Study "City in the Arctic." International Publications, 1971. 56 pp. $7.50

A05-1878
Stuttgart Universität, Institute of Lightweight Structures. Shadow in the Desert. International Publications, 1973. 88 pp. $7.50

A05-1879
Sutherland, H. and Reese, R. C. Reinforced Concrete Design.
    John Wiley, 1943. 559 pp. $11.95

A05-1880
Szilard, R. Theory and Analysis of Plates: Classical and Numerical Methods. Prentice-Hall, 1973. 672 pp. $25.00
    Serves as an introduction to the classical theory and analysis of plates. Covers the static aspects of linear plate theory, non-linear plate theory and stability of plates. Dynamic behavior is not restricted to linearly elastic plates, but plastic behavior and method of analysis such as yield-line method are also included. More than 170 readily usable plate formulas.

A05-1881
Takabeya, F. Calculation and Moment-Tables: The Methods of Cross, Kani, and Takabeya. Adler's Foreign Books. 276 pp. $28.50
    Prof. Takabeya classified 337 kinds of rigid frames as well as loading cases and illustrated moment-tables, from which, without further calculation, one can obtain necessary moments. The moment-tables are the appendices of the book and the tables give material for actual design of such frames. The book is divided into three parts. Part A contains multi-story frames subjected to vertical and horizontal loads; Part B gives applications of moment-tables for horizontal and vertical loading; Part C discusses moment-tables as related to symmetrical rigid frames and girders.

A05-1882
Tall, L. Structural Steel Design. Ronald Press, 1974. 875 pp. $18.50
    The aim of the second edition is to further understanding of the basic behavior of structures. Its preparation has been prompted by changes in design specifications and by advances in the knowledge of structural behavior. The material is arranged so that basic behavior and the basis for design rules may be studied first, without references to a specific code. However, various specifications and the intent of their rules are cited frequently throughout the book.

A05-1883
Teng, W. C. Foundation Design. Prentice-Hall, 1962. 466 pp. $17.20
    Treatise of foundation design encompassing theory and practice. Reviews basic principles underlying foundation engineering.

A05-1884
Terzaghi, K. and Peck, R. B. Soil Mechanics in Engineering Practice. John Wiley, 1967. 729 pp. $17.25
    Directed specifically to those aspects of civil engineering dealing with foundations, deep excavations, landslides, earth fills, and earth or rockfill dams. Recent findings of research workers in soil mechanics that bear on practice are included. Consideration

Design--Civil and Structural

is given to the implications of the inherent variability of natural deposits of soil, and procedures are developed for safe, economical design.

A05-1885
Thadani, B. N. Modern Methods in Structural Engineering. Asia Publishing House, 1964. 619 pp. $17.00
The first part analyzes determinate structures and calculates deflections. The second part explains modern methods for the analysis of indeterminate structures including a complete exposition of Kani's method for the analysis of multi-storeyed structures, together with the author's contributions to the subject. A chapter on arches, multi-arched and multi-gables frames is included, and the book concludes with an introduction to the plastic methods of structural analysis.

A05-1886
Timoshenko, S. Strength of Materials. Part 1, Elementary Theory and Problems. Van Nostrand Reinhold, 1955. 434 pp. $10.95
Covers tension and compression, analysis of stress, strain, shear and bending moment, and beam deflection. Statically indeterminate problems in bending are given as are beams of variable cross section and of two materials.

A05-1887
Timoshenko, S. Strength of Materials. Part 2, Advanced Theory and Problems. Van Nostrand Reinhold, 1956. 572 pp. $13.50
Contents: Beams on Elastic Foundations. Beams with Combined Axial and Lateral Loads. Special Problems in Bending of Beams. Thin Plates and Shells. Buckling of Bars, Plates and Shells. Deformations Symmetrical about an Axis. Torsion. Stress Concentration. Deformations Beyond the Elastic Limit. Mechanical Properties of Materials.

A05-1888
Timoshenko, S. P. and Gere, J. M. Mechanics of Materials. Van Nostrand Reinhold, 1972. $14.95
Topics covered are analysis of structural members subjected to axial load, torsion and bending, basic concepts of mechanics of materials, strain energy, stress and strain transformations, and inelastic behavior.

A05-1889
Timoshenko, S. and Woinowsky-Krieger, S. Theory of Plates and Shells. McGraw-Hill, 1959. 480 pp. $21.00
The text offers discussion of deflection of plates due to transverse shear, stress concentrations, and bending of plates.

A05-1890
Timoshenko, S. and Young, D. H. Elements of Strength of Materials. Van Nostrand Reinhold, 1968. 388 pp. $9.50

The same approach is used as in the earlier edition. This consists of proceeding gradually from simpler cases to more complex ones and relying on physical and geometrical considerations of deformation to establish the patterns of stress distribution under various types of loading.

A05-1891
Timoshenko, S. and Young, D. Theory of Structures. McGraw-Hill, 1965. 629 pp. $19.50
This edition develops an understanding of the basic principles underlying structural analysis methods--primarily for truss and frame type structures.

A05-1892
Tonge, B. Y. The Indeterminate Beam: Theory and Examples. Butterworth, 1972. 302 pp. £4.50 (paperback £2.60)
A major step is the progression from the analysis of simple structural problems to the solution of problems which are statically indeterminate. This has many methods of analysis which are introduced by their application to the statically indeterminate beam. A complete understanding of the beam problem is necessary for the development of the theories for the solution of rigid frame problems. This book aims to provide this understanding by confining the various methods of analysis to beam problems only.

A05-1893
Tschebotarioff, G. Foundations, Retaining and Earth Structures: The Art of Design Construction and Its Scientific Basis in Soil Mechanics. McGraw-Hill, 1973. 704 pp. $24.50
An extensive revision of the classic, Soil Mechanics, Foundations and Earth Structures, this new edition has been completely updated. New cases and examples are included to insure the usefulness of the book among soils and structural engineers as well as architects and administrators.

A05-1894
Tsytovich, N. A. Mechanics of Frozen Ground. McGraw-Hill, 1975. 448 pp. $29.50
Co-published with Scripta Technica, this translation of the 1973 Soviet edition will be of use to geologists, architects, civil engineers, builders and anyone whose work entails a consideration of the peculiar characteristics of frozen ground.

A05-1895
Tuma, J. J. Structural Analysis: Including 200 Solved Problems. McGraw-Hill, 1969. 304 pp. $5.95
Contents include statics of simple structures, stresses in beams, arches, frames and trusses, elastic curve, virtual work, energy, and more.

A05-1896
Ural, O. Finite Element Method: Basic Concepts and Applications. Intext, 1973. 250 pp. $13.00

Design--Civil and Structural

A complete theoretical derivation of the finite element method using very basic mathematical notation and operations. Several types of finite elements are discussed and an engineering evaluation of their applications is presented. Computer and long-hand solutions to problems are given and new areas of possible applications of the finite element method are suggested.

A05-1897
Ural, O. Matrix Operations and Use of Computers in Structural Engineering. Intext, 1971. 271 pp. $7.00
Illustrates the use of matrices and digital computers in structural analysis and presents available computer programming systems in structural analysis and design.

A05-1898
Vawter, J. and Clark, J. G. Elementary Theory and Design of Flexural Members. John Wiley, 1950. 215 pp. $7.95

A05-1899
Venkatraman, B. and Patel, S. Structural Mechanics with Introductions to Elasticity and Plasticity. McGraw-Hill, 1970. 704 pp. $19.50
Methods for analyzing the mechanical behavior of structures are developed in this book which discusses all aspects of elasticity, structural mechanics, and plasticity.

A05-1900
Vinson, J. R. Structural Mechanics: The Behavior of Plates and Shells. John Wiley, 1974. 173 pp. $14.25
Contents--Derivation of the Governing Equation for a Rectangular Plate. Solutions to Problems of Rectangular Plates. Thermal Stresses in Plates. Circular Plates. Buckling of Thin Plates. Energy Methods. Cylindrical Shells. Elastic Stability of Shells.

A05-1901
Vinson, J. R. and Chou, T. Composite Materials and Their Use in Structures. John Wiley, 1975. 438 pp. $47.50
Provides a fundamental understanding of the physical and mathematical aspects of the materials system, and structures comprised of composite materials, at a level that can be understood by students and practicing engineers alike.

A05-1902
Vlasov, V. Z. and Leont'ev, N. N. Beams, Plates and Shells on Elastic Foundations. Israel Program for Scientific Translations, 1960. 368 pp.
The book presents a new theory of the design and calculation of constructions created on elastic foundations which is much simpler than earlier theories. The theory advanced is based on V. Z. Vlasov's variational methods.

A05-1903
Volterra, E. and Gaines, J. H. Advanced Strength of Materials.
    Prentice-Hall, 1971. 522 pp. $18.95
    Offers balanced presentation of advanced subjects in the field of strength of materials while emphasizing practical solutions to engineering problems. Its principle object is to provide fundamental bases for theories in order to prepare for advanced study.

A05-1904
Wah, T. and Calcote, L. R. Structural Analysis by Finite Difference Calculus. Van Nostrand Reinhold, 1970. 264 pp. $14.95
    This book shows the use of finite difference calculus to analyze complex engineering structures and the many practical applications inherent in its analytical capabilities. Applications in problem areas include elastic, elastoplastic, vibration, and stability analyses of various structural types.

A05-1905
Walker, A. C. Design and Analysis of Cold Formed Sections.
    John Wiley, 1975. 174 pp. $14.95
    The methods of analysis relevant to cold formed sections are introduced and include local buckling, warping torsion, and torsional instability. The mathematics has been kept to a minimum and worked examples of design analysis are given.

A05-1906
Walmer, M. E. and Baron, S. L. Manual of Structural Design and Engineering Solutions. Prentice-Hall, 1975. 1089 pp. $49.95
    The book is divided into nine sections. The first is on the mechanics of beams and includes wind loading. The second topic is structural steel design and covers sections, beams, columns, trusses, decks, welding and fasteners. The third section is on timber design including framed dome design. Concrete design follows and encompasses slabs, beams, columns, stairs, foundations and retaining walls. Other sections include properties of sections, rigid frame design, high rise design and pile driving.

A05-1907
Wang, C. K. Applied Elasticity. McGraw-Hill, 1953. 357 pp. $18.50
    The basic knowledge of theory is given so the reader can formulate any problem occurring in the classical theory of elasticity and solve it by using analytical or numerical methods. Topics covered include analysis of stress and strain, torsion of shaped bars; energy principles and variation methods; theory of thin shells and curved plates, etc.

A05-1908
Wang, C. K. Computer Methods in Advanced Structural Analysis.
    Intext, 1973. 398 pp. $14.00
    Introduction to structural dynamics, stability, finite element

Design--Civil and Structural 315

analysis and minimum-weight design techniques. Ten basic computer programs--two or three in each area--are used to teach advanced students to apply similar methods to the solution of a wide range of analysis problems. Numerical examples and answers are provided within each chapter.

A05-1909
Wang, C. K. Introductory Structural Analysis with Matrix Methods. Prentice-Hall, 1972. 240 pp. $14.95
Integrates matrix methods with conventional methods throughout. Demonstrates the methods of structural analysis and inserts at appropriate points, numerical exercises that make the book useful as a workbook for programmed study. Develops energy theorems in simple terms so that they may form the basis for advanced study of "finite-element" techniques.

A05-1910
Wang, C. K. Matrix Methods of Structural Analysis. Intext, 1970. 406 pp. $14.50
This second edition begins with a chapter on matrix multiplication, inversion, and transportation and then covers the displacement method of analyzing trusses, continuous beams, and rigid frames with and without sidesway. The force method is presented for the analysis of trusses and rigid frames.

A05-1911
Wang, C. K. Statically Indeterminate Structures. McGraw-Hill, 1953. 424 pp. $14.50
Covers deflection of statically determinate beams, frames, and trusses, consistent deformation method, the three moment equation, the slope deflection method, analysis of fixed arches, and more.

A05-1912
Wang, C. K. and Eckel, C. L. Elementary Theory of Structures. McGraw-Hill, 1957. 387 pp. $13.50
Covers building bents, equilibrium of coplanar force systems, influence diagrams, shears and bending moments in beams, stresses in trusses, and criteria for maximums.

A05-1913
Wang, C. K. and Salmon, C. G. Reinforced Concrete Design. Intext, 1973. 934 pp. $18.50
This new edition emphasizes the ultimate strength method, and features expanded coverage of concrete and reinforcement steel materials and properties, shear strength, reinforcement, two-way and flat slabs, and torsion--in each case incorporating and amplifying the provisions of the 1971 ACI Code.

A05-1914
Wang, P. C. Numerical and Matrix Methods in Structural Mechanics with Applications to Computers. John Wiley, 1966. 426 pp. $17.25

Contents include finite difference method in structural mechanics, numerical integration, matrix algebra, the force and relaxation methods in structural mechanics.

A05-1915
Wass, A. Manual of Structural Details for Building Construction. Prentice-Hall, 1968. 386 pp. $16.95

Contents include material on drawings, specifications and cost accounting. Construction details include post and beam construction, masonry, curtain walls, flashing and steel details. The building types comprise a parking garage, old age homes, shop building and county office building, among others.

A05-1916
Weaver, W., Jr. Computer Programs for Structural Analysis. Van Nostrand Reinhold, 1967. 320 pp. $4.95

This book contains flow charts for the analysis of structures by the stiffness method. Each program can accommodate all types of framed structures subjected to any number of loading systems. They are independent of any particular machine or computer language, and are presented in a sequence of increasing difficulty corresponding to the size and complexity of the structures to be analyzed. Background theory, example problems, computer results and supplementary programming topics are also included.

A05-1917
Westergaard, H. M. Theory of Elasticity and Plasticity. Dover, 176 pp. $2.00

Compact account of fundamental ideas and mathematical techniques of elasticity theory, containing much original material. Useful to anyone with background in mechanics of materials, especially graduate students in civil or mechanical engineering.

A05-1918
Whitaker, T. The Design of Piled Foundations. Pergamon, 1970. 200 pp. $12.00 (paperback $5.50)

An introduction to piled foundation design based on examination of practical methods. Piles are classified as displacement or non-displacement and methods of installation, their merits and faults are discussed.

A05-1919
White, R. N., et al. Structural Engineering. Vol. 1, Introduction to Design Concepts and Analysis. John Wiley, 1972. 260 pp. $10.50

Evolution of a structure, objectives of structural design, man-made and natural loads, structural form, etc.

A05-1920
White, R. N., et al. Structural Engineering. Vol. 2, Indeterminate Structures. John Wiley, 1972. 330 pp. $11.50

Displacement calculations, moment-area method, virtual

Design--Civil and Structural

force method, direct stiffness method, influence lines, slope-deflection equations and more.

A05-1921
White, R. N., et al. Structural Engineering. Vol. 3, Behavior of Members and Systems. John Wiley, 1974. 597 pp. $16.95

Introduction to structural behavior, structural properties of engineering materials, steel tension members, behavior of steel structures, concrete and timber structures, and more.

A05-1922
White, R. N., et al. Structural Engineering. Vol. 4, Design of Structures. John Wiley, 1976.

Design philosophy and codes, building design in steel and concrete. Highway bridge design.

A05-1923
Wiegel, R. L. Earthquake Engineering. Prentice-Hall, 1970. 518 pp. $26.95

A collection of articles on the causes of earthquakes, the factors regarding damage, and the requirements and methods for building lasting structures in earthquake zones. Many illustrations are included to clarify the material and to dramatize the extent of destruction caused by earthquakes.

A05-1924
Wilby, C. B. Pre-Stressed Concrete Beams-Design and Logical Analysis. Applied Science, 1969. 97 pp. £3.00

This book presents a new theory for an accurate and logical design in pre-stressed concrete beams. Large numbers of post-tensioned pre-stressed concrete bridges are being designed and this book contains computer programmes for all the theories developed so that the accurate design proposed, although complex, can be effected rapidly and inexpensively on a computer.

A05-1925
Wilby, C. B. and Khwaja, I. Concrete Shell Roofs. Applied Science, 1974. 390 pp.

Part I of the book gives design tables for a large number of ordinary and north-light cylindrical shells. The use of the tables in the design of shells and the detailing of reinforcement in the shells are illustrated by numerical examples. Part II of the book deals with the analysis and design of hyperbolic paraboloidal shells. The governing equations of such shells are derived and a new method developed for the solution of these equations in the case of hyperbolic paraboloidal shells of the inverted umbrella type.

A05-1926
Wilby, C. B. and Nagvi, M. M. Reinforced Concrete Conoidal Shell Roofs--Flexural Theory Design Tables. Cement and Concrete Assn., 1973. 80 pp. £3.00

A05-1927
Willems, N. and Lucas, W. M., Jr. Matrix Analysis for Structural Engineers. Prentice-Hall, 1968. 364 pp. $17.20
Emphasizes the relation between conventional methods of analysis and matrix methods, and their formulation in familiar concepts (e.g., energy, slope deflection, consistent deformation). Applications include beams, plane and space structures. Notation is directly applicable to computer programming.

A05-1928
Williams, A. Analysis of Indeterminate Structures. Hart Publishing Co., 1968. 336 pp. $12.00
Contents include statistical indeterminacy, virtual work methods, indeterminate pin-jointed frames, conjugate beam methods, influence lines, elastic center and column analogy methods, moment distribution methods, model analysis, ultimate load analysis, matrix methods, and elastic instability.

A05-1929
Williams, C. D. and Harris, E. C. Structural Design in Metals. Ronald Press, 1957. 655 pp. $9.50
Emphasis is placed on basic training in structural design through application of statics and strength of materials to details of design rather than to a relatively few complex structures. The design procedure followed is a sequence of correlated steps leading from the selection of sections through their connection into integrated structures. Welded and riveted details are analyzed.

A05-1930
Winter, G. and Nilson, A. H. Design of Concrete Structures. McGraw-Hill, 1973. 640 pp. $19.00
This book covers the field of reinforced and prestressed concrete design, from basic information on material and member behavior through the most fundamental methods of members design and analysis to at least the simpler aspects of the design of buildings, bridges, arches and shells, spread foundations and precast construction. This edition has been brought into conformance with the 1971 American Concrete Institute Building Code. Chapter on precast concrete buildings, material on design for torsion, two-way floor systems, on light-weight concrete, and on slender columns.

A05-1931
Wolfer, K. H. Elastically Supported Beams. Adler's Foreign Books, 1971. 576 pp. $40.00
Tables of coefficients for bending moment shear force and bearing pressure diagrams based on the modulus of subgrade reaction. For low bearing capacity of the soil, heavy loads on the foundations, rule out the use of individual footings under piers and columns. The designer is often tempted to use over-simplified assumptions for the bearing pressure distribution. The results obtained in this way are frequently inaccurate. These tables have been compiled to provide the designer with a reliable and quick

Design--Civil and Structural

means of dealing with these problems. It gives ready-to-use coefficients for the analysis and design of strip footings, spread footings and raft-type foundations. The tabulated coefficients have been determined on the basis of the theory of the modulus of subgrade reaction.

A05-1932
Woodward, R., et al. Drilled Pier Foundations. McGraw-Hill, 1972. 288 pp. $18.95
 This is the first book written on the subject of drilled piers. It discusses design considerations, techniques, specification engineering supervision, and inspection, and contains many specific examples.

A05-1933
Yamada, Y. and Gallagher, R., eds. Theory and Practice in Finite Element Structural Analysis. Proceedings of the 1973 Tokyo Seminar on Finite Element Analysis. Int'l. Scholarly Book Services (Peter Peregrinus, Ltd.), 1973. 754 pp. $25.00
 42 contributions from eight countries covering basic theory, nonlinear analysis, dynamics, applications, and program development. Also papers on fluid flow, diffusion, and equation solving.

A05-1934
Yu, W. Cold-Formed Steel Structures. McGraw-Hill, 1973. 512 pp. $22.50
 This is the first complete book written on the subject of thin-wall, cold-formed steel structures. Written for practicing engineers, consultants, and fabricators, this unique professional's reference discusses material design criteria, the designing of building elements, connections, corrugated sheets, plus a number of other important and practical topics.

A05-1935
Zilly, R. G., ed. Handbook of Environmental Civil Engineering. Van Nostrand Reinhold, 1976. 1029 pp. $37.50
 This handbook emphasizes the environmental aspects of civil engineering while presenting a basic text on such civil engineering subjects as structures, foundations, water supply, and waste treatment and disposal. Engineering management is also stressed both as a professional employee and in private practice. Soils and foundation engineering are covered including construction in difficult sites, such as swamps.

Section   A06

DESIGN: MECHANICAL AND ELECTRICAL

Electrical Equipment
Heating, Ventilation, Airconditioning
Lighting
Plumbing
Vibration

A06-1960
ACGIH. Industrial Ventilation. American Conference of Governmental Industrial Hygienists, 1972. 312 pp. $8.00 (paperback $5.00)

The manual is sufficiently complete so that an industrial ventilating system can be designed without reference to other text. It deals with air pollution design, dilution ventilation, hood design, fans, make-up and recirculated air, plus a section on testing of ventilating systems to assist ventilation engineers and other personnel in the field in measuring discharge stacks, air flows and in testing and evaluating industrial exhaust systems.

A06-1961
AIChE. Cooling Towers. American Institute of Chemical Engrs., 1972. 145 pp. $15.00 (for AIChE members, $7.00)

Twenty-nine papers describe the ecological problems associated with waste heat rejection systems, future national water needs and sources, dry-type cooling towers, the use of lakes and ponds, and blow down disposal.

A06-1962
AIChE. Industrial Process Design for Pollution Control. American Institute of Chemical Engrs., 1972. 118 pp. $15.00 (for AIChE members, $7.50)

This volume is an expansion of the series to encompass articles on the design of plants for minimal air as well as water pollution. The designer is challenged in providing facilities for clean operation during all phases of the operation of a chemical plant--startup, shutdown, upset conditions, normal operations and emergencies. This volume presents case histories of some successful pollution control designs. Control facilities, and the monitoring and prevention of spills are discussed. Federal effluent standards are evaluated.

A06-1963
Air Conditioning. Howard W. Sams. 464 pp. $5.95

Home, commercial, and automobile air conditioning explained. The material has been organized to provide a practical understanding of the construction, operation, and basic fundamentals so important in diagnosing operating faults in an air conditioning system. A description of the purpose and function of each operating component, along with coverage of refrigerants used in various applications is included.

A06-1964
Alden, J. L. and Kane, J. M. Design of Industrial Exhaust Systems. Industrial Press, 1970. 244 pp. $17.00
 This book contains all information required to design, purchase and operate an exhaust system which will comply with state regulations for the protection of employees and which will minimize atmospheric pollution. The same data apply to systems for the recovery of valuable air-borne solid products. Design data for low pressure pneumatic conveying systems are also included.

A06-1965
Allphin, W. Primer of Lamps and Lighting. Addison-Wesley, 1973. 256 pp. $11.50
 A book that lets anyone entering the lighting industry gain a working knowledge of lamps and the basics of lighting layouts. It requires no background in electricity. It is a book for the electrical salesman, the electrical contractor, the building and maintainence superintendent and the electrical foreman. And it is also a reference for practicing architects who want to take advantage of the possibilities of light as an architectural tool.

A06-1966
Ambrose, E. R. Heat Pumps and Electric Heating: Residential, Commercial, Industrial Year-Round Air Conditioning. John Wiley, 205 pp. $12.25
 Covers thermodynamics, design and application of heating and cooling systems, energy consumption computation, air conditioning, and more.

A06-1967
Andrews, F. T. Architect's Guide to Mechanical Systems. Van Nostrand Reinhold, 1966. 256 pp. $14.95
 This reference describes all significant mechanical systems including heating, air conditioning, cooling, ventilating, plumbing, and fire protection, for all kinds of buildings. It presents to the specialist information needed to ensure adequate consideration of these systems in designing a building. Simple rules have been developed to determine the space requirements and costs of various mechanical systems.

A06-1968
Angus, T. C. The Control of Indoor Climate. Pergamon, 1968. 130 pp. $7.00
 Emphasizes the importance of securing conditions which do not impose stresses on the physiological processes of man. Of interest to all concerned in the relevant subjects of architecture and building science, public health engineering, pharmacy, psychiatry and mental health, and marine engineering.

A06-1969
ASHRAE. Equipment Volume 1972. American Society of Heating, Refrigerating and Air Conditioning Engrs., 1972. $30.00
 The equipment volume is divided into six sections; the first

Design--Mechanical and Electrical

involving air handling equipment and the second, refrigeration equipment. This is followed by heating equipment, general components, unitary equipment and codes and standards.

A06-1970
ASHRAE. 1974 Applications Volume. American Society of Heating, Refrigerating and Air Conditioning Engrs., 1974. $40.00

The seven sections of the book begin with air conditioning and heating applications for comfort in residences, stores, public and educational buildings, and for transportation facilities. This is followed by special cases of air conditioning and heating such as industrial buildings, laboratory, clean areas, hospitals, etc. Other sections include food refrigeration and distribution of chilled and frozen foods. Low temperature applications such as cryogenics, environmental test equipment and biomedical applications follow. The final sections are the industrial application of refrigeration and a catalog data section.

A06-1971
ASHRAE. 1973 Systems Volume. American Society of Heating, Refrigerating and Air Conditioning Engrs., 1973. $40.00

Four sections include air-conditioning and heating systems, industrial ventilation, refrigeration systems practices and a general section that comprises odor, sound and vibration control, snow melting, testing, and costs to own and operate.

A06-1972
ASHRAE. 1972 Handbook of Fundamentals. American Society of Heating, Refrigerating and Air Conditioning Engrs., 1972. $32.00

There are seven sections in this handbook beginning with theory of thermodynamics, heat and mass transfer, fluid flow, psychrometrics and sound control. Other sections include general engineering data, basic materials, load calculations, duct and pipe sizing, general terminology and basic tables.

A06-1973
ASME. AMD. Vol. 1, Isolation of Mechanical Vibration, Impact, and Noise. American Society of Mechanical Engrs., 270 pp. $25.00 (for ASME members, $20.00)

Presents an account of new analytical results and recently developed practical techniques and applications in the field. Includes reduction of vibrations from complex structures and flow-induced structural vibrations, isolation of building structures and machinery vibrations, acoustic enclosures, and high performance shock isolation.

A06-1974
ASME. ASME Handbook--Engineering Tables. McGraw-Hill, 1956. 692 pp. $22.50

Contains tabulations of essential data on dimensions, standards, etc.

A06-1975
ASME. Dry and Wet/Dry Cooling Towers for Power Plants.
American Society of Mechanical Engrs., 154 pp. $14.50
(for ASME members, $11.60)
Contains 13 papers on the technological and economic development of dry and wet/dry cooling concepts covering: design optimization of the dry cooling tower; design concepts for combined wet/dry cooling systems; economic optimization of plant loading due to climatic variation; effect of climatic variation on turbine design and operations; and practical design and operational requirements.

A06-1976
ASSE. The Principles of Residential Plumbing Inspection. American Society of Sanitary Engineering, 138 pp. $3.00 (for ASSE members, $2.25)
The book covers the basics of plumbing installation in residential one and two family buildings. Also contains a section on swimming pools. Illustrations.

A06-1977
ASTM. Metallic Electrical Conductors. American Society for Testing and Materials, 1975. 582 pp. $18.25 (for ASTM members, less 20%)
A compilation of 89 standards covering all ASTM electrical conductor standards including copper, aluminum, and steel clad wire.

A06-1978
AWS. Resistance Welding--Theory and Use. American Welding Society, 1956. 163 pp. $6.00 (for AWS members, $4.50)
This is a handbook for welding thinner gauges of all kinds of metals. Every aspect of resistance welding from basic principles to most effective use is presented.

A06-1979
Babbitt, H. Plumbing. McGraw-Hill, 1960. 649 pp. $26.00
Discusses applications of plumbing--from hydraulics, mechanics, and pneumatics to design, materials, installation, and repairs involved.

A06-1980
Baker, W. E., ed. Use of Models and Scaling in Shock and Vibration. American Society of Mechanical Engrs., 1963. 78 pp. $7.00 (for ASME members, $5.60)
Papers from the 1963 Winter Meeting include the modeling of air blast, modeling of large elastic and plastic deformations of structures subject to transient loading, and dynamic modeling with similar materials.

A06-1981
Banham, R. Architecture of the Well-Tempered Environment.

Design--Mechanical and Electrical 325

Univ. of Chicago Press, 1969. 296 pp. $15.00 (paperback $5.95)

Assesses the impact of environmental engineering on the design of buildings and on the minds of architects. Even though mechanical and electrical utilities constitute as much as fifty percent of building costs and require a major part of architectural and engineering effort to integrate them into the structure and furnishings of a building, Banham argues that innovations in mechanical environment control may be more valid tests of technological modernity than the commonly accepted touchstones, such as steel or concrete construction.

A06-1982
Barkan, D. Dynamics of Bases and Foundations. McGraw-Hill, 1962. 496 pp. $16.00

Examines the basics of soil mechanics, the elasticity theory, and the dynamics of vibratory motion to develop usable solutions for machine foundation design problems.

A06-1983
Barton, J. J. Domestic Heating and Hot Water Supply. Applied Science, 1970. 310 pp. £7.00

This book deals with the fundamental principles of low pressure hot water central heating, and the application of those principles to the design of domestic gravity and pump operated systems, and of hot water supply. During the period 1970-72 the British building industry was required to change to construction and service installations based on metric drawings and documents. To accord with this change, tables, formulae, equations, and worked examples are expressed in both British and metric units.

A06-1984
Baturin, V. V. Fundamentals of Industrial Ventilation. Pergamon, 1972. 756 pp. $36.00

A treatise on ventilation in industrial buildings, covering model techniques, jet theory, air movement problems etc. For post-graduate students and researchers in the field of industrial ventilation; practising heating and ventilating engineers and consultants.

A06-1985
Baumelster, T. and Marks, L. Standard Handbook for Mechanical Engineers. McGraw-Hill, 1967. 2,456 pp. $35.00

This monumental book has been updated and revised to include recent information on cryogenics, aerospace, and computers plus essential data on iron and steel coatings, paints and protective coatings, fuels, gearing, pipes and fittings, steam turbines, materials handling, and cost accounting. The volume offers a wealth of mathematical tables.

A06-1986
Bean, A. R. and Simons, R. H. Lighting Fittings--Performance and Design. Pergamon, 1969. 340 pp. $12.00

Gives an account of the new materials and processes which have been introduced to lighting fittings performance and design. Suitable for students of electrical engineering, lighting engineering, environmental engineering or architecture, photometerists, lighting fitting designers, consulting engineers, architects, and lighting engineers.

A06-1987
Beeman, D.   Industrial Power Systems Handbook.   McGraw-Hill, 1955.   948 pp. $32.75

This handbook outlines industrial power system practices in a simple, practical manner, using many examples and a wide range of design data. Industrial and commercial installations are thoroughly covered.

A06-1988
Beranek, L. L.   Noise and Vibration Control.   McGraw-Hill, 1971.   864 pp. $32.50

The behavior of sound waves, levels, decibels, spectra; traducers; field measurements; data analysis; sound propagation in small and large places; acoustical properties of porous and solid materials; mufflers; wrappings; duct linings and the like; hearing risk; and criteria for noise and vibration in various sites. Written for the engineer with noise problems that must be solved, this book examines the "engineering" of quiet into buildings, transportation systems, products, factories, power plants, air-conditioning systems, and farm and road building equipment. This book should aid the engineer in making cost-effective designs that work the first time, thus improving competitive advantages in the market place.

A06-1989
Beranek, L. L.   Noise Reduction.   McGraw-Hill, 1960.   752 pp. $26.00

Designed for the engineer with no special training in acoustics, this practical volume on noise control treats the nature of sound and its measurement, fundamentals of noise control, criteria, and case histories.

A06-1990
Billington, N. S.   Building Physics: Heat.   Pergamon, 1967.   248 pp. $7.00 (paperback $5.50)

Shows the ways in which environments can be designed by emphasizing the underlying physical principles on which design calculations must be based to provide a satisfactory thermal environment in which to live and work. Contents include heat flow, transfer of moisture, human comfort and ventilation.

A06-1991
Blake, M. P. and Mitchell, W. S., eds.   Vibration and Acoustic Measurement Handbook.   Hayden, 1972.   656 pp. $30.00

This volume discusses the value of vibration measuring in terms of living situations and actual machine problems rather than in terms of unrelated theory or design problems. The keynote of

Design--Mechanical and Electrical   327

the handbook is the well-being, the safety, the economy, and the
general control of the cost of performance of existing machinery,
through vibration measurement techniques.

A06-1992
Blendermann, L. Design of Plumbing and Drainage Systems. Industrial Press, 1963. 456 pp. $15.00
   Encyclopedic coverage of modern American plumbing practice. Contents include drywells, manholes, drainage for parking
and planting areas, septic tank systems, vent stack terminals, preventing sewage backflow, storm drainage, water supply, water heaters, piping systems, and more.

A06-1993
Bond, H. NFPA Inspection Manual--Third Edition. National Fire
   Protection Association, 1974. 352 pp. $6.75
   Tells what to look for in inspecting all types of properties.
Describes common and special hazards, building features, exits,
inspection of extinguishing equipment, detection systems, report
writing and other essentials. Standard plan symbols in full color
and many other illustrations.

A06-1994
Bosich, J. F. Corrosion Prevention for Practicing Engineers.
   Cahners, 1970. 250 pp. $7.95 (paperback $4.95)
   Corrosion mechanisms, types of corrosion, prevention techniques, testing techniques and interpreting corrosion tests are subjects dealt with in this book.

A06-1995
Boud, J. Lighting Design in Buildings. Peter Peregrinus, 1973.
   220 pp. $17.00 (available from International Scholarly Book
   Services)
   This book bridges the gap between text books on illumination engineering and albums of lighting schemes. Although primarily addressed to architects, it has much to say to all practitioners
involved in building design, students, and services engineers.
There are line drawings and documented illustrations, mostly photographs of completed interiors.

A06-1996
Bredahl, C. Home Wiring Manual. McGraw-Hill, 1957. 216 pp.
   $9.50
   Here is a training and reference manual on the planning of
home wiring systems. Emphasizing methods of analyzing electrical requirements and planning adequate systems, rather than commonly-known installation methods, the book shows how to determine
the right number of outlets, the proper number and size of circuits, and the capacity of the electrical service needed to carry
the load.

A06-1997
Brinkworth, B. J. Solar Energy for Man. John Wiley, 1972.

251 pp. $9.95

The author reviews first principles of solar energy and relates this to heating and converting solar energy to electricity and work. Photo-electricity, photo-chemistry and photo-biology are also discussed.

A06-1998
Burch, M. Basic House Wiring. Harper and Row, 1976. 280 pp. $9.95

Designed for the beginner but broad enough in scope and information to be used as a reference guide by the more advanced. Contents include: The Basics of Electricity; Determining Your Electrical Needs; Planning Your Wiring and Reading Wiring Blueprints; Electrical Tools and How to Use Them; Electrical Materials and How to Use Them; Basic Wiring; Circuits; Wiring a New Home; Rewiring an Old Home; Wiring for Heavy-Duty Appliances; Wiring Your Shop; Wiring Outbuildings and Rural Wiring; Indoor Low-Voltage Wiring; Outdoor Wiring; Testing; Trouble Shooting and Repair Wiring.

A06-1999
Burkhardt, C. Residential and Commercial Air Conditioning. McGraw-Hill, 1959. 324 pp. $16.00

A manual on the sizing, installation, and servicing of commercial and residential air conditioners. Explains the theory of refrigeration and air conditioning, and provides methods for calculating heat gains; sizing the cooling system; and installing, operating, servicing, and troubleshooting the unit, its components, and controls. Compressors, condensers, expansion valves, cooling towers, and other components are covered, with many installation diagrams and a chart method of service procedure.

A06-2000
Burton, R. Vibration and Impact. Dover, 310 pp. $3.00

An introductory treatment of vibrations for advanced undergraduate engineers, covering a variety of kinds of vibration plus numerical computations, measurement, analysis of control systems.

A06-2001
Carmichael, C., ed. Kent's Mechanical Engineers' Handbook. John Wiley, 1950. 1,611 pp. $21.50
Design and production.

A06-2002
Carrier Air Conditioning Co. Handbook of Air Conditioning System Design. McGraw-Hill, 1965. 832 pp. $32.50

Prepared by the foremost firm in the field, this practical guide covers all important design steps from load estimation and air distribution to analysis of all-air, air-water, and water and direct expansion systems.

A06-2003
Caudill, W. W., et al. A Bucket of Oil: The Humanistic Approach

Design--Mechanical and Electrical

to Building Design for Energy Conservation. Cahners, 1974. 88 pp. $10.95

This book reports on 27 years of experience using natural means to design buildings that will conserve on resources, materials, and energy and meet human needs at the same time. Natural means of economy covered include working with the climate, working with the site, considering the orientation of buildings, and shaping buildings. Economy in interior design includes energy-efficient systems for heating and cooling, lighting, fenestration, and wall/roof components.

A06-2004
Central Electricity Generating Board. Modern Power Station Practice. Vol. 1, Planning and Layout. Pergamon, 1971. 536 pp. $18.00

Covers the planning and layout of power stations. Deals with the site investigation and design to meet the site conditions including foundations, superstructure, and construction methods. The use of scale models is discussed.

A06-2005
Central Electricity Generating Board. Modern Power Station Practice. Vol. 2, Mechanical (Boilers, Fuel- and Ash-Handling Plant). Pergamon, 1971. 425 pp. $15.00

Deals with the design and selection factors of boilers and associated plant. The influence of fuels on the type and size of boiler plant is discussed. Sections are devoted to ancillary plant, such as economisers, air and feed water heaters, fuel, dust and ash handling arrangements.

A06-2006
Chalkley, J. N. and Cater, H. R. Thermal Environment. Architectural Press, 1968. 224 pp. £2.10

An introduction to the field of environmental engineering and the associated practices of heating, ventilating and air-conditioning. All calculations in the book are given in both S.I. units and Imperial units.

A06-2007
Chasis, D. A. Plastic Piping Systems. Industrial Press, 1976. 224 pp. $17.50

Plastic pipe, valves and fittings are described in terms of design, applications, and installation techniques. Contents include physical characteristics, pressure ratings, product selection, cost comparisons, and more.

A06-2008
Church, A. H. Centrifugal Pumps and Blowers. John Wiley, 1944. 308 pp. $10.50

A06-2009
Commercial Refrigeration. Howard W. Sams. 448 pp. $6.50

Gives information on the installation, operation, servicing,

A06-2010
Crocker, M., ed. Noise and Vibration Control Engineering: Proceedings of the Purdue Noise Control Conference, 1971, Lafayette, Ind. Purdue University, 1971. $16.50

Widely varying topics discussed at the conference have been loosely grouped into nine sections: surface transportation noise, machinery noise, industrial noise criteria and control, vibration control and biodynamics, legislation and city planning, aircraft and rotor noise, jet noise and sonic boom, noise in buildings, and general topics.

A06-2011
Crocker, S. and King, R. Piping Handbook. McGraw-Hill, 1967. 1526 pp. $38.50

Data and procedures on the most effective uses of piping in industrial, municipal, and building piping systems. Contains new sections on nuclear, cryogenic, and other piping applications.

A06-2012
Croft, T., et al. American Electricians' Handbook. McGraw-Hill, 1970. 1548 pp. $25.80

An efficient and helpful guide to the selection, installation, operation and maintenance of electrical equipment, the new ninth edition reflects the heavy revisions made in the 1968 edition of the National Electrical Code. Thoroughly revised and updated, this work now covers the use of latest wiring devices, ground fault circuit interrupters, all types of electrical space heating equipment, new light sources (e.g., metal hallide and quartz halogen lamps) and more.

A06-2013
CTI. Cooling Tower Performance Curves. Cooling Tower Institute, 800 pp. $50.00 (for CTI members $35.00)

A collection of curves relating the variables which affect water cooling tower performance to a common index. This index provides a characteristic, a "degree of difficulty," for the various combinations of cooling ranges, approaches, wet bulb temperatures, and water-to-air loadings. It may be used by the cooling tower owner to evaluate performance of his cooling tower at actual, or projected, conditions.

A06-2014
Daniels, G. Solar Homes and Sun Heating. Harper and Row, 1976. 176 pp. $8.95

The book includes such information as how to plan for solar heating if you are thinking of building a new home; how to add solar heating to an existing structure; insulation; and seasonal data for the entire United States to help predict system capabilities. More than ten different systems are illustrated and described in

Design--Mechanical and Electrical

detail. For the do-it-yourselfer Mr. Daniels has complete step-by-step instructions and for the not-so-confident he tells you where to find the professional.

A06-2015
Davis, P. Plumbing, Heating, and Piping: Estimator's Guide. McGraw-Hill, 1960. 213 pp. $15.00
   This practical time-saver presents tested, workable standards for jobs of all sizes and allows the estimator to handle his job more effectively and profitably. It provides 75 "labor factor" tables pinpointing various elements of work involved and specific methods of cost analysis.

A06-2016
Den Hartog, J. P. Mechanics. Dover, 1948. 462 pp. $3.00
   Hundreds of applications and design problems are used to illuminate fundamentals of trusses, jacks, hoists, loaded beams and cables, gyroscopes, etc.

A06-2017
Dossat, R. J. Principles of Refrigeration. John Wiley, 1961. 544 pp. $14.00

A06-2018
Down, P. G. Heating and Cooling Load Calculations. Pergamon, 1969. 272 pp. $10.00
   A general study of heat transfer combined with the specialized application of the theory to environmental engineering producing an introduction to new techniques and a handbook of design principles. Suitable for environmental engineering students, and practising engineers.

A06-2019
Draffin, J. O. and Collins, W. L. Statics and Strength of Materials. Ronald Press, 1950. 398 pp. $8.50
   In statics, coplanar force systems are covered in detail, and a chapter devoted to systems in space. The treatment of materials and working stresses includes consideration of modern lightweight alloys, high-strength steel, and plastics. Resistance of materials to energy, creep, and repeated loading, and stress concentrations are treated. The presentation emphasizes geometric interpretation and physical visualization of problems rather than purely mathematical consideration, yet with few exceptions all equations are rigidly derived.

A06-2020
Ebeling, A. M. Basic Guide to Electric Heating. Business News Publishing, 1971. 171 pp. $7.95
   There are illustrations of the latest equipment, examples of the most recent approaches to system wiring and the recommended methods for calculating energy requirements and costs. These methods will give a realistic projection of actual operating costs. There are guidelines that will show how to apply electric heating

for cost-saving advantage in homes, apartments, offices, schools and commercial buildings.

A06-2021
Eck, B. Fans: Design and Operation of Centrifugal, Axial Flow and Cross Flow Fans. Pergamon, 1973. 616 pp. $45.00
The book gives the fundamental fluid dynamics of turbomachinery and applies this to fans. It continues with experimental investigations of fans and includes a comprehensive account of all work done in this field. Both theoretical and empirical results are applied to problems of design.

A06-2022
Egan, M. D. Concepts in Thermal Comfort. Prentice-Hall, 1974. 224 pp. $10.95
Basics of thermal comfort principles illustrated and explained with numerous illustrations, charts, graphs, etc. Analysis of climate and shelter design strategies, use of materials, and mechanical system selection.

A06-2023
Emerick, R. Heating Handbook. McGraw-Hill, 1964. 522 pp. $19.50
A guide to good practice and the many regulatory codes and standards governing space heating. Covers everything from fuel storage to smoke control--all types of fuels and heating systems-- how to convert from one system to another--rating systems, contracts, codes, etc.

A06-2024
Emerick, R. Troubleshooters' Handbook for Mechanical Systems. McGraw-Hill, 1969. 524 pp. $16.95
Here is a practical, field-tested manual for checking out and repairing defects in heating, ventilating, air conditioning, piping, and related systems in buildings. It gives the troubleshooter a clear cause-and-effect guide to the remedy for each condition, and the solutions to the most frequent problems. The handbook covers incinerators, swimming pools, compressed air systems, and air pollution.

A06-2025
Eshbach, O. W. and Souders, M. Handbook of Engineering Fundamentals. John Wiley, 1975. 1520 pp. $29.95
A revised and updated version of this handbook which embodies in a single volume those fundamental laws and theories of science basic to engineering practice.

A06-2026
Evans, F. L. Equipment Design Handbook for Refineries and Chemical Plants. Vol. 1. Gulf Publishing Co., 1971. 192 pp. $13.95
This two-volume handbook provides a reference for equipment design in the oil refining and chemical industries. The first

Design--Mechanical and Electrical

volume describes, by example, all the design procedures needed for major rotating equipment items: drivers, compressors, ejectors, pumps and process refrigeration.

A06-2027
Evans, F. L. Equipment Design Handbook for Refineries and Chemical Plants. Vol. 2. Gulf Publishing Co., 1973. 280 pp. $13.95

Volume 2 covers major nonrotating equipment items: fired heaters, heat exchangers, cooling towers, vessels, separators and accumulators, flare stacks, piping, sewers and valves.

A06-2028
Fanger, P. O. Thermal Comfort: Analysis and Applications in Environmental Engineering. McGraw-Hill, 1973. 244 pp. $15.50

Based on the results of recent research work in the U.S. and Denmark, where it was originally published, the book applies these findings to conditions for thermal comfort, methods of assessing thermal environments, and principles for performing thermal analyses of enclosures, based on comfort criteria. Featured within the text are a series of comfort diagrams showing all combinations of the environmental variables which will provide thermal comfort for man. New physiological comfort criteria are established including a discussion of the influence of age, sex and national-geographic origin.

A06-2029
Federal Energy Administration. Energy Conservation: Lighting and Thermal Operations. U.S. Gov't. Printing Office, 1975. 286 pp. $3.55. Order No. FE 1.22:18 S/N 041-018-00084-4

This book studies energy conservation principles as they apply to office lighting. It includes the relationship of illumination to visual performance and of lighting energy to heating and cooling energy, and provides a summary of newer techniques and equipment for energy conservation.

A06-2030
Federal Energy Administration. Guidelines for Saving Energy in Existing Buildings. U.S. Gov't. Printing Office, 1975. Part 1, 304 pp. $5.25. Order No. FE 1.22:20 S/N 041-018-00079-8; Part 2, 465 pp. $5.05. Order No. FE 1.22:21 S/N 041-018-00080-1

Many of the ideas outlined in this two-part manual can be executed by building managers and engineers with little or no capital outlay, but can still achieve significant energy savings. This manual describes specific ways energy can be saved in the fields of heating, ventilation, cooling, domestic hot water, commercial refrigeration, lighting, and power. A bibliography, selected references, and specific examples of energy conservation are also included.

A06-2031
Fink, D. and Carroll, J. Standard Handbook for Electrical Engineers. McGraw-Hill, 1968. 2596 pp. $38.50

The revision of this handbook now includes modern and sophisticated techniques for computer control of power generation and distribution along with conventional data on wiring and switch-gear. New sections cover nuclear power plants; direct conversion of heat to electricity; d-c power plants; power system instrumentation; EDP and its applications to engineering design, control, and communications; and industrial electronics.

A06-2032
Fitzgerald, R. W. Strength of Materials. Addison-Wesley, 1968. 418 pp. $11.50 (instructor's guide $2.95)

Contents include: Stress and strain. Torsion. Shear and moment diagrams. Stresses in beams. Deflection of beams. Statically indeterminate beams. Columns. Additional applications of axial stress and strain. Connections. Combined stresses and eccentric loading. Combined stresses. Properties of materials.

A06-2033
Flügge, I. Handbook of Engineering Mechanics. McGraw-Hill, 1962. 1632 pp. $39.50

Presents all the mathematical tools needed by practicing engineers, gives data on mechanics of rigid bodies, including material on variational principles and gyroscopes, and details the theory of elasticity and its applications, plasticity, and viscoelasticity. Covers mechanical vibrations, surge tanks, wing flutter, stochastic loads, acoustics, etc., plus the field of fluid mechanics from basic concepts to special topics such as supersonic flow, lubrication, waves, and seepage flow.

A06-2034
Flynn, J. E. and Mills, S. M. Architectural Lighting Graphics. Van Nostrand Reinhold, 1962. 224 pp. $19.50

By text, photographs and more than 1500 illustrations the authors discuss light distribution, exterior lighting, lighting equipment, light control, color, lighting design, etc.

A06-2035
Flynn, J. E. and Segil, A. W. Architectural Interior Systems: Lighting, Air Conditioning, Acoustics. Van Nostrand Reinhold, 1970. 306 pp. $13.95

Here are guidelines for choosing acoustical, air conditioning, and lighting systems that best satisfy human sensory and behavioral needs.

A06-2036
Geiringer, P. L. High Temperature Water Heating: Its Theory and Practice for District and Space Heating Applications. John Wiley, 1963. 333 pp. $21.25

Design--Mechanical and Electrical

A06-2037
General Electric. Heat Transfer & Fluid Flow Data Books. Vol. 1, Heat Transfer. Vol. 2, Fluid Flow. General Electric. Vols. 1 and 2, 500 pp. $290.00 (annual updating, $75.00)

For 25 years, General Electric heat transfer authorities have collected and condensed heat transfer and fluid flow data from internal and world-wide sources. Information on conduction, forced and free convection, radiation, condensation, evaporation, combined and transient heat flow, heat exchangers and properties of gases, liquids and solids are given in volume 1. The second volume covers duct design and fan selection and design.

A06-2038
Giachino, J. W., et al. Welding Technology. American Technical Society, 1973. $9.25 (Study Guide $2.90)

Contains a revised chapter on automated welding and a section on safety in welding which meets OSHA standards. Covers metallurgy of welding, testing of weldments, strength of materials, joint design, and estimating.

A06-2039
Giles, R. L. Layout of E.H.V. Substations. Peter Peregrinus, 1972. 223 pp. $14.50 (available from International Scholarly Book Services)

The theory behind substation layout, showing many examples of how a proper application of basic principles can result in sound practical designs. Several novel techniques are described, and a world survey of existing practice and experience is included.

A06-2040
Goldstern, W. Steam Storage Installations. Pergamon, 1970. 160 pp. $10.75

After a short outline of the history and fundamentals of the equipment a description is given of the main parts which together make up most steam storage installations. The two main chapters are devoted to varying-pressure and constant-pressure accumulators. The description of the principle of the accumulator leads to the design calculations, to the arrangement of storage within the steam plant as a whole and to its method of operation. Ten installations are illustrated with comparative load graphs, arrangement diagrams and explanatory notes.

A06-2041
Gosling, C. T. Applied Air Conditioning and Refrigeration. Applied Science, 1974. 250 pp.

This book is aimed at all levels of personnel and skills. It will be of value to those engineers who have not received formal training, and to the overworked competent engineer in dealing with his problems and bringing him up to date with the latest thinking on air conditioning.

A06-2042
Graham, K. C. Industrial and Commercial Wiring. American

Technical Society, 290 pp. $6.50

Offering a resume of all facets of industrial and commercial wiring, this edition takes into account the changes and progress made in fixture design, equipment, and new materials. There is new material on lighting and wiring problems for special purposes such as swimming pools, area lighting, display case lighting, etc.

A06-2043
Graham, K. C. Interior Electric Wiring--Residential. American Technical Society, 312 pp. $5.25

This edition explains practical wiring procedures and considers wiring safety, alternative wiring methods or forms and their merits, designs, and estimating. New equipment, current usage and needs are discussed.

A06-2044
Griffin, C. W. Energy Conservation in Buildings: Techniques for Economical Design. The Construction Specifications Institute, 1975. 183 pp. $20.00

Life-cycle costing is stressed in chapters on glass wall design, HVAC, insulation, conservation by waste heat reclamation and architectural design, lighting, solar energy, and computer-controlled HVAC systems. The book is intended to use architectural and engineering techniques to conserve energy.

A06-2045
Gunther, R. C. Refrigeration, Air Conditioning, and Cold Storage: Principles and Applications. Chilton, 1969. 1398 pp. $22.95

This book explains the fundamental principles and applications of thermal energy, pressure, gases, refrigerants, thermodynamic functions, equipment, controls, prime movers, refrigeration calculations, heat transfer, indirect refrigeration, and lubrication.

A06-2046
Haines, J. Automatic Control of Heating and Air Conditioning. McGraw-Hill, 1961. 389 pp. $19.50

The basic principles of automatic control, and today's methods of applying these controls to heating, ventilating, and air-conditioning, are thoroughly discussed in this volume.

A06-2047
Haines, R. W. Control Systems for Heating, Ventilating, and Air Conditioning. Van Nostrand Reinhold, 1971. $12.95

This volume provides a non-mathematical overview, of the subject from the elements of control systems, basic control devices for elementary systems, and more complex systems.

A06-2048
Harris, C. Handbook of Noise Control. McGraw-Hill, 1957. 1184 pp. $33.50

Here is an authoritative exposition of noise, its nature, measurement, and techniques for its control in buildings, industry,

Design--Mechanical and Electrical

transportation, and the community. The text provides fundamentals, practical engineering techniques, and reference data in connection with topics such as the effect of noise on human efficiency, bearing, and speech; vibration problems and control; the control of industrial noise, and treatment of special types of noise sources.

A06-2049
Harris, C. M. Shock and Vibration Handbook. McGraw-Hill, 1976. 1344 pp. $29.50

Combined into one volume (from the original three), the updated second edition contains new material including expanded coverage of practical engineering problems. There are new chapters on vibration standards, and the effects of vibration on life. Provides engineers, especially those in the fields of mechanical, civil, acoustical, aeronautical and electrical engineering, with a unified and definitive treatment of instrumentation and measurement, data analysis and testing, practical methods of control, solutions to engineering problems, equipment design and packaging.

A06-2050
Harris, N. Modern Air Conditioning Practice. McGraw-Hill, 1974. 521 pp. $14.50

This text covers theory and practice, and discusses physics, psychrometry, load calculations, refrigeration theory, summer and winter air conditioning system design, and sales engineer work.

A06-2051
Hemeon, W. C. L. Plant and Process Ventilation. Industrial Press, 1963. 485 pp. $20.00

This book places emphasis on estimating ventilation quantities required in various industrial situations, while the subject of duct design has been simplified by condensation of design data. Data provide the engineer, as well as the student, with design factors for any situation that may be found in a plant.

A06-2052
Herkimer, H. Cost Manual for Piping and Mechanical Construction. Chemical Publishing, 1958. 176 pp. $11.25

This book of 115 tables--all calculated in man hours--will save time and effort to estimators in arriving at reliable figures. It will enable engineers to preestimate--at least roughly--how much a job should cost and to check the estimates submitted them.

A06-2053
Hicks, T. Standard Handbook of Engineering Calculations. McGraw-Hill, 1972. 1206 pp. $19.50

This collection of worked-out problems most frequently encountered by working engineers constitutes a "cookbook" for the practitioner. Step-by-step solutions to problems give the reader an understanding of the basic methods and an ability to apply them. The handbook covers all of the major engineering areas because engineers must frequently handle garden-variety design problems outside of their nominal specialities.

A06-2054
Hicks, T. and Edwards, T.   Pump Application Engineering.
    McGraw-Hill, 1970.   432 pp. $21.00
       This is a completely revised and updated edition of Pump
Selection and Application. Because the authors have broadened the
scope and depth of coverage, the title has been changed to signify
this edition's new look. This edition reflects recent advances in
pump technology.

A06-2055
Hill, P. H.   The Science of Engineering Design.   Holt, Rinehart
    and Winston, 1960.   550 pp. $11.50
       Science involves methods which, once applied to a problem
will produce an answer, and when reapplied will again produce a
comparable answer. The prime purpose of the text is to reveal to
the reader through reason, explanation, example, and case history
the science of engineering design so that it may be learned and
practiced.

A06-2056
Home Refrigeration and Air Conditioning.   Howard W. Sams.
    576 pp. $6.95
       A service guide that offers information covering all phases
of modern household mechanical refrigeration and air conditioning.
The text covers specialized tools used in servicing refrigeration
and air conditioning equipment, plus trouble charts for diagnosis
of common troubles.

A06-2057
Hopkinson, R. G. and Kay, J. D.   The Lighting of Buildings.
    Praeger or Wittenborn, 1969.   366 pp. $10.00
       The authors stress that the criterion of good lighting is
visual satisfaction rather than strict adherence to numerical specifications. They explain basic architectural and engineering principles of day lighting and artificial lighting and show how they can
be integrated. They also discuss special lighting requirements for
schools, hospitals, offices, laboratories, factories, private houses,
homes for the aged, and buildings in the tropics.

A06-2058
Housing and Urban Development Dept.   Multifamily Housing, Final
    Report.   U.S. Gov't. Printing Office, 1974.   177 pp.
    $2.50.   Order No. HH1.2:En 2/7 S/N 023-000-00282-5
       Identifies the components of residential energy consumption
and evaluates technologies or practices that could lead to the conservation of energy. Considers type, size, geographic location,
and structural properties of multifamily housing, and the kinds of
appliances in such housing. This is the final of two reports. The
first is entitled "Residential Energy Consumption, Phase 1 Report," and is available for $1.20.

A06-2059
Housing and Urban Development Dept.   Single-Family Housing,

Design--Mechanical and Electrical 339

Final Report. U.S. Gov't. Printing Office, 1973. 184 pp. $2.10. Order No. HH 1.2:En 2/5 S/N 2300-00258
A technically oriented study that discusses how to get greater efficiency in residential energy use.

A06-2060
Hovanessian, S. and Pipes, L. Digital Computer Methods in Engineering. McGraw-Hill, 1969. 320 pp. $17.50
Digital computer programs, written in FORTRAN and BASIC programming languages, are given for numerical examples to familiarize the reader with computer logic.

A06-2061
Hutchinson, F. W. Design of Refrigeration Systems for Air Conditioning. Industrial Press, 1963. 286 pp. $16.00
Treatment simplified by restricting refrigeration to air conditioning applications.

A06-2062
IEEE. National Electrical Safety Code, 1973 edition. John Wiley, 1973. 352 pp. $11.00
Consists of the parts of the National Electrical Safety Code (NESC) currently in effect.

A06-2063
IEEE. Recommended Practice for Grounding of Industrial and Commercial Power Systems. John Wiley, 1972. 95 pp. $8.75

A06-2064
IEEE. Standard and American National Standard Graphic Symbols for Electrical and Electronics Diagrams. John Wiley, 1971. 88 pp. $12.75
Including Reference Designation Class Designation Letters.

A06-2065
IEEE Recommended Practice for Protection and Coordination of Industrial and Commercial Power Systems. The Institute of Electrical and Electronics Engineers, Inc., 1975. 312 pp. $19.95 ($14.95 for IEEE members)
This standard is an authoritative document on the subjects of calculation of short-circuit currents, use of overcurrent relays and circuit breakers, ground fault protection and protection for motors, transformers, conductors, bus and switchgear, generators and service supply lines.

A06-2066
IEEE Standard Dictionary of Electrical and Electronics Terms. John Wiley, 1972. 716 pp. $23.25

A06-2067
IES Lighting Handbook. Illuminating Engrg. Society, 1972. 772 pp. $37.50 (for IES members, $30.00)
Basic principles of light and vision through description and

illustration of interior and exterior lighting installations of all
types. Includes section on color; detailed listing of recommended
lighting levels. Photographs, charts, graphs, and tables.

A06-2068
Inglis, C. Applied Mechanics for Engineers. Dover, 404 pp.
$3.00
Specific problems are used to illustrate methods in working
with taut wires, framework stress, harmonic motion, periodic vibrations, coupled vibrating systems, gyroscopic principles, other
areas involving statics and dynamics.

A06-2069
International Institute of Refrigeration (Paris). Bibliographic Guide
to Refrigeration 1960-64. Pergamon, 1966. 1000 pp.
$20.00
Contents: Physical phenomena relating to refrigeration;
Production and distribution of cold; Refrigerating plants; Transport
and packaging; Air conditioning and heat pumps; Industrial and miscellaneous applications of refrigeration; Biological and medical applications of refrigeration; Cold storage of agricultural and food
products.

A06-2070
Jacobson, I. D. Plumbing Dictionary. American Society of Sanitary Engineering. $7.50 (for ASSE members, $5.65)
Contains over 2,500 plumbing words and terms, crossreferenced and illustrated. Compiled by Plumbing Nomenclature
Committee. Supply limited.

A06-2071
Javitz, A. E., ed. Materials Science and Technology for Design
Engineers. Hayden, 1972. 560 pp. $23.95
An interdisciplinary volume in the series, this work presents an advanced approach to the subject. Authorities contribute
both practical and far-reaching ideas in such areas as: the basic
structure and molecular behavior of all relevant materials, the nature and effects of critical environments, and new techniques in
reliability studies.

A06-2072
Johnson, J. R. Johnson's Guide to the Welding Section of the Los
Angeles Building Code. Anchor Neptune, 1971. $2.00

A06-2073
Johnson, R. C. Electrical Wiring: Design and Construction.
Prentice-Hall, 1971. 356 pp. $13.95
Fundamentals of electrical theory; concentrates on theory
and practical applications of the design problems associated with
wiring systems in buildings. Studies three-phase power supplies,
delta-and wye-connected loads, voltage, current and power measurements. Offers an operational study of motors and transformers.
Covers voltage, current, power, and speed variations with respect

Design--Mechanical and Electrical

to mechanical or electrical loading. Provides diagrams that show principles of grounding.

A06-2074
Jones, F. D. and Schubert, P. B. Engineering Encyclopedia. Industrial Press, 1963. 1431 pp. $18.50

A combined encyclopedia and dictionary for everyone in any kind of mechanical work who can use essential facts about thousands of standard and special engineering subjects. Contains various important mechanical laws, rules and principles; physical properties and compositions of a large variety of standard and special metals used in machine construction and in other engineering structures.

A06-2075
Kalff, L. C. The Creative Light. Philips Books, 1971. 132 pp. £6.00

This book discusses such topics as design of the visual field, development of lighting techniques, and the function of the human eye.

A06-2076
Kallen, H. Handbook of Instrumentation and Controls. McGraw-Hill, 1961. 750 pp. $24.50

This handbook gives practical data on instrumentation and controls for mechanical services used in institutional and commercial buildings and industrial plants. It reflects modern advances in its coverage of control systems for high-pressure high-temperature steam power plants, controls for large central air conditioning and other systems used in office buildings, schools, hospitals, etc. The book includes data on boiler controls selection, flame-failure safeguards, and instrument system design.

A06-2077
Karassik, I., et al., eds. Pump Handbook. McGraw-Hill, 1976. 1000 pp. $32.50

Prepared by over fifty experts, this handbook provides practical data and techniques needed for the design application, specification, purchase and maintenance of all types of pumps. The handbook covers design methods, noise and vibration control, materials or construction, and specialized designs (such as pumps for nuclear service, solids handling, marine service, and pulp and paper stock pumps). A particularly valuable section provides helpful information on the economic usage of pumping equipment in every major industry.

A06-2078
Kasuda, T. Use of Computers for Environmental Engineering Related To Buildings. U.S. Gov't. Printing Office, 1971. 813 pp. $7.75. Order No. C 13.29/2:39

Symposium at NBS, Nov. 30-Dec. 2, 1970. Building Science Series No. 31.

A06-2079
Katz, D., et al. Handbook of Natural Gas Engineering. McGraw-Hill, 1959. 802 pp. $46.50

This handbook provides all the information required by an engineer practicing in the natural gas industry. It gives design calculations and quantitative charts for determining equipment needed and methods required for handling natural gas from its occurrence through its marketing.

A06-2080
Kell, J. R. and Martin, P. L. Faber and Kell's Heating and Air Conditioning of Buildings. Architectural Press, 1971. 562 pp. £6.75

This new edition is the tenth printing of a book which has come to be regarded as a classic in heating and air-conditioning practice in England. Much fresh information has been incorporated, such as current thought on the forecasting of building reactions to solar gains and new chapters on the topical subjects of district heating and total energy.

A06-2081
Kellogg Company. Design of Piping Systems. John Wiley, 1964. 385 pp. $22.75

A06-2082
King, G. Modern Refrigeration Practice. McGraw-Hill, 1971. 512 pp. $14.95

Provides a description of the principles of refrigeration including the uses of refrigeration in modern society, types of jobs available, physics, chemistry and thermodynamics. Most up-to-date methods of producing refrigeration are covered.

A06-2083
King, R. C. Piping Handbook. McGraw-Hill, 1975. 1652 pp. $36.85

For designing, specifying, fabrication, installing or testing piping systems. Included in this handbook are chapters on materials, expansion and flexibility, pipe hangers and supports, insulation, corrosion, piping systems for cryogenics, fire-protection, gas, and underground steam. Water supply, plumbing, sewerage and other types are also covered.

A06-2084
Kolousek, V. Dynamics in Engineering Structures. Butterworth, 1973. 580 pp. £12.00

The author of this book has modified existing methods of analysing dynamic effects in engineering structures and has set out new ways of approaching individual problems. Of importance is his development of the theory of the frequency functions. An appendix contains the tabulated values of these functions, with which many vibration problems can be solved using a desk-top calculator. The analytical techniques here developed are useful to the structural and the mechanical engineer.

Design--Mechanical and Electrical

A06-2085
Kuljian, H. A. Nuclear Power Plant Design. A. S. Barnes, 1968. 272 pp. $40.00

This book is a guide in the design of nuclear power plants for mechanical, electrical and structural engineers who are already familiar with the design of conventional, fossil-fuel-fired power systems. The first part deals with basic principles of nuclear fission, and the general aspects of reactor design and operation. The second section discusses basic design calculations and features of all the nuclear reactor power systems which have been developed to date.

A06-2086
Kurtz, E. B. and Shoemaker, T. M. The Lineman's and Cableman's Handbook. McGraw-Hill, 1975. 1024 pp. $22.50

A basis for understanding the electric power system, construction practices, maintenance procedures, and safety guidelines is provided, along with expanded coverage of underground electric distribution and new material on protective equipment, safety rules, heart-lung resuscitation, and pole-top rescue. New emphasis has been placed on the National Electrical Safety Code and the 1970 National Power Supply, Federal Power Commissions.

A06-2087
Kut, D. Heating and Hot Water Services in Buildings. Pergamon, 1968. 434 pp.

Intended for all concerned with the study, design or practice of heating and hot water services in buildings. The subject matter has been deliberately concentrated on heating and hot water supply only.

A06-2088
Kut, D. Warm Air Heating. Pergamon, 1970. 400 pp. $18.00

Describes the underlying principles of heating by warm air and shows how these are carried into practice. The book is presented in both British and metric units. For advanced students and researchers in the fields of heating and ventilating science; practising heating and ventilating engineers.

A06-2089
Laub, J. M. Air Conditioning and Heating Practice. Holt, Rinehart and Winston, 1963. 784 pp. $14.00

Both theory and practice of all phases of air conditioning, ventilation, and heating are covered. Contents include types of air conditioners, load calculations, refrigeration cycle, room air conditioners, water-cooled packaged air conditioners, cooling towers, air distribution, estimating, and more.

A06-2090
Laube, H. L. How to Have Air Conditioning and Still Be Comfortable. Business News Publishing, 254 pp. $7.95

The author explains what comfort is, and how it comes about. His contention is that often the people responsible for building design do not take into account the comfort of its future occupants.

A06-2091
Lefax Pub. Co., eds. Home Heating. Lefax Publishing Co., Data Book 610. $2.00

A06-2092
Lefax Pub. Co., eds. Illumination. Lefax Publishing Co., Data Book 614. $2.00

A06-2093
Levinson, I. J. Introduction to Mechanics. Prentice-Hall, 1968. 346 pp. $12.95

Covers elements of theory and problems drawn from contemporary engineering situations. Featured are sections on fluid statics, beam analysis, free-body diagrams, components of acceleration, the Coriolis effect, balancing, variable forces, work energy principles, momentum, and mechanical vibrations.

A06-2094
Lewis, B. Facilities and Plant Engineering Handbook. McGraw-Hill, 1974. 1024 pp. $33.00

This handbook provides the facilities engineer with the latest in proven management and engineering techniques needed to run a cost effective engineering program. It describes the complete, "cradle-to-grave" sequence of planning, design, construction, maintenance, management, and related functional tasks required to support production.

A06-2095
Lindeke, W. Technical Dictionary of Heating, Ventilation and Sanitary Engineering. Pergamon, 1970. 182 pp. $21.00

A multilingual dictionary presenting some 4,300 terms in English, German, French and Russian in the field of heating, ventilation and sanitary engineering. The work is arranged in four sections: the main section providing a side-by-side comparison of the terms in all four languages, and the following three sections the terms in separate language indexes.

A06-2096
Littleton, C. Industrial Piping. McGraw-Hill, 1962. 349 pp. $20.00

Offers coverage of the latest developments, materials, and techniques in industrial piping systems. Studies flow sheets and design procedures, steam piping and condensate systems, valves and more.

A06-2097
Loftness, R. L. Nuclear Power Plants: Design, Operating Experience, and Economics. Van Nostrand Reinhold, 1964. 548 pp. $12.50

Provides data on the various types of reactors that have

Design--Mechanical and Electrical

been or are being constructed--their characteristics, problems, design operating experience, and development trends. Discusses nuclear power costs and presents data on predicted power costs.

A06-2098
Lyons, S. Management Guide to Modern Industrial Lighting. Applied Science, 1972. 202 pp. £3.00 (paperback £1.00)
Covers lighting equipment to suit industrial environments, outdoor lighting, installation design, how electric lamps work, and more.

A06-2099
McGuinness, W. J. and Stein, B. Mechanical and Electrical Equipment for Buildings. John Wiley, 1971. 1011 pp. $23.95
A reference book for the selection, planning, and development of the essential mechanical and electrical services in buildings. Coverage includes drainage, plumbing, sewage treatment and disposal, heat loss and gain, principles of heating and cooling, space conditioning of housing and large buildings, basic electricity, electrical wiring and lighting, and elevators and escalators.

A06-2100
McPartland, J. How to Design Electrical Systems. McGraw-Hill, 1968. 208 pp. $13.50
This manual explains, in easy-to-understand terms, how to design safe reliable industrial, commercial, and residential electrical systems.

A06-2101
McPartland, J. and Novak, W. Electrical Design Details. McGraw-Hill, 1960. 231 pp. $12.00
This is a complete design file of electrical wiring diagrams, schematics, and detail drawings and of actual electrical installations in operation in the United States.

A06-2102
McPartland, J. and Novak, W. Electrical Equipment Manual. McGraw-Hill, 1965. 284 pp. $11.50
Here is a practical reference for anyone seeking a background in the fundamentals of electrical equipment.

A06-2103
McPartland, J. and Novak, W. Practical Electricity. McGraw-Hill, 1964. 104 pp. $7.95
This book discusses, in an essentially nonmathematical approach, the basics of electricity, and the operation and application of practical circuits to electrical design.

A06-2104
Manas, V. National Plumbing Code Handbook. McGraw-Hill, 1957. 503 pp. $22.50
Offers guidance to the National Plumbing Code, explaining and illustrating meaning and intent, paragraph by paragraph.

Includes much technical information to aid in the design and installation of systems that will meet code standards.

A06-2105
Manas, V. T. National Plumbing Code Illustrated. Manas, 1965.
$8.00

A06-2106
Marshall, J. L. Lightning Protection. John Wiley, 1973. 190 pp. $18.25
Contents--The Toll of Lightning. The Nature of Lightning. Magnitude of the Lightning Discharge. The Earth as a Discharge Terminal. Protective Grounding Systems: General. Safety of Life. Grounding of Communications, Towers and Systems. Protection Systems for Buildings. Protection of Electric Power-Transmission Systems.

A06-2107
Meirovitch, L. Elements of Vibration Analysis. McGraw-Hill, 1975. 480 pp. $19.50
This book offers a fundamental overview of vibration theory and application for control. Sample problems help the reader develop his own understanding of vibration dynamics. Mechanical, civil, and aeronautical engineers will find this volume particularly useful.

A06-2108
Merritt, F. Mechanical and Electrical Design of Buildings for Architects and Engineers. McGraw-Hill, 1966. 144 pp. $4.95
Material on environmental control in buildings selected from the editor's Building Construction Handbook, 2nd ed.

A06-2109
Miliaras, E. S. Power Plants with Air-Cooled Condensing Systems. MIT Press, 1974. 240 pp. $12.95
Air-cooled condensing systems can reduce environmental thermal pollution if air-cooled equipment is included from the outset as an integral component in the entire power plant design. The technology for building air-cooled condensing systems is reviewed in this book.

A06-2110
Mischke, C. R. An Introduction to Computer-Aided Design. Prentice-Hall, 1968. 211 pp. $3.95
Introduces the reader to the elements of computer-aided design and exposes him to Computer-Aided Design Engineering Technique. While FORTRAN is the language of the volume, the approach employed is independent of the language used and the computer utilized. Fundamentals learned in the first four chapters are applied to a sequence of problems of increasing dimensional complexity.

Design--Mechanical and Electrical

A06-2111
Morris, J. L. Welding Principles for Engineers. Prentice-Hall, 1951. 511 pp. $12.95
Contains information on welding processes, applications and testing and inspection of welds. Includes data on brazing, soldering, stress and distortion, and weld design.

A06-2112
Morris, N. M. Essential Formulae for Electrical Engineers. John Wiley, 1974. 26 pp. $2.95
Contains the essential formulae in the fields of electronics, electrical engineering, control systems, measurement, logic, telecommunications, and mathematics.

A06-2113
Morrison, D. Engineering Design: The Choice of Favorable Systems. McGraw-Hill, 1969. 200 pp. $9.95
The strategical rules which govern design decisions are accented in this book which details major steps in design--steps concerned with how the device to be designed will interact with features of its environment. Covers machines, energy and power, load carrying elements and structures, etc., and shows how the application of concepts discussed leads to results.

A06-2114
Morse, F. T. Power Plant Engineering. Van Nostrand Reinhold, 1953. 687 pp. $9.95
The emphasis is on small and medium sized plants. Includes discussions of gas turbines and the fundamentals of nuclear energy; engineering principles of power plants and their operation as well as design; the operating engineer's problems, such as making tests and identifying sources of difficulties. Incorporates a review of applied heat theory nomenclature and calculations; many solved examples illustrating the principles of the subject.

A06-2115
Nadeau, G. Introduction to Elasticity. Holt, Rinehart and Winston, 1964. 288 pp. $10.00
A modern introduction to elasticity with a minimum of mathematical complications. Vectors and dyadics are used for developing theory and solving problems.

A06-2116
National Science Foundation. Demand Analysis, Solar Heating and Cooling of Buildings. U.S. Gov't. Printing Office, 1974. 169 pp. $2.45. Order No. NS 1.33:So 4/3 S/N 038-000-00207-4
This report focuses on two separate subjects relating to the use and acceptance of solar heating and cooling by the general public. Part one is a summary of the solar water heater industry in Miami, Florida. It describes products and technology of the solar water heater industry, economics of marketing solar water heaters, users' costs, and opinions of Miami residents on solar water

heaters. Part two reviews the attitudes of lending institutions and financiers throughout the country toward the solar heating and cooling of buildings.

A06-2117
National Science Foundation. Proceedings of the Solar Heating and Cooling for Buildings Workshop, 1973. Part 1, Technical Sessions. U. S. Gov't. Printing Office, 1974. 226 pp. $2.50. Order No. NS 1.2:So 4/3pt.1 S/N 3800-00171

Summaries of presentations concerning solar collectors, energy storage, heating, solar air conditioning, and combined solar heating/cooling systems. Part 2, "Panel Sessions," is not yet available.

A06-2118
National Science Foundation. Solar Cooling for Buildings. U. S. Gov't. Printing Office, 1974. 231 pp. $3.00. Order No. NS 1.2:So 4/8 S/N 3800-00189

The proceedings of a workshop on solar cooling held February 6-8, 1974, and sponsored by the National Science Foundation.

A06-2119
NESCA. Basic Refrigeration, and Student Workbook for Basic Refrigeration. National Environmental Systems Contractors Assn., $50.00 (Student Workbook, $4.00)

Basic Refrigeration is an instructor's course outline in basic refrigeration theory and principles. Contains lesson plans, demonstrations, student worksheet samples, illustrations, experiments, and examinations, in loose-leaf form. Suitable for inshop training of service mechanics or for technical and vocational school courses. Student Workbook for Basic Refrigeration exactly parallels coverage and content of Instructor's Course Outline--contains over 100 pages of worksheets and experiments for the student.

A06-2120
Neville, A. M. and Kennedy, J. B. Basic Statistical Methods for Engineers and Scientists. Intext, 1964. 325 pp. $9.25

A treatment of engineering applications of statics which does not dwell on mathematical derivations. Includes many examples and solutions based on actual test data.

A06-2121
NFPA. Electrical Code for One and Two-Family Dwellings. National Fire Protection Assn., 1975. 167 pp. $4.50

Vital information for electricians and those in the construction industry. Covers those wiring methods and materials most often encountered in the construction of new one and two-family dwellings. Precise, informative excerpts from the current National Electrical Code without modification and with minimum editorial changes.

A06-2122
NFPA. National Electrical Code 1975. National Fire Protection

Design--Mechanical and Electrical 349

Assn., 1975. 640 pp. $5.50
Authoritative requirements for safe design and installation of wiring and electrical equipment. Approved by the American National Standards Institute.

A06-2123
NFPA. National Fuel Gas Code. National Fire Protection Assn., 1974. 150 pp. $3.00
Offers general criteria for the installation, operation and maintenance of gas piping, gas appliances and nonspecialized gas equipment on residential, commercial and industrial premises for use with fuel gases, such as natural gas, manufactured gas, LP-Gas in the vapor phase, LP-Gas-air mixtures, or mixtures of these gases (including mixtures within the flammable range) at pressures up to and including 60 psi.

A06-2124
Nielsen, L. Standard Plumbing Engineering Design. McGraw-Hill, 1963. 312 pp. $19.75
This reference for architects, engineers, plumbing contractors, and builders gives modern engineering design of plumbing systems for buildings.

A06-2125
Olivieri, J. B. How to Design Heating-Cooling Comfort Systems. Business News Publishing, 1973. 320 pp. $14.95
The book defines comfort and gives a method for calculating year-round heat loss and gain. The reader is shown how to select equipment from manufacturer's catalogs. There follow chapters on psychrometrics, water circulation and environmental control. Cost estimates are given.

A06-2126
Osborne, W. C. Fans. Pergamon, 1966. 166 pp. $9.00
This text explains the simple physical principles necessary for the proper application, control and design of fans.

A06-2127
Pansini, A. J. Basic Electrical Power Distribution. Vol. 2. Hayden, 1971. 128 pp. $3.95
Underground construction, cables, ducts, risers, service equipment. Substations, street lighting, essentials of electricity, transformers.

A06-2128
Parrish, A., ed. Mechanical Engineer's Reference Book. CRC Press, 1973. 1800 pp. $41.20
Covers theory and design data, drawing practice and metrology, screw threads, metallurgy and non-metals technology, welding and surface finishes, fabrication of materials, pipework, lubricants and lubrication, plastics, and more.

A06-2129
Patton, A. R.   Solar Energy for Heating and Cooling of Buildings.
    Noyes Data Corp., 1975.   328 pp. $24.00
        Solar energy can be used for indirect heating purposes in
many ways. The information in this book has been limited to so-
called low temperature solar thermal processes which are obtain-
able from collectors that are fairly simple to construct. Large
scale applications designed for schools and similar building are beginning to appear or are in the planning stage. This book describes in detail several large scale feasibility studies with designs
suitable for institutions and industrial plants.

A06-2130
Pesco, C., ed.   Solar Directory.   Ann Arbor Science, 1976.
    624 pp. $20.00
        Gives information on research, manufacturing, and construc-
tion projects involving solar heating, cooling and energy. Subjects
covered include collector systems, water and space heating, retro-
fitting, incentives for energy conservation and new construction
technology.

A06-2131
Peters, M. and Timmerhaus, K.   Plant Design and Economics for
    Chemical Engineers.   McGraw-Hill, 1968.   704 pp. $18.50
        This edition studies profitability techniques, cost-estimation
data, and process-design evaluation methods in chemical engineering.

A06-2132
Petrusewicz, S. A. and Longmore, D. K.   Noise and Vibration
    Control.   American Elsevier, 1974.   300 pp. $18.50
        This book is aimed primarily at people working in industry
who have little or no knowledge of noise and vibration control, but
who are confronted from time to time with the task of trying to
keep levels of noise and vibration within acceptable limits.

A06-2133
Pfluger, A. R. and Lewis, R. E., eds.   Weld Imperfections.
    Addison-Wesley, 1968.   652 pp. $15.00
        The contents include welding design, imperfections, frac-
ture mechanics and weld evaluation. Weld inspection standards are
discussed. The book is comprised of papers from a symposium
on the subject.

A06-2134
Phillips, D.   Lighting in Architectural Design.   McGraw-Hill,
    1964.   312 pp. $18.50
        A presentation of all aspects of lighting.   385 illustrations.

A06-2135
Plant Engineering Electrical Library.   Plant Engineering, about
    400 pp. $18.50
        Nine booklets in a loose-leaf folder cover the application

Design--Mechanical and Electrical

and specification of electric motors, motor maintenance, electric power systems, troubleshooting industrial controls, electrical estimating, plant utilities control, designing industrial lighting and protecting electrical equipment.

A06-2136
Porges, F. The Design of Electrical Services for Buildings. Chapman and Hall, 1974. 275 pp. £4.00

This book explains the electrical equipment used in buildings, the types of cable employed, installation practice and the methods of distributing electricity within a building. The chapter devoted to protection against faults develops the subject in a logical manner. Every service which an electrical consultant may be called upon to design or specify is explained, including communal TV systems, call systems, low voltage systems, fire alarms, lightning conductors and lifts. The theoretical knowledge needed for design is given and the practical details of installation are also described.

A06-2137
Porges, F. The Design of Electrical Services for Buildings. John Wiley, 1975. 264 pp. $18.75

Explains fully the electrical equipment used in buildings, the types of cable employed, installation practice and the methods of distributing electricity within a building.

A06-2138
Porges, J. and Porges, F. Handbook of Heating Ventilating and Air Conditioning. Butterworth, 1971. 304 pp. £4.00

Providing the data regularly needed by design engineers, this book is accepted as a standard work of reference. It consists mainly of formulae and tables rather than theoretical explanations. The new edition gives most data and tables in SI units while retaining their Imperial equivalents. Although written primarily for heating and ventilating engineers, much of the information is applicable to other branches of mechanical and chemical engineering.

A06-2139
Power Magazine. Plant Energy Systems. McGraw-Hill, 1967. 480 pp. $16.00

Based on the special reports published by Power Magazine, this book discusses such topics as pumps, compressing air, fans, refrigeration, building heat, piping, cooling towers, lubricants, corrosion, and more.

A06-2140
Price, S. G. Air Conditioning for Building Engineers and Managers: Operation and Maintenance. Industrial Press, 1970. 136 pp. $17.50

This handbook is written specifically for building engineers, superintendents and property managers concerned with the operation and maintenance of air conditioning equipment in commercial buildings, large apartment buildings and institutions. Topics include basic refrigeration, self-contained units, central plants, cooling towers, filters, etc. Sample maintenance programs for heating and cooling plants are also given to provide economical plant operation.

A06-2141
Ramsey, M. A. Tested Solutions to: Design Problems in Air Conditioning and Refrigeration. Industrial Press, 1966. 178 pp. $20.00

Here are recommendations for designs which will reduce or eliminate the frequently occurring problems that arise in air conditioning. The book analyzes problems on the basis of fundamental principles, and the resulting solutions preclude such considerations as "usual" or "customary" practices, which are often incorrect or inappropriate. Instead, when the author proposes any design principles or procedures, the theoretical reasons behind them are given.

A06-2142
Rase, H. F. Piping Design for Process Plants. John Wiley, 1963. 295 pp. $18.25

A06-2143
Richter, H. P. Practical Electrical Wiring. 10th ed. McGraw-Hill, 1976. 672 pp. $17.50

An instruction manual rather than a reference volume, this book is for anyone who wants a practical understanding of the "how's" and "why's" of electrical wiring. With this book anybody can learn how to wire homes and farms, as well as industrial and commercial buildings, schools and churches, all in strict accordance with the latest National Electrical Code.

A06-2144
Rogers, A. and Connolly, T. Analog Computation in Engineering Design. McGraw-Hill, 1960. 450 pp. $17.50

Guidance on using and applying analog computers to solve engineering design problems. Includes examples of computer applications.

A06-2145
Rogers, T. S. Thermal Design of Buildings. John Wiley, 1964. 196 pp. $12.25

A06-2146
Rossi, B. Welding Engineering. McGraw-Hill, 1954. 786 pp. $13.95

A study of welding processes, design and fabrication methods, testing and inspection of welds.

A06-2147
Rothbart, H. A. Mechanical Design and Systems Handbook. McGraw-Hill, 1964. 1594 pp. $39.50

Design--Mechanical and Electrical

Provides a scientific basis for the dynamic analysis of mechanical systems and for machine design, and covers topics such as system analysis and synthesis, dynamics of contacting bodies, dynamics of materials, fasteners, and power control components and subsystems.

A06-2148
Sauer, H. J., Jr. and Howell, R. H. Environmental Control Principles. American Society of Heating, Refrigerating and Air-Conditioning Engineers, 1975. 432 pp. $8.00

A supplemental workbook for use with ASHRAE's Handbook of Fundamentals permits self-instruction in such subjects as heating and cooling loads, design calculations, air-conditioning systems, etc. More than 400 problems are given. A separate manual of solutions to the problems is available for $2.50.

A06-2149
Schenck, H., Jr., ed. Introduction to Ocean Engineering. McGraw-Hill, 1975. 384 pp. $18.50

The book covers all important aspects of engineering for ocean applications. It provides integration of oceanographic theory, especially in the chapters on waves, soil mechanics, underwater acoustics, and water quality. With many examples and over 150 problems this volume offers practical applications of basic engineering theory to the in-water environment.

A06-2150
Schweitzer, P. A. Handbook of Corrosion Resistant Piping. Industrial Press, 1969. 358 pp. $30.00

The wide range of corrosion resistant pipe and fittings presently available are placed in one source to provide a means of rapid evaluation and comparison. The book contains pertinent design, installation, corrosion resistant, and economic factors necessary to determine the optimum system to handle specific corrodents. Each of the materials is discussed individually, and suitable construction materials are indicated for over 500 corrodents. Available sizes, weights and types of fittings are given for each material.

A06-2151
Schweitzer, P. A. Handbook of Valves. Industrial Press, 1972. 180 pp. $20.00

This volume aids the valve user and or specifier in making a correct choice of a valve for any particular application. Explaining that valve selection starts with an understanding of a valve's function, and the factors which affect its performance, the author then discusses and explains these factors: the properties of fluids going through the valves; fluid friction losses; operating conditions; materials of construction; and size.

A06-2152
Sculthorpe, W. J. Design of High Pressure Steam and High Temperature Water Plants. Industrial Press, 1972. 103 pp.

$20.00

Here is information for the engineer which is completely integrated for both kinds of plants. Almost all of the literature is outdated in respect to this field, and the book takes the designer from start to completion on the basis of current practices.

A06-2153
Segeler, C. G., ed. Gas Engineers Handbook. Industrial Press, 1965. 1550 pp. $47.50

This reference, sponsored by the American Gas Association and written by a staff of 150 specialists, answers any general or specific engineering information requirement in regard to natural, liquefied petroleum, and manufactured gases. It presents all "working" facts and data on fuel gases needed by engineers, industry and government personnel. The Handbook brings together in one volume and 125 chapters all conceivable engineering methods and operating data of the entire gas industry, from source to burner.

A06-2154
Seto, W. W. Mechanical Vibrations: Including 225 Solved Problems. McGraw-Hill, 1967. 203 pp. $3.95

Covers the single degree of freedom system, two and several degrees of freedom, torsional vibration, nonlinear vibration, electrical analogies, analog computer and vibration and sound.

A06-2155
Severns, W. H. and Fellows, J. R. Air Conditioning and Refrigeration. John Wiley, 1958. 563 pp. $12.75

A06-2156
Sherratt, A. F. C., ed. Air Conditioning System Design for Buildings. Applied Science, 1969. 249 pp. £6.00

This book contains the eleven papers from a conference on the relationship of air conditioning to building design. The interdependence of the building and the air-conditioning system in providing the best environment at minimum cost is brought out in the papers and, ranging as they do from human requirements to computerised design, they form a textbook of present-day thinking on the implications of air conditioning on building design.

A06-2157
Sherratt, A. F. C., ed. Integrated Environment in Building Design. John Wiley, 1975. 281 pp. $32.50

Proceedings of a conference on integrated environment in building design. It includes an assessment of design criteria, examination of the techniques by which these criteria could be implemented, and discusses the cost implications.

A06-2158
Sherwood, D. R. and Whistance, D. J. The Piping Guide. Syentek Books, 1976. 210 pp. $16.95 (paperback $11.95)

This handbook covers the design of piping systems and

Design--Mechanical and Electrical

drafting techniques. Pipe, valves and fittings, pumps and compressors and pipe hangers are described and illustrated in numerous drawings and tables. The methods and procedures of organizing piping drawings are also discussed.

A06-2159
Sheilds, C. Boilers. McGraw-Hill, 1961. 566 pp. $24.50
 Explaining the terminology of this subject, the book also contains a wealth of data for buying, designing, selling, specifying, installing, operating, and maintaining all types of boilers for heat or power use.

A06-2160
Simmons, D. M. Wind Power 1975. Noyes Data Corp., 1975. 300 pp. $24.00
 This book is based on international studies by industrial, engineering and university groups. The various phases for developing wind power are considered in the first three chapters of this book. The next three describe wind power developments in the U.S., Canada, U.S.S.R., Germany, Denmark, France, U.K., Sweden and other countries in Asia and Africa. The last chapter describes commercially available apparatus.

A06-2161
Snowdon, J. C. Vibration and Shock in Damped Mechanical Systems. John Wiley, 1968. 486 pp. $21.50
 This book describes how the steady-state and transient vibration of damped mechanical systems can be analyzed by the modern and concise methods of complex algebra and the Laplace transformation.

A06-2162
Steidel, R. F. An Introduction to Mechanical Vibrations. John Wiley, 1971. 393 pp. $15.75
 Presents mechanical vibrations with three major objectives: stresses problems and examples; emphasizes idiomatic concepts in the field; discusses energy dissipation or damping in mechanical engineering systems.

A06-2163
Stephens, R. W. B., ed. Sound: In Eight Languages. John Wiley, 1974. $37.50
 A multilingual handbook which brings together the concepts of many fields of science in order to provide complete coverage of all the ways in which sound is important in modern technology. Includes a general coverage of the physical concepts relating to mechanical vibrations and an exhaustive treatment of topics ranging from architectural acoustics to audiology and phonetics.

A06-2164
Stoecker, W. F. Principles for Air Conditioning Practice. Industrial Press, 1968. 160 pp. $16.00
 The material included here enables the practicing engineer

to gain insight into the solution of non-routine air conditioning problems. The book supplements handbooks and reference books by delving deeper into the topics covered. The fifteen chapters are in three major sections: flow in confined spaces, flow in unconfined spaces, and heat and mass-transfer processes. A chapter on sound and acoustics is included.

A06-2165
Stoecker, W. F. Refrigeration and Air Conditioning. McGraw-Hill, 1958. 397 pp. $15.00
Methods, types, and applications of refrigeration and air conditioning, the vapor compression cycle, multipressure systems, heat pumps, psychometry, cooling towers and evaporative condensers, piping and accessories, and more.

A06-2166
Strock, C. and Koral, R. L. Handbook of Air Conditioning, Heating and Ventilating. Industrial Press or R. S. Means, 1956. 1472 pp. $35.00
Section headings include: Climatic data. Load calculation. Air Conditioning systems and components. Refrigeration for air conditioning. Air handling and ventilation. Fuels and combustion. Space heating. Piping. Plumbing and drainage. Motors and controls. Building types. Noise and radiation control.

A06-2167
Thomson, W. T. Vibration Theory and Applications. Prentice-Hall, 1965. 384 pp. $15.95
Discusses single degree of freedom systems in free, forced, and transient conditions. Supplements this material with analog computer techniques and nonlinear vibrations. A parallel approach to the problem of formulation with generalized coordinates and Lagrange's Method leading to mode summation methods.

A06-2168
Threlkeld, J. L. Thermal Environmental Engineering. Prentice-Hall, 1970. 495 pp. $16.15
Covers refrigeration, psychometrics, solar radiation, and heat transmission in buildings, with emphasis on theory and analysis throughout the text. New material on thermoelectric cooling, air separation, properties of moist air, psychometry, humidity measurement, and the estimating of thermal loads.

A06-2169
Timoshenko, S. Vibration Problems in Engineering. Van Nostrand Reinhold, 1955. 468 pp. $9.95
Systems with one degree of freedom. Systems with nonlinear and variable spring characteristics. Systems with two degrees of freedom. Systems with several degrees of freedom. Vibrations of elastic bodies.

A06-2170
Traister, J. E. Electrical Design for Building Construction.

Design--Mechanical and Electrical

McGraw-Hill, 1976. 192 pp. $15.00

Written primarily for electrical designers and engineers working for architectural/consulting engineering firms. Topics covered include: lighting with incandescent lamps, power wiring for convenience outlets, power wiring for special outlets, power wiring for heating, ventilating, and air conditioning, service equipment, panelboards and feeders, special electrical systems, and energy conservation methods. In addition, actual design problems for lighting a church, bank, laundry, and several other buildings are given.

A06-2171
Udin, H., et al. Welding for Engineers. John Wiley, 1954. 430 pp. $10.00

A06-2172
Van Straaten, J. F. Thermal Performance of Buildings. Applied Science, 1967. 311 pp. £6.00

Many of the unsatisfactory thermal and ventilation conditions found in buildings can be avoided if the approved principles of functional planning and design are applied. In this book, the author outlines these principles, as far as thermal and ventilation conditions in buildings are concerned, in terms that will be understood by architects, engineers, building inspectors and building scientists.

A06-2173
Vilbrandt, F. C. and Dryden, C. E. Chemical Engineering Plant Design. McGraw-Hill, 1959. 534 pp. $16.50

Discusses development of the project, process design, selection of process equipment and materials, plant layout, site selection, preparation, structures, and more.

A06-2174
Walker, R. Pump Selection--A Consulting Engineer's Manual. Ann Arbor Science, 1975. 128 pp. $10.00

Written by a pump specialist, this book tells about fluid pump selection, application, design and performance characteristics. From it you can specify pumping equipment required to suit your specific installations, with essential design parameters to insure reliable and efficient service with minimum annual costs.

A06-2175
Waller, R. A. and Atkins, W. S. Building on Springs. Pergamon, 1969. 98 pp. $6.75

Describes a new way of protecting buildings and machinery from vibration by means of springs. Existing information is not quoted in detail and the newness of the subject is emphasized together with its advantages and drawbacks. Although intended for architects, engineers and those responsible for property development this book could be of interest to anyone interested in town planning and the construction industry.

Construction Information

A06-2176
Watt, J. and Stetka, F. NFPA Handbook of the National Electrical Code. McGraw-Hill, 1972. 748 pp. $13.75
   This edition of the annotated and illustrated text of the National Electrical Code continues the tradition begun by Arthur Abbott in 1932. The Code is now on a three-year revision cycle, and this revision, based on the 1971 Code, contains much new illustrative material and explanatory commentary which clarifies all aspects of the Code including about 500 changes that have occurred in it.

A06-2177
Watt, J. H. and Summers, W., eds. NFPA Handbook of the National Electrical Code. National Fire Protection Assn., 1975. 800 pp. $15.95
   The complete text of the 1975 NEC with explanatory comments. Diagrams, photos, formulas and tables help you understand and apply the code. Emphasizes new provisions. Latest information on new materials and methods. A valuable reference for electrical work in all types of buildings.

A06-2178
Weaver, M. K. and Kirkpatrick, J. M. Environment Control: Air Conditioning and Refrigeration Theory and Application. Harper and Row, 1974. $13.95
   An air conditioning and refrigeration textbook incorporating behavioral objectives. Divided into six parts: basic refrigeration principles, domestic refrigeration, commercial refrigeration, domestic air conditioning and heating, commercial air conditioning and heating, and automobile air conditioning.

A06-2179
Weaver, R. Process Piping Design. Gulf Publishing Co. Vol. 1, 1973. $11.95; Vol. 2, 1973. $11.95
   Tells how to lay out a grass roots refinery, how to orient nozzles on horizontal and vertical vessels, defines process terms. Chapters cover general piping, plant arrangement and storage tanks, process unit plot plans, piping systems and details, pipe fabrication, vessels and instrumentation.

A06-2180
Westinghouse Lighting Handbook. Westinghouse Electric Corp., 1974. 264 pp. $7.50
   This practical manual informs the designer about light sources and luminaires and includes recent data from the Illuminating Engineering Society on criteria on visual comfort probability and equivalent sphere illumination. There are chapters on interior wiring for lighting, illumination levels, floodlight design, industrial lighting and cost factors for lighting.

A06-2181
Williams, J. R. Solar Energy, Technology and Applications. Ann Arbor Science, 1975. 144 pp. $9.95

Design--Mechanical and Electrical 359

Gives current facts about collectors, heat for buildings, water heaters, air conditioning, power generation and other uses.

A06-2182
Wolberg, J. Application of Computers Engineering Analysis. McGraw-Hill, 1971. 288 pp. $14.50
Written to help the user close the gap between a sketchy knowledge of programming and the ability to apply the computer to engineering analysis problems.

A06-2183
Woolrich, W. R. Handbook of Refrigerating Engineering. Vol. 1, Fundamentals. AVI Publishing Co., 1965. 460 pp. $17.00
The 4th edition of the Handbook of Refrigerating Engineering is published in two volumes. Volume I contains basic refrigeration and food processing data. New charts of the thermodynamic properties of the principal refrigerants, and a previously unpublished chart of the recently adopted azeotrope Refrigerant 500 are featured.

A06-2184
Woolrich, W. R. Handbook of Refrigerating Engineering. Vol. 2, Applications. AVI Publishing Co., 1966. 434 pp. $17.00
Makes available information on processing, storage, transportation, protection and refrigeration of perishable foods and other products.

A06-2185
Woolrich, W. R. and Hallowell, E. R. Cold and Freezer Storage Manual. AVI Publishing Co., 1970. 338 pp. $21.50
The principles of refrigeration are presented with a consideration of the advantages and disadvantages of refrigerants, both those presently used and those formerly employed. The book also discusses the construction, materials and equipment to be used in modern types of refrigerated warehouses. Information is presented concerning the optimum temperature and humidity which should be maintained in rooms in which all kinds of fresh and frozen foods, furs, etc. are stored.

A06-2186
Yerges, L. F. Sound, Noise and Vibration Control. Van Nostrand Reinhold, 1969. 204 pp. $9.95
This book utilizes the designer's own idiom to explain the methods, procedures, and materials that have proven to be the most effective deterrents to troublesome acoustical situations.

Section   A07

DESIGN: TRANSPORTATION, HEAVY CONSTRUCTION

　　　　　　Airports
　　　　　　Bridges
　　　　　　Highway Engineering

A07-2200
ACI. First International Symposium on Concrete Bridge Design.
Amer. Concrete Institute, 1969. 800 pp. $27.50 (for ACI members, $22.50)
This symposium volume contains 45 papers authored by over 64 eminent exponents on concrete bridge design from all over the world. Classified into 13 distinct groups, the areas covered are general and aesthetic, concrete slab decks, load distribution in multibeam and box-girder concrete bridges, skew and curved bridges, analysis of concrete box-girder bridges, dynamic response and seismic design, high-strength reinforcement and crack control, ultimate strength and limit design, design and construction of prestressed concrete bridges, precast concrete bridges, long-span concrete bridges, and sub-structures.

A07-2201
ACI. Second International Symposium on Concrete Bridge Design.
University Microfilm, 1971. 1300 pp. $45.00
New methods of design, fabrication and erection, construction, unconventional methods of aggregate grading, innovative cementing materials, and exciting treatment of concrete are being rapidly evolved. This volume is aimed at disseminating the new methods and techniques compiled in this massive work to bridge engineers throughout the world. 44 papers, 63 authors.

A07-2202
Antoniou, J. Environmental Management: Planning for Traffic.
McGraw-Hill, 1972. 200 pp. $22.50
This book examines the elements of environmental management and their practical application in order to assess the principal effects of the use of such techniques in Great Britain and the USA. The quality of an environment in relation to the needs of pedestrians is a major consideration in the methods proposed for the control of vehicular traffic.

A07-2203
APWA. Feasibility of Utility Tunnels in Urban Areas. American Public Works Assn., 1971. 167 pp. $10.00 (for APWA members, $8.00)
This study of the technical, legal, and economic aspects concludes that utility tunnels are justified in high density urban areas. Although current technology can settle the unresolved technical problems, further research is needed to prove economic feasibility and general benefit to the public. Much of the research was

conducted under contract by Stanford Research Institute. Engineers, lawyers, planners, administrators, and educators will find the report a valuable reference.

A07-2204
APWA. History of Public Works in the United States, 1776-1976. American Public Works Assn., 1976. $15.00

Scheduled for publication in 1976 as part of the U.S. Bicentennial celebration, this book has been supported by the Federal Government in the form of Public Law 92-54, enacted by Congress and signed by the President. It documents the role played by public works in U.S. development during the nation's 200-year history. The book is being produced by a special APWA committee with several prestigious public and private organizations cooperating.

A07-2205
APWA. Proceedings, Engineering Utility Tunnels in Urban Areas. American Public Works Assn., 1971. 135 pp. $10.00

Representatives of utilities, government agencies, engineering firms, and contractors met to discuss what must be done to promote the construction of utility tunnels. Organized by APWA and its Institute for Municipal Engineering, the conference covered systems compatibility and safety, design problems, foreign experience, operating and maintenance procedures; and financing, legal, and right-of-way considerations.

A07-2206
APWA. Public Works Computer Applications. American Public Works Assn., 1970. 143 pp. $10.00 (for APWA members, $8.00)

This report is designed to extend the use of computers beyond their application in the public works field. It weighs the problems, approaches, and opportunities of further computer use in terms of feasibility; process character; equipment design, acquisition, and placement; economic factors; and training of personnel.

A07-2207
APWA. Streetscape Equipment Sourcebook. American Public Works Assn., $15.00 (for APWA members, $12.00)

Authored by the Center for Design Planning, this is the first American publication featuring top-rated street furniture for streets, sidewalks, plazas, malls, and other public spaces. The equipment listed, which was judged excellent by a knowledgeable jury, includes lighting, traffic control, information signs, public safety/security, amenities and good housekeeping.

A07-2208
APWA. A Survey of Urban Arterial Design Standards. American Public Works Assn., 1969. 91 pp. $5.00 (for APWA members, $4.00)

Municipal engineers will find good use for this book as a basis for local standards, a reference, and a guide to further information. It reports the standards and practices of 24 urban

Design--Transportation, Heavy Construction 363

areas, including lateral clearance, curvature and grade, corner radii, lane width, medians, curb and gutter, roadway width, driveways, sidewalk and border, and right-of-way.

A07-2209
APWA Directory. American Public Works Assn., $20.00
    This is an encyclopedia of information on the APWA, listing all members, past and present officers, chapters, award recipients, institutes, committees, foundations, and much more.

A07-2210
ARBA Directory of Transportation and Agency Personnel. American Road Builders' Assn., 1973. 149 pp. $3.50
    A listing of the names and addresses of federal, state and county transportation agency officials and engineers. Included also are the various Congressional committees and other highway-related groups.

A07-2211
Army Department. Planning and Design of Roads, Airbases, and Heliports, in the Theater of Operations. U.S. Gov't. Printing Office, 1974. 792 pp. $9.80. Order No. D 101.11:5-330/2/rep S/N 0820-00482
    Presents common factors to be considered in road, airbase, and heliport construction, and outlines responsibilities for construction of these facilities. An Army field manual that has useful information for those in the heavy construction field. Looseleaf.

A07-2212
Army Engineers Corps. Roads and Airfields. U.S. Gov't. Printing Office, 1975. 392 pp. $4.50. Order No. D 103.111/2:64-9 S/N 008-022-00086-1
    Gives information on planning, building, and maintaining military roads and airfields. Such information can be useful for construction and care of any kind of road. The course consists of seven lessons, and its presentation is programmed. That is, the ideas are presented in a logical learning sequence and the text often repeats ideas in different forms to reinforce correct responses.

A07-2213
Arnison, J. H. Roadwork Technology. Butterworth, 1967. Vol. 1, 156 pp. £2.35; Vol. 2, 164 pp. £2.20; Vol. 3, 172 pp. £2.35
    These three volumes on the subject of roadwork, which cover basic and practical applications, are intended to serve as a standard text book. This is a comprehensive and well planned study which students and teachers of the subject will find both useful and stimulating.

A07-2214
ASCE. Airport Terminal Facilities. American Society of Civil Engineers, 1967. $9.00 (for ASCE members, $4.50)

ASCE-AOCI Specialty Conference. 1967.

A07-2215
ASCE. Airports--Challenges of the Future. American Society of Civil Engineers, 1973. $5.00 (for ASCE members, $2.50) Proceedings, Air Transport Division Conference, 1973.

A07-2216
ASCE. Airports--Key to the Air Transportation System. American Society of Civil Engineers, 1971. 296 pp. $6.00 (for ASCE members, $3.00)
This publication contains the papers, and speeches presented at the biennial Airports Specialty Conference held April 14-16, 1971 at Atlanta, Georgia. The book is of interest as a state of the art for airport operators, urban and transportation planners, city managers, and all engineers concerned with the siting, maintenance, and construction of airport facilities.

A07-2217
ASTM. An Analysis of the Literature on Tire-Road Skid Resistance. American Society for Testing and Matls., 1973. 152 pp. $5.50 (for ASTM members, less 20%)
A comprehensive review dealing with tires, road surfaces, vehicles and safety. The most comprehensive book of its type thus far published.

A07-2218
ASTM. Skid Resistance of Highway Pavements. American Society for Testing and Matls., 1973. 165 pp. $12.25 (for ASTM members, less 20%)
Nine papers explore in depth the analysis of slipperiness of various types of pavements, the analysis of skid resistance and relationship to tires. Of interest to auto designers, tire technologists, and pavement designers.

A07-2219
ASTM. Surface Texture Versus Skidding. American Society for Testing and Matls., 1975. 154 pp. $12.00 (for ASTM members, less 20%)
Ten papers cover measurement systems, frictional aspects, and safety features. Particularly useful to civil engineers, highway designers, automotive engineers, tire technologists and others concerned with highway safety.

A07-2220
Baerwald, J., et al., eds. Institute of Traffic Engineers Transportation and Traffic Engineering Handbook. Prentice-Hall, 1975. 992 pp. $24.50
Handbook is a basic reference work on transportation and traffic engineering. Topics include vehicle, highway and travel facts, driver and pedestrian characteristics, urban travel characteristics, mass transportation characteristics, highway capacity, traffic accident analysis, parking, loading and terminal facilities.

Design--Transportation, Heavy Construction

A07-2221
Barenberg, E. J., et al. Pavement Distress Identification and Repair. National Technical Information Service, 1975. 133 pp. $5.45. Order No. AD-758-447

Used for field identification of pavement problems, their causes, and ways to repair them. The pavements that are covered include asphalt, concrete and portland cement concrete.

A07-2222
Baumann, D. M. B. and Wilson, D. G., eds. Urban Engineering and Transportation. American Society of Mechanical Engrs., 1969. 109 pp. $8.75 (for ASME members, $7.00)

Includes following topics: mass production housing, cities and transportation, urban transportation, feasibility and cost of urban transportation systems, the guideway systems for automated transportation, and technical and cost-effectiveness considerations.

A07-2223
Bruun, P. Port Engineering. Gulf Publishing Co., 1973. 448 pp. $21.95

The book provides information on planning and layout of ports, port navigation, hydraulics, breakwaters, jetties, piers, wharves, quays, dolphins and mooring devices; harbor transport; littoral drift and sedimentation problems at ports; coastal geomorphology vs. port engineering; tidal inlets on alluvial shores; dredging and fishing ports.

A07-2224
Cedergren, H. R. Drainage of Highway and Airfield Pavements. John Wiley, 1974. 285 pp. $18.00

A07-2225
Cornick, H. F. Dock and Harbour Engineering. Vol. 1, The Design of Docks. Charles Griffin, 1968. 338 pp. £9.00

This volume deals with general aspects of the planning of docks, enclosed or tidal, and their detailed design, including that of wharves of various kinds. The historical approach is favoured, since dock and harbour engineers of today can sometimes learn much from designs, constructional methods, and expedients adopted in the past to meet particular problems.

A07-2226
Cornick, H. F. Dock and Harbour Engineering. Vol. 2, The Design of Harbours. Charles Griffin, 1969. 352 pp. £10.00

Covers the design of artificial harbours and their component parts and the improvement of natural harbours, together with a consideration of tides, the analysis of wave action and its influence on the design of breakwaters, the theory of piled structures, and the subject of the regulation and rectification of entrance channels.

A07-2227
Cornick, H. F. Dock and Harbour Engineering. Vol. 3, Buildings and Equipment. Charles Griffin, 1960. 320 pp. £7.50

Devoted to dock and harbour layout, buildings, sheds, and storage accommodation, communications, power distribution, and working equipment, with special reference to cargo handling--both of general cargo and of materials in bulk, including grain, coal, and ore, oil, and timber. "Working equipment" chapter gives a full treatment of dredgers and dredging. "Auxilliary services" considers wreck-raising and salvage, diving appliances and underwater work, radar, fire-prevention, port welfare, oil pollution control, etc.

A07-2228
Cornick, H. F. Dock and Harbour Engineering. Vol. 4, Construction. Charles Griffin, 1962. 407 pp. £9.00

Deals with the theoretical and practical aspects of constructional work, beginning with the relevant facts of geology and then discussing, in turn, the principles of soil mechanics, site investigation, calculation of earth pressures, and the construction of flexible walls and of graving docks. Materials in all their variety, and the formation and use of reinforced--concrete piles, are dealt with in detail, as are constructional plant and appliances. Temporary works and construction methods are described. Section on ground-water lowering.

A07-2229
Drew, D. Traffic Flow Theory and Control. McGraw-Hill, 1968. 447 pp. $18.50

Traffic engineering is presented as a science and practicing traffic engineers are provided with an insight into contemporary traffic research approaches. The book integrates experimental psychology, mathematics, physics, and statistics as a means of determining traffic flow.

A07-2230
Halprin, L. Freeways. Van Nostrand Reinhold, 1966. 160 pp. $15.00

The author, who is personally involved in the design of high speed roads, criticizes the existing concepts of design and offers many constructive suggestions for improvement.

A07-2231
Heggle, I. Transport Engineering Economics. McGraw-Hill, 1973. 265 pp. $25.00

Written to help the professional transport planner evaluate and reconcile the conflicting engineering, social, and economic characteristics of major transport decisions, this book provides a means of evaluating engineering projects for all kinds of transport. It is useful to engineers, economists, and landuse planners, as well as research workers.

A07-2232
Hennes, R. and Ekse, M. Fundamentals of Transportation Engineering. McGraw-Hill, 1969. 624 pp. $18.50

The latest design and operation information is incorporated

Design--Transportation, Heavy Construction

in this authoritative survey. Undersea transportation, urban rail transit, roads and pavements, airports, railroads, rivers and harbors, pipelines, and beltlines--all are discussed with reference to the engineering problems they present, how existing systems may be evaluated, and how new, improved systems can be designed and constructed.

A07-2233
Henry, D. and Jerome, J. A. Modern British Bridges. Applied Science, 1965. 189 pp. £2.50

Nearly one hundred bridges, large and small, significant and functional, are illustrated by means of photographs and line drawings: these feature almost every type of modern design and construction technique used. A short introductory text provides an historical background to bridge building generally and a simple explanation of construction techniques.

A07-2234
Hickerson, T. Route Location and Design. McGraw-Hill, 1967. 626 pp. $16.00

This reworked book contains new information on route surveying, general surveying, and covers newer topics such as city and regional planning. It conforms with the latest concepts and practices with reference to the interstate system and the AASHO Design Standards for highways other than freeways.

A07-2235
Holland, H. Travellers' Architecture. Wittenborn, 1971. 223 pp. $14.50

Architecture on the highways. Railway stations in Europe, United States, Canada, Argentina, Africa, India and Australia. Railway stations in the 20th century. European seaports. North American seaports. Airports in Great Britain. European, American and Canadian airports. Travel architecture in the future.

A07-2236
Hopkins, H. J. A Span of Bridges: An Illustrated History. Praeger, 1970. 272 pp. $12.50

An account of the history of bridges from ancient times to the 1960's. Beginning with the first tree trunk laid across a stream, the author traces the development of techniques up to the completion of the breathtaking span of New York's Verrazano-Narrows Bridge. He describes the origins and evolution of arches, suspensions, beams and girders, iron railway bridges, modern steel or reinforced concrete, cantilevers, and long spans. American bridge-designers, Eads, Roebling, and Amman are fully treated.

A07-2237
Horonjeff, R. Planning and Design of Airports. McGraw-Hill, 1975. 500 pp. $22.50

This new edition covers all aspects of the planning and design of airports, including the design of pavements and drainage facilities. Attention has been given to environmental and economic planning, heliports and Stolports, airport capacity and delay, and aircraft noise. The processes involved in the preparation of airport master plans and environmental impact statements are covered.

A07-2238
Legault, A. R. Highway and Airport Engineering. Prentice-Hall, 1960. 483 pp.
Information on the development, planning, engineering, financing and administration of highway and airport facilities. Covers plans, specifications and contracts, drainage, pavements, bituminous and concrete materials, etc.

A07-2239
Li, S. T. Bibliography on Airport Engineering. American Society of Civil Engineers. $8.00 (for ASCE members, $4.00) Covers period between 1941 and 1960.

A07-2240
Malt, H. Furnishing the City. McGraw-Hill, 1970. 264 pp. $19.95
This is the first book ever published to thoroughly discuss the subject of street furniture (traffic signs, mailboxes, light fixtures, directional signals, and similar items). The author discusses the urban product environment with its problems, goals, and systems design approach--and explores man's relationship to this synthetic setting.

A07-2241
Marcus, H. S. The Challenge of Deepwater Terminals. Lexington Books, 1974. 144 pp. $10.00
In this largest of trading nations and the world's leading consumer of energy, there are no terminals large enough to accept a ship of 200,000 tons and more. Unless it acts quickly to build deepwater terminals, the United States must face a serious challenge from foreign countries for the bulk-cargo business of the future. The author discusses the technical, economic, environmental, governmental, and legal aspects of this problem.

A07-2242
Minikin, R. R. Winds, Waves and Maritime Structures. Charles Griffin, 1963. 294 pp. £3.50
This edition includes new matter on ship locks, the sand by-passing of harbors, the use of wind-protection screens in harbors, and the design of structures for the open sea. Contents include details of wave pressure against obstructions, foundations, and stability and stresses on breakwaters.

A07-2243
Moses, R. Public Works: A Dangerous Trade. McGraw-Hill, 1970. 864 pp. $14.50
Covers topics such as Parks, Public Works and Transportation, Power, Planning, Housing, United Nations, Lincoln Center, World's Fairs, the Depression and Work Relief, Domestic and

Design--Transportation, Heavy Construction

Foreign Surveys, as well as the behind-the-scenes activities of politicians and others.

A07-2244
Myers, J., et al. Handbook of Ocean and Underwater Engineering. McGraw-Hill, 1969. 800 pp. $33.50
This work meets the need for a collection of tested methods, procedures, and facts for use in underwater construction and engineering projects. The subject matter ranges from basic concepts in oceanography and fluid properties, through facts on tools and techniques, cable technology, underwater power sources, fixed and floating structures, diving and other operations.

A07-2245
Navy Department. Design Manual, Airfield Pavements. U.S. Gov't. Printing Office, 1973. 147 pp. $2.10. Order No. D 209.14/2:21/3 S/N 0850-00108
Basic criteria regarding airfield pavements, for use by architects and engineers. Contents include airfield orientation, structural design of pavements, evaluation, aircraft facilities and loadings, and more.

A07-2246
Navy Department. Design Manual, Dry-Docking Facilities. U.S. Gov't. Printing Office, 1974. 308 pp. $4.50. Order No. D 209.14/2:29/3 S/N 0850-00111
Design criteria and construction information for drydocks, marine railways, and lifts. Includes foldout plans, along with many diagrams and charts.

A07-2247
Odier, L., et al. Low Cost Roads: Design, Construction and Maintenance. Butterworth, 1971. 166 pp. £3.50
Promoted by UNESCO as part of their programme to provide standards, codes and guides for engineering works in developing countries, this book has been written for the guidance of road planners and builders in tropical and sub-tropical countries. Its purpose is to provide up-to-date information on techniques of determining priorities in road planning, on standards of road layout and design, and the design of pavements.

A07-2248
Oglesby, C. H. Highway Engineering. John Wiley, 1975. 783 pp. $18.95
Contents include highway systems and organizations, highway transportation planning, economics and finance, highway design and construction, and a description of highway materials including gravel and crushed rock, macadam, bituminous pavements, and Portland-cement concrete pavements.

A07-2249
Paquette, R. J., et al. Transportation Engineering--Planning and Design. Ronald Press, 1972. 760 pp. $16.00

This book covers the planning and design of facilities for land, sea, and air transportation. Analyses of economic and operating characteristics encourage students to view the transportation problem in the context of a complex system for moving goods and people. Urban transportation is given special emphasis, using mathematical planning models and computer techniques for traffic forecasting. The design of streets, highways, and railroads is presented in a unified discussion. Specific planning procedures and design criteria are given for air transportation facilities. Material on the design of harbors and port facilities includes up-to-date considerations of the coastal environment. Pipelines and belt conveyor systems are described, emphasizing their role in moving freight.

A07-2250
Pignataro, L. J. Traffic Engineering: Theory and Practice. Prentice-Hall, 1973. 502 pp.
Designed to meet the needs of students, practitioners and researchers in traffic engineering. Presents material on planning, safety, economics and traffic theory. Bibliography contains 1622 entries. Worked out problems on statistical application, capacity and signalization are included.

A07-2251
Quinn, A. Design and Construction of Ports and Marine Structures. McGraw-Hill, 1972. 608 pp. $29.50
The revised edition of this manual is for use by design engineers, consulting engineers, port authority personnel, oil and bulk materials corporations, project managers, chief engineers, students--anyone involved in port planning, construction or maintenance. Every chapter has been rewritten and updated to reflect the latest on container ports, fenders for docks, new information on breakwaters and piles, up-to-date tabulation of ship sizes (including the new supertankers), plus an additional chapter on marinas.

A07-2252
Reclamation Bureau. Design of Small Dams. U.S. Gov't. Printing Office, 1974. 852 pp. $14.25. Order No. I 27.19/2:D 18/974, S/N 024-003-00089-2
Intended primarily as a guide to safe practices in public works programs in the United States. Provides engineers with information and data necessary for the proper design of small dams, provides specialized and highly technical knowledge in a form that can be used readily by engineers who do not specialize in this field, and simplifies design procedures for earthfill dams.

A07-2253
Ritter, L. J., Jr. and Paquette, R. J. Highway Engineering. Ronald Press, 1967. 782 pp.
Standard textbook on highway engineering, now includes the latest technological information: the 1966 geometric design criteria and traffic capacity data, new thickness design methods for flexible

and rigid pavements, new hydraulic design methods for culverts, revised asphalt specifications, and a new description of bridge types.

A07-2254
Robinson, J. Highways and Our Environment. McGraw-Hill, 1971. 320 pp. $30.50
   This illustrated book describes just how the motor car contributed to the transformation of our country's landscape. Using vintage and modern photographs, the author illustrates the book's underlying philosophy--that although we can never return our country to one of quiet country roads, we can still restore beauty to our highways and our land.

A07-2255
Rodgers, L. M. and Sands, L. G. Automobile Traffic Signal Control Systems. Chilton Book Co., 1969. 200 pp. $7.95
   Intended for those concerned with the need for and the design, installation, and timing of traffic-control devices, this text describes various designs, specific equipment, control circuits and maintenance problems.

A07-2256
Rowe, R. E. Concrete Bridge Design. Applied Science, 1972. 372 pp. £8.00
   This book deals with the problems in bridge design created by modern traffic requirements. It is based on extensive experimental research both on laboratory model bridges and on actual bridges. The methods of analysis and design procedures are presented in detail, and attention is focused exclusively on the bridge superstructure and its load distribution properties. Substructure or foundation problems have not been considered.

A07-2257
Sargious, M. Pavements and Surfacings for Highways and Airports. John Wiley, 1975. 619 pp. $48.50
   Deals with both modern and conventional types of pavements, including recent theories and approaches to pavement design and construction. Some topics include stabilized soils, the materials used and the design of flexible pavements for highways and airports. Rigid pavement design is covered along with reinforcement expansion joints and anchorage. Pavement construction and maintenance are also discussed.

A07-2258
Schriever, B. A. and Seifert, W. W. Air Transportation 1975 and Beyond: A Systems Approach. MIT Press, 1968. 516 pp. $20.00
   In 1967 a Transportation Workshop was organized to look at this whole problem. The organizers did not confine their attention to any one aspect of it for improvements must be made from all directions at once. To mount this systems approach, the Workshop established six panels to look at socioeconomic trends, air

vehicles technology, air traffic control, airports and terminals, collection and distribution of passengers, and government policies.

A07-2259
Sharp, O. R.   Concrete in Highway Engineering.   Pergamon, 1970. 172 pp. $7.40
　　Deals with the use of cement and concrete in the construction of road pavings. Design and constructional problems, problems of obtaining good appearance of the finished highway, maintenance and repair are treated.

A07-2260
Sherard, J. L., et al.   Earth and Earth-Rock Dams: Engineering Problems of Design and Construction.   John Wiley, 1963. 725 pp. $33.25

A07-2261
Steinman, D. B. and Watson, S.   Bridges and Their Builders. Dover.   401 pp. $2.50
　　Foremost modern bridge builder gives history, engineering ideas, techniques of bridges, from earliest times to present. Stories of many great bridges given in detail: Brooklyn Bridge, work of Eads, many others.

A07-2262
Thomas, H. H.   The Engineering of Large Dams.   John Wiley, 1975.
　　A book on the engineering of large dams which covers the major factors of investigation including hydrology, geology and aspects of design including mathematical and model analyses and the various techniques of construction.

A07-2263
Thorn, R. B. and Simmons, J. C. F.   Sea Defence Works: Design, Construction and Emergency Works.   Butterworth, 1971.   128 pp.   £3.00
　　Basing their account on extensive experience and on the principles and findings of coastal hydraulics, the authors deal with all aspects of sea defences and emergency works. Recommendations put forward stem from the authors' personal knowledge of works carried out on over 230 kilometres of sea defences in south east England, including wide variations in foreshore conditions and works.

A07-2264
Urban Mass Transportation Dept.   Subway Environmental Design Handbook.   Vol. 1, Principles and Applications.   U.S. Gov't. Printing Office, 1975.   416 pp. $4.65.   Order No. TD 7.8:Su 1/v.1 S/N 050-014-00005-3
　　A valuable guide and reference for the planning, construction, and operation of underground rapid transit systems. Prepared for those primarily responsible for environmental control, the handbook follows the logical flow path from criteria through

Design--Transportation, Heavy Construction 373

load analysis, and from system conceptual design to selection of equipment.

A07-2265
Wallace, H. and Martin, J. Asphalt Pavement Engineering. McGraw-Hill, 1967. 351 pp. $17.95
This complete paving manual and reference is valuable for practicing engineers and contractors. It presents material on petroleum asphalts, mix-design methods, and more.

A07-2266
Wetteland, S. S. and Bruun, P., eds. Proceedings POAC--Norway 1971. Vols. 1 and 2. Gulf Publishing Co., 1972. 1457 pp. $37.50
This is a collection of papers on the International Conference on Port and Ocean Engineering Under Arctic Conditions. A few of the important subjects are: Arctic Ocean environments; data on waves and currents; observations of ice; land and ice slides; forces due to winds; waves, currents and ice; littoral drift and sediment transport; special soil mechanics problems; and instrumentation. The applied sciences include: surveying; coastal protection; pipelines under Arctic conditions; transportation and pollution problems.

A07-2267
Wiegel, R. L. Oceanographical Engineering. Prentice-Hall, 1964. 532 pp. $24.75
Information on how to overcome problems encountered in the design, installation and use of facilities and equipment in the ocean. Describes mixing processes theoretically, including mixing by waves of waste dumps in the ocean.

A07-2268
Wohl, M. and Martin, B. Traffic System Analysis for Engineers and Planners. McGraw-Hill, 1967. 500 pp. $19.50
This text gives a suitable framework for traffic planning and engineering, and fundamental reasons for highway or traffic decision making.

A07-2269
Woods, K. Highway Engineering Handbook. McGraw-Hill, 1960. 1696 pp. $40.25
All aspects of financing, planning, traffic engineering, design, construction, maintenance, landscaping and other important topics are fully discussed in this comprehensive handbook on modern highway engineering.

A07-2270
Yang, N. Design of Functional Pavements. McGraw-Hill, 1973. 480 pp. $26.50
This book correlates both the empirical and theoretical methods of pavement design. Practical in his approach, the author provides a review of basic material, discusses the existing design methods, and then proceeds to chapters on pavement support

conditions, quality control, environmental effects, vehicle-pavement interaction, systems of pavement design analysis, and redesign of existing pavements.

A07-2271
Yoder, E. J. and Witczak, M. W.   Principles of Pavement Design.
    John Wiley, 1975.   711 pp. $27.95
    Presents a complete coverage of all aspects of the theory and practice of pavement design including the latest concepts. Topics include pavement types, wheel loads, design factors, stresses in asphalt and concrete pavements, traffic considerations, soil, climate and environmental factors, and pavement materials.

Section   A08

BUILDING MATERIALS, COMPONENTS, FINISHES

        Adhesives and Sealants
        Concrete, Brick, Masonry
        Metal Properties, Corrosion
        Paints, Coatings, Finishes
        Plastics and Elastomers
        Wood and Wood Products

A08-2280
ACI. Behavior of Concrete Under Temperature Extremes. American Concrete Institute, 1973. 216 pp. $9.00 (for ACI members, $6.50)

A discussion of Canadian research into effects of below normal and above normal temperatures on properties of concrete. The reaction of temperature and age on thermal expansion and modulus of elasticity of concrete; vacuum processing; pulse velocity; field curing and protection are covered. Also included are effects of heat applied during curing and after placement, effects of steam injection and review of research by National Research Council of Canada on fire endurance of concrete masonry walls.

A08-2281
ACI. Causes, Mechanism, and Control of Cracking in Concrete. American Concrete Institute, 1968. 246 pp. $15.00 (available from University Microfilms)

Virtually all aspects of the subject on the causes, mechanism, and control of cracking in concrete are covered. Contrasts are presented to show that in some plain concrete structures, cracking is to be avoided. In mass concrete, elaborate precautions are taken to insure that during the early life of the structure, thermal stresses remain less than the tensile strength of the concrete. However, in reinforced concrete, cracking is inevitable; at times cracking is induced.

A08-2282
ACI. Cement and Concrete Terminology. American Concrete Institute, 1967. 144 pp. $3.00 (for ACI members, $2.00)

This first edition is the result of 10 years of work by ACI Committee 116 to provide the industry with a glossary of most used terms. Alphabetically arranged, the glossary contains over 1400 terms relating to cement manufacturing and construction, design and research in concrete.

A08-2283
ACI. Commentary on Building Code Requirements for Reinforced Concrete. American Concrete Institute, 1965. 91 pp. $2.00 (for ACI members, $1.00)

Gives insight into the reasons behind the provisions of the Code and explains what is intended to be accomplished by the requirements. It is not an all-inclusive resume of the committee's studies, but an explanation of some of the less common provisions.

Building Materials, Components, Finishes 377

A08-2284
ACI. Concrete Thin Shells. American Concrete Institute, 1971.
442 pp. $15.75 (for ACI members, $11.75)
Seventeen papers given at ACI's 1970 Convention. Specific papers cover: cylindrical shells and folded plates, hyperbolic paraboloids, cable-supported concrete shells, specialized computer programs. Conceptual design and construction, historical view of shell construction, outstanding recently completed examples, and problems of analysis and design.

A08-2285
ACI. Designing for Effects of Creep, Shrinkage and Temperature in Concrete Structures. American Concrete Institute,
445 pp. $15.75 (for ACI members, $11.75)
Much has been written on the failures resulting from designs without adequate provision for the effects of creep, shrinkage and temperature. The object of this symposium volume is to combine the laboratory and actual field experience into a working procedure for estimating the effects of volumetric changes. By comparing the experiences of researchers and design engineers this publication aims for an eventual mathematical procedure that could be as well agreed upon as moment distribution.

A08-2286
ACI. Durability of Concrete. American Concrete Institute, 1975.
385 pp. $14.95 (for ACI members, $11.00)
This book gives an insight to those factors which determine the durability of concrete; how concrete is attacked, means of preventing the effects of this attack, and specific case histories of problems and their solutions. Emphasis is placed on freeze-thaw problems.

A08-2287
ACI. Epoxies with Concrete. American Concrete Institute, 1968.
140 pp. $10.00 (available from University Microfilms)
Includes papers on the uses, application techniques, and methods of determining physical and bonding characteristics of epoxies and on important facets of preparing application specifications. Prepared by ACI Committee 503, this symposium volume contains 12 papers, each identified in the table of contents.

A08-2288
ACI. Expansive Concrete. American Concrete Institute, 1973.
504 pp. $12.50 (for ACI members, $9.50)
A study of the use of expansive cements and its effects on cracking in concrete. Covers research, design, field applications and self-stressing. 20 papers, including reports on use of expansive cement in walls, reinforced concrete pipe, pavements, grout, taxiways, and slabs.

A08-2289
ACI. Fiber Reinforced Concrete. American Concrete Institute, 1974. 570 pp. $12.75 (for ACI members, $9.75)

A selection of 29 papers on the properties and applications of fiber reinforced concrete by authors from six countries. Arranged in two general categories, the first emphasizes the mechanics and properties of steel or glass reinforced concrete and mortar. The second describes developments in the application to pavements, block walls, and prefabrication of structural units.

A08-2290
ACI. Lightweight Concrete. American Concrete Institute, 1971. 296 pp. $15.75 (for ACI members, $11.75)
Emphasis is placed on the importance and acceptance of structural lightweight concrete on the construction scene. There is increased interest in the use of lower density concretes in some of the building systems developed to alleviate the housing shortage. The peculiar properties of these materials, such as energy absorption, make them attractive for uses other than conventional building construction.

A08-2291
ACI. Mechanical Fasteners for Concrete. American Concrete Institute, 1969. 200 pp. $15.00 (available from University Microfilms)
Covers research, design, and practical applications. Among fasteners discussed are those for precast wall panels, for composite construction, for precast prestressed members, and for anchorage to concrete foundations. Coverage includes rock bolts, powder actuated fasteners, stud shear connectors, anchor bolts grouted in drilled holes, and anchorages set in place before concrete is cast.

A08-2292
ACI. Polymers in Concrete. American Concrete Institute, 1973. 368 pp. $11.00 (for ACI members, $8.25)
Covers process technology for preparation of concrete which contains polymer; material properties; fundamental studies; work performed on development of applications. Methods discussed are monomer impregnation of normal hardened concrete followed by in-situ polymerization (PIC); mixing with aggregate and polymerizing after placement (PC); addition of a polymer during mixing of fresh concrete (PCC).

A08-2293
ACI. Precast Concrete Wall Panels. American Concrete Institute, 1965. 143 pp. $10.00 (available from University Microfilms)
A collection of 7 papers that pinpoint problem areas of concern and their solutions to architects, engineers, material suppliers, and manufacturers of precast concrete panels.

A08-2294
ACI. Shotcreting. American Concrete Institute, 1966. 224 pp. $15.00 (available from University Microfilms)
A 13-paper symposium volume which presents the most

Building Materials, Components, Finishes

current information on applications, properties, construction practices, and equipment for pneumatically applied mortars and concretes.

A08-2295
ACI. Temperature and Concrete. American Concrete Institute, 1971. 312 pp. $15.00 (for ACI members, $11.00)
This publication reports new research on the effect of exposures up to 1600°F on comprehensive strength of concrete; designing prestressed concrete reactor vessels; design of mass concrete dams; laboratory and field studies on the effect of temperature differentials on slabs; temperature expansion in continuous span bridges; temperature effect on the curing process; concrete construction in hot climates; water-reducing admixtures effect on temperature rise in mass concrete.

A08-2296
ACI and Natl. Concrete Masonry Assn. Menzel Symposium on High Pressure Steam Curing. American Concrete Institute, 1972. 250 pp. $12.00 (for ACI members, $9.00)
Contains 10 papers representing the latest information available in the field of high pressure steam curing and 3 classics written by Carl Menzel and published in the ACI Journal in 1934, 1935, and 1936. The 10 papers discuss the benefits autoclaving contributes to: concrete block and brick, asbestos-cement products and cellular concrete.

A08-2297
Addleson, L. Materials for Building. Vol. 1, Physical and Chemical Aspects of Matter and Strength of Materials. Butterworth, 1972. 182 pp. £3.50
This series of books provides a detailed survey of how protection is best afforded to building materials affected by being continuously exposed to the weather, to the atmosphere with its corrosive elements, to accidental damage by fire and flooding, and to the effects of heat and vibration. The books, with their use of many practical examples, will be invaluable to surveyors, architects and engineers in their efforts to make the maximum use of existing materials and to counteract, to the fullest possible extent, the depredations of the elements. Part 1 provides an appreciation of the structure and behaviour of materials by presenting the relevant basic physics and chemistry. Part 2 includes a description of the resistance offered by materials to cracking and other failures caused by movements, local damage, shaping or fixing.

A08-2298
Addleson, L. Materials for Building. Vol. 2, Water and Its Effects-1. Butterworth, 1972. 172 pp. £5.50
The second volume of the series discusses the basic problems associated with water, and its effect on materials and buildings.

A08-2299
Addleson, L.   Materials for Building.   Vol. 3, Water and Its Effects-2.   Butterworth, 1972.   136 pp.   £4.50

Volume 3 deals with the deleterious chemical effects of water on materials, and includes a consideration of the effects of the weather.

A08-2300
Addleson, L.   Materials for Building.   Vol. 4.   Butterworth.   (In preparation)

Contents: Heat. General Considerations. Exposure. Thermal Movement. Thermal Insulation. Thermal Response. Condensation and Staining. Staining and Degradation. Fire. Nature of Fire Hazard. Performance of Materials. Planning and Precautions.

A08-2301
Adler, R. R.   Vertical Transportation for Buildings.   American Elsevier, 1970.   228 pp. $27.80

This book analyzes vertical traffic demands for principal building types and outlines system designs and operating practices for satisfying those demands. It compares relative costs and performance for alternative means of vertical transport.

A08-2302
AISC.   Iron and Steel Beams 1873 to 1952.   American Institute of Steel Construction, 142 pp. $7.00

The American Institute of Steel Construction often receives requests for information on the properties of beam and column shapes, including those which are no longer rolled. Through the co-operation of rolling mills, steel fabricating companies, engineers, railroads and libraries, additional historical material was obtained. With the consolidation of this data, it is believed that information on practially all beam and column sections that have been produced in this country is provided in this one reference book.

A08-2303
Akers, L. E.   Particle Board and Hardboard.   Pergamon, 1966.   192 pp. $6.00

Deals with manufacturing methods in broad outline, with physical and strength properties and test methods in detail, as well as the practical aspects of utilization as required by the designer and craftsman.

A08-2304
Akroyd, T. N. W.   Concrete: Its Properties and Manufacture.   Pergamon, 1962.   336 pp.

A reference work for the site engineer and a guide for the student and graduate. It describes the properties of concrete, its manufacture and use in building and civil engineering construction.

Building Materials, Components, Finishes

A08-2305
Aluminum Assn. Aluminum Finishing Seminar Papers. Aluminum Assn., 1973. 400 pp. $15.00
   Set of three books containing the 30 papers presented at the 1973 seminar. The three books cover the Anodizing, Applied and General Sessions. Illustrated with photographs, charts and drawings. Individual session papers may be purchased at $10.00 each for the Anodized and Applied and $5.00 for the General Session papers.

A08-2306
Aluminum Assn. Aluminum Standards and Data. Aluminum Assn., 1973. 204 pp. No charge
   Issued biennially, this is a comprehensive reference book containing data on mechanical, physical and other properties, tolerances and other useful information on aluminum mill products in general use. Includes separate sections on sheet and plate, rolled rod and bar, extruded rod, bar, tube and shapes, forging, electrical conductors and other aluminum forms and shapes.

A08-2307
Aluminum Assn. and American Welding Society. Aluminum Welding Seminar Papers. Aluminum Assn., 1973. $10.00
   Set of five papers presented at a seminar jointly sponsored by the Aluminum Association and the American Welding Society. Papers cover metallurgy selection of alloys, MIG and TIG welding, quality control and design data.

A08-2308
American Chemical Society. Fire Retardant Paints. American Chemical Society, 1954. 91 pp. $4.50
   Theory of flame-proofing, effectiveness and formulation, and testing of fire-retardant paints, aircraft coatings, and flame-resistant mastics.

A08-2309
Amrhein, J. E. Reinforced Masonry Engineering Handbook. Masonry Institute of America, 1973. 320 pp. $14.50
   Latest code requirements based on the 1973 Uniform Building Code. Includes theory of masonry design and wind, earth and seismic lateral force design. There are design examples of a retaining wall, industrial building and a multi-story building. 170 pages of diagrams and tables for ten masonry strengths and two steel strengths.

A08-2310
Anderson, E. A. and Earle, G. F., eds. Design and Aesthetics in Wood. State Univ. of N.Y. Press, 1967. 222 pp. $15.00
   From a symposium at State Univ. of N.Y., College of Environmental Science and Forestry, Syracuse, 1967. Fifteen articles stress wood as art and artifact, wood as architectural material, design in a dynamic technology and wood as a material.

A08-2311
Annett, F. Elevators. McGraw-Hill, 1960. 388 pp. $23.50
   A practical treatment of vertical-transportation equipment and types of buildings where it is used.

A08-2312
Architectural Precast Concrete. Prestressed Concrete Institute, 1973. 173 pp. $15.00
   The manual includes design, detailing, and specifying information in a practical, easy-to-use book.

A08-2313
Arnison, J. H. Floor and Structural Surfaces. Butterworth, 1969. 164 pp. £2.80 (paperback £1.80)
   All types of floors, floor coverings and wall surfaces are dealt with in this book. Their basis, construction, structure and maintenance are outlined and their cleaning is described. A guide to the best surfaces to use according to conditions, in the construction of buildings, is given.

A08-2314
Arnold, L. K. Introduction to Plastics. Iowa State Univ. Press, 1968. 205 pp. $6.95
   Features a brief presentation of chemistry common to the whole field, methods of producing plastics, their practical applications, and properties and uses of various plastics.

A08-2315
ASCE. Structural Plastics--Properties and Possibilities. American Society of Civil Engineers, 1969. 244 pp. $2.50
   Offers an opportunity to update knowledge of the use of plastic materials as structural elements. This volume contains 8 papers presented in Louisville, Kentucky in April 1969.

A08-2316
ASCE Structural Div. Wood Structures. The American Society of Civil Engineers, 1975. 416 pp. $12.00 (for ASCE members, $6.00)
   More than 30 authors have contributed to this state-of-the art book on structural lumber, plywood, connections, fire resistance, and structural design. Such structural subjects as metal reinforcement, prestressed wood, trusses and glulam elements are described as are a number of structural wood systems.

A08-2317
Ashton, J. E., et al. Primer on Composite Materials: Analysis. Technomic, 1969. 124 pp. $10.00
   The fundamental principles of fiber-reinforced laminated materials are presented.

A08-2318
ASM. Metals Handbook. Vol. 1, Properties and Selection. American Society for Metals, 1975. 1300 pp. $45.00

Building Materials, Components, Finishes

Principal subjects include definitions and reference tables, carbon steel, cast iron, stainless steel, nonferrous metals and properties of pure metals.

A08-2319
ASM. Metals Handbook. Vol. 2, Heat Treating, Cleaning and Finishing. American Society for Metals, 1975. 668 pp. $42.50

Topics include heat treating, cleaning and finishing of carbon steel, cast iron, tool steel, stainless steel and nonferrous metals. Electroplating and nonmetallic coatings are described.

A08-2320
ASM. Metals Handbook. Vol. 3, Machining. American Society for Metals, 1975. 512 pp. $42.50

The machining processes are described as well as their application to steel and cast iron.

A08-2321
ASM. Metals Handbook. Vol. 4, Forming. American Society for Metals, 1975. 496 pp. $42.50

The shearing, bending and forming of carbon steel, stainless steel, aluminum, copper and other nonferrous materials are described.

A08-2322
ASM. Metals Handbook. Vol. 5, Forging and Casting. American Society for Metals, 1975. 448 pp. $37.50

The book describes the hammers and presses and the dies used for forging stainless steel, aluminum, copper and other alloys. Melting and casting methods for grey, ductile and malleable iron, steel, aluminum, copper and other alloys are given.

A08-2323
ASM. Metals Handbook. Vol. 6, Welding and Brazing. American Society for Metals, 1975. 734 pp. $45.00

Principal subjects included are welding, electroslag, electro gas and resistance welding, flash and friction welding, gas welding, and electron beam welding. Furnace and torch brazing are also covered.

A08-2324
ASM. Metals Handbook. Vol. 7, Atlas of Microstructures; Vol. 8, Metallography, Structures and Phase Diagrams; Vol. 9, Fractography and Atlas of Fractographs. American Society for Metals, 1975. $99.00 for set

These three volumes complement each other to provide comprehensive coverage on metal structure. For use in metallographic laboratories.

A08-2325
ASM. Metals Handbook. Vol. 10, Failure Analysis and Prevention. American Society for Metals, 1975. 580 pp. $47.50

An overview of preventive measures in design, materials selection, assembly, operation and maintenance that contribute to improved product performance.

A08-2326
ASME Handbook--Metals Engineering: Design.   McGraw-Hill, 1965.
605 pp. $28.75
Discusses machinability, strength, and other characteristics of metals, and explains their significance to engineers and others who design metal products.

A08-2327
ASME Handbook--Metals Properties.   McGraw-Hill, 1954.   445 pp. $22.50
Considers the properties of more than 500 metals, and offers tabulated data on their strength, hardness, machinability, electrical conductivity, thermal conductivity, and composition.

A08-2328
Asphalt Institute.   Asphalt in Hydraulic Structures.   The Asphalt Institute, 1961.   152 pp.   No charge
Information to assist engineers in design and construction with asphalt of canal, reservoir and storage pond linings, revetments, beach erosion control structures, and dams.

A08-2329
ASTM.   Corrosion in Natural Environments.   American Society for Testing and Materials, 1974.   352 pp. $29.75 (for ASTM members, less 20%)
This publication contains 19 papers dealing with atmospheric corrosion, metal corrosion in seawater, and statistical planning and analysis of corrosion experiments.

A08-2330
ASTM.   Environmental Effects on Advanced Composite Materials. American Society for Testing and Materials, 1976.   102 pp. $10.00
Program definition and preliminary results of a long-term evaluation program of advanced composites for supersonic cruise aircraft applications; flight simulation testing equipment for composite material systems; effects of thermal cycling environment on graphite/epoxy composites; effects of graphite/epoxy composite materials on the corrosion behavior of aircraft alloys; effect of natural weathering on the mechanical properties of graphite/epoxy composite materials; influence of outdoor weathering on dynamic mechanical properties of glass/epoxy laminate.

A08-2331
ASTM.   Heat Transmission Measurements in Thermal Insulations. American Society for Testing and Materials, 1974.   319 pp. $30.75 (for ASTM members, less 20%)
Nineteen papers deal with definitions and thermal modelling,

Building Materials, Components, Finishes

techniques, results and applications. A must for anyone concerned with the energy crisis.

A08-2332
ASTM. Localized Corrosion--Cause of Metal Failure. American Society for Testing and Materials, 1972. 322 pp. $22.50 (for ASTM members, $18.00)
Fifteen papers cover local attack including exfoliation, intergranular corrosion, crevice corrosion, pitting and dealloying. Covered are: conditions of attack and prevention, mechanisms, and testing.

A08-2333
ASTM. Masonry: Past and Present. American Society for Testing and Materials, 1975. 295 pp. $30.00 (for ASTM members, less 20%)
Fifteen papers give the background on existing specifications and current activities in this field. An extremely useful book for architects, masonry specifiers and engineers.

A08-2334
ASTM. Materials Performance and the Deep Sea. American Society for Testing and Materials, 1969. 146 pp. $9.50 (for ASTM members, $7.60)
The effects of deep seas on service life, performance, and assessing corrosion and degradation of materials in the marine environment are covered in 10 outstanding papers. A truly critical analysis.

A08-2335
ASTM. Metal Corrosion in the Atmosphere. American Society for Testing and Materials, 1968. 396 pp. $27.00 (for ASTM members, $21.60)
This book is one of the most comprehensive in the field dealing with 37 aluminum, 19 copper, 2 lead, 5 magnesium, 5 nickel, 8 titanium and 3 other alloys and some steels. The material includes calibration of sites, handling data, statistical interpretation and effects of stress and exposure conditions.

A08-2336
ASTM. Paint Testing Manual. American Society for Testing and Materials, 1972. 600 pp. $27.50 (for ASTM members, less 20%)
The most complete manual of its type ever published. Eleven parts cover optical, physical, mechanical, and chemical properties; weather, film, and whole paint testing, raw materials, specific products, instrumentation, and specifications. 43 contributors.

A08-2337
ASTM. Significance of Tests and Properties of Concrete and Concrete Making Materials. American Society for Testing and Materials, 1966. 580 pp. $12.00 (for ASTM members,

$9.60)

The four parts cover general principles, tests and properties of concrete, tests and properties of concrete aggregates, and tests and properties of other materials. The 45 papers review the present status of concrete testing.

A08-2338
ASTM. Stress Corrosion Cracking of Metals--A State of the Art. American Society for Testing and Materials, 1972. 172 pp. $11.75 (for ASTM members, $9.40)

Ten papers containing information on steels, aluminum, copper, titanium and nickel high strength alloys. Current test methods are described. Aids in the selection of materials, fabrication and maintenance, and failure analysis from the point of view of stress corrosion.

A08-2339
ASTM. Window and Wall Testing. American Society for Testing and Materials, 1974. 75 pp. $6.25 (for ASTM members, less 20%)

High rise structures with extensive use of glass and decorative panels present architectural problems discussed in seven extensive papers.

A08-2340
Austin, R. and Ueda, K. Bamboo. Weatherhill, 1970. 216 pp. $16.50

This book explores the world of bamboo in nature and in the hands of man. 162 photos.

A08-2341
Baer, E., ed. Engineering Design for Plastics. Van Nostrand Reinhold, 1964. 1350 pp. $32.00

Contents include the chemical composition, mechanical behaviour, thermal stability, and electrical and optical properties of plastics. Reinforced plastics and foams are among the forms discussed.

A08-2342
Banov, A. Paints and Coatings Handbook. Structures Publishing Co., 1972. 320 pp. $20.00

Information on surface protection, including paint specifications, surface preparation and coatings application. Procedures for cost cutting in new construction and maintenance painting.

A08-2343
Barer, R. D. and Peters, F. B. Why Metals Fail. American Society for Nondestructive Testing, 1970. 345 pp. $18.50 (for ASNT members, $16.00)

The book is aimed at the non-metallurgist engineer who has to contend with failures of everyday metals and metal components. The contents include 134 case histories, illustrated with 350 photographs--all chosen to demonstrate the type of reasoning and diagnostic ability essential in solving metal problems.

Building Materials, Components, Finishes 387

A08-2344
Bate, S. C. C., et al. Handbook on the Unified Code for Structural Concrete. Cement and Concrete Assn., 1972. 153 pp. £3.00

A08-2345
Battelle Memorial Institute. Engineering Properties of Selected Ceramic Materials. American Ceramic Society, 674 pp. $16.00 (for ACS members, $12.00)
A plastic ring-bound materials selection data book, prepared at the Columbus Labs of Battelle Memorial Institute under contract to the U.S. Air Force.

A08-2346
Becker, W. E. U.S. Sandwich Panel Manufacturing/Marketing Guide. Technomic, 1968. 146 pp. $5.00
The design, production, on-site handling, and marketing of sandwich panels is presented. A directory of manufacturers and suppliers is included.

A08-2347
BIA. Recommended Practice for Engineered Brick Masonry. Brick Institute of America, 1970. 350 pp. $5.00
A summary of knowledge compiled into a manual for structural engineers, architects and advanced students. Designed to assist the practicing structural engineer in the design of brick masonry structural elements and systems.

A08-2348
BIA. Reinforced Brick Masonry--Lateral Force Design. Brick Institute of America, 1953. 271 pp. $1.50
Data on the performance of brick masonry, recommended design and construction procedures and a review of currently accepted design criteria as related to lateral forces (i.e., wind, earthquake or blast).

A08-2349
Biczok, I. Concrete Corrosion and Concrete Protection. Chemical Publishing, 1967. 543 pp. $25.25
Topics include cements and their corrosion resistance, aggregates, groundwater, thin walled concrete objects, and factors increasing or reducing corrosion.

A08-2350
Bodnar, M. J., ed. Structural Adhesives Bonding. John Wiley, 1966. 495 pp. $20.50
31 chapters are authored by different experts in the field. Subjects include physical and chemical properties of adhesives, design and performance of joints between similar and dissimilar materials and methods of fabricating bonded joints.

A08-2351
Boyne, D. A. C. A. and Wright, L., eds. Architects' Working

Details. Architectural Press, Each volume contains 160 pp. and costs £2.50. Vol. 1, 1953. Vol. 2, 1954. Vol. 3, 1955. Vol. 4, 1957. Vol. 5, 1958. Vol. 6, 1959. Vol. 7, 1960. Vol. 8, 1961. Vol. 9, 1962. Vol. 10, 1964. Vol. 11, 1965. Vol. 12, 1968. Vol. 13, 1969. Vol. 14, 1971. Vol. 15, 1973.

The details in these fifteen volumes are grouped under the headings of windows, doors, staircases, walls and partitions, roofs and ceilings, furniture and fittings, balconies, covered ways and canopies, heating and lighting. They show the recent work of leading architects and are selected from the working details which have appeared in the Architects' Journal.

A08-2352
Brady, G. Materials Handbook. McGraw-Hill, 1971. 1024 pp. $26.50

The book is written in a non-technical manner, giving information on properties and applications of metals and alloys, abrasives, plastics, woods, synthetic resins, industrial chemicals, petroleum products, fuels, refractories, minerals, and many other materials.

A08-2353
Brasunas, A. de S. and Stansbury, E. E., eds. Symposium on Corrosion Fundamentals. University of Tennessee Press, 1957. 262 pp. $5.00

In this abundantly illustrated book sixteen recognized authorities discuss the major aspects of corrosion, beginning with general considerations and continuing through specialized topics. The volume is an outgrowth of lectures given at The University of Tennessee Corrosion Conference of 1955.

A08-2354
Brick, R., et al. Structure and Property of Alloys. McGraw-Hill, 1965. 505 pp. $19.50

The authors seek to relate the basic engineering properties of metals to their metallurgical structure. The book displays an outstanding use of photomicrographs.

A08-2355
Bricks: Their Properties and Use. The Construction Press. £6.95

This volume brings together for the first time, in a convenient hardback format, the publications of the Brick Development Association which contain both basic information on clay and calcium silicate bricks and detailed recommendations for their choice and use to meet varying requirements. In addition there is a comprehensive selection of the most relevant technical data on related aspects of bricks and brickwork.

A08-2356
Brownell, A. H. Architectural Hardware Specifications Handbook. American Institute of Architects, 1971. 171 pp. $14.95

Building Materials, Components, Finishes   389

(for AIA members, less 10%)

A reference on writing hardware specifications. Some topics are: six systems for writing hardware specifications, details of various hardware items, a number system adaptable for computer use, security, and BHMA standards for finishes.

A08-2357
Brydson, J. A. Plastics Materials. Van Nostrand Reinhold, 1970. 576 pp. $19.50

The chemical, mechanical, electrical and optical properties of many polymers are given, including polyethylene, diene rubbers, vinyl chloride, acrylics, styrene, polycarbonates, epoxide resins, silicones, and others.

A08-2358
Bucksch, H. Dictionary of Wood and Woodworking Practice. Adler's Foreign Books. Vol. 1, German-English, 461 pp. $20.00; Vol. 2, English-German, 536 pp. $23.00

More than 40,000 terms are supplemented where necessary by definitions or explanations in English or German. The terminology ranges from sapling to finished product, with special attention given to forestry, timber conversion and woodworking machinery. Related fields, such as adhesives, have also been taken into account.

A08-2359
Building Research Establishment, ed. The Strength Properties of Timber. Herman Publishing, Inc., 1974. 208 pp. $24.50

Brings together in bound and indexed form those publications from the Forest Products Research Laboratory of the UK's Building Research Establishment which are concerned with the strength properties of non-U.S. timbers. Concise and comprehensive guide to those strength properties which are commonly required by members of the various construction and engineering professions.

A08-2360
Bureau of Reclamation, U.S. Dept. of the Interior. Concrete Manual. Concrete Construction Publications, 1966. 501 pp. $4.75

This volume presents a discussion of techniques, materials and equipment for all aspects of construction with concrete. Some subjects discussed are: properties of concrete, aggregates, mix design, batching, handling, forming, placing, finishing, curing, reinforcement, prestressed concrete, hot and cold weather concreting and repair of concrete.

A08-2361
Burns, R. M. and Bradley, W. Protective Coatings for Metals. Van Nostrand Reinhold, 1967. 768 pp. $25.00

Contents include corrosion control, surface preparation, types of coatings and methods of application, and coatings of zinc, cadmium, tin, nickel, chromium, copper, lead, aluminum, etc.

A08-2362
Butler, G. and Ison, H. C. K. Corrosion and Its Prevention in
    Waters. Van Nostrand Reinhold, 1966. 312 pp. $14.95
    Subjects covered include corrosion principles, types of
water, corrosion forms, ferrous and non-ferrous metal corrosion,
flow, temperature and heat transfer, water treatment, and protective coatings and cathodic protection.

A08-2363
[No entry]

A08-2364
Cagle, C. V., et al., eds. Handbook of Adhesive Bonding.
    McGraw-Hill, 1973. 754 pp. $27.50
    Describes ways to solve virtually any problem in adhesive
bonding. The book draws on the experience of 27 experts whose
contributions add up to a complete overview of the use of adhesives
for maximum effect and efficiency.

A08-2365
Ceiling Systems Handbook. Ceilings and Interior Systems Contractors Assn., 1967. 352 pp. $5.95
    The latest edition contains four new chapters including concealed suspension systems, luminous ceilings, integrated ceilings
and safety regulations. The original fourteen chapters cover various ceiling systems, installation tools, building plans, installation
conditions and has two special chapters on sound control and architectural terms. The handbook is widely used by ceiling contractors' installation crews and in carpenter apprenticeship training
programs.

A08-2366
Childe, H. L. Concrete Finishes and Decoration. Cement and
    Concrete Assn., 1963. 138 pp. £0.90

A08-2367
Childe, H. L. Everyman's Guide to Concrete Work. George Godwin Ltd., 1969. 170 pp. £0.75
    As concrete is now the only building material made on site,
it is important that all engaged in its production and use should
have knowledge of its properties, its possibilities and its limitations. This book makes available the information needed by builders, foremen, craftsmen, students and other members of the
building team.

A08-2368
Clauser, H., et al., eds. Encyclopedia of Engineering Materials
    and Processes. Van Nostrand Reinhold, 1963. 798 pp.
    $29.50
    All engineering materials of any consequence are included.
More than 300 articles of a descriptive nature include tabulation of
data wherever needed.

Building Materials, Components, Finishes   391

A08-2369
Close, P. D.   Sound Control and Thermal Insulation of Buildings.
    Van Nostrand Reinhold, 1966.   510 pp. $18.50
    This book includes essential design data and product information, together with application details. Some subjects included in the book are noise abatement in homes, apartments, schools, hospitals and commercial buildings, thermal insulation, heating and cooling, and condensation control.

A08-2370
Comber, A. W.   Composition Flooring and Floorlaying.   Charles
    Griffin, 1950.   118 pp.   £1.00

A08-2371
Composite Materials in Engineering Design.   Technomic, 1972.
    500 pp. $35.00
    Researchers have been seeking new concepts and methods by which to apply composites in the solution of design problems. Since composite materials are used primarily for structural applications, many well-developed concepts and methods of mechanics and applied mathematics can be adapted to solve important problems in composite materials. This book brings the latest design concepts and techniques, applied to a wide range of products, to practitioners in the field.

A08-2372
Concrete Society.   Polymers in Concrete.   The Construction Press,
    1975.   £15.00
    The use of polymers in concrete is a subject of outstanding importance to members of the construction industry as well as people in the chemistry and plastics industries. This volume is based on the proceedings of the First International Congress on Polymer Concretes (1975), organized by The Concrete Society. Both theoretical and practical aspects of the subject are covered in this comprehensive and up to date volume.

A08-2373
Conway, J. B., et al.   Fatigue, Tensile, and Relaxation Behavior
    of Stainless Steels.   U.S. Atomic Energy Commission, 1974.
    (Now Energy and Research Development Administration)
    A detailed treatment of short-term tensile, relaxation, and low-cycle fatigue behavior with special emphasis on stainless steels at elevated temperatures. Some new test procedures are highlighted and discussed fairly thoroughly to focus on the types of information that can be obtained. Data generated by use of these procedures are summarized in a fairly comprehensive manner along with similar data reported in other studies.

A08-2374
Cook, J. P.   Construction Sealants and Adhesives.   John Wiley,
    1970.   269 pp. $18.25
    Assembles the available information on materials and methods of sealants and adhesives in construction in an organized form.

Emphasizes performance rather than chemistry. Discusses properties, testing, stresses and cost of materials.

A08-2375
Cordon, W. A. Freezing and Thawing of Concrete: Mechanisms and Controls. American Concrete Institute, 1966. 100 pp. $4.50 (for ACI members, $3.00)
Since freezing and thawing of concrete when moisture is present is a major cause of deterioration this monograph focuses on the behavior of various concrete structures exposed to these conditions. Published jointly by the ACI and Iowa State University Press, Monograph No. 3 explains theories regarding mechanisms which produce this behavior in concrete structures and recommends methods for overcoming these mechanisms and thereby preventing deterioration.

A08-2376
Corrosion Abstracts Yearbooks. National Assn. of Corrosion Engrs., 1962, 1963. $50.00 each (for NACE members, $30.00 each); 1964-70, $75.00 each (for NACE members, $50.00 each)
Thoroughly cross indexed, approximately 4000 abstracts of world's corrosion control literature per issue.

A08-2377
Critchell, P. L. Joints and Cracks in Concrete. Applied Science, 1968. 228 pp. £3.50
A survey of jointing materials and methods of construction, correct and economical jointing techniques and the diagnosis and treatment of faults.

A08-2378
Croome, D. J. and Sherratt, A. F. C., eds. Condensation in Buildings. Applied Science, 1972. 271 pp. £7.50
The book contains 15 papers presented at a Conference held at the University of York in January 1972, and organised in consultation with the Royal Institute of British Architects, the Institute of Building, the Institution of Heating and Ventilating Engineers and the Department of the Environment. These papers deal with many aspects of the condensation problem, and taken together provide an authoritative and comprehensive statement of the present situation and point, in a practical way, to future possibilities for solving and preventing condensation in both new and existing buildings.

A08-2379
Dalzell, J. Simplified Masonry Planning and Building. McGraw-Hill, 1955. 362 pp. $7.95
A manual on the planning and construction of all common types of concrete, concrete block, stucco, and similar structures.

A08-2380
Damusis, A., ed. Sealants. Van Nostrand Reinhold, 1967.

Building Materials, Components, Finishes

352 pp. $17.50

This book provides both producers and users with information on every type of polymeric binder known in 1967.

A08-2381
Danz, E.   Sun Protection.   Praeger, 1967.   152 pp. $15.00

The author discusses the aesthetic, structural, and economic aspects of this problem. He treats the comparative costs of various systems, the heat absorptive and retentive qualities of different materials, the construction and service of mechanical devices, and the possible combinations of systems. Separate chapters present every means of regulating sunlight, from a well-placed shade tree to mechanically controlled louvers. Among the methods discussed are cantilevered roofs, balconies and loggias, structural framing, shutters and blinds, awnings and marquees, and shaded glass. The international selection of examples illustrated includes works of such architects as Breuer, Rudolph, and Neutra.

A08-2382
David Litter Labs.   Paints and Protective Coatings.   U.S. Gov't. Printing Office, 1974.   225 pp. $7.25.   Order No. TM 5-618 or Navfac MO-110

Developed for use by U.S. military agencies, this book covers the materials and techniques of painting buildings and structures. Paint and coating failures are described as are measures to prevent them.

A08-2383
Delollis, N. J.   Adhesives for Metals: Theory and Technology. Industrial Press, 1970.   230 pp. $16.50

A survey of the factors involved with adhesion, adhesive chemistry, and materials and bonding practice. It also covers the concepts of cleanliness, joint design, stress relief and vibration damping.

A08-2384
Diamant, R. M. E.   Chemistry of Building Materials.   Beekman, 1970.   258 pp. $17.50

An up-to-date and comprehensive treatment of building materials from a strictly chemical-scientific point of view.

A08-2385
Diamant, R. M. E.   Insulation of Buildings: Thermal and Acoustic. Butterworth, 1965.   256 pp.   £3.50 (paperback £2.25)

A book which places the question of the insulation of buildings on a scientific and technological foundation. It shows how the quality of insulation required can be calculated from first principles, and how the thickness needed is evaluated from economic considerations.

A08-2386
Diamant, R. M. E.   Prevention of Corrosion.   Beekman, 1971. 199 pp. $21.50

This volume covers the electrochemical theory of corrosion, as well as the actual practical methods of preventing corrosion under specific conditions. Contents include the Nature of Corrosion, Galvanic and Differential Aeration Corrosion of Ferrous Metals, Corrosion of Non-Ferrous Metals, Corrosion-Resistant Alloys, etc.

A08-2387
Dietz, A. G. H. Plastics for Architects and Builders. MIT Press, 1970. 144 pp. $7.95
The objective of this "primer" on plastics is to acquaint the architect, designer, builder, and contractor with polymers and their potentialities in building, covering not only structural aspects, but interior, decorative, and lighting applications as well. The user of this book will obtain basic, essential knowledge of the nature and the applicability of these versatile materials.

A08-2388
DuBois, J. H. and John, F. W. Plastics. Van Nostrand Reinhold, 1967. 352 pp. $11.75
Here is a simplified presentation of plastics materials, processing procedures, selection, and essential design data. Procedures for material selection are included with tables of properties of important modern materials. Stresses fabrication and end-use properties.

A08-2389
Duck, E. W. Plastics and Rubbers. Philosophical Library, 1972. $15.00
This liberally illustrated volume is a sourcebook in the plastics and rubber industries. It not only relates in detail the theoretical bases for both these sciences, but it also provides indepth presentations of the hundreds of different polymerization processes involved in the manufacture of rubber and plastic.

A08-2390
Evans, L. S. Selecting Engineering Materials for Chemical and Process Plant. John Wiley, 1974. 164 pp. $14.95
Contains information on carbon steels, cast iron, nickel alloys, copper, lead, aluminum, glass, cements, bricks and tiles, plastics and metallic and organic coatings. Pressure vessel design codes are discussed.

A08-2391
Fibre Reinforced Cement and Concrete. The Construction Press, 1975. £15.00
This comprehensive volume of up-to-date information covering every type of reinforcing fibre is a reference work for those who make it their business to keep abreast of current developments in concrete technology. The book contains all the papers presented at the 1975 RILEM Symposium (September 1975).

A08-2392
Fintel, M., ed. Handbook of Concrete Engineering. Van Nostrand Reinhold, 1975. 801 pp. $42.50
This handbook gives authoritative information on the design and construction of reinforced concrete buildings and other structures. It is based on the ACI "Building Code Requirements for Reinforced Concrete."

A08-2393
Fontana, M. and Green, N. Corrosion Engineering. McGraw-Hill, 1967. 416 pp. $17.50
Simplifies and blends theory with practice while treating all materials, including nonmetals.

A08-2394
Forest Products Laboratory. Wood Handbook: Wood as an Engineering Material. U.S. Gov't. Printing Office, 1974. 432 pp. $7.95. Order No. A 1.76:72/973 S/N 0100-03200
Contains information on the physical and mechanical properties of wood, and how these properties are affected by variations in the wood itself. Also, includes chapters on wood based products, and the principles of how wood is dried, fastened, finished, and preserved.

A08-2395
Frisch, K. C. and Saunders, J. H., eds. Plastic Foams. Marcel Dekker. Vol. 1, 1972. 464 pp. $36.50; Vol. 2, 1973. 704 pp. $47.50
Gives an integrated picture of the fundamental principles, technology, and applications of foams, and offers a thorough treatment of specific types of plastic foams. Thirty-two authors contribute to 21 chapters, including one on the architectural uses of foam plastics and the thermal decomposition and flammability of foams.

A08-2396
Gage, M. Guide to Exposed Concrete Finishes. Concrete Construction Publications. 161 pp. $8.45
Using over 270 photographs as illustrations, this book covers the factors affecting the selection of exposed concrete finishes, precast and in situ. Following are 70 pages of informations sheets, each describing a particular type of finish and how it is created. Over 21 finishes are included. Specifications and costs.

A08-2397
Gage, M. and Kirkbride, T. Design in Blockwork. Architectural Press, 1972. 120 pp. £2.50
The first part of this book is a revised version of Guide to Concrete Blockwork. A new section consists of a guide to the design and specification of blockwork for structural purposes and

includes an appraisal of the available structural design data and a comparative review of UK and foreign codes of practice.

A08-2398
Gage, M. and Newman, K. Guide to Ready Mixed Concrete. Architectural Press, 1972. 60 pp. £0.75
 This book describes the factors the architect, engineer or contractor must take into account when using ready mix. It describes the properties of the material, gives specimen specification clauses and check lists and offers information and advice on ordering methods and supervising the use of ready mix on site.

A08-2399
Gage, M. and Newman, K. Specification and Use of Ready Mixed Concrete. Architectural Press, 1973. 66 pp. £1.50
 This book describes the factors the architect, engineer and building contractor must take into account when using the material. It covers properties, gives typical specification clauses and offers advice on such matters as ordering methods and supervising the use of ready mixed concrete on site.

A08-2400
Garthwaite, C. H. Metric Handbook for Reinforced Concrete. Butterworth, 1969. 64 pp. £1.00
 A book for engineers which contains nomograms scaled in metric and imperial units representing the compound units used in calculations for reinforced concrete design. They will be useful for checking calculations in metric units at all stages of a design, during the period when engineers are getting the feel of the new units.

A08-2401
Gatz, K. and Thierry, J., eds. Architect's Detail Library. Vol. 1, Wrought Iron Railings, Doors and Gates. Butterworth, 1966. 120 pp. £2.50
 The selection of examples of wrought and forged ironwork used outdoors or as part of building constructions and interior structures is based on what is considered necessary for everyday application in this field.

A08-2402
Gatz, K. and Thierry, J., eds. Architect's Detail Library. Vol. 2, Ceilings in Wood. Butterworth, 1966. 120 pp. £2.50
 This book concentrates on modern ceiling design in wood, showing the many possible applications in present-day buildings. Examples have been chosen to prove the cohesion and sympathy of the wood ceilings with the entire interior design concept.

A08-2403
Gatz, K. and Thierry, J., eds. Architect's Detail Library. Vol. 3, Windows and Window Walls. Butterworth, 1966. 120 pp. £2.50
 This collection of designs aims to provide architects and

Building Materials, Components, Finishes

window specialists with the challenge to attempt a new solution for the new problems posed by the special conditions of building where the windows act visually and technically within the harmonic building unit.

A08-2404
Gatz, K. and Thierry, J., eds. Architect's Detail Library. Vol. 5, Entrances and Staircases. Butterworth, 1967. 120 pp. £2.50

The examples in the first part illustrate the effect of using metal for doors and other entrance fittings. The stairs reproduced in the second part demonstrate how cost of construction and architectural effect are related, provided that full advantage is taken of the ability of steel to withstand heavy stresses and loads.

A08-2405
Gatz, K. and Thierry, J., eds. Architect's Detail Library. Vol. 6, Exterior Detailing in Concrete. Butterworth, 1967. 120 pp. £2.50

Current trends in the development of building shapes based on the plastic properties of concrete demand more than mere calculating design methods and craftsmen for their construction. Appearance depends on the detailing and enlivening possible within the limits imposed by the material. Examples include masonry and wall structures, walls in precast concrete units, balconies, staircases and screens.

A08-2406
Gerwick, B. C., Jr. Construction of Prestressed Concrete Structures. John Wiley, 1971. 411 pp. $13.75

Presents general principles and specific techniques for practical utilization, manufacture and construction of prestressed concrete. Discusses means of ensuring quality, economy, durability and desired performance in service for various types of buildings and engineering construction.

A08-2407
Gibbs and Cox, Inc. Marine Design Manual for Fiberglass Reinforced Plastics. McGraw-Hill, 1960. 376 pp. $26.00

This manual explains the engineering and structural principles involved in building commercial, military, and pleasure vessels from these plastics.

A08-2408
Glanvill, A. B. Plastics Engineer's Data Book. Industrial Press, 1973. 216 pp. $10.00

First published in England, this book brings together a wealth of reference material applicable to U.S. plastics practice. Because of its practical value and usefulness, this reference is recommended to those who design and process plastics. The processing and technical data it contains, which is intended for engineers and designers in the plastics industry is thorough.

A08-2409
Glanville, W. and Thomas, F. G. Explanatory Handbook on the BS Code of Practice for Reinforced Concrete CP 114:1957 with Metric Appendix. Cement and Concrete Assn., 1973. 203 pp. £2.50

A08-2410
Gratwick, R. T. Dampness in Buildings. 2nd ed. John Wiley, 1975. 360 pp. $23.00
  Dampness is a root cause of a great proportion of all building deterioration. This new edition includes chapters covering condensation, electrodampproofing applications and the commonly neglected interrelationship between dampness and the thermal effects of combinations of building materials.

A08-2411
Griffin, C. W. Manual of Built-Up Roof Systems. American Institute of Architects, 1970. 241 pp. $17.50 (for AIA members, less 10%)
  A book of practical information and fundamental theories of built-up roofs, discussing each of the built-up roofing components. Illustrated with diagrams, tables and photos.

A08-2412
Gross, W. Applications Manual for Paint and Protective Coatings: A Guide to Types of Coatings, Methods of Surface Preparation, and Hand Application Techniques. McGraw-Hill, or Machine Tool Publications, 1970. 320 pp. $19.50
  This manual presents the essential facts required by either a professional or amateur painter to obtain successful application of a paint or a protective coating.

A08-2413
Gurfinkel, G. Wood Engineering. Southern Forest Products Association, 1975. 540 pp. $9.95
  The physical properties of wood are covered in great detail as the basis of the sections on design which follow. These include information on connectors, beams, trusses, columns and glulam.

A08-2414
Guy, A. G. Essentials of Materials Science. McGraw-Hill, 1976. 416 pp. $18.00
  This introduction to materials science covers all the essential aspects of the behavior of metals, ceramics, semi-conductors, and polymers. Using everyday examples of materials, the author describes their internal structure, their electrical, magnetic, and optical behavior, and the effects of high temperature, fracture, and deterioration.

A08-2415
Guyon, Y. Limit-State Design of Prestressed Concrete. Vol. 1, The Design of the Section. Applied Science, 1972. 485 pp. £12.00

The author is an authority on prestressed concrete and a member of the Comité Européen du Béton and the Fédération Internationale de la Précontrainte. The recommendations of these two bodies now form the basis for the Codes of Practice adopted or in preparation in most European countries, including Great Britain.

A08-2416
Hanson, A. and Parr, J. G. The Engineer's Guide to Steel. Addison-Wesley, 1965. 406 pp. $15.00

As a practical reference and for courses in industrial metallurgy, this book covers what steel is--how to select steel--why a specification takes a particular form--and what are the limitations of a certain steel and its advantages.

A08-2417
Harper, C. Handbook of Plastics and Elastomers. McGraw-Hill, 1975. 950 pp. $29.50

This handbook provides data, performance application information and guidelines for the entire range of plastics and elastomers, and for every area of plastic and elastomer product use. Coverage progresses from fundamentals and basic electrical, physical, mechanical, and chemical properties, to material and product categories such as laminates, reinforced plastics, fibers, foams, liquid resin systems, coating and adhesives, then to standards and specifications.

A08-2418
Heinle, E. and Bacher, M. Building in Visual Concrete. British Book Centre. 208 pp. $22.75

The term visual concrete is used in the book to define any concrete surface planned to remain visible in the finished building. The term is intended to embrace both exposed concrete whose surface has been left untouched after dismantling of the formwork, and concrete whose surface has been subjected to raking, hammering, sandblasting or tooling of any other kind. It covers also concrete whose surface has been removed altogether, by washing, brushing or etching with acid, so as to leave exposed the aggregate used in its manufacture. The original German text has been translated and edited by a consultant to Britain's largest cement-manufacturing organization.

A08-2419
Hoffmann, K., et al. Building with Wood: Form, Structural Design, and Preservation. Praeger, 1969. 180 pp. $15.00

Building with Wood surveys examples of the successful use of wood in modern buildings in the United States, Japan, Scandinavia, England, and Central Europe. There are single-family houses, row houses, and vacation houses, as well as schools, offices, shops, and churches. Structural details treated include stairs, sun decks, screens, benches, fences, and roofs. The importance of wood in modern engineering and new possibilities for its use are also noted.

A08-2420
Hornung, W. J. Reinhold Data Sheets. Van Nostrand Reinhold, 1965. 256 pp. $18.50

This book presents reference and drawing data in the field of architecture, construction, and design. Information is given on site planning, kitchen and bath equipment, thermal insulation, wood, concrete block and poured concrete construction, structural steel details, heating, and a large number of additional building elements.

A08-2421
Jayne, B. A., ed. Theory and Design of Wood and Fiber Composite Materials. Syracuse Univ. Press, 418 pp. $20.00

Scientists concerned with research and experimentation on composite systems--paper fiberboard, particleboard, laminates, the many new composites that have come into existence in recent years will find this volume invaluable. Heretofore, each has been studied largely independently of the others, but recent years have seen the development of concepts and methods of analysis applicable to all of these materials. These methods not only make possible a better understanding of composites now manufactured; they are also the key to future progress in creative design of new wood and fiber systems. Book is from the Proceedings of a Conference held at the University of Washington.

A08-2422
Johnson, F. B., ed. Designing, Engineering and Construction with Masonry Products. Gulf Publishing Co., 1969. 500 pp. $35.00

Architects and engineers from government, education, research and industry met at the University of Texas in 1967 to discuss the state of research, education, and practice relating to masonry structural systems. This book contains the proceedings of this conference. The book includes discussions of the creative challenges of masonry, materials science relevant to structure, structural performance, design methodology, and construction.

A08-2423
Jones, R. Mechanics of Composite Materials. McGraw-Hill, 1975. 450 pp. $21.00

This work covers the exciting new area of development in design and physical properties of materials used in building and aircraft design. It covers the metallurgical and engineering aspects of advanced strength properties and stress analysis, and examines the various approaches to materials use.

A08-2424
Katz, I. and Cagle, C. Adhesive Materials: Their Properties and Usage. Foster Publishing Co. 536 pp. $18.00

Covers adhesives, binders, cements, mortars, mucilages, pastes, gummed and presensitive adhesives. The book gives a picture of the different types of joining materials and their precise chemical, physical and performance characteristics.

Building Materials, Components, Finishes

A08-2425
Keyser, C. A. Materials Science in Engineering. Charles Merrill, 1974. 448 pp.
 This is a clear, complete and practical book intended to introduce students to why materials behave as they do, and how they can be used to the best advantage in engineering applications. Theory is treated to the extent that it contributes to understanding. Topics include metals, concrete, plastics, wood and clay products, their physical properties, failure modes, and their behaviour at elevated temperatures and under fatigue. Corrosion and its prevention is also discussed.

A08-2426
Kinniburgh, W. Dictionary of Building Materials. Applied Science, 1966. 285 pp. £2.00
 The dictionary is not confined to materials which are actually employed directly in buildings, but includes substances which are used in the preparation of building materials, where reference to these leads to a better understanding of the nature of the building materials themselves. Where applicable, references are made to British Standards, Codes of Practice, and Building Research Station Digests.

A08-2427
Kissin, G. H., ed. Finishing of Aluminum. Van Nostrand Reinhold, 1963. 243 pp. $13.50

A08-2428
Knowles, P. R. Composite Steel and Concrete Construction. John Wiley, 1973. 200 pp. $18.75

A08-2429
Koster, W. Expansion Joints in Bridges and Roads. Applied Science, 1969. 333 pp. £8.00
 This book for engineers is packed with valuable data and design criteria for expansion joints in bridges and concrete roads. It deals in detail with all the movements at expansion joints; basic properties of the various types of joints (open joints, sealed joints, tooth joints, trailing plate and floating plate joints and others) and their suitability for particular conditions; design of bridge joints; design of road joints; precautions in construction of various types of joints and faults and failures.

A08-2430
Krebs, R. and Walker, R. Highway Materials. McGraw-Hill, 1971. 448 pp. $18.50
 Introduces concepts underlying the characterization and use of soil, aggregate, portland cement, and bituminous materials in major engineering works such as highways, airfields, and hydraulic structures. Emphasis is placed on current practice and test methods, with reference made to standard ASTM and AASHO test procedures and definitions.

A08-2431
Labahn, O. and Kaminsky, W. A.   Cement Engineers Handbook.
 Adler's Foreign Books, 1971.   250 pp. $21.50
 The book features convenient presentation of information and numerical data, graphs and formulas.

A08-2432
LaLonde, W. and Janes, M.   Concrete Engineering Handbook.
 McGraw-Hill, 1962.   1172 pp. $45.50
 Presents essential methods, standards, and data of concrete engineering--covering the entire field from planning and design of reinforced concrete structures and elements to proved construction practices for buildings, bridges, pavements, and other concrete work.

A08-2433
Launchbury, W., ed.   AJ Handbook of Fixings and Fastenings.
 Architectural Press, 1971.   80 pp.   £1.00
 This book provides a guide to the properties, uses and methods of fixing of nails, wood screws, light plugs, heavy masonry fixings, in situ fixings, rivets and screws, roofing and cladding fixings, masonry clamps, pipe and conduit fixings, power tools, welding, brazing and soldering equipment, and adhesives.

A08-2434
Lea, F. M.   The Chemistry of Cement and Concrete.   Chemical
 Publishing Co., 1970.   760 pp. $39.50
 The book deals with the chemical and physical properties of cements and concretes, and their relation to the practical problems that arise in manufacture and use. Attention is given to problems arising in the use of concrete, to the suitability of materials, to conditions under which concrete may deteriorate, and to precautionary or remedial measures that can be adopted.

A08-2435
Lee, H. and Neville, K.   Handbook of Epoxy Resins.   McGraw-
 Hill, 1966.   960 pp. $39.50
 Almost everything known about epoxy resins has been compiled and summarized in this handbook which includes thousands of sophisticated industrial applications.

A08-2436
Lesley, R., et al.   History of the Portland Cement Industry in the
 United States.   Arno Press, 1924. $17.00
 This authorized history of the Portland cement industry describes the improvements in production methods and chronicles the increasing use of Portland cement in water works, tunnels, and buildings. The author shows how cement aided the development of rapid construction techniques and gave new directions to American architecture. The appendices cover progress of the industry by years and an outline of the organization and activities of the Portland Cement Association.

Building Materials, Components, Finishes

A08-2437
Lloyd, W. B. Millwork--Principles and Practices. Cahners, 1966. 426 pp. $8.95
   Covers manufacturing, distribution and uses of all woodwork items in all phases of the construction industry. Millwork construction, marketing techniques, sizes, designs, grades, uses, specifications, industry standards, future trends. Includes an 800-word glossary of woodwork terms.

A08-2438
Logan, H. L. The Stress Corrosion of Metals. John Wiley, 1966. 306 pp. $18.75
   Contents--The Phenomena and Mechanism of the Stress Corrosion Cracking of Metals. Stress-Corrosion Cracking of Low Carbon Steels. Stress-Corrosion Cracking of Stainless Steels. Stress-Corrosion Cracking of Copper Base Alloys. Stress-Corrosion of Aluminum Alloys. Magnesium Alloys. Titanium and Titanium Alloys. Methods of Evaluating the Resistance of Materials to Stress Corrosion Cracking.

A08-2439
Love, T. W. Construction Manual: Concrete and Formwork. Craftsman Book Co., 1973. 176 pp. $3.75
   Provides solutions to such problems as the best type of mix for a job, depth of the footing and quantity of material needed. Contains man hour tables, charts, and illustrations.

A08-2440
Lydon, F. D. Concrete Mix Design. Applied Science, 1972. 148 pp. £3.50
   This monograph presents an approach to the mix design process using most of the conventional methods. Background coverage is given on the properties of fresh and hardened concrete so that the principles underlying the choosing of mix proportions can be seen.

A08-2441
McIntosh, J. D. Concrete and Statistics. Applied Science, 1968. 139 pp. £3.00
   An engineer's approach to the testing of concrete and its constituent materials to check compliance with specification requirements, using statistical concepts to give a sense of proportion in drafting, checking and enforcing these requirements.

A08-2442
McMillan, F. R. and Tuthill, L. H. Concrete Primer. American Concrete Institute, 1973. 96 pp. $3.00 (for ACI members, $2.00)
   A question and answer format developing the principles governing concrete mixtures. It shows how a knowledge of these principles and of the properties of cement can be applied to the production of permanent structures in concrete.

A08-2443
Mallinson, J.   Chemical Plant Design with Reinforced Plastics.
  McGraw-Hill, 1969.   416 pp. $29.50
    A distillation of theory and practical experience, this source
reference is the first book to cover the use of chemical reinforced
plastic material as applied to the chemical process industries.
Discusses the safe use and fabrication of reinforced plastic materials, procurement methods, stocking programs, and corrosion resistance data.

A08-2444
Malloy, J. F.   Thermal Insulation: Integrated Design Manual.
  Van Nostrand Reinhold, 1969.   688 pp. $21.50
    This book provides a solid grasp of the methods used to
make a complete economic evaluation of thermal insulation installations, products, and applications. Topics include heat transfer,
functions and properties of thermal insulation, weather and vapor
barriers, indoor coverings and finishes, design of industrial insulation systems and more.

A08-2445
Mantell, C.   Engineering Materials Handbook.   McGraw-Hill, 1958.
  1960 pp. $41.50
    This reference work presents information on engineering ma-
terials with respect to design, structure, and serviceability. Covers
all materials with emphasis on their fabricated forms, their physical and mechanical properties, their adaptability, advantages, limitations, protection against deterioration, their stability, etc.

A08-2446
Mark, H. F., et al., eds.   Encyclopedia of Polymer Science and
  Technology: Plastics, Resins, Rubbers, Fibers. Vol. 1,
  Ablative Polymers to Amino Acids.   John Wiley, 1964.
  893 pp. $50.00
    This encyclopedia, comprising 16 volumes, including the in-
dex, brings together progress and developments in the polymer
field. The volumes contain comprehensive treatments of all mono-
mers and polymers, their properties, methods and processes for
their preparation and manufacture as well as broad treatments of
theoretical fundamentals.

A08-2447
Marsh, P.   Concrete as a Visual Material.   Cement and Concrete
  Assn., 1974.   £2.00

A08-2448
Marsh, P. and Beckett, D.   Mechanical Fixing Devices in the Construction Industry. The Construction Press.   £6.75
    To ensure that the safest and most economical fixing is al-
ways achieved it is essential that architects and builders should be
aware of the range of devices available and also of the specific
purpose for which each type was designed. This book deals with
fixings in masonry, brickwork, timber, metal, plaster-board and

Building Materials, Components, Finishes 405

lightweight bases; with "through" and "anchor," restraint and load-bearing, friction and threaded, expansion and toggle devices. Adhesives are referred to in one of the appendices.

A08-2449
Martens, C. R., ed. The Technology of Paints, Varnishes and Lacquers. Van Nostrand Reinhold, 1968. 732 pp. $28.50
This book offers technical data relevant to the preparation and use of paints, varnishes, and lacquers. It explains the raw materials required for the formulation, production, testing, and application of protective coatings. Provides information on the performance of specific coatings for trade, industrial, and maintenance use.

A08-2450
Masonry Institute of America. Masonry Design Manual. Masonry Institute of America, 1972. 384 pp. $14.50
Provides technical assistance to architects, engineers, designers, developers, contractors and building officials. Includes structural design information, construction details, outline specifications and ideas for masonry systems. Brick, concrete masonry, stone, veneer and arches are treated.

A08-2451
May, C. and Tanaka, Y., eds. Epoxy Resins: Chemistry and Technology. Marcel Dekker, 1973. 704 pp.
Brings together the contributions of a number of outstanding researchers in the field of epoxy resins. Not only emphasizes the chemistry and technology of epoxy resins, but also deals with many industrial applications.

A08-2452
Meinecke, E. A. and Clark, R. C. Mechanical Properties of Polymeric Foams. Technomic, 1973. 175 pp. $20.00
This book contains design information for working with foams.

A08-2453
MIA. Marble Design Manual. Marble Institute of America, 178 pp. $12.95
A comprehensive guide to the architectural use of marble organized to conform with the Uniform System for Construction Specifications. Covered are such subjects as the physical and structural properties of marble, and its installation and maintenance. The uses of marble in roads, precast concrete, and as building stone are covered, along with specialized subjects as curtainwall systems, masonry restoration and veneer stone.

A08-2454
Mills, A. P., et al. Materials of Construction. John Wiley, 1955. 650 pp. $18.25

A08-2455
Moffat, D. W. Plant Engineer's Handbook of Formulas, Charts

and Tables. Prentice-Hall, 1976. $24.95

Over 500 formulas, charts, tables and nomograms provide data on soil mechanics and foundation design, concrete blocks and wood, structural steel and aluminum, security and electrical power. Other subjects include piping, HVAC, elevators and conveyors, roadways and parking, and outdoor signs.

A08-2456
Mohr, J. G., et al. SPI Handbook of Technology and Engineering of Reinforced Plastics/Composites. Van Nostrand Reinhold, $29.95

A reliable, documented source for use in calculating configurations for reinforced plastics products, it contains full information on molding methods, tool and product design, and properties of materials. The subjects discussed include processes involving low-temperature cure; processes for intermediate cure (architectural and industrial paneling); high-temperature curing processes; miscellaneous thermoset molding processes; reinforced thermoplastics; unique composite materials; and potential and future growth.

A08-2457
Moselle, M. Practical Lumber Computer. Craftsman Book Co., 1956. 121 pp. $2.00

This handy book quickly and easily gives the board footage for all standard sizes and lengths of lumber from one to one thousand pieces. All the work is done for you--no mathematics needed. You arrive at the precise answer in seconds. A table is also included which allows anyone to rapidly determine the board feet per linear foot of lumber.

A08-2458
NACE. Control of Pipeline Corrosion. National Assn. of Corrosion Engrs., 200 pp. $6.50 (for NACE members, $5.50)

A08-2459
NACE. Fundamental Aspects of Stress Corrosion Cracking. National Assn. of Corrosion Engrs., 1967. 711 pp. $36.50
From 1967 meeting at Ohio State University.

A08-2460
NACE. Industrial Maintenance Painting. National Assn. of Corrosion Engrs. $6.00 (for NACE members, $5.00)
Illustrated, tabular data.

A08-2461
NACE. 1960 Bibliographic Survey of Corrosion. National Assn. of Corrosion Engrs., 1960. $35.00 (for NACE members, $25.00)

Contains 3400 abstracts of 1960 world literature on corrosion control, arranged according to NACE Abstract Filing Index, alphabetical subject and author index.

Building Materials, Components, Finishes

A08-2462
National Bureau of Standards. A Compilation and Evaluation of Mechanical, Thermal and Electrical Properties of Selected Polymers. U.S. Gov't. Printing Office, 1973. 843 pp. $15.80. Order No. C 13.44:132 S/N 0303-01082

This compilation abstracts original experimental data on the mechanical, thermal, and electrical properties of six commercially available polymers. A summary of property data is included.

A08-2463
Nesbit, J. K. Structural Lightweight-Aggregate Concrete. Cement and Concrete Assn., 1967. 282 pp. £1.40

A08-2464
Neumann, J. A. and Bockhoff, F. J. Welding of Plastics. Van Nostrand Reinhold, 1959. 288 pp. $10.00

With the increasing use of plastics in construction, this book gives important data on welding by hot-gas, heated-tool and friction.

A08-2465
Neville, A. M. Creep of Concrete: Plain, Reinforced, and Prestressed. North-Holland, 1970. 622 pp. $45.50

Provides an assessment of the effects of creep on stresses and deformations and gives information on prediction of creep under any conditions. Both metric and British units used throughout.

A08-2466
Neville, A. M. Hardened Concrete: Physical and Mechanical Aspects. American Concrete Institute, 1971. 260 pp. $15.00 (available from University Microfilms)

Intended as an aid for designers, construction engineers and students. The author presents a review of the major physical and mechanical properties of hardened concrete, with emphasis on deformation, environmental influences, and understanding the phenomena involved. A list of references is included.

A08-2467
Neville, A. M. High Alumina Cement Concrete. John Wiley, 1975. 201 pp. $22.00

Explains how and when high alumina cement concrete can be used advantageously and safely, and why defects occur; shows how to identify it and test it, and describes possible courses of remedial action; analyzes case studies.

A08-2468
Neville, A. M. Properties of Concrete. John Wiley, 1973. 686 pp. $16.00

Contents include Portland cement, properties of aggregate, fresh concrete, and the strength, elasticity, shrinkage, creep and durability of concrete. There is also information on lightweight and heavy concrete.

A08-2469
NFPA. Tentative Guide for Plastics in Building Construction. National Fire Protection Assn., 1973. $2.00

The purpose of this guide is to discuss the characteristics of plastics used in building construction which should be considered by designers, prospective users, and regulatory authorities concerned about fire safety. The chapters cover general fire considerations of plastics, foamed plastics, and fire test procedures. Appendix A discusses the history and development of the thermal barrier concept while Appendix B provides data on representative fire research selected to illustrate specific fire problem areas.

A08-2470
Nicholas, D. D., ed. Wood Deterioration and Its Prevention by Preservative Treatments. Vol. 1, Degradation and Protection of Wood. Syracuse Univ. Press, 1973. 380 pp. $20.00

In the first volume the history of wood preservation precedes the detailed consideration of the major causes of wood deterioration and the principles of protection from those causes.

A08-2471
Nicholas, D. D., ed. Wood Deterioration and Its Prevention by Preservative Treatments. Vol. 2, Preservatives and Preservative Systems. Syracuse Univ. Press, 1973. 402 pp. $22.00

The second volume of the series deals primarily with the treatment of wood, the effectiveness of preservatives, treating processes, properties of preservatives, and, because of its increasing importance, the problem of water-pollution abatement and control.

A08-2472
Nicholls, R. Composite Construction Materials Handbook. Prentice-Hall, 1976. 608 pp. $24.00

The book integrates chemical structure, mechanical behavior, and design optimization of composite materials used in the construction industry--aggregate-binder, fiber-reinforced, laminar, and structural form systems. Features include a fusion of analysis and design procedures for construction composites that were previously available only from scattered sources, and an emphasis throughout that relates the dependence of mechanical properties upon chemical structure.

A08-2473
Norton, F. Fine Ceramics: Technology and Applications. McGraw-Hill, 1970. 512 pp. $29.95

This overall picture of the fine ceramics industry, both domestic and foreign, discusses in detail the general principles involved in the production of fine ceramics, the manufacturing methods for specific products, and the overall economic background. Emphasis is on economic aspects, production methods, raw materials and applications.

Building Materials, Components, Finishes   409

A08-2474
Norton, F. Refractories. McGraw-Hill, 1968. 450 pp. $24.75
    Provides information on the manufacture, properties, and uses of heavy refractories for the metallurgical, chemical, and glass industries.

A08-2475
Nylen, P. and Sunderland, E. Modern Surface Coatings: A Textbook of the Chemistry and Technology of Paints, Varnishes and Lacquers. John Wiley, 1965. 750 pp. $25.50

A08-2476
Ogorkiewicz, R. M., ed. Engineering Properties of Thermoplastics. John Wiley, 1970. 318 pp. $18.75
    A compilation of information about the engineering characteristics and behavior of a wide range of thermoplastics in a form which will facilitate the design of plastics components.

A08-2477
Orchard, D. F. Concrete Technology. Vol. 1, Properties of Materials. Applied Science, or John Wiley, 1973. 375 pp. $26.00
    Contents include the kinds of cements and their properties, aggregates, lightweight concrete, mix design and quality control. Deterioration of concrete and its resistance to chemical attack are also covered.

A08-2478
Orchard, D. F. Concrete Technology. Vol. 2, Practice. Applied Science or John Wiley, 1973. 440 pp. $32.50
    The volume covers the use and testing of concrete and concrete materials. Covers the testing of materials before mixing, immediately after mixing and after hardening of the concrete, and the problem of compacting the concrete into moulds and curing it. Methods of transporting and storing the materials required for concrete making and of transporting the mixed concrete by mechanically propelled vehicles, by using a concrete pump or by pneumatic means are dealt with.

A08-2479
Parker, E. Materials Data Book. McGraw-Hill, 1967. 416 pp. $11.50
    Here is a convenient source of reliable engineering design data for metals, alloys, wood, ceramics, concrete, and plastics. It is of equal value to students and practicing engineers.

A08-2480
Parker, M. E. Pipe Line Corrosion and Cathodic Protection. Gulf Publishing Co., 1962. 166 pp. $5.95
    This practical field manual, explains how to utilize field data once it has been obtained. The contents include soil resistivity surveys, line currents, current requirement surveys, rectifier systems for coated lines, ground bed design and installation,

galvanic anodes on coated lines, hot spot protection, stray current electrolysis, interference on cathodic protection, operation, maintenance coating inspection and testing.

A08-2481
Parkyn, B., ed. Glass Reinforced Plastics. CRC Press, Inc., 1970. 295 pp. $24.00
The reinforced plastics industry has grown enormously. Parallel to the growth has been the development of better resin systems, curing systems and glass fiber reinforcement. This book covers the whole field of glass reinforced plastics with the emphasis on end products and design criteria.

A08-2482
Patton, W. J. Construction Materials. Prentice-Hall, 1975. 416 pp. $13.95
Discussing both the traditional and the modern synthetic construction materials, this book explores their properties, applications, advantages, and limitations for specific uses. Patton examines the pertinent properties of these construction materials such as strength, thermal conductivity, and fire resistance. In addition, the author covers the applications of the materials ranging from rock to plastic foams, including such modern practices as water-filled building columns, sandwich construction, foamed urethane roofs and urethane floors.

A08-2483
Patton, W. J. Materials in Industry. Prentice-Hall, 1975. 496 pp. $16.95
Here is a basic and thorough analysis of engineering materials, ceramics, metals, organics, biomaterials, and semifinished products that explains the complex properties of each. The author stresses the need for responsible appraisal and compromise when selecting such materials for specific applications and particular products. He examines a wide range of materials-- fuels, industrial radioisotopes, adhesives, refrigerants, foods, paints, plastic foams, prosthetic materials, bone, laminates, sandwich materials, and industrial gases, as well as such traditional materials as stone, wood, steel, and concrete.

A08-2484
Patton, W. J. Plastics Technology: Theory, Design and Manufacture. Prentice-Hall, 1975. 320 pp. $15.95
In this book, the author explains the usual and unusual properties of plastics and how to select, design and apply them to the particular task at hand. While the author devotes the main emphasis to the influence of polymer chemistry and manufacturing method, he also gives more than ample treatment to fillers and reinforcements, additives and even devotes a separate chapter to foamed plastics.

A08-2485
Pawley, M. Garbage Housing. Architectural Press, 1974. £3.00

Building Materials, Components, Finishes

In spite of the growing energy crisis and the world-wide shortage of raw materials, billions of tons of packaging and similar products are wasted annually. This book shows how they could be used or adapted for building.

A08-2486
Penn, W. S. Plastics-in-Building Handbook. Applied Science, 1964. 326 pp. £4.00

There are hundreds of applications of plastics, but unfortunately the average user can only become aware of this through journals, manufacturers' literature, and the visits of manufacturers' representatives. The amount of literature is great and biased. This handbook is designed to overcome these difficulties. In the first place a systematic classification of products into groups and sub-groups has been made. In the second place the technical advantages of using plastics instead of other materials are given.

A08-2487
Peter, J. P. Design with Glass Materials in Modern Architecture. Van Nostrand Reinhold, 1965. 160 pp. $14.00

Demonstrates the design potential of glass as a building material. Each detail in the book has been redrawn from the original blueprints by a trained architectural draftsman to bring out the glass design detail. Photographs show the glass detail as well as the entire building. Analysis of 36 buildings shows the various uses of glass design.

A08-2488
Petzold, A. and Rohrs, M. Concrete for High Temperatures. American Elsevier, 1970. 220 pp. $16.50

During the last decade progress has been made in developing fire-resisting concrete which does not need to be baked before going into service. For such concrete, attention must be given to the cement and the aggregates, but the mixing and casting process is analogous to that of ordinary concrete. Thus, there is the same flexibility of choice in the shape which can be made, the concrete can be cast in place when required and the resultant product has been cheapened. The temperatures considered are chiefly those above 1500°C.

A08-2489
Pindar, N. T. Engineering Wall and Partition Components for Prefabrication. Herman Publishing, Inc., 1976. 128 pp. $13.50

Contents: general description of principles and process; principles of panel engineering: orientation, conversion, mensuration, modulation; development of steps in panel engineering: computation forms, panel dimensioning plans; openings identification print, window/door tally form, panel work sheets, cutting list.

A08-2490
Plant Engineering Directory and Specifications Catalog. Technical Publishing Co. $25.00

Product specifications and manufacturer catalogs are given for more than 15,000 plant engineering and maintenance products. Names, addresses and telephone numbers of local sales offices are listed.

A08-2491
Plummer, H. Brick and Tile Engineering. Brick Institute of America, 1962. 460 pp. $3.50

For the engineer, with some chapters on architectural applications, this is a volume concerning all aspects of brick and tile construction.

A08-2492
Pollack, H. W. Materials Science and Metallurgy. Prentice-Hall, 1973. 412 pp. $13.50

Includes discussions of manufacture, structure and physical properties of metals; interpretation of equilibrium diagrams and TTT curves, steel, tool steel, stone, wood, plastics, rubber, etc. Features problems and questions with answers included.

A08-2493
Portland Cement Assn. Notes on ACI 318-71 Building Code Requirements with Design Applications. Portland Cement Assn. 1972. 560 pp. $8.25

Reference manual on the proper application of the ACI 318-71 Building Code. More than 75 design examples show how to apply code provisions in design. A total of 26 topics are included, 25 concerned directly with specific design provisions of the Code and one concerned with PCA computer program developments for the analysis and design of concrete structures.

A08-2494
Portland Cement Assn. Principles of Quality Concrete. John Wiley, 1975. 312 pp. $14.95

Compiled by the Portland Cement Association (PCA). Covers the fundamentals required to produce quality concrete. Includes data on concrete materials such as cement, water, aggregates, air entrainment and admixtures. Other subjects covered are quality concrete and proportioning and mixing concrete.

A08-2495
Portland Cement Assn. Proceedings of the PCA-ACI Teleconference on ACI 318-71 Building Code Requirements. Portland Cement Assn., 1972. 135 pp. $3.15

Consists of papers presented at televised conference on January 18, 1972, on the provisions of the new edition of ACI design standard: Building Code Requirements for Reinforced Concrete.

A08-2496
Portland Cement Assn. Special Concretes, Mortars and Products. John Wiley, 1975. 482 pp. $19.95

Describes the various concretes available and ways to create

Building Materials, Components, Finishes

projects to meet new standards and designs. Includes sections on lightweight, insulating and heavy concrete, decorative concretes, reinforced concrete, concrete masonry, concrete pipe, soil cement, and fire resistance of concrete.

A08-2497
Powers, T. C.   The Properties of Fresh Concrete.   John Wiley, 1968.   664 pp. $31.75
   Describes the properties of fresh concrete and explains the nature of concrete mixtures in terms of structure and interparticle forces.

A08-2498
Preston, H.   Prestressed Concrete for Architects and Engineers. McGraw-Hill, 1964.   204 pp. $14.50
   Past, present, and future uses of prestressed concrete are what this book is all about.   Also gives guidance on new structural design.

A08-2499
Preston, H. and Sollenberger, N.   Modern Prestressed Concrete. McGraw-Hill, 1967.   352 pp. $17.50
   This reference furnishes information on the design of safe, economical prestressed concrete.

A08-2500
Probert, S. D., ed.   Thermal Insulation.   Applied Science, 1968. 121 pp.   £4.00
   Too often thermal insulation is still an afterthought whereas it should be included in the design stage.   This book describes actual examples of building and engineering systems in which this approach has been adopted, thus minimizing energy wastage and providing optimum economic working and operating conditions.

A08-2501
Quarmby, A.   Plastics in Architecture.   Praeger, 1973.   224 pp. $18.50
   The increasing sophistication of the plastics industry is producing materials and fabrication technologies that are revolutionizing the concept of building.   Until Plastics in Architecture, no work written especially for the architect has covered thoroughly the infinite uses and adaptability of plastic as a new constructional material.

A08-2502
Rabinowicz, E.   Friction and Wear of Materials.   John Wiley, 1965.   244 pp. $15.50
   Topics include material properties which influence interactions, surface interactions, friction, adhesive and abrasive wear and lubrication.

A08-2503
[No entry]

A08-2504
Reboul, P. and Mitchell, R. G. Plastics in the Building Industry.
  Butterworth, 1968. 238 pp. £2.50

A08-2505
Rice, P. F. User's Guide to the ACI Building Code. Van Nostrand Reinhold. (In preparation)
  This guide to the 1971 American Concrete Institute Building Code interprets the code and provides formulas, examples of acceptable designs or construction methods, tables of data and other helpful information. This book is of value to structural engineers, concrete engineers and construction engineers.

A08-2506
Robson, T. D. High Alumina Cements and Concretes. John Wiley, 1963. 263 pp. $7.95

A08-2507
Ross, R. B. Metallic Materials Specification Handbook. John Wiley, 1973. 833 pp. $25.00
  This book lists all known specifications, trade names, and symbols for metals, and under each is shown the analysis, condition, supplier, and mechanical properties where known. The materials are divided into 130 groups each of which contains similar materials with a single index referring the reader to the appropriate group. A table of physical properties is given with each group under the headings of general metallurgical characteristics, thermal treatment, welding and brazing, flaw detection methods, corrosion protection, machinability and uses. Covers the increasingly important modern materials such as titanium, cobalt, and nickel alloys.

A08-2508
Rostron, M. R. Light Cladding of Buildings. Architectural Press, 1964. 352 pp. £3.15
  Provides a summary of available knowledge and gives guidance on technical data.

A08-2509
Ryan, T. Gunite, A Handbook for Engineers. Cement and Concrete Assn., 1973. 63 pp. £1.50

A08-2510
Sahlin, S. Structural Masonry. Prentice-Hall, 1971. 290 pp. $17.25
  Gives the reader an understanding of the physical behavior of structural masonry. It also provides a basis for predicting the behavior of elements or structures composed of masonry units.

A08-2511
Salter, W. L. Floors and Floor Maintenance. Applied Science or John Wiley, 1974. 360 pp. $24.75
  The types of flooring discussed range through carpet, timber,

concrete, quarry and other hard tiles, linoleum and cork, rubber, asphalt and vinyl asbestos and flexible vinyl. The design, choice of materials, equipment for installation and maintenance are described.

A08-2512
Sarvetnick, H. A. Polyvinyl Chloride. Van Nostrand Reinhold, 1968. 300 pp. $13.50
 Here are the facts about PVC, the most widely compounded plastic material in use today. Emphasis is placed on PVC problems and their solutions in terms of technical data and economics.

A08-2513
Schweitzer, P. A. Handbook of Corrosion Resistant Piping. Industrial Press, 1969. 358 pp. $25.00
 For the first time, the wide range of corrosion-resistant pipe and fittings that are available are compared. The book contains factors necessary to determine the optimum system to handle specific corrodents. Each material is discussed individually, and suitable construction materials are indicated for over 500 corrodents.

A08-2514
Scott, W. L., et al. Explanatory Handbook of the BS Code of Practice for Reinforced Concrete CP 114 (1957) (including Amendment No. 1:1965). Cement and Concrete Assn., 1968. 172 pp. £1.00

A08-2515
Seiffert, K. Damp Diffusion and Buildings. Applied Science, 1970. 209 pp. £6.00
 The author shows how to determine quantitatively the behaviour of a wall or roof under various conditions of temperature and humidity, both internally and externally. He shows how moisture condensation can be either avoided altogether or kept within acceptable limits. The tables of data which have been gathered from long-term tests on various building materials, including a wide range of modern synthetic sheeting and insulating products, are an important part of the book.

A08-2516
Sementsov, S. A. and Kameiko, V. A., eds. Designer's Manual: Masonry, Including Reinforced Masonry in Industrial, Residential and Communal Buildings and Structures. Peter Peregrinus, 1971. 236 pp. $20.00 (available from Intl. Scholarly Book Services.)
 A manual on all aspects of brickwork and other masonry construction including: analysis, designs, structure types, materials and procedures. Chapters on prefabrication and on construction in earthquake areas. Translated from Russian.

A08-2517
Shacklock, B. W. Concrete Constituents and Mix Proportions. Cement and Concrete Assn., 1974. 120 pp. £3.00

A08-2518
Shand, E.   Glass Engineering Handbook.   McGraw-Hill, 1958.
 471 pp. $22.50
 This encyclopedic handbook on the composition, manufacture, properties, and applications of glass as an engineering material provides practical data on the use of glass and glass products in engineering, industry and research, and covers less common topics such as photosensitive glass, glass-ceramics, electrically conducting glass, glass reinforced plastics, etc.

A08-2519
Shields, J.   Adhesives Handbook.   CRC Press, 1970.   355 pp.
 $43.50
 A chapter on adhesives selection helps the user to define his bonding problem and view it in terms of the service conditions anticipated for the bonded assembly. The mechanical, physical and chemical properties of adhesives are summarized in tables to assist in the selection of adhesives for bonding a variety of materials. More specific information on the properties, processing characteristics and applications of basic chemical types of adhesive is covered in an encyclopedic section of adhesive materials and properties. Commercially available adhesives, representative of the basic types, are described in a 157-page Adhesives Products Directory.

A08-2520
Short, A. and Kinniburgh, W.   Lightweight Concrete.   Applied
 Science, 1968. 368 pp.   £5.00
 A text dealing with all known types of lightweight aggregates and concretes, including aerated concrete, properties of materials and application, British, European and American practice and comparison of respective Standards and Recommendations. The results of original research throughout the world are discussed and used in framing recommendations for design and application.

A08-2521
Siau, J. F.   Flow in Wood.   Syracuse Univ. Press, 1971.   131 pp.
 $15.00
 Unifying and analyzing recent research, this book provides a theoretical background for a better understanding of wood preservation processes and other treatments involving transport in wood.

A08-2522
Simonds, H. R. and Church, J. M.   Concise Guide to Plastics.
 Van Nostrand Reinhold, 1963.   410 pp. $13.50
 This guide instructs in the selection, use and forms of plastics, and discusses which ones best suit particular products. Includes basic data on strength, properties, processes, production and prices. Lists the most important plastics producers with their addresses.

A08-2523
Simpson, J. W. and Horrobin, P. J., eds.   The Weathering and

Building Materials, Components, Finishes

Performance of Building Materials. John Wiley, 1970. 286 pp. $12.50

A collection of information on weathering and performance of building materials. Includes new and previously unpublished information.

A08-2524
Skaar, C. Water in Wood. Syracuse Univ. Press, 1972. 218 pp. $12.50

This fundamental book on wood-water relationships provides a unified treatment of all the important physical phenomena relating to the behavior of wood in response to changes in its moisture content.

A08-2525
Skeist, I. Epoxy Resins. Van Nostrand Reinhold, 1958. 308 pp. $10.50

Covers resins, curing agents, modifiers, fillers and fiber reinforcement. Explains methods of casting and potting, including the tools used.

A08-2526
Skeist, I., ed. Handbook of Adhesives. Van Nostrand Reinhold, 1962. 700 pp. $25.00

A compilation of adhesive materials and bonding technology with detailed information on adhesives chemistry, manufacture, and application by more than 50 authors.

A08-2527
Skeist, I., ed. Plastics in Building. Van Nostrand Reinhold, 1966. 480 pp. $20.00

Twenty-three of the world's leading architects, plastics engineers, chemists, and building-code specialists have contributed articles to this illustrated book. Thorough discussions of coatings, adhesives, and sealants are included. The importance of building codes and specifications is emphasized.

A08-2528
Smith, R. C. Materials of Construction. McGraw-Hill, 1973. 448 pp. $12.50

Here is an updated version of a well-illustrated, descriptive text covering all of the major materials in the building construction industry. Well-suited for self-study or for classroom use, this edition covers recent advances including the uses of reflective glass, plastics, glass brick, and weathering steel. Smith describes the origin, manufacturing, uses, and engineering aspects of the building products available in the field. He then shows how these materials can best be used in modern building practices.

A08-2529
Snow, F. Formwork for Modern Structure. Chapman and Hall, 1965. $6.75 (available from Barnes and Noble)

A08-2530
Society of Manufacturing Engineers. Surface Preparation and Finishes for Metals. McGraw-Hill, 1971. 544 pp. $21.50
Here is a comprehensive guide to the many processes currently being used for cleaning and finishing metals, giving the advantages and disadvantages of the coatings available which can be applied to the clean metal.

A08-2531
Spring, S. Preparation of Metals for Painting. Van Nostrand Reinhold, 1965. 318 pp. $13.50
Covers pickling and removal of oxides from ferrous and nonferrous metals, abrasive removal of oxide, paint-bond treatments, acidic surface conditioning, and coatings for steel, aluminum and other metals.

A08-2532
Steel Structures Painting Council. Steel Structures Painting Manual. Vol. 1, Good Painting Practice. Steel Structures Painting Council, 1972. 432 pp. $15.00
A summarization of good practice in the painting of steel structures.

A08-2533
Steel Structures Painting Council. Steel Structures Painting Manual. Vol. 2, Systems and Specifications. Steel Structures Painting Council, 1971. 350 pp. $19.00 (vols. 1 and 2, set $30.00)
Complete guides and specifications of recommended painting systems for most steel structures in most exposures. Guides and specifications on surface preparation, paint application, paint thickness measurement, paints and paint systems.

A08-2534
Strakosch, G. R. Vertical Transportation: Elevators and Escalators. John Wiley, 1967. 365 pp. $17.25
A view of the relation of elevators and escalators to people and to buildings. A complete appraisal of all the factors related to vertical transportation.

A08-2535
[No entry]

A08-2536
Svec, J. J. and Jeffers, P. E. Modern Masonry Panel Construction Systems. Cahners, 1972. 130 pp. $12.50
In this collection of articles originally published in Brick and Clay Record, the authors describe the various methods of panel construction and building systems with panels. Also included are scores of ideas used in many parts of the world, as well as a description of the various sizes and composition of brick panels and methods to reduce construction time and costs. Many of the panel systems are available for licensing.

Building Materials, Components, Finishes

A08-2537
Swenson, E. G. Performance of Concrete. Univ. of Toronto Press, 1968. 280 pp. $10.00
 This book is a compilation of papers at a symposium held in honor of the Canadian scientist, the late Dr. T. Thorvaldson. It is a memorial in recognition of his contribution to the development of sulphate-resistant cement and concrete.

A08-2538
Taylor, W. H. Concrete Technology and Practice. American Elsevier, 1969. 697 pp. $15.00
 This handbook details all aspects of concrete production and its behavior in structures. Design of concrete mixes, formwork, concrete erosion and concrete construction are covered. Air entrained, lightweight aggregate, and polymer concretes are also described as are the various means of concrete placement.

A08-2539
Terrington, J. S. and Turner, F. H. Design of Non-Planer Roofs. Cement and Concrete Assn., 1964. 108 pp. £0.75

A08-2540
Traxler, R. N. Asphalt: Its Composition, Properties and Uses. Van Nostrand Reinhold, 1961. 300 pp. $12.50
 The chemical composition and physical properties of asphalt are covered, as well as its durability and manufacture. The uses of asphalt in paving, roofing and in hydraulics are also topics.

A08-2541
Tretyakov, A. Concrete and Concreting. Beekman, 1976. 312 pp. $11.00
 This book is translated from the Russian language. Gives the methods of proportioning and mixing concrete, considers the methods of curing, and discusses the production of reinforced concrete structures.

A08-2542
Troxell, G., et al. Composition and Properties of Concrete. McGraw-Hill, 1968. 513 pp. $17.50
 Studies the uses of ordinary concrete, problems involving special concretes, new materials and types of mechanical equipment, and the chemistry of cement.

A08-2543
Tsai, S. W., et al., eds. Composite Materials Workshop. Technomic, 1968. 345 pp. $5.00
 Thirteen authorities present the physical aspects of high-performance composites. Topics include mechanics, fracture behavior, physical and mathematical theories, design, and synthesis.

A08-2544
Uhlig, H. H. Corrosion and Corrosion Control: An Introduction to Corrosion Science and Engineering. John Wiley, 1971.

419 pp. $15.95
Systematic description and basic presentation of the electrochemical principles underlying corrosion reactions. Discusses the scientific principles upon which control of corrosion is based. Cites practical and engineering applications illustrating how the basic science is employed usefully.

A08-2545
Uhlig, H. H., ed. Corrosion Handbook. John Wiley, 1948. 1188 pp. $25.25

A08-2546
UNESCO. Reinforced Concrete: An International Manual. Butterworth, 1971. 430 pp. £9.00
This text is a combination of code of practice and design manual. It will become a standard text for libraries and those consultants with interests in the developing countries of the world.

A08-2547
van Amerongen, C. Dictionary of Cement: Manufacture and Technology. (German-English/English-German.) Adler's Foreign Books, 202 pp. $20.50
A dictionary solely devoted to cement manufacture and technology. It contains terms from mechanical and electrical engineering, chemistry, physics, etc. Terms relating to the pit and quarry industry, excavation and haulage, mechanical handling, crushing and grinding automation, fuel technology and heat engineering, refractories, chemical and physical testing are all included.

A08-2548
Van Vlack, L. H. A Textbook of Materials Technology. Addison-Wesley, 1973. $14.50
Designed to introduce materials to the student who does not have a rigorous background in science, this text at the same time retains the problem-solving approach of the technologist and engineer. To provide a broad approach to the technology of materials, he examines processing principles in addition to the structure and properties of metals, polymers, and ceramics, which are studied in sequence.

A08-2549
Volkart, K. Gypsum and Plaster Dictionary: German-English-French. International Publications, 1971. 192 pp. $30.00
Contains a total of 1,452 terms. Intended for the gypsum and gypsum board industry, the gypsum processer and others who deal with these products. French and English indexes are numerically keyed to the main part.

A08-2550
Waddell, J. Concrete Construction Handbook. McGraw-Hill, 1974. 978 pp. $32.50
The specifics of mixing, pouring, and curing of specification-quality concrete are clearly outlined in this new edition of the

Building Materials, Components, Finishes

practical guide for field inspectors, construction superintendents, field engineers, and contractors. In addition to serving as a quality control manual, this revision includes new data on Portland cement, admixtures, steel, testing and inspection, pumping, fusion and curing, and lift slabs. Among the recent advances covered are those in geophysical and aerial surveying methods, concrete toughness and creep studies, use of plastics in formwork, aggregate beneficiation, etc.

A08-2551
Waddell, J. Practical Quality Control for Concrete. McGraw-Hill, 1962. 396 pp. $16.50
   Emphasizes the prevention and cure of defects, and offers guidance on the processing, classification, stockpiling, and inspection of concrete materials.

A08-2552
Walley, F. and Bate, S. C. C. A guide to the B.S. Code of Practice for Prestressed Concrete CP 115. Cement and Concrete Assn., 1961. 110 pp. £2.00

A08-2553
Watson, D. Construction Materials and Processes. McGraw-Hill, 1972. 512 pp. $13.95
   This survey covers basic materials, assemblies of materials, and trade practices in heavy and light construction. It includes the latest developments in the field, emphasizing assemblies of materials and installation of prefabricated units. Special topics are high rise masonry-bearing wall construction, vacuum processing of steel, and illumination and light distribution curves. Material is grouped according to the CSI format for Building Specifications and the Uniform Filing System.

A08-2554
The Weathering and Performance of Building Materials. The Construction Press. £4.95
   A detailed analysis of how the major building materials behave in use, this important book is the first which adequately describes the weathering and performance of the five most commonly encountered building materials. The editors and authors have collected together in one volume all the relevant information on concrete, clay products, timber, metal and plastics, and have added much previously unpublished information.

A08-2555
Whitehurst, E. A. Evaluation of Concrete Properties from Sonic Tests. American Concrete Institute, 1966. 94 pp. $4.50 (for ACI members, $3.00)
   This monograph traces the history and development of the two general methods of sonic testing: Resonant Frequency Techniques and Pulse Velocity Techniques. Use of both types of test to calculate "dynamic" properties is explained and the degree of correlation of these properties with conventional test results is discussed.

A08-2556
Whittington, L. R. Whittington's Dictionary of Plastics. Technomic, 1968. 261 pp. $10.00

Authoritative scientific, commercial, and legal definitions of over 3,100 terms basic to plastic technology.

A08-2557
Wilson, C. L. and Oates, J. A. Corrosion and the Maintenance Engineer. Hart Publishing Co., 196 pp. $15.00

This book endeavors to present and identify many causes of corrosion and how such corrosion can be eliminated. The practical and economic aspects of the subject are discussed.

A08-2558
Wilson, J. G. Exposed Concrete Finishes. Vol. 2, Finishes to Precast Concrete. John Wiley, 1964. 170 pp. $9.95

A08-2559
Woods, H. Durability in Concrete Construction. American Concrete Institute, 1968. 190 pp. $6.50 (for ACI members, $5.00)

Published jointly by ACI and the Iowa State University Press. Contains references to studies of practical means for achieving durability and presents the results of extensive research. Prepared especially for engineers and those who are interested in making or specifying durable concrete and who realize the need for information on possible deterioration under various conditions.

A08-2560
Zakar, P. Asphalt. Chemical Publishing Co., 1971. 212 pp. $13.50

The book is divided into three parts, respectively: Production and Properties of Asphalt; Manufacture of Asphalt-distilled, extracted, blown and cracked; Uses of Asphalt, road construction, roofing, building etc.

Section   A09

CONSTRUCTION, FABRICATION, INSTALLATION

        Construction Equipment
        Excavation
        House Construction
        Renovation and Rehabilitation

A09-2580
ACI. Concrete Construction. American Concrete Institute, 1968.
220 pp. $15.00 (available from University Microfilms)
A collection of some of the top construction articles which have appeared in the ACI Journal since 1952. Especially prepared to give the builder an assortment of the best in concrete construction. As a supplement, three often referred to ACI Standards on formwork, shotcreting, and cold weather concreting are included.

A09-2581
ACI. Symposium on Concrete Construction in Aqueous Environments. American Concrete Institute, 1964. 116 pp.
$10.00 (available from University Microfilms)
A collection of 10 papers on various aspects of the construction of concrete waterfront structures, concrete waterholding structures, and concrete construction in aqueous environments. Covers tremie concrete, prestressed tanks, concrete piles, joints and cracks, concrete pontoons, effect of sea water, and pier design.

A09-2582
ACI Manual of Concrete Inspection. American Concrete Institute, 1967. 270 pp. $5.00 (for ACI members, $3.00)
Fifth edition provides information on settlement of concrete, proportioning of lightweight concrete, determination of yield, shoring and formwork, strength requirements, cold weather concreting, and shotcrete.

A09-2583
Alerich, W. N. Electrical Construction Wiring. American Technical Society, 476 pp. $7.75 (Study Guide $2.60)
The author covers the recent changes and developments that have modified materials and procedures and the changes in the National Electrical Code that have caused many of the previous methods to become obsolete. This text includes the latest accepted changes in wiring, plus the traditional wiring procedures that are still recognized by the NEC.

A09-2584
Althouse, A. D., et al. Modern Welding. Goodheart-Wilcox, 1970. 712 pp. $10.96
Designed to help students, apprentices, adult workers become competent welders. Arranged so student may start welding experiences with any process. Covers properties of metals, metal identification, heat treatment, metallurgy, welding tools, equipment,

Construction, Fabrication, Installation

welder certification requirements. Each chapter contains a section on welding safety. Review questions cover important points in chapter.

A09-2585
Ambrose, J. E. Building Structures Primer. John Wiley, 1967. 123 pp. $9.50
A non-mathematical, non-technical presentation of the basic principles of structural behavior and the vocabulary of contemporary building structural materials and systems.

A09-2586
American Concrete Pipe Assn. Concrete Pipe Installation Manual. American Concrete Pipe Association, 128 pp. $3.75
The Concrete Pipe Installation Manual is a manual for engineers, inspectors, contractors and all personnel associated with the design or construction of sewers and culverts. This manual presents a guide for the proper installation of concrete pipe and discusses manufacturing methods, material specifications, test requirements and design principles. Numerous explanatory illustrations are included to emphasize all phases of pipe installation so that design criteria regarding specified pipe strength, type of installation, excavation and dimensional limitations and bedding and backfilling requirements are realized in actual construction.

A09-2587
American Institute of Architects. Manual of Built-up Roofing Systems. McGraw-Hill, 1970. 256 pp. $17.50
Presents information on the design and construction of built-up roofing systems, with the focus on those roof types most commonly used.

A09-2588
American Institute of Timber Construction. Timber Construction Manual. John Wiley, 1974. 799 pp. $16.50
This manual, prepared by the American Institute of Timber Construction, covers the physical and mechanical properties of wood, design loads, structural connections, detailing and erection. For use by architects, contractors, laminators and fabricators concerned with engineered timber buildings.

A09-2589
American Society of Civil Engrs. Underground Rock Chambers. American Society of Civil Engineers, 1971. 616 pp. $14.00 (for ASCE members, $7.00)
Contains the 16 papers presented at the ASCE Meeting in Phoenix, Arizona in January 1971. This volume provides a ready reference to new developments and current practice in the design and construction of underground powerhouses and similar rock chambers.

A09-2590
Anderson, L. How to Build a Wood-Frame House. Dover, 1973. 223 pp. $3.00

A guide to building a wood frame house, using established methods. Deals with everything from selecting a site to ventilation, protection against fire and methods of holding down building costs.

A09-2591
Anderson, L. O.   Wood Frame House Construction.   U. S. Gov't. Printing Office, 1970.   223 pp. $2.60

This publication presents sound principles for wood frame house construction and suggestions for suitable materials to insure a well-constructed house. It sets forth acceptable practices in assembling and arranging the parts of a well-designed wood frame house. The book is arranged in chronological order of house construction, covering everything from site location and excavation to home maintenance and repair.

A09-2592
Anderson, L. O.   Wood Frame House Construction.   Craftsman Book Co., 1971.   232 pp. $2.75

A source book and guide to sound wood-frame house construction. Includes material on framing, exterior trim, roofing, windows and doors, siding, plumbing and heating, insulation and vapor barrier, finishes, floor covering, millwork, cabinets, chimneys and fireplace, driveways and walks.

A09-2593
Army Department.   Carpenter, U. S. Army Technical Manual. TM 5-551B.   U.S. Gov't. Printing Office, 1973.   196 pp. $2.40.   Order No. D 101.11:5-551B S/N 0820-00487

While written for the training of the Army carpenter, this manual is a valuable source of information for anyone interested in carpentry and wood working. Information, supplemented by hundreds of illustrations, is provided on construction techniques, building layout, forming for concrete, frame carpentry, roofing, and many other related subjects.

A09-2594
ASTM.   Performance Monitoring for Geotechnical Construction. American Society for Testing and Materials, 1975.   204 pp. $14.00 (for ASTM members, $11.20)

Eleven papers give case history of monitoring excavations. Civil engineers, soil and foundation engineers will find this a valuable addition to their library.

A09-2595
ASTM.   Sampling of Soil and Rock.   American Soc. for Testing and Materials, 1971.   198 pp. $8.00 (for ASTM members, $6.40)

Thirteen papers deal with analytical methods and testing techniques essential for site evaluation. Emphasis on appropriate sampling.

Construction, Fabrication, Installation

A09-2596
ASTM. Special Procedures for Testing Soil and Rock for Engineering Purposes. American Soc. for Testing and Materials, 1970. 630 pp. $15.75 (for ASTM members, $15.75)
  Contains 64 suggested methods and numerous references to standard methods for testing soil and rock. Provides a comprehensive picture of current practice.

A09-2597
ASTM. Underwater Soil Sampling, Testing, and Construction Control. American Soc. for Testing and Materials, 1972. 240 pp. $15.50 (for ASTM members, $12.40)
  This book summarizes techniques for soil sampling and testing in deep water and describes procedures used to control underwater construction. Contains 13 pages including a state-of-the-art review.

A09-2598
Auslander, L. Domestic Oil Burners and Oil Heat. Holt, Rinehart and Winston, 1958. 400 pp.
  This manual covers installation and repair of hot water, warm air, and steam systems. Subject matter includes types of heating systems, principles of combustion, types of burners, controls, chimneys, wiring, troubleshooting, and heat loss surveys.

A09-2599
AWS. Brazing Manual. American Welding Society, 1963. 290 pp. $7.50 (for AWS members, $5.62)
  First published in 1955, this manual is updated to provide coverage for today's methods and materials.

A09-2600
AWS. Soldering Manual. American Welding Society, 1959. 180 pp. $6.50 (for AWS members, $4.50)
  This manual gives comprehensive data on all phases of soldering.

A09-2601
Badzinski, S., Jr. Carpentry in Residential Construction. Prentice-Hall or R. S. Means, 1972. 308 pp. $17.25
  Provides carpenter apprentices and architectural or construction technology students with an up-to-date discussion of carpentry in home and apartment building, written from the tradesman's viewpoint. Includes chapters on construction for forms for concrete footings, walls, columns, beams, floor slabs and stairways. Also included are important chapters on storefront construction and finishing; and movable partitions, cabinet and fixture work.

A09-2602
Badzinski, S., Jr. Light Frame House Construction. Prentice-Hall, 1975. 256 pp. $10.50
  For the home owner, the prospective home buyer, or the real estate salesperson, here is a book that explains the basics of

home and apartment construction. Among other topics, this book describes and illustrates: foundation work; floor, wall, and roof construction; plumbing, heating and electrical facilities; finish floors, interior trim, and hardware; and more.

A09-2603
Barber, G. Builders' Plant and Equipment. Butterworth, 1973. 216 pp. £4.40 (paperback £2.80)
The second edition of Builders' Plant and Equipment has been revised and brought up to date with much new material covering management and safety, plant legislation and plant output. SI units are used throughout, with imperial units where necessary.

A09-2604
Barry, R. Construction of Buildings. 4 vols. Beekman, 1969-71. 508 pp. $30.00 set
Presents information on everything necessary to construct a building. Collection includes: foundations, masonry, lattice truss construction, steel frames, fireplaces and flues, insulation, waterproofing, internal finishes, and external renderings.

A09-2605
Benjamin, A. The American Builder's Companion. Dover, 1827. 114 pp. $3.00
The most widely used early 19th-century architectural style and source book, for colonial up into Greek Revival periods. Extensive development of geometry of carpentering, construction of sashes, frames, doors, stairs; plans and elevations of domestic and other buildings. Hundreds of thousands of houses were built according to this book, now invaluable to historians, architects, restorers, etc.

A09-2606
Berne, J. How to Make Money in the Remodeling Business. Cahners, 1964. 165 pp. $9.95
Provides a management training course for dealers who want to go into the field or who want to increase their remodeling business and make it more profitable.

A09-2607
Blake, L. S., ed. Civil Engineer's Reference Book. Butterworth. (In preparation)
This edition has been revised to bring together in one volume all aspects of civil engineering. The 42 sections have been contributed by engineers who are specialists in their particular field. The information contained in this volume in addition to providing engineers with background theory and theoretical data, covers present-day practices in all aspects of construction and management.

A09-2608
Bolz, H. A. and Hagemann, G. E., eds. Materials Handling Handbook. Ronald Press, 1958. 1740 pp. $22.50
This book presents principles, procedures, methods, engineering data, and examples from the operating experience of broad

segments of industry, and applicable to virtually all industrial situations. Efficient materials handling methods, procedures, and systems that have led to significant reductions in both manufacturing and distribution costs are explained. Analyzes the engineering specifications and compares the operating capabilities of each type of equipment.

A09-2609
Boudreau, E. H. Making the Adobe Brick. Fifth Street Press, 1971. 87 pp. $2.95
The book acquaints the reader with all information necessary to make adobe bricks that conform to the Uniform Building Code. The author made 7000 bricks and built a nine room home with them. It takes the reader through the construction stages, from testing the brick, foundation, walls, and roof to plumbing and electrical wiring.

A09-2610
Bouwcentrum, ed. Modern Steel Construction in Europe. North-Holland, 1963. 279 pp. $21.80
Though steel elements find many applications in the building trades, recognition of the decisive role which steel must come to play in the building industry is still largely wanting. This volume illustrates the potentialities of constructional steel by reference to existing applications in a number of European countries, to recent trend reports, and to details of model developments in the use of steel.

A09-2611
Bradzinski, S. Stair Layout (Design and Building). American Technical Society. $2.75
By applying the information outlined in this text, it is possible to design and build straight stairs, stairs with landings, and winding stairs--all with proper headroom and unit rise-to-unit-run ratio. Design, layout, and building are each discussed to serve high-school level students, apprentices, and builders. A method is given for laying out, cutting, and installing stair carriages, cleated or dadoed stringers, housed stringers, and mitered stringers--all under widely varying requirements.

A09-2612
Branden, V. D. and Hartsell, T. L. Plastering Skill and Practice. American Technical Society, 543 pp. $10.75 (Study Guide $2.25)
This text provides a practical groundwork for apprentices at all levels. For journeymen, it provides a coverage of the entire subject, either as a refresher or reference text. Photographs and illustrations show proven means and methods. The text includes 40 pages on math for plasterers and a list of 140 sources for tools, equipment, and materials.

A09-2613
Brighty, S. G. Setting Out: Guide for Site Engineers. Beekman,

1975.   264 pp. $17.50
Practical handbook to teach the skills of translating plans and drawings into marks, pegs, profiles, and other devices which construction teams use to carry out and control their work.

A09-2614
Brooks, H.   Illustrated Encyclopedic Dictionary of Building and Construction Terms.   Prentice-Hall, 1975.   320 pp. $22.95
More than 2,400 terms are explained and examples given of their application. Terms are indexed by function in 23 subject areas including professional services construction management, real estate, financing, building codes, architectural and structural parts of a building, foundations, concrete work, masonry, HVAC, electrical work, etc.

A09-2615
Bucksch, H.   Dictionary of Civil Engineering and Construction Machinery and Equipment.   Adler's Foreign Books.   Vol. 1, German-English, 1971.   1184 pp. $70.00; Vol. 2, English German, 1971.   1220 pp. $70.00
A dictionary of civil engineering, construction machinery, construction equipment and related fields. The author has refrained from a mere repetition of general expressions in order to concentrate more on terms and definitions, which are difficult to trace.

A09-2616
Builder's Encyclopedia.   Howard W. Sams.   608 pp. $7.95
This book will help bridge the gap existing between the various skilled and unskilled workers in the construction trades. Inexperienced home builders will find it useful in communicating with plumbers and electricians. Many terms common to one trade but meaningless to others are explained.

A09-2617
The Building Research Establishment.   Essential Data for Construction.   Vol. 1, Building Construction.   Cahners, 1974.   250 pp. $17.50
Each volume, taken from the monthly publication of the Building Research Establishment, has over 40 digests. Volume 1 contains information on foundations, floors, walls and windows, roofs, joints, site control, and domestic dwellings.

A09-2618
The Building Research Establishment.   Essential Data for Construction.   Vol. 2, Building Materials.   Cahners, 1974.   250 pp. $17.50
Volume 2 contains information on concrete, brickwork, timber, metals, plastics, plaster and design and appearance.

A09-2619
The Building Research Establishment.   Essential Data for

Construction, Fabrication, Installation

Construction. Vol. 3, Services and Environmental Engineering. Cahners, 1974. 250 pp. $17.50
Volume 3 contains information on piped services, heating, ventilating and air conditioning, acoustics, vibration and noise control, lighting, electrical and external environment.

A09-2620
The Building Research Establishment. Essential Data for Construction. Vol. 4, Building Defects and Maintenance. Cahners, 1974. 240 pp. $17.50
Volume 4 contains information on design and appearance, foundations and walls, floors, roofs and joinery, painted surfaces and services.

A09-2621
Building with Tilt-Up. 2nd ed. Know-How Publications, 1958. 160 pp. $10.00
This book gives the necessary technique to construct tilt-up buildings with economical methods that will produce structures with high structural integrity and low yearly maintenance costs. It is not confined just to panel fabrication, but includes such subjects as soil and fill, floors and walls, panel footings, cold weather concrete, pick-up, erection, hot weather concrete, and prestressing.

A09-2622
Burgess, R. A., et al., eds. The Construction Industry Handbook. Cahners, 1971. 464 pp. $22.50
An account of recent developments in every important aspect of the construction industry. Includes a guide to properties of building materials and a selection of environmental design data on lighting, heating, sound insulation and acoustics. A reference for architects, contractors, building technologists, surveyors, engineers, students and teachers.

A09-2623
Burgess, R. A., et al., eds. Progress in Construction Science and Technology. Barnes & Noble. Vol. 1, 1971. 322 pp. $18.50; Vol. 2, 1973. 223 pp. $18.50
Collection of research reports by British authors on recent developments in the construction industry. Is of interest to architects, contractors and structural designers.

A09-2624
Burkhardt, C. Domestic and Commercial Oil Burners. McGraw-Hill, 1969. 538 pp. $12.50
A practical aid for technicians who must install and service automatic oil heating equipment and its components, this book deals with all types of oil burners and their accessories; installation procedures and control systems; service and maintenance. New material covers central piping, systems, modern combustion chambers, water level and feed controls, new types of oil fired burners, and recent advances in control mechanisms.

A09-2625
Bush, V. G. Handbook for the Construction Superintendent. R. S. Means, 1973. $13.75

The superintendent is led through each phase of the project, whatever it may be, by some theory and many suggestions for managing the work and anticipating the problems. Emphasis is placed on practices and procedures that are not found in textbooks or manuals but that are usually learned by experience. This handbook can aid the superintendent in managing labor, in communicating effectively with staff, architect, and owner, and in dealing with the public. It covers all the responsibilities and relationships of the modern, efficient superintendent.

A09-2626
Campbell, M. D. and Lehr, J. H. Water Well Technology: Field Principles of Exploration and Drilling for Ground Water. McGraw-Hill, 1973. 600 pp. $27.50

A definitive reference, this work reviews the current well construction methods and techniques presently used by the petroleum, mining and ground water industries. Since the drilling for ground water and the drilling for minerals are both directly interconnected, they should be treated together. This merger of ground water and minerals exploration practices will reduce expensive duplication of effort by industry and government.

A09-2627
Carpenters and Builders Library. Vol. 1, Tools, Steel Square, Joinery. Howard W. Sams. 464 pp. $4.95

Contents include fasteners, and marking, measuring, holding and cutting tools in addition to tools for facing and boring. Cabinetmaking and kitchen cabinet construction are illustrated.

A09-2628
Carpenters and Builders Library. Vol. 2, Builders Math, Plans, Specifications. Howard W. Sams. 336 pp. $4.95

Contents include mathematics for carpenters and builders; surveying; strength of timbers; practical drawing; how to read plans; architectural drawing; specifications; barn construction; small house construction; motels and home workshop layout.

A09-2629
Carpenters and Builders Library. Vol. 3, Layouts, Foundations, Framing. Howard W. Sams. 320 pp. $4.95

Contents include laying out; foundations; concrete forms; concrete block construction; framing; girders and sills; skylights; porches and patios; chimneys and fireplaces; insulation of building; scaffolding and staging; hoisting apparatus.

A09-2630
Carpenters and Builders Library. Vol. 4, Millwork, Power Tools, Painting. Howard W. Sams. 368 pp. $4.95

Contents include roofing; cornice construction; miter work; doors; windows; sheathing and siding; stairs; flooring; walls and ceiling; millwork; lathes; planers, jointers, and shapers; mortisers and tenoners; standing; boring; power operated hand tools; termite protection; painting; maintenance and repair; and wood.

A09-2631
Carpentry and Building. Howard W. Sams. 448 pp. $5.95
Deals with problems in the residential or light construction field. Answers questions most frequently asked about layout, foundations, framing, ceilings and walls, insulation and moisture, condensation, acoustics, concrete, masonry, painting and finishing, roofing, floors, plumbing, and windows.

A09-2632
Carson, A. Foundation Construction. McGraw-Hill, 1965. 407 pp. $22.50
A practical, illustrated survey of underground and underwater construction methods. Provides a background in the preliminaries of any job.

A09-2633
Carson, A. General Excavation Methods. McGraw-Hill, 1961. 392 pp. $20.00
Examines procedures and equipment used in excavation work, for those involved in earth and rock excavation, ground water control, and bank stabilization.

A09-2634
Cassiday, B. Vacation Houses. Dodd-Mead, 1973. 214 pp. $6.95
A guide to choosing and building the right vacation house at the right price. Comparative costs, photographs and floor plans are presented for each house discussed. Included are cabins that can be built by simply putting prepared logs together; plywood houses designed for a minimum of cutting and shaping by the amateur carpenter; prefabricated houses that come with sections prebuilt for immediate assembly; predesigned vacation houses that can be purchased in a package; and custom-designed homes to be built by a professional contractor.

A09-2635
Chandler, F. J. Structures for Building Technicians. Beekman, 1975. 74 pp. $12.50
Covers timber, steel, and concrete structural design. Each chapter ends with a brief learning program to help the student consolidate information.

A09-2636
Christian, M. Negro Ironworkers in Louisiana. Pelican, 1972. 61 pp. $4.50 (paperback $2.50)
This documented work outlines the notable contributions by Negro craftsmen to the distinctive ironwork of New Orleans, as well as innovations in ironwork throughout the state. The author also furnishes an interesting refutation to the "Toledo Sources."

A09-2637
Clarke, F. W. Installing Small Pipe Central Heating. Applied Science, 1967. 117 pp. £2.00
   Small pipe central heating is a young and rapidly expanding industry. Many of the principles and methods used are new and although special equipment and controls have been designed, there is a great need for competent men to install this form of central heating. This book with its coverage of available heating systems for various fuels, will act as a guide.

A09-2638
Colby, J. P. Building Wrecking: The How and Why of an Important Business. Architectural Book Publishing, 1973. 96 pp. $5.95
   Jean Colby explains when and why certain techniques and machinery are used, how risks are minimized and how demolition relates to the environment.

A09-2639
Collins, F. T. Manual of Tilt-Up Construction. 2nd ed. Know-How Publications, 1958. 148 pp. $10.00
   This sixth edition has been updated. The chapter on panel fabrication is revised to include late developments on erection and bracing. Also new is an outline of needs for a new look at tilt-up joinery. New information on drying shrinkage and sandwich wall panels completes the revision. Other subjects include bond breaking, painting, foundations, exposed aggregate, sandwich panels and critical path techniques.

A09-2640
Collymore, P. Altering, Repairing and Extending Houses. Architectural Press, 1974. £4.00
   The majority of smaller scale works undertaken by architects and builders are concerned with various aspects of house conversion. This handbook fills a need for a guide of what is involved.

A09-2641
A Complete Guide to Demolition. Construction Press. £6.75
   As building technology has changed over the past two decades, so the techniques for demolishing today's buildings will, in due course, also have to change. This book looks at the different ways of pulling down existing structures and sets out recommended courses of action that should be taken in the case of most building types existing today, including the various types of stressed and tensioned concrete buildings.

A09-2642
Crimmins, R., et al. Construction Rock Work Guide. John Wiley, 1972. 241 pp. $16.75
   Provides a reference book of practical information and illustrations on the business of constructing foundations in rock. The material is organized to help the contractor and engineer recognize the obligations of their agreement.

Construction, Fabrication, Installation   435

A09-2643
Cutler, L. and Cutler, S. Handbook of Systems Housing for Designers and Developers. Van Nostrand Reinhold, 1973. 160 pp. $17.50
   Here is a view of industrial buildings in which the living units, when joined, offer themselves as total units. These buildings can be composed in an almost infinite series of arrangements, as compared with other building systems, which offer precut units with a limited number of unit types of models. This book shows how to take advantage of these situations to rationalize construction and speed up design while satisfying the user's needs. The book also covers areas such as spatial alternatives, unit types and building types.

A09-2644
Dahl, A. and Wilson, J. D. Cabinetmaking and Millwork: Tools, Materials, Layout, Construction. American Technical Society, 352 pp. $7.15 (Study Guide $1.40)
   This book covers construction and layout methods for cabinets, sashes, windows, doors, interior trim, exterior woodwork, and stairway building.

A09-2645
Dalzell, J. R. Repairing and Remodeling Guide for Home Interiors: Planning, Materials, Methods. McGraw-Hill, 1973. 216 pp. $8.95
   This edition makes numerous changes from the first edition because of differences in homebuilding and remodeling practice. Because of the increasing demand by homeowners for controlled indoor environment, chapters have been added on electrical outlets and lighting, and on heating and airconditioning. The new edition reflects the changes in standard sizes of materials and describes new interior finish materials.

A09-2646
Dalzell, J. R. Simplified Concrete Masonry Planning and Building. McGraw-Hill, 1972. 384 pp. $9.95
   This up-dated edition of a book for novice masons has been completely revised--with new illustrations. The book serves as an ideal guide for all those who need to know basic principles involved in building common types of concrete, concrete block, stucco, and similar structures. It takes even the most inexperienced reader step by step through every type of concrete job in home building.

A09-2647
Dalzell, J. R. and Townsend, G. Concrete Block Construction for Home and Farm. American Technical Society. 228 pp. $5.50
   To conform with the latest designations, codes, and standards, this text has been revised and updated. For students, apprentices, tradesmen, or foremen, it gives fundamentals, plus information on pre-fab masonry panels and modular components.

Other features include details of mortar and concrete mixing, safety ways and means, and metric conversion.

A09-2648
Dalzell, J. R. and Townsend, G. Masonry Simplified. Vol. 1, Tools, Materials, Practice. 3rd ed. American Technical Society, 1974. 408 pp. $8.50 (Study Guide $1.85)

This revised edition provides a learning sequence in bricklaying and cement trades. It is a useful learning tool for the vocational student or the apprentice as well as a valuable reference work for the tradesman. It covers the various changes in tools and materials and explains how these changes affect working methods.

A09-2649
Dalzell, J. R., et al. Masonry Simplified. Vol. 2, Practical Construction. 2nd ed. American Technical Society, 1957. 438 pp. $6.95 (Study Guide $1.50)

A comprehensive study of masonry construction. Emphasizes the techniques underlying the design of a structure and the factors controlling good construction.

A09-2650
Daniels, G. Home Guide to Plumbing, Heating and Air Conditioning. Popular Science Publishing Co., 1967. 186 pp. $4.95 (paperback $3.50)

Principles, tools, techniques, trouble shooting for each.

A09-2651
Day, D. A. Construction Equipment Guide. John Wiley or R. S. Means, 1973. 563 pp. $30.25

Contents include choosing construction equipment, costs and control, compressors and pumps, earthwork equipment, trenching and dredging, power excavators and cranes, foundation and excavation equipment, material handling, concreting equipment, compactors and paving machinery.

A09-2652
de Sousa Ferreira, K. Glossary of Building and Bridge Construction. Adler's Foreign Books. 216 pp.

The book is a collection of German and English-American technical terms as used in the building and construction industries. The usual listing of equivalent terms in different languages is generally replaced by an arrangement where each term is given in its context.

A09-2653
Diamant, R. M. E. Industrialised Building. Butterworth. Vol. 1, 50 International Methods. 1964. 214 pp. £5.00; Vol. 2, 50 International Methods. 1965. 208 pp. £5.00; Vol. 3, 70 International Methods. 1968. 216 pp. £5.00

These volumes describe industrial building techniques selected from Great Britain, Scandinavia, France, Holland, the U.S.A.

Construction, Fabrication, Installation 437

West Germany, Eastern Europe and Italy. Materials are discussed as are methods used in the construction of houses and small blocks of flats. Constructional data, and costs where available, allow one method to be compared with another.

A09-2654
Dietz, A. G. H. Dwelling House Construction. Van Nostrand Reinhold, 1954. 396 pp. $7.50
Details established methods of construction and the assembly of a house from the first inspection of a site to the last coat of paint. Information on panelized construction of houses includes developments adapted from standard details, housing based upon standard dimensions, stressed-skin panels and sandwich panels.

A09-2655
Douglas, J. Construction Equipment Policy. McGraw-Hill, 1975. 320 pp. $16.50
This book offers an integrated understanding of the uses of computers, the modeling of economic resources and machines, and engineering technology.

A09-2656
Drake, G. The Complete Handbook of Power Tools. Prentice-Hall, 1976. 544 pp. $13.95
For the home handyperson or the professional, this book tells all you need to know about bench powered tools--what the tools are, how to use them, accessories, and how to care for the tools. The author covers all types of hand power tools. Useful as a handy shop reference, this book also describes workshop floor plans, lighting, shop and tool safety, workbenches, tool stands, storage, and more.

A09-2657
Dunham, C. W. Foundations of Structures. McGraw-Hill, 1962. 722 pp. $16.50
Covers soils as foundation materials, exploration of site, principles of foundation action, spread footings, foundation walls, mats, foundations subjected to overturning forces, piles, pile foundations, cofferdams and other aids for open excavations, caissons, bridge piers, bridge abutments, underpinning, and machinery foundations.

A09-2658
Durbahn, W., et al. Fundamentals of Carpentry. Vol. 1, Tools, Materials, Practices. American Technical Society, 416 pp. $6.00 (Study Guide $1.65)
Includes: new information on tools, hardware, fasteners, adhesives, and building materials. Also covers safety and accident prevention.

A09-2659
Eastwick-Field, J. and Stillman, J. Design & Practice of Joinery. Herman Publishing, Inc., 1972. $17.50

The beauty of timber for finishings, inside and out, is enjoying a new appreciation. For architects, students, all who use wood. Covers construction, detailing, data for achieving economy.

A09-2660
EPA. Processes, Procedures, and Methods to Control Pollution Resulting from all Construction Activity. U.S. Environmental Protection Agency, 1975. 234 pp. $2.30. Order No. EPA 430/9-73-007

This book examines the methods of controlling erosion, sediment and pollution that results from construction activities. Useful for site planning and evaluation.

A09-2661
Equipment Guide-Book. Compaction Equipment. Equipment Guide-Book Co., 1976. 900 pp. $150.00 annual subscription

Specifications are given for the compaction equipment of 46 manufacturers. Includes vibratory, pneumatic-tired, steel wheel, sheepsfoot, segmented and grid type rollers both towed and self-propelled.

A09-2662
Equipment Guide-Book. Construction Equipment Cost Reference Guide. Equipment Guide-Book Co., 1976. 750 pp. $85.00 annual subscription

A handbook to help estimate construction equipment ownership and operating costs for more than 6,000 items. Each listing describes make, model or size for identification, and shows basic data such as capacity, weight, horsepower, and ship option tons. Ownership costs per day, overhauling and costs per hour and operating costs per hour are broken out in detail. Some factors are presented as daily charges, others as hourly, and some as lot information. All figures are based on current list prices, or offer a reasonable operating cost schedule where no actual job data is available.

A09-2663
Equipment Guide-Book. Crawler Cranes. Equipment Guide-Book Co., 1976. Vol. 1, 320 pp. $75.00 annual subscription; Vol. 2, 1050 pp. $125.00 annual subscription; Vol. 3, 1000 pp. $125.00 annual subscription

Published in 3 volumes according to size of bucket. Vol. 1, currently 7 manufacturers, 3/8 to 3/4 cu. yd. capacity. Volume 2, currently 10 manufacturers, 7/8 to 2 1/2 cu. yd. capacity, Vol. 3, currently 9 manufacturers, 2 3/4 cu. yd. capacity and larger. Current data on engines, working weights, crawler length and shoe sizes, lifting capacities, booms, jibs, line pulls and speeds, bucket options, etc.

A09-2664
Equipment Guide-Book. Electric Lift Trucks. Equipment Guide-Book Co., 1976. 1600 pp. $150.00 annual subscription

Specifications are given for electric lift trucks of currently

Construction, Fabrication, Installation 439

27 manufacturers, 1600 pages capacity, lifting height, dimensions, drive and pump motors, batteries and chargers for walkies, riders, narrow aisle models from 1,000 to 100,000 lbs. capacity.

A09-2665
Equipment Guide-Book. <u>Fork Lifts</u>. Equipment Guide-Book Co., Vol. 1, 1976. 2000 pp. $150.00 annual subscription; Vol. 2, 1976. 1100 pp. $115.00 annual subscription
   Specifications are given for gas and diesel fork lifts. Volume 1 contains lifts of 28 manufacturers, shows weights, dimensions, lifting height and capacity, tires, etc. for self-propelled lift trucks under 10,000 lbs. capacity, gas, diesel, LPG. Volume 2 contains fork lifts of 21 manufacturers; gas and diesel, self propelled fork lifts over 10,000 lbs. capacity: lfiting height, power, tires, dimensions, options.

A09-2666
Equipment Guide-Book. <u>Green Guide</u>. Volumes 1 and 2. Equipment Guide-Book <u>Co., 1976.</u> 3800 pp. $85.00 each or $145.00 the set for annual subscription
   The Green Guide handbooks are designed for use as a general market reference for new and used prices on U.S. construction equipment manufactured during the past 10 years. Each volume is divided into sections according to type of equipment, and alphabetically by manufacturer for quick reference. Each machine listing shows type, make, model, capacity, power, weight and factory-built options. New price show is for the last year of manufacture or current list price, while average resale and wholesale values are by year of manufacture.

A09-2667
Equipment Guide-Book. <u>Green Guide: Off-Highway Trucks and Trailers</u>. Equipment Guide-Book Co., 1976. 625 pp. $60.00 annual subscription
   This is a guide to heavy hauling units and their optional attachments produced during the past 10 years. It gives realistic values of off-highway trucks, haul units, prime movers and trailers, and their factory built-extras and tires.

A09-2668
Equipment Guide-Book. <u>Green Guide's Older Equipment</u>. Equipment Guide-Book Co., 1974. 1500 pp. $85.00 annual subscription
   The average auction and resale values of older equipment is available in this book. When used with the Green Guides, values and basic specifications for 14 categories of construction equipment for the past 20 years may be seen. The reader can check serial numbers to determine year of manufacture, check the last published list price, look up manufacturer's options and accessories, and have immediate reference to current average market values.

A09-2669
Equipment Guide-Book. <u>Hydraulic Cranes</u>. Equipment Guide-Book

Co., 1976. 1350 pp. $150.00 annual subscription

Specifications are given for the hydraulic cranes of 25 manufacturers. Includes optional mountings and carriers, lifting capacities, booms, jibs, line pulls and speeds, outriggers, engines and options.

A09-2670
Equipment Guide-Gook. Hydraulic Excavators. Equipment Guide-Book Co., 1976. 1600 pp. $150.00 annual subscription

Specifications are given for the hydraulic excavators of 36 manufacturers. Shows hydraulic system, capacities, bucket capacities, optional dippers, hoe and shovel attachments, ranges, carriers, etc.

A09-2671
Equipment Guide-Book. Lift Trucks Green Guide. Equipment Guide-Book Co., 1976. 800 pp. $60.00 annual subscription

Here is a reference for comparison of resale and trade-in values for gas and diesel powered fork lift trucks from 31 manufacturers. Each manufacturer has his own section. A brief description of specifications is shown--capacity, lifting height, type of power and drive, tires (both cushion and pneumatic), and other pertinent data to help identify the exact model. A separate section lists available serial numbers by year of manufacture.

A09-2672
Equipment Guide-Book. Motor Graders. Equipment Guide-Book Co., 1976. 426 pp. $70.00 annual subscription

Specifications are given for the motor graders manufactured in the U.S. Currently 11 manufacturers are included. Data such as blade lift mechanisms, ranges and pressures, circle, moldboard, frames, tires, weights, engines, special equipment.

A09-2673
Equipment Guide-Book. Motor Scrapers. Equipment Guide-Book Co., 1976. 660 pp. $75.00 annual subscription

Current specifications for scrapers manufactured in the U.S. Currently 11 manufacturers are included. Both the tractor and scraper units described; engine model and horsepower, transmissions, tires, hydraulic systems, capacities, etc.

A09-2674
Equipment Guide-Book. Off-Highway Haul Units. Equipment Guide-Book Co., 1976. 850 pp. $110.00 annual subscription

Specifications are given for the off-highway haul units of 27 manufacturers. Capacities, loading heights, engine specifications, etc. for bottom dump, end dump, and prime mover when integral part of unit.

A09-2675
Equipment Guide-Book. OSHA Requirements for Construction. Vol. 1, Equipment. Equipment Guide-Book Co., 1976.

$130.00 annual subscription

Reference for OSHA requirements and standards for construction equipment. Equipment is listed by type. With each are given specifications, summary of standards, and recommended safety practices.

A09-2676
Equipment Guide-Book. OSHA Requirements for Construction. Vol. 2, Operations Standards. Equipment Guide-Book Co., 1976. $110.00 annual subscription

OSHA standards for construction cover such operations as materials handling, storage, use of tools, welding, excavation, trenching and shoring, steel erection and demolition. General safety, personal protective equipment and fire safety are also covered.

A09-2677
Equipment Guide-Book. Rental Rate Blue Book. Equipment Guide-Book Co., 1974. 1020 pp. $75.00 annual subscription

Lists more than 22,000 monthly, weekly, daily and hourly rates by make, model capacity. Covers such equipment types as air tools, crushing and conveying, asphalt and bituminous, compaction, concrete, tractors and earthmoving, excavating, generators and motors, hoists and derricks, pile driving, pumping, road maintenance, etc.

A09-2678
Equipment Guide-Book. Trenching Equipment. Equipment Guide-Book Co., 1976. 750 pp. $105.00 annual subscription

Specifications are given for the trenching equipment of 21 manufacturers. Every size and make of ladder or wheel type trencher. Digging depths and widths, capacities, weights, engines, speeds, etc.

A09-2679
Equipment Guide-Book. Truck Cranes. Equipment Guide-Book Co., 1976. 1500 pp. $170.00 annual subscription

Specifications are given for the truck cranes of 16 manufacturers. Working weights, lifting capacities, drums, booms, jibs, outriggers, etc. for 12 1/2 to 250 ton mechanical truck cranes.

A09-2680
Equipment Guide-Book. Wheel & Crawler Loaders. Equipment Guide-Book Co., 1976. 2350 pp. $180.00 annual subscription

Specifications are given for the wheel and crawler loaders of 51 manufacturers. Shows capacity of buckets both struck and heaped, loading reach and height, breakout force, digging depth, engines, weights, etc.

A09-2681
Equipment Guide-Book. Wheel & Crawler Tractors. Equipment

Guide-Book Co., 1976. 1025 pp. $140.00 annual subscription

Specifications are given for the wheel and crawler tractors of 26 manufacturers. Specifications include turning radius, speed, drawbar pull, lubricants and fuel capacities, general dimensions, weights, options.

A09-2682
Federal Highway Administration. Construction Manual. U.S. Gov't. Printing Office, 1975. 496 pp. $5.75. Order No. TD 2.8:C 76 S/N 050-001-00084-9

Designed to assist persons and firms performing highway construction projects under contract to the Federal Highway Administration. Examines the responsibilities and authority of the project engineer, general relationships between contractor and agency, records and orders, permits and licenses, sanitary provisions, final payments, and more.

A09-2683
Feirer, J. L. and Hutchings, G. R. Carpentry in Building Construction. Bennett Books.

A basic text for students in industrial education classes at the high school, vocational school, and the community junior college levels, written to prepare the learner to do quality work on the job in building construction. It will also be helpful in apprenticeship training and for do-it-yourselfers in building and remodeling. This text contains much current information not found in similar books, including a unit on manufactured (industrial) housing and recreational vehicles.

A09-2684
Feld, J. Construction Failure. John Wiley, 1968. 399 pp. $14.50

Intended to provide the construction practitioner with up-to-date guides that cover theory, design and practice. The author offers a collection of experiences which point out the procedures that lead to construction failure. Topics include surface and below surface construction, building materials such as timber, steel, concrete, and masonry, formwork and temporary structures. Other subjects of interest include responsibility and negligence in construction failures.

A09-2685
Fitchen, J. The Construction of Gothic Cathedrals: A Study of Medieval Vault Erection. Oxford University Press, 1961. 366 pp. $17.00

This book treats the entire erectional process, including the building equipment and falsework requirements, the actual operations undertaken, and the sequence of these operations, as specifically as they can be deduced today.

A09-2686
Foster, N. Practical Tables for Building Construction. McGraw-

Construction, Fabrication, Installation 443

Hill, 1963. 248 pp. $9.95

A digest of vital numerical data for immediate use to estimators, contractors, and construction foremen.

A09-2687
Gatz, K., ed., et al. Curtain Wall Construction. Praeger, 1968. 174 pp. $15.00

This book counteracts the design-impoverished treatment of external walls by demonstrating that no one type of wall is the solution for every kind of building. All types of facings are considered, including precast concrete, steel, marble, asbestos, aluminum, brick, granite, and timber, as well as window walls and balconies.

A09-2688
Geary, R. Work Study Applied to Building. George Godwin, 1965. 143 pp. £1.40

Drawing on practical experience in the industry, the author shows how the application of work study techniques can help to obtain maximum efficiency from labour, plant, equipment, materials and supervision.

A09-2689
Giachino, J. W., et al. Welding Skills and Practices. American Technical Society, 426 pp. $7.50 (Study Guide $1.90)

Coverage is given on standard methods and new practices. Beginners and veterans can profit from the technical information and explanations of manipulative skills.

A09-2690
Gladstone, J. Air Conditioning Testing and Balancing, A Field Practice Manual. Van Nostrand Reinhold, 1975. 110 pp. $7.95

This book is designed for the person responsible for balancing and testing an air conditioning system. He is brought up to date on instrumentation, fan laws and air flow so that he can be familiar with startup adjustments to speed, pressure and volume.

A09-2691
Gregory, C. E. Explosives for North American Engineers. Trans-Tech Publications, 1973. 300 pp. $20.00

The intelligent use of explosives is an important tool in many fields of engineering, yet the principles of explosives are generally not part of an engineer's formal education. This book is a study of explosives and their uses to achieve the desired results. Laws relating to explosives and specifications for their storage are also given.

A09-2692
Groneman, C. General Woodworking. McGraw-Hill, 1965. 353 pp. $8.36

This text shows how woodworking processes are performed, emphasizes safety precautions, and reinforces teaching points. Offers simple language and defines new terms.

A09-2693
Groneman, C. and Glazener, E. Technical Woodworking. McGraw-
    Hill, 1966. 474 pp. $10.60
    Reports on current practices and techniques employed in the
operation of modern woodworking machinery and equipment. A
how-to-do-it book with 1,550 illustrations.

A09-2694
Guild of Architectural Ironmongers. Timber Joinery and Applied
    Ironmongery. Oxford University Press, 1969. 114 pp.
    $4.00
    This text provides information on timbers used by joiners
and carpenters. It contains numerous detailed drawings of joints
and associated ironmongery. The subject is treated mainly through
illustrations, with the text kept to a minimum. Throughout the
joinery section which deals with doors, windows and stairs, suitable ironmongery is recommended.

A09-2695
Halperin, D. A. Building with Steel. 2nd ed. American Techni-
    cal Society, 1966. 268 pp. $6.60 (Study Guide $2.25)
    Covers all primary elements of buildings constructed with
steel, and secondary elements such as prefabricated curtain walls,
doors, and door frames, etc. Current practices in riveted, bolted,
and welded connections are covered. This revision includes the
use of all new steels, and tables from the American Institute of
Steel Manual for use in solving problems.

A09-2696
Hammond, R. Earthmoving and Excavating Plant. Applied Science,
    1964. 315 pp. £4.00
    This book deals with the principles of earthmoving and exca-
vating and the plant and equipment used. It advises on the selection of the best type of plant for the work envisaged and includes
a bibliography.

A09-2697
Hammond, R. Mobile and Movable Cranes. Applied Science, 1963.
    251 pp. £4.50
    Handling adds to the cost of a building product without add-
ing anything to its value, which means that particular care should
be taken to obtain the best equipment for mechanical handling.
This is particularly important in the selection of cranes.

A09-2698
Hammond, R. Modern Foundation Methods. Applied Science,
    1967. 176 pp. £4.00
    The book concentrates on practical methods and avoids high-
er mathematics. Special reference has been made to machine
foundations, which are becoming increasingly important due to the
need for avoiding noise and vibration in built-up areas. Contents
include soil mechanics, grouting, pile driving, cast-in-place piling
and large diameter piles.

Construction, Fabrication, Installation 445

A09-2699
Hanna, T. H.   Foundation Instrumentation.   Trans-Tech Publications, 1973.   400 pp. $30.00
   This book shows how the foundation engineer may benefit from the use of field instrumentation and establishes it as a necessary part of design, construction and performance evaluation. Such parameters as load, pore water pressure, earth pressure and ground movement are discussed.

A09-2700
Hansen, H. J., ed.   Architecture in Wood.   American Institute of Architects, 1971. $40.00 (for AIA members, less 10%)
   A study tracing the evolution of wooden buildings, with emphasis on construction techniques. Discusses practical and esthetic uses of wood in architecture.

A09-2701
Havers, J. and Stubbs, F.   Handbook of Heavy Construction. McGraw-Hill, 1971.   1440 pp. $39.50
   This source book codifies the basics of heavy construction, from tunneling to skyscraper erection. It incorporates a manual of construction management, a handbook of construction equipment and contains sections on basic operations like concreting and rock excavation along with special-purpose sections on pipelines and pavements.

A09-2702
Hodge, J. C.   Brickwork for Apprentices.   Herman Publishing, Inc., 1971.   248 pp. $6.50

A09-2703
House Wiring.   Howard W. Sams.   192 pp. $5.95
   This book offers answers to many often-asked questions about all phases of house wiring. Containing concise interpretations of the NEC, it will be valuable to contractors and electricians.

A09-2704
Huntington, W. C. and Mickadeit, R. E.   Building Construction: Materials and Types of Construction.   R. S. Means or John Wiley, 1975.   768 pp. $18.50
   Survey of traditional and contemporary materials of building construction, their uses, and the terms of building construction. Featuring: Complete modernization of the chapter on structural steel--consistent with the new AISC Steel Handbook--chapter on concrete amended to include new technology in ACI codes--new section on hardware--new references to timber construction. Included are drawings and photos of new products and techniques--expanded to adequately cover both light and heavy construction.

A09-2705
Hurd, M. K.   Formwork for Concrete.   American Concrete Institute, 1973.   376 pp. $18.50 (for ACI members, $11.50)

Prepared under the direction of ACI Committee 347. Contains revised design information reflecting changes in standard lumber dimensions, and in allowable lumber stresses. A how-to-do-it handbook for designers, contractors, engineers, and architects following guidelines established in the ACI Standard "Recommended Practice for Concrete Formwork (ACI 347-68)." Included are tables, diagrams, and formulas for form design loads. Planning, building, and using formwork are dealt with extensively.

A09-2706
Hurst, A. E. Painting and Decorating. Charles Griffin, 1963. 490 pp. £3.60
This book discusses techniques and materials used today. The contents include a description of equipment and materials in addition to painting and paperhanging techniques.

A09-2707
Huston, J. Hydraulic Dredging: Theoretical and Applied. Cornell Maritime, 1970. 352 pp. $12.50
Today's dredge is an unparalleled excavating plant. When sufficient water is available it has no economic competitor. Here is a handbook on the theory, applications, mechanics, business and other aspects of the subject of hydraulic dredging.

A09-2708
Hutchinson, B. D. and Barton, J. Physical Aspects of Building Science. Butterworth, 1970. 212 pp. £1.80
This book brings together the relevant basic facts required by students in the construction industry. It will also be of value to architectural students and technicians requiring a basic knowledge of the scientific principles involved in construction.

A09-2709
Illingworth, J. Movement and Distribution of Concrete. McGraw-Hill, 1972. 256 pp. $18.00
This practical book begins with a discussion of the limitations which the site and the specification impose on the choice of methods and equipment, and a detailed examination of what constitutes a valid cost analysis and the methods by which such analysis should be carried out.

A09-2710
Imrie, R. S. Building Site Organization. Beekman, 1972. 180 pp. $14.00
Guidance and instruction in the numerous aspects of site supervision. A special feature of the book is the use of drawings, charts, and photographs designed to present a clear understanding of the subject.

A09-2711
Innocent, C. F. The Development of English Building Construction. David and Charles, 1916. 320 pp. $15.00
Reprint of classic 1916 survey of architectural development of the workingman's house. Contents: primitive forms of building;

Construction, Fabrication, Installation

curved tree principals; fully-developed timber buildings; details of timber buildings; the carpenter; walls; floors; slated roofs; thatching; doors; windows and chimneys.

A09-2712
Inott, F. A. Carpentry & Joinery: Multi-Question Course. Beekman, 1974. 96 pp. $5.95
 Designed to assist students of carpentry and joinery to assess their progress in technical studies so that specific weaknesses can be rectified.

A09-2713
Insall, D. The Care of Old Buildings Today: A Practical Guide. Watson-Guptill, 1974. 197 pp. $13.95
 The book provides administrative advice: the owner, architect and contractor; the preservation plans; preservation and public; new uses for old buildings; cost control; etc. The second part deals with specific restoration techniques; repair or renewal; sheet roof coverings; timber structures, fungi and pests; stonework decay and repair; glass in old buildings; etc.

A09-2714
Intl. Soc. of Soil Mechanics and Foundation Engrg. Grouts and Drilling Muds in Engineering Practice. Butterworth, 1963. 242 pp. £6.75
 Symposium organized by the British National Society of the International Society of Soil Mechanics and Foundation Engineering. Contains the papers presented at the symposium, together with discussion.

A09-2715
Jay, P. and Hemsley, J. Electrical Services in Buildings. Applied Science, 1968. 180 pp. £2.00
 This book describes electrical installation practice, the application of all the common types of wiring systems, distribution equipment and protective devices. It gives an account of fundamental principles so that the reader can understand exactly why certain practices are necessary.

A09-2716
Kaberlein, J. J. Air Conditioning Metal Layout. Bruce Publishing Co., 1967. $6.75

A09-2717
Kidder, F. E. and Parker, H. Architects' and Builders' Handbook. John Wiley, 1931. 2315 pp. $27.00

A09-2718
Knowles, C. C. and Pitt, P. H. The History of Building Regulation in London 1189-1972. Architectural Press, 1973. 140 pp. £3.95
 This history of building regulation in London from the twelfth century to the present is of considerable importance. Building

regulations had a powerful effect on construction methods and hence on architectural style. The book also covers the complex legislation that regulates building in London today.

A09-2719
Knowles, P. R. Composite Steel and Concrete Construction. Butterworth, 1974. 208 pp. £5.00
Provides theoretical background and practical design information. For the engineer who wishes to use composite construction in either bridge or building structures. A historical survey of the development of composite construction and some discussion of its economics is followed by a detailed treatment of the theoretical aspects. All numerical work is carried out in SI units.

A09-2720
Koenigsberger, F. and Adair, J. R. Welding Technology. Hart Publishing Co., 1969. 424 pp. $12.50
A view of the potentiality of welding, descriptions and annotated diagrams and photographs, and bias towards modern economic production methods characterize this work. The book includes new techniques such as electron beam, ultra-sonic, friction and dip transfer welding.

A09-2721
Komendant, A. E. Contemporary Concrete Structures. McGraw-Hill, 1972. $22.50
This guide covers the most advanced theories, material characteristics, and construction methods relating to poured-in-place and prefabricated concrete structures.

A09-2722
Koncz, T. Manual of Precast Concrete Construction. Vol. 1. Adler's Foreign Books, 1968. 304 pp. $40.00
Design, analysis and construction of reinforced and prestressed concrete roof and floor units, and wall panels. 556 illustrations.

A09-2723
Koncz, T. Manual of Precast Concrete Construction. Vol. 2. Adler's Foreign Books, 1971. 427 pp. $55.00
Design, analysis and construction of reinforced and prestressed concrete industrial shed-type and low-rise buildings, and special structures. 881 illustrations.

A09-2724
Koncz, T. Manual of Precast Concrete Construction. Vol. 3. Adler's Foreign Books, 1970. 380 pp. $45.00
Design, analysis and construction of reinforced and prestressed concrete multi-story industrial and administrative buildings, school and university buildings, and residential buildings. 522 illustrations.

Construction, Fabrication, Installation 449

A09-2725
Kratfel, E. Introduction to Modern Sheet Metal. Prentice-Hall, 1976. 288 pp. $12.95
Here is a basic introduction to the definition and characteristics of modern sheet metals. The author presents a basic set of hand tools, and provides layout, constructions, and precise sheet metal bend calculations. Also, the book offers chapters dealing with the types, use, and operations of metal forming and shearing machines with emphasis on safe and proper use. The book equips the reader with a working knowledge of fastening by welding, brazing, soldering, rivets, sheet metal screws, and insert hardware.

A09-2726
Labor Department. Training and Entry into Union Construction. U.S. Gov't. Printing Office, 1975. 207 pp. $2.80. Order No. L 1.39/3:39 S/N 029-000-00239-8
Procedures and standards that building trades unions use to admit craftsmen to journeyman status (obtaining full union wage rates).

A09-2727
Lafever, M. The Modern Builder's Guide. Dover, 1969. 146 pp. $5.00
Historically important book in the development of the Greek revival in American domestic architecture. 90 plates show architectural features, floor plans, etc. Text describes carpentry, materials, historical documents of Greek architecture, etc. Reprint of 1833 edition.

A09-2728
Lair, E. Carpentry for the Building Trades. McGraw-Hill, 1953. 310 pp. $8.60
Covers trade data, reference material, standard methods of construction, and how-to-do-it information on practical carpentry.

A09-2729
Leonards, G. Foundation Engineering. McGraw-Hill, 1962. 1146 pp. $29.50
The current status of foundation engineering, with emphasis on design and construction.

A09-2730
Leonhardt, P. Prestressed Concrete Design and Construction. William Heinman, 1964. $35.00

A09-2731
Lewicki, B. Building with Large Prefabricates. North-Holland, 1966. 460 pp. $30.90
This is a revised English translation of the corresponding Polish book. It describes techniques for the construction of buildings from large prefabricated units. Examples are used from a number of countries developing the building system in which large

A09-2732
Libby, J. R. Modern Prestressed Concrete: Design Principles and Construction Methods. Van Nostrand Reinhold. (In preparation)
   This book stresses practical information for solving fabrication and design problems. Much of the material appears in book form for the first time, the erection of precast members, a methodology for estimating creep and shrinkage when test data are not available, and data on cracking. The new ACI (American Concrete Institute) Code is discussed.

A09-2733
Lindsey, F. R. Pipefitters Handbook. Industrial Press, 1967. 464 pp. $11.00
   Provides answers to problems involving pipe bending and fabricating in the shop and field. A reference work for pipefitters, steamfitters, layout men, contractors, plumbers, apprentices, and anyone else concerned with the layout and installation of pipe.

A09-2734
Lippsmeier, G. Building in the Tropics/Tropenbau. International Publications, 1969. 282 pp. $32.00
   A compendium for planners-architects, engineers and also clients with a basic knowledge of building. Particular emphasis upon the problems of developing countries.

A09-2735
Lloyd, N. A History of English Brickwork: With Examples and Notes of the Architectural Use and Manipulation of Brick from Medieval Times to the End of the Georgian Period. Benjamin Blom, 1925. 449 pp. $28.50
   Examines the remains of Roman work, the reappearance of brick in the 13th century, the manufacturing process, statutes and ordinances, and some accounts of the lives of the workers and builders in brick. Techniques and aesthetics of bricklaying of all kinds. Over 450 photographs.

A09-2736
Love, T. W. Stair Builders Handbook. Craftsman Book Co., 1974. 400 pp. $5.95
   If the floor to floor rise is known, this handbook will give everything else: the number and dimension of treads and risers, the total run, the correct well hole opening, the angle of incline, the quantity of materials and settings for the framing square. Tables give over 3,500 code approved rise and run combinations--several for each 1/8 inch interval from a 3 foot to a 12 foot floor to floor rise.

Construction, Fabrication, Installation 451

A09-2737
Lytle, R. J., et al. Farm Builders Handbook. Structures Publishing Co., 1973. 266 pp. $20.00
   Data on metal buildings and storage structures, including silos, new lumber standards. Occupational Safety and Health Administration (OSHA) standards for both the farm builder and his customers.

A09-2738
Lytle, R. J., et al. Industrialized Builders Handbook. Structures Publishing Co., 1971. 248 pp. $24.95
   The Industrialized Builders Handbook is designed as a source book for builders and manufacturers of components, packaged and modular buildings. Covers modular building construction, explaining manufacturing techniques, materials and equipment. Ranges from selection of materials and equipment through delivery and installation of the completed units, and future advances.

A09-2739
McCue, G. M., et al. Creating the Human Environment. Univ. of Illinois Press, 1971. $4.95
   The authors look at the building industry today and determine how it will shift in structure to accommodate change. The future of the architectural profession is examined. The year 2000 is used as a focus.

A09-2740
McGregor, K. Drilling of Rock. Applied Science, 1967. 306 pp. £6.00
   Although there is a varied literature of technical papers and articles on the subject of rock drills and drilling, this volume is probably the first determined attempt to integrate the subject and to make available in easily readable form the latest developments and techniques.

A09-2741
McKaig, T. Building Failures. McGraw-Hill, 1962. 261 pp. $12.50
   Case studies of over 200 building failures in recent years--failures due to defective building materials, over loading, wind action, etc.

A09-2742
Maguire, B. Carpentry in Commercial Construction. Prentice-Hall, 1976. 320 pp. $10.50
   Bridging the gap between residential carpentry and light commercial carpentry, this book explains why and how tasks are performed differently, and prepares readers to participate in and understand their role in commercial construction, be it as contractor, foreman, scheduler, or worker. Including many useful tables, graphs, and figures, the book covers all aspects of building as they relate to commercial carpentry, from concrete form work and framing to exterior and interior tasks. The author fills readers in

on such cost-saving techniques as the single wall siding system, truss design, fire control, and sound control techniques, and shows readers how to plan for effective man-hour use.

A09-2743
Makowski, Z. S. Steel Space Structures. Fernhill, 1965. $9.50

A09-2744
Manning, G. P. Concrete Water Towers, Bunkers, Silos and Other Elevated Structures. Cement and Concrete Assn., 1973. 372 pp. £4.00

A09-2745
Manning, G. P. Design and Construction of Foundations. Cement and Concrete Assn., 1972. 305 pp. £4.00

A09-2746
Manning, G. P. Reinforced Concrete Reservoirs and Tanks. Cement and Concrete Assn., 1967. 382 pp. £2.50

A09-2747
Masons and Builders Library. Vol. 1, Concrete, Block, Tile, Terrazzo. Howard W. Sams. $5.95
Covers proportions and mixing, tools, concrete slabs, walks, and driveways, reinforced concrete, concrete forms, hot and cold weather handling, pouring, finishing, and curing. Treats fancy finishes, stucco, concrete block, concrete anchors, repairing and patching and projects with concrete.

A09-2748
Masons and Builders Library. Vol. 2, Bricklaying, Plastering, Rock Masonry, Clay Tile. Howard W. Sams. $5.95
Covers brick and mortar; tools and bonding. How to lay brick; wall types, thicknesses, and anchoring; corners, openings, and arches, brick surfaces and patterns. Chimneys and fireplaces; hollow tile and glass block; brick terraces, walls, and floors. Brick repairs and maintenance; masonry patio projects; plastering; stone and rock masonry and how to read blueprints.

A09-2749
Matthias, A. J., et al. How to Design and Install Plumbing. American Technical Society, 446 pp. $5.50 (Study Guide $1.65)
This text explains standard practices for students, architects, apprentices, journeymen, and other construction tradesmen. The coverage is from small details to complete systems of water supply, pumping, drainage, sewage, ventilation.

A09-2750
Mead, H. T. and Mitchell, G. L. Plant Hire for Building and Construction. Butterworth, 1972. 196 pp. £4.50
Few building contractors could operate as efficiently as they do without the aid of plant-hire companies. The object of this book

Construction, Fabrication, Installation

is to give those in the construction industry better understanding of the plant-hire system and to stimulate more interest in the economics of hiring contractors' plant, so that contractors can make the best decision when faced with the choice of hiring or buying.

A09-2751
Medlycott, A. Applied Building Construction. John Wiley, 1967. 138 pp. $5.50

A09-2752
Meleka, A. Electron-Beam Welding: Principles and Practice. McGraw-Hill, 1971. 328 pp. $21.00
Compiled with the combined knowledge and experience of the editor and eleven authors who are authorities on particular aspects of electron beam welding, this book has been published at a time when industry is beginning to exploit the considerable possibilities of the process.

A09-2753
Merritt, F. Building Construction Handbook. McGraw-Hill, 1965. 842 pp. $25.00
Information on building design and construction. Fundamental data on building materials, structural design, and electrical and mechanical services are covered as well as CPM scheduling, the new concrete and steel codes, and folded-plate and thin-shell designs.

A09-2754
Modular Building Standards Association. Modular Practice. John Wiley, 1962. 198 pp. $12.75

A09-2755
Morris, R. C. Air Conditioning Cutter's Ready Reference. Business News Publishing, 1971. 363 pp. $16.50
361 illustrations make it easy to follow the step-by-step instructions given for every fitting and piece of duct used in air conditioning work.

A09-2756
Mulligan, J. Handbook of Brick Masonry Construction. McGraw-Hill, 1942. 526 pp. $18.50
A reference work with practical technical data covering all aspects of masonry construction.

A09-2757
Navy Department. Equipment Operator. U.S. Gov't. Printing Office, 1973. 465 pp. $4.25. Order No. D 207.208/2:Eq 5 S/N 0847-00162
This U.S. Navy Rate Training Manual provides the technical information and skill requirements necessary to prepare personnel for jobs involving the handling of automobiles and heavy equipment. It contains information on earth moving, road building, asphalt paving, and related subjects.

Construction Information

A09-2758
Nichols, H. L., Jr. Heavy Equipment Repair. North Castle Books, 1964. 640 pp. $12.50
   This book provides thorough and specific background information on the design of all major earthmoving machines and their sub-assemblies, and on the tools and methods used to lubricate, adjust, and overhaul them. It emphasizes the importance of careful work and of preventive maintenance.

A09-2759
Nichols, H. L., Jr. How to Operate Excavation Equipment. North Castle Books, 1954. 164 pp. $4.00
   A thorough set of instructions on operation of the principal types of heavy equipment, abridged from the book "Moving the Earth."

A09-2760
Nichols, H. L., Jr. Moving the Earth (The Workbook of Excavation). North Castle Books, 1962. 1488 pp. $25.00 (first section available separately, 608 pp. $12.50)
   The first section contains practical instructions for most earthmoving jobs. Explains what equipment to use and the most efficient ways to use it. Covers costs and estimates from the contractor's standpoint. There are chapters on surveying, explosives, and mining. The second section describes every important type of machine used in digging, transporting, and grading soil and rock.

A09-2761
Oberg, F. R. Heavy Timber Construction. American Technical Society, 373 pp. $8.75 (Study Guide $2.40)
   This book covers all phases of heavy timber construction. It covers dwellings and commercial structures, trestles, decks, bridges, tunnels, tanks, towers, waterfront structures, dams, and retaining walls.

A09-2762
Oppenheimer, S. Erecting Structural Steel. McGraw-Hill, 1960. 256 pp. $12.75
   Describes the process of building all types of steel structures, from beginning plans to the final installation of the roof, beams, etc.

A09-2763
[No entry]

A09-2764
Parker, H. and MacGuire, J. W. Simplified Site Engineering for Architects and Builders. John Wiley, 1954. 250 pp. $11.00

A09-2765
Parker, H., et al. Materials and Methods of Architectural Construction. John Wiley, 1958. 724 pp. $16.00

Construction, Fabrication, Installation 455

A09-2766
Paxton, J. M. Manual of Civil Engineering Plant and Equipment. Applied Science, 1971. 592 pp. £25.00
This manual contains full details and specifications of mechanical plant and equipment manufactured in Great Britain and used in connection with all civil engineering, building, quarrying, and general construction.

A09-2767
Perkins, P. H. Floors--Construction and Finishes. Cement and Concrete Assn., 1973. 132 pp. £4.00

A09-2768
Peurifoy, R. Formwork for Concrete Structures. McGraw-Hill, 1964. 330 pp. $18.50
Explains methods and materials vital to the design, erection, use, and removal of forms for most concrete structures built today.

A09-2769
Plant Engineering Construction Library. Plant Engineering. 229 pp. $15.00
Three booklets in a loose-leaf folder give data on construction briefs (79 pp.), specifications (106 pp.), and civil engineering for the plant engineer (44 pp.).

A09-2770
Plumbers and Pipe Fitters Library. Vol. 1, Materials, Tools, Calculations. Howard W. Sams. 320 pp. $4.95
For use by master plumbers, journeymen, apprentice pipe fitters, sheet metal workers, draftsmen and building trade students. Covers physical principles, materials, sheet metal, pipe, soldering, lead work, safe water pressures, pipe threading and copper tubing.

A09-2771
Plumbers and Pipe Fitters Library. Vol. 2, Drainage, Fittings, Fixtures. Howard W. Sams. 320 pp. $4.95
For use by master plumbers, journeymen, apprentice pipe fitters, sheet metal workers, draftsmen and building trade students. Treats water supply, drainage, sewage disposal, pipe joints and fittings, roughing-in, valves and faucets, fixtures, percolation tests, septic tanks, ventilation stacks, etc.

A09-2772
Plumbers and Pipe Fitters Library. Vol. 3, Installation, Heating, Welding. Howard W. Sams. 288 pp. $4.95
For use by master plumbers, journeymen, apprentice pipe fitters, sheet metal workers, draftsmen and building trade students. Includes gas piping, steam and hot water heating systems, boiler fittings and fuel oil tank installation. Also covered are air conditioning, gravity systems, return systems and pipe supports. Brazing and welding are additional subjects of interest.

A09-2773
Portland Cement Association. Administrative Practices in Concrete Construction. John Wiley, 1975. 230 pp. $18.95
Provides a working knowledge of the legal obligations faced by the concrete technologist. Covers the various aspects of estimating--adequate planning and scheduling. Bidding for contracts, and operating at a profit. Deals with the collection, organization, and presentation of data, characteristics of frequency distribution, and various limits.

A09-2774
Portland Cement Association. Basic Concrete Construction Practices. John Wiley, 1975. 468 pp. $18.95
Presupposing a working knowledge of how quality concrete is made, the book deals with its purposes, construction, and use, details the handling of concrete, and provides information for performing concrete work under special conditions.

A09-2775
Portland Cement Association. Concrete Inspection Procedures. John Wiley, 1975. 146 pp. $12.95
Contains information on the concrete inspector and his job function, quality control inspection and testing of concrete in street construction, floor construction, and soil cement, concrete field problems, and repair patching of concrete and safety requirements.

A09-2776
Portland Cement Association. Laboratory and Exercise Manual on Concrete Construction. John Wiley, 1975. 267 pp. $11.95
This laboratory manual, divided into 43 laboratory sessions, deals with testing procedures for cement and concrete. Major topics include: principles of quality concrete, basic concrete construction practices, and special concretes, mortars and products.

A09-2777
Progress in Construction Science and Technology. Construction Press. £7.50
This series of volumes provides a medium for the publication of major reviews, at an advanced level, of developing areas of building science and technology. The authorship is both international and authoritative and each volume represents the highest standard of scholarship.

A09-2778
Purbahn, W. E. and Sundberg, E. W. Fundamentals of Carpentry. Vol. 2, Practical Construction. American Technical Society, 544 pp. $7.40 (Study Guide $1.80)
The text is illustrated with floor plans, detailed drawings, line cuts, and photographs. The coverage includes preparing for the job, leveling instruments, foundation formwork, wall and floor framing, roof framing, exterior finishing, and interior finishing. Information on new materials, new trends, and constructional variations due to regional and climate differences.

Construction, Fabrication, Installation

A09-2779
Putnam, R. Bricklaying Skill and Practice. 3rd ed. American Technical Society, 1974. 272 pp. $5.75 (Study Guide $1.85)
  This edition presents the techniques, practices, and materials now used in bricklaying. It meets the needs of today's apprentice tradesmen, and the journeymen, foremen, and supervisors who need refresher training. Included is new information on epoxy mortars, pre-fab panels, bricklaying in heavy construction, and use and handling of welding equipment, plus a chapter on concrete block.

A09-2780
[No entry]

A09-2781
Questions and Answers for Plumbers Examinations. Howard W. Sams. 256 pp. $5.95
  This is a review of fundamental principles underlying answers to questions usually asked on plumbers license examinations. Many questions are answered as to types of fixtures to use, size of pipe to install, design of systems, size and location of septic tank systems, and procedures used in installing material.

A09-2782
Rampaul, H. Pipe Welding Procedures. Industrial Press, 1973. 238 pp. $15.00
  This book is intended for the welder and it is written from his viewpoint. Although primarily devoted to procedures used to weld both thick-wall and thin-wall pipe by the shielded metal-arc process, there is a chapter on root bead welding by the gas tungsten-arc process. Other subjects include: essentials of shielded metal-arc technology, heat input and distribution, essentials of welding metallurgy, distortion, pipe welding defects, qualification of the welding procedure and the welder, and welding safety.

A09-2783
Randall, F. A. History of the Development of Building Construction in Chicago. Arno Press, 1949. $20.00
  This volume traces building techniques in Chicago from the beginning of the city to the date of publication.

A09-2784
Rapp, W. G. Construction of Structural Steel Building Frames. John Wiley, 1968. 340 pp. $15.00
  The author focuses on the work of erecting tier, column-core, and mill buildings, hangars, pier sheds, and power houses.

A09-2785
Reband, P. Related Mathematics for Carpenters. American Technical Society, 218 pp. $3.65
  This revised edition acquaints the student with math fundamentals as related to carpentry. The student is provided many practice applications.

A09-2786
Reid, D. A. G. Construction Principles 1: Function. George Godwin, 1973. 198 pp. $11.95 (available from Construction Publishing Co.)

This is the first book in a series designed to complement and extend the work of Principles of Modern Building. The volume is in metric throughout. Whilst adopting the same approach as Principles of Modern Building, the material has been updated, condensed and presented in a form appropriate to the graduate level student in the professions that serve the construction industry.

A09-2787
Reiner, L. E. Methods and Materials of Construction: A Guide for Builders, Owners, Architects and Engineers. Prentice-Hall, 1970. 353 pp. $12.95

Survey of the construction process from site selection to the occupation of the structure by the tenant. Considers the latest developments in construction methods and materials and gives informed opinions on comparative costs and quality. Indicates ways to avoid costly errors and demonstrates that an attractive viable structure need not be expensive.

A09-2788
Reynolds, C. E. Concrete Construction. Cement and Concrete Assn., 1967. 512 pp. £3.00

A09-2789
Richardson, J. G. Formwork Notebook. Cement and Concrete Assn., 1972. 94 pp. £2.00

A09-2790
Richardson, J. G. Precast Concrete Production. Cement and Concrete Assn., 1973. 232 pp. £3.50

A09-2791
Ritchie, T. Canada Builds, 1867-1967. Univ. of Toronto Press, 1968. 456 pp. $12.50 (Canadian Building Series No. 1)

The first part of the book focuses on the state of building in the mid-1860's, and on the construction of the Parliament Buildings, a project which represented a number of major innovations that were soon to become commonplace in the developing trade. The remaining three parts survey the evolution of the Canadian building industry.

A09-2792
Rose, J. Questions and Answers on Pipework and Pipe Welding. Butterworth, 112 pp. £0.75

Subjects covered are pipework systems, metals and materials used in pipe construction, joining, pipe welds, solderless joints, connectors and couplings, flexible hoses, bending of tubing, pressure testing, etc.

Construction, Fabrication, Installation

A09-2793
Rossnagel, W.   Handbook of Rigging.   McGraw-Hill, 1964.   375 pp.
    $18.50
    A practical manual for construction and industrial rigging operations. Slings, scaffolds, ladders; cranes, jacks, and other methods for supporting and hoisting weights are detailed, with emphasis on their safe and efficient management.

A09-2794
Rufus, G.   Practical Rafter Calculator.   Craftsman Book Co.,
    1972.   124 pp. $3.00
    Gives the exact length for common, valley, hip and jack rafters for any span up to 50 feet and for any rise from 1/2 in 12 to 30 in 12. Provides the correct rafter length to the nearest 1/16-inch. Angle, plumb and level cuts are included.

A09-2795
Salzman, L. F.   Building in England.   Oxford University Press,
    1952.   654 pp. $13.00
    The book is concerned with the contemporary written accounts of medieval buildings and the processes of their construction. It covers the period from Saxon times down to the dissolution of the monasteries and deals with the organization of the building industry, such matters as hours and wages, and the methods and materials used.

A09-2796
Sandstrom, G.   Man the Builder: Origins of Civil Engineering.
    McGraw-Hill, 1970.   280 pp. $17.50
    Recounts the history of man's architectural and engineering feats from Stonehenge to modern power dams. This reference work is illustrated with more than 200 sketches, engravings, maps, drawings, and photographs.

A09-2797
Schmidt, J. L., et al.   Construction: Principles, Materials and
    Methods.   The Interstate, 1972.   698 pp. $20.00
    This book has the purpose of providing a better understanding of construction principles as well as gathering into a single source many recognized industry standards and specifications. A wide spectrum of building material is covered and with each is given the most important design criteria and construction recommendations. However, the book is not intended to be a catalog of past methods or a manual of minimum standards. It has been written for use in reviewing home construction plans and making on-site inspections.

A09-2798
Schofield, C. W.   Mathematics for Construction Students.   Herman
    Publishing, Inc., 1975.   200 pp. $9.95

A09-2799
Searles, W. H., et al.   Field Engineering.   John Wiley, 1949.
    836 pp. $14.00

A09-2800
Sedwell, N. A. Calculations for Bricklayers. Herman Publishing, Inc., 1966. 128 pp. $2.95

A09-2801
Shank, M. E., ed. Control of Steel Construction to Avoid Brittle Failure. American Welding Society, 1957. $4.50

A09-2802
Siegele, H. H. Roof Framing. Drake Publishers, 1972. $5.95

A09-2803
SMACNA. Architectural Sheet Metal Manual. Sheet Metal and Air Conditioning Contractors' National Assn., 1972. 282 pp. $17.00 (50% discount to architects, engineers, schools)
A standards reference for proper design and installation of architectural sheet metal. Detailed drawings and text material reflect industry practice. Contains rainfall data and drainage factors as well as details of roof drainage systems, expansion tables, weight and gage tables and details of gravel stops, facia, flashings, copings, expansion joints, metal roofs, skylights, louvers, sunshapes, snow guards, metal column covers, metal decking, siding and many miscellaneous specialty sheet metal items.

A09-2804
Smith, R. C. Principles and Practices of Heavy Construction. Prentice-Hall, 1967. 343 pp. $13.95
Outlines blasting techniques, gives an account of types, uses of excavation equipment; describes modern soil testing methods, and concrete design.

A09-2805
Smith, R. C. Principles and Practices of Light Construction. Prentice-Hall, 1970. 368 pp. $13.95
Covers various phases of light construction beginning with site layout, excavating, form work and foundations. Considerable detail is given to the types and applications of exterior and interior finishes.

A09-2806
Stewart, D. A. Design and Placing of High Quality Concrete. John Wiley, 1962. 162 pp. $8.00

A09-2807
Stone, P. A. Building Economy: Design, Production and Organization--A Synoptic View. Pergamon, 1966. 264 pp. $8.50
Describes and assesses the forms of construction used today, the methods of production, and the way the industry is organized in England. Considers the impact which new methods, materials, forms of organization and contractual relationships might have on industry.

Construction, Fabrication, Installation

A09-2808
Swearer, H. Installing and Servicing Electronic Protective Systems. Tab Books, 1975. 256 pp. $7.95 (paperback $4.95)
Although this book is intended to serve as an installation and maintenance manual for intrusion alarm systems, it is also useful to familiarize others in security systems. A glossary and list of manufacturers are included, as are the regulations which must be met in order to obtain Federal crime insurance.

A09-2809
Taylor, J. B. Plastering. George Godwin, 1970. 159 pp. $11.95 (available from Construction Publishing Co.)
Traditional as well as contemporary practices and techniques of the trade are fully explained in this book on plastering. A comprehensive overview as well as a working aid, it includes a chapter on special finishes. It covers a wide variety of solid and fibrous plaster work with appropriate methods and application technique.

A09-2810
Technical Group for Nuclear Explosives Engrg. Proceedings of the Special Session on Nuclear Excavation. American Nuclear Society, 1968. 144 pp. $7.50
The Proceedings present some of the most important thought on this new area of technology. Some papers include: Nuclear Excavation: Theory and Applications; Results of the Cabriolet Excavation Experiment; Project Buggy: A Nuclear Row Excavation Experiment; The Corps of Engineers Nuclear Construction Research Program, etc.

A09-2811
Teller, E., et al. The Constructive Uses of Nuclear Explosives. McGraw-Hill, 1968. 320 pp. $12.75
A summary of the results of research aimed at determining the earth-moving capabilities of nuclear explosives. The book is written from an engineering standpoint, and the data presented are adequate for making preliminary estimates of size and cost of excavation projects.

A09-2812
Terry-Smith, W. H. Building Illustrated. Charles Griffin, 1949. 141 pp. £1.90

A09-2813
Testa, C. The Industrialization of Building. Reinhold Books, 1973. 200 pp. $19.95
Describes this emerging technology as a total process, giving due consideration to functions of the various participants. Explains the conditions that have allowed the development of industrialized building projects utilizing ultra-modern methods.

A09-2814
Thames Valley Group. Students Project Work in Construction.

George Godwin, 1971.   131 pp.   £1.85

The eight projects range over the whole of the contract period and cover: road and public drainage; preliminaries; foundations; plant and scaffolding; superstructure; roofing; cladding, and services and internal finishes. Realistic documents are provided and the tasks set for each project present the student with construction problems which can be solved by applying the knowledge he has gained.

A09-2815
Timber Engineering Co., Timber Design and Construction Handbook. McGraw-Hill, 1956.   622 pp. $18.50

This practical field handbook offers the information needed to develop and construct wood structures.

A09-2816
Tomlinson, M. J.   Foundation Design and Construction.   John Wiley, 1975.   785 pp. $40.00

Covers site investigations and soil mechanics, general principles of foundation design, ground movements, spread and box foundations, piles, construction methods, control of ground water, etc. This new edition has been updated to include SI units. A revision has been made in that part of the text which deals with the skin friction of piles into clays.

A09-2817
Trill, J. and Bowyer, J. T.   Problems in Building Construction: A Scientific Method Approach.   Architectural Press.   Vol. 1, 1972.   100 pp.   £1.00; Vol. 1, Instructor's Manual, 1972.   220 pp.   £3.00; Vol. 2, 1974.   100 pp.   £1.00; Vol. 2, Tutor's Guide, 1974.   128 pp.   £3.00

Each book examines 20 problems concerned with typical failures of the sub and superstructure of domestic buildings. The method used is one step-by-step investigation in which the student participates by being given clues from which he can arrive at the way the problem was finally solved. In this way, the book combines an introduction to practical investigative techniques with an extension of the teaching of the principles of construction which makes clear their relationship to building science.

A09-2818
Turner, L. and Lakeman, A.   Concrete Construction Made Easy. Cement and Concrete Assn., 1958.   118 pp.   £0.30

A09-2819
Urquhart, L.   Civil Engineering Handbook.   McGraw-Hill, 1959. 1148 pp. $29.50

Principles, methods, practical pointers, and helpful data are contained in this easy-to-use, comprehensive reference guide to all branches of civil engineering. It covers everything from surveying water supply to highway and airport engineering.

Construction, Fabrication, Installation

A09-2820
Vallings, H. G. Mechanisation in Building. Applied Science, 1964. 95 pp. £2.00
　　The building industry in Britain is faced with a tremendous demand for new construction. To meet this demand with the labour force available the industry will have to make the fullest use of machines. This book describes the plant available and deals with its application and performance. It is intended as a guide to mechanization.

A09-2821
Wachsmann, K. The Turning Point of Building. Van Nostrand Reinhold, 1961. 240 pp. $19.00
　　Presents an exposition of today's mass prefabrication techniques and the new concepts of planning and building that must be mastered to realize their potential.

A09-2822
Waddell, J. Concrete Construction Handbook. McGraw-Hill, 1968. 918 pp. $28.50
　　The specifics of mixing, pouring, and curing of specification-quality concrete are outlined in this practical guide for field inspectors, construction superintendents, field engineers, and contractors. Designed to serve as a quality control manual, this handbook also includes sections on form work, shoring, aggregates, and additives.

A09-2823
Walton, H. How to Build your Cabin or Modern Vacation Home. Popular Science Publishing, 1973. 160 pp. $1.95
　　Picking the site, selecting plans, building the house.

A09-2824
Wass, A. Methods and Materials of Residential Construction. Prentice-Hall, $13.50
　　Features erecting housing units on their correct locations and elevations in space, the implementation of latest building techniques, and the assembly of men, materials and machines to safely and profitably erect residential units. Highlights job control with guide lists of overhead expenses, progress schedules, including an introduction to critical path method, subtrades and change orders.

A09-2825
Wass, A. and Sanders, G. A. Building Construction: Roof Framing. Prentice-Hall, 1960. 170 pp. $14.60
　　Comprehensive treatment of roof framing is illustrated with shop drawings.

A09-2826
Weaver, L. English Leadwork: Its Art and History. Benjamin Blom, 1909. 283 pp. $22.50
　　A major work on the evolution of the craft, styles and techniques, and use in buildings. Illustrated.

A09-2827
Welders Guide. Howard W. Sams. 928 pp. $9.95
   This new edition is a manual on the theory, practical operation, and maintenance of all welding machines. Covers both electric and oxy-gas welding.

A09-2828
West, A. S. Piling Practice. Butterworth, 1972. 122 pp.
   £3.20 (paperback £1.80)
   A book which gives guidance on the selection of the most appropriate technique in any given set of circumstances, and which also details the organization and documentation of piling work. The first part describes the various main types of pile available, their method of construction, the plant required for their construction and the advantages and disadvantages of each type. The second considers in detail all the steps necessary for construction of a piled foundation.

A09-2829
Whittick, A. R. Questions and Answers on Carpentry and Joinery.
   Butterworth. 160 pp. £0.75
   The basic aspects of the work of the carpenter and joiner are discussed. Covers a wide range of topics such as power tools, timber defects, joints, temporary work, carcassing, first and second fixings, woodworking machinery and timber preserving and finishing.

A09-2830
Wilson, J. Practical House Carpentry. McGraw-Hill, 1973.
   424 pp. $11.25
   A guide to housebuilding showing the reader how to do each carpentry job in foundation work, framing, exterior and interior finishing. The second edition of this standard guide to the subject of how-to-do-it house construction contains new material on power hand tools, a new chapter on reading house plans, a new section on basements, and a complete updating of lumber standards.

A09-2831
Wilson, J. D. and Werner, S. O. Simplified Roof Framing. McGraw-Hill, 1948. 161 pp. $6.50
   This manual on roof framing treats common types of roofs and tells how to lay out all types of rafters.

A09-2832
Winslow, T. F. Construction Industry Production Manual. Craftsman Book Co., 1972. 176 pp. $6.00
   Contains man-hour tables for all types of construction. Charts and tables give the estimator the information he needs to compile an accurate estimate.

A09-2833
Woodward, W. S., et al. Drilled Pier Foundations. McGraw-Hill, 1972. 304 pp. $16.50

This book covers the application, design and construction of drilled pier foundations. It discusses safety requirements, specifications, and legal and contractual problems. Drilling and easing techniques and the tools used on the job are described.

A09-2834
Yardley, D. H., ed. Rapid Excavation--Problems and Progress: Proceedings of the Tunnel and Shaft Conference, 1968. Society of Mining Engrs. of AIME, 1970. 410 pp. $23.00 (for AIME members, $17.00)

With emphasis on tunnels in rock, this volume covers the current technology of tunneling and shaft sinking. Since preliminary engineering studies are important to the success of a tunnel project, stress is placed on present methods of site investigation and exploration and how to improve them. Case histories of recent tunnel and shaft projects are presented; there is a section on various types of moles and discussion of "mole tunneling." Support and linings, the legal aspects of tunnel contracting, and the environmental problems encountered underground are dealt with.

A09-2835
Zboinski, L. and Tyszynski, L., eds. Dictionary of Architecture and Building Trades: English/German/Polish/Russian. Pergamon, 1963. 500 pp. $28.50

Contains nearly 8000 entries from all branches of architecture and building, as well as other scientific terms commonly used in books on these subjects.

A09-2836
Zeevaert, L. Foundation Engineering for Difficult Subsoil Conditions. Van Nostrand Reinhold, 1973. 320 pp. $22.95

Information on the laws, theories and working hypotheses available in soils and foundation engineering, for the purpose of designing more successful building foundations under difficult environmental conditions. The book covers areas such as dewatering of excavations to reduce heave, injection of water outside excavations to reduce settlements, and application of plastic theory to estimate friction in piles.

A09-2837
Zinngrabe, C. J. Mathematics for the Sheet Metal Technician, Part 1. American Technical Society, 1969. 146 pp. $3.00

Practical information for the sheet metal worker, technician, draftsman, designer. In workbook form, Part 1 and Part 2 make a complete course to develop better skills in this field.

Section   A10

CONSTRUCTION INDUSTRY

>Building Trades
>Information Sources
>Industrialized Buildings

A10-2850
AIA. Glossary of Construction Industry Terms. American Institute of Architects, 1970. $1.25
Definitions of terms with special meaning or connotation in the construction industry.

A10-2851
ASCE. A Biographical Dictionary of American Civil Engineers. American Society of Civil Engineers, 1972. 163 pp. $5.00 (for ASCE members, $2.50)
Contains many hitherto unpublished photos and illustrations of some of the outstanding civil engineers and their works of the 18th, 19th and early 20th century, along with their biographies.

A10-2852
Bender, R. Selected Technological Aspects of the American Building Industry. The Industrialization of Building. National Technical Information Service, 1969. 98 pp. $3.00. Order No. PB-185211
The report prepared by the National Commission on Urban Problems, describes the state of industrialization in building, suggests ways that new materials and methods can be used to produce better buildings and communities, and outlines a framework within which these developments may take place.

A10-2853
Berne, J. How to Make Money in the Remodeling Business. Cahners, 1964. 165 pp. $9.95
Provides a management training course for dealers who want to go into the field or who want to increase their remodeling business and make it more profitable.

A10-2854
Building Blocks: Design Potentials and Constraints. Cornell Univ., Center for Urban Devel. Research, 1971. $6.00
Construction with factory-produced modules is a concept with far-reaching implications. Misused, however, the concept could have debilitating effects on the housing environment. Indeed, experience has revealed serious problems: to the environment, endless repetition and monotony; to the individual consumer, constricted floor plans resulting from the 12-foot maximum over-the-road width; to the producer, inability to accommodate changing and heterogeneous markets because of expensive, fixed manufacturing processes and equipment. In Building Blocks, the pitfalls and

potentials of the modular concept are examined. Differences of opinion are discussed, recommendations are made, and a new module system is presented which should provide ideas to anyone interested in the factory-produced housing field.

A10-2855
Building Science Directory.   Building Research Institute, annual.
$42.15
Listings of approximately 500 American building industry associations and societies, with names and addresses of officials; 300 private research and testing facilities, detailing fields of specialization and executives; 100 colleges and universities conducting building research, listing types of projects and contact personnel; 100 federal, state and local public agencies engaged in building research, with a breakdown of contact personnel. Also contains data sheets on 250 selected associations' and societies' organizational structure, research and educational programs, publications and standards or codes, arranged under 100 major subject headings and 3,000 subdivisions.

A10-2856
Bureau of the Census.   New One-Family Homes Sold and For Sale: 1963-1967.   U.S. Gov't. Printing Office, 1967.   300 pp. $4.75.   Order No. C3.2:H 75/2/963-67
This volume brings together all data on number of new one-family homes sold and for sale during this 5 year period into a single reference book covering sales price, type of financing, physical characteristics, installed appliances and other special features. It provides continuity of data for current C25 Census publications, Sales of New One-Family Homes. Selected annual and quarterly data give totals for the United States broken down by inside and outside standard metropolitan statistical areas and by region. The monthly data present United States totals only.

A10-2857
Burgess, R. A., et al., eds.   The Construction Industry Handbook. Cahners, 1973.   452 pp. $22.50
This book is a source of information for architects, contractors, surveyors, and other members of the construction industry. It includes a comprehensive analysis of the construction industry in Great Britain, reviews of developments in specialized building, a major listing of the properties of commonly used materials, essential data for environmental design, and a catalogue of international information sources.

A10-2858
Carreiro, J., et al.   The New Building Block: A Report on the Factory-Produced Dwelling Module.   Cornell Univ., Center for Urban Devel. Research, 1968.   278 pp. $6.00

A10-2859
Construction Labor Report.   Construction Craft Jurisdiction Agreements.   Bureau of National Affairs, 1974. $6.00

A collection of the building trades jurisdictional agreements which are recognized by the National Joint Board for the Settlement of Jurisdictional Disputes, but which are not printed in the "Green Book." This revised edition, originally published as a Special Supplement to BNA's Construction Labor Report, contains each agreement in large facsimile, including photographic and diagrammatic exhibits.

A10-2860
Davidson, H. A. Housing Demand: Mobile, Modular or Conventional? Van Nostrand Reinhold, 1973. 442 pp. $17.50

This book identifies and evaluates determinants of the demand for mobile homes in relation to the demand for conventional housing. It depicts major problem areas, as well as new developments, relevant to the current and future total housing market; analyzes historical growth of the mobile home sales; forecasts the demand for mobile homes through 1980 based on the key determinants of demand; and evaluates the effect of changes in demand on the financial structure of the industry.

A10-2861
Dietz, A. G. The Building Industry. National Technical Information Service, 1969. 291 pp. $3.00. Order No. PB-185 208

Prepared for the National Commission on Urban Problems, this report describes the building industry in the United States as a large loose aggregation of small to large units, each performing some part of the myriad operations that result in the production of buildings. In the usual pattern a team of the necessary elements is drawn together to carry out one building operation for an owner; the team is then dispersed and may not be reassembled in precisely the same form for another building. Although this procedure is fairly typical on a world-wide basis, the vertically-integrated construction firm performing many or all of the functions of design and construction, plus, if necessary, land development, financing, and management, is more common abroad than here.

A10-2862
Dubinsky, I. Reform in Trade Union Discrimination in the Construction Industry. Praeger, 1973. 332 pp. $18.50

Conceived in response to protest by an urban black militant group, Pittsburgh's Operation Dig, created in 1968, evolved into a government-sponsored manpower program to remedy trade union discrimination.

A10-2863
Electrical World Directory of Electric Utilities, 1975-1976. McGraw-Hill, 1975. 1000 pp. $95.00

The names, addresses and phone numbers of top utility personnel are given as well as the company statistics on fuels, customers, substation interconnections, generating capacities, etc. Also listed are national, state and industry associations and commissions that regulate and manage the industry.

A10-2864
Engineering News Record. Probing the Future. McGraw-Hill, 1974. 538 pp. $10.00

For over a year Engineering News Record editors researched the field and interviewed authorities in an attempt to forecast the course of the next hundred years in the engineering and construction industry. Their report was published as the April 30th 1974 issue of ENR. This book will make that report available to those who do not subscribe to ENR. The problems and issues of water, waste, energy resources, transportation, tools and methods, materials, and environmental design are topics covered.

A10-2865
Europe Design Intelligence Unit. Europe's Building Industries United Kingdom. George Godwin. 64 pp. $16.95 (available from Construction Publishing Co.)

Invaluable for any firm contemplating construction involvement in Great Britain, this reference covers structure of the construction industry, legislation, unions, governmental departments, and other statistical and analytical information.

A10-2866
Foster, H. G. Labor Supply in the Construction Industry: A Case Study of Upstate New York. National Technical Information Service, 1969. 415 pp. $3.00. Order No. PB 184 045

The purpose of the investigation that resulted in this publication was to describe, analyze and evaluate the manpower resources in the construction industry; to depict and dissect the extent and nature of the problems pertaining to labor supply and to appraise how and with what degree of success the industry copes with them.

A10-2867
Ghosh, D. The Economics of Building Societies. Lexington Books, 1974. $12.00

This is a timely and penetrating probe into the affairs of a section of the British world of finance which are the subject of fierce controversy. The study sets out to explain the structure and behavior of building societies, using a theoretical model which examines the aggregate building society balance sheet behavior during the 1960s.

A10-2868
Godel, J. B. Guide to Information Sources in the Construction Industry. Construction Publishing Co., 1975. 128 pp. $18.75

The book tells you what professional associations and periodicals serve the industry, where to find lists of buyers for your services, potential sources of new business, where to look for answers to your questions about standards and codes, how to find reports on specific construction subjects. A list of industry directories is included.

Construction Industry

A10-2869
Gordon, J. B., et al. Year-Round Employment in the Construction Industry: A Systems Analysis. Praeger, 1973. 148 pp. $16.50

Prepared for the Joint Study Group on Construction Seasonality of the U.S. Departments of Labor and Commerce. An econometric study of the costs and benefits of year-round planning in heavy industry, based on the U.S.-Canadian Barnhart Island Dam Project (1954-58). Investigates the comparative expenditure patterns and technology of the two nations, as well as the social impact on the workers involved.

A10-2870
Gramlich, E. and Jaffee, D. M. Savings Deposits, Mortgages and Housing. Lexington Books, 1974. $12.00

A work which brings up to date the work started in 1966 on the FRB-MIT Penn quarterly model of the U.S. economy. Concentrating on the areas of savings deposits, mortgage market and residential construction the book analyzes how the workings of monetary policy and monetary forces effect the real economy. Examines in depth models for each sector, the dynamics of sector interaction and the development of a summary model.

A10-2871
Grebler, L. Large Scale Housing and Real Estate Firms: Analysis of a New Business Enterprise. Praeger, 1973. 206 pp. $15.00

Details the emergence of large business in housing production and urban development. Analyzes four types of firms: on-site builders, home manufacturers, land developers, and builder-investors. Studies the growth of public corporations, acquisitions of real estate firms, and the internally generated real estate activities of major diversified corporations.

A10-2872
Grubb, C. A. and Phares, M. I. Industrialization: A New Concept for Housing. Praeger, 1972. 140 pp. $12.50

A10-2873
Guide to Architectural Information. Design Data Center. $4.95

This is a source book for information and references in the subjects of housing, urban design and architecture.

A10-2874
Havighurst, C. C., ed. Regulating Health Facilities Construction. American Enterprise Institute, 1974. 314 pp. $8.50 (paperback $4.00)

This volume contains the proceedings of a conference co-sponsored by the AEI and the Committee on Legal Issues in Health Care at Duke University School of Law. Over the years, government regulation of the health care has been reduced. This conference focused on the effect the trend toward government regulation and reduced competition could have on future health care delivery.

A10-2875
Housing and Urban Development, Dept. of. Bibliography on Housing Building and Planning. U.S. Dept. of Housing and Urban Development, Div. of International Affairs, 150 pp.
no charge
Lists approximately 500 recent books and periodicals available in the U.S. on housing, building and planning.

A10-2876
Housing and Urban Development, Dept. of. Operation Breakthrough. Volume 3. U.S. Gov't. Printing Office, 1974. 166 pp. $2.95. Order No. HH 1.2:B 74/9/v. 3 S/N 023-000-00272-8

The Department of Housing and Urban Development started Operation Breakthrough in 1969 to illustrate how all income groups can get quality housing. HUD selected 21 housing systems producers to build models. This volume describes the transportation, handling, storage, and erection systems employed by each of the producers, and discusses certain methods that could reduce logistics and construction costs for builders.

A10-2877
Housing and Urban Development, Dept. of. Registry of Minority Contractors and Housing Professionals. Vol. 1, A Listing of Minority Construction Contractors and Housing Professionals for Cities in: Connecticut, Maine, Massachusetts, New Hampshire, Rhode Island, and Vermont. U.S. Gov't. Printing Office, 1973. 111 pp. $2.85. Order No. HH 1/2:M 66/4/v.1 S/N 2300-00226

A listing of minority construction contractors and housing professionals in selected cities. The Registry is divided into ten separate regional volumes, with each volume listing, by metropolitan areas, names of contractors in alphabetical sequence. Part two provides additional background information on each of the businesses listed.

A10-2878
Housing and Urban Development, Dept. of. Registry of Minority Contractors and Housing Professionals. Vol. 2, A Listing of Minority Construction Contractors and Housing Professionals for Cities in: New Jersey, New York, and Puerto Rico. U.S. Gov't. Printing Office, 1973. 424 pp. $4.85. Order No. HH 1.2:M 66/4/v.2 S/N 2300-00227

A10-2879
Housing and Urban Development, Dept. of. Registry of Minority Contractors and Housing Professionals. Vol. 3, A Listing of Minority Construction Contractors and Housing Professionals for Cities in: Delaware, District of Columbia, Maryland, Pennsylvania, Virginia, and West Virginia. U.S.

Gov't. Printing Office, 1973. 626 pp. $6.55. Order No. HH 1.2:M 66/4/v.3 S/N 2300-00228

A10-2880
Housing and Urban Development, Dept. of. Registry of Minority Contractors and Housing Professionals. Vol. 4, A Listing of Minority Construction Contractors and Housing Professionals for Cities in: Alabama, Georgia, Florida, Kentucky, North Carolina, South Carolina, Tennessee, and Mississippi. U.S. Gov't. Printing Office, 1973. 1325 pp. $12.45. Order No. HH 1.2:M 66/4/v.4 S/N 2300-00229

A10-2881
Housing and Urban Development, Dept. of. Registry of Minority Contractors and Housing Professionals. Vols. 5A & 5B, A Listing of Minority Construction Contractors and Housing Professionals for Cities in: Illinois, Indiana, Michigan, Minnesota, Ohio, and Wisconsin. U.S. Gov't. Printing Office, 1973. 1842 pp. $16.70. Order No. HH 1.2:M 66/4/v.5A & 5B S/N 2300-00230

A10-2882
Housing and Urban Development, Dept. of. Registry of Minority Contractors and Housing Professionals. Vol. 6, A Listing of Minority Construction Contractors and Housing Professionals for Cities in: Arkansas, Louisiana, Oklahoma, New Mexico and Texas. U.S. Gov't. Printing Office, 1973. 1424 pp. $13.05. Order No. HH 1.2:M 66/4/v.6 S/N 2300-00231

A10-2883
Housing and Urban Development, Dept. of. Registry of Minority Contractors and Housing Professionals. Vol. 7, A Listing of Minority Construction Contractors and Housing Professionals for Cities in: Iowa, Kansas, Missouri, and Nebraska. U.S. Gov't. Printing Office, 1973. 109 pp. $1.75. Order No. HH 1.2:M 66/4/v.7 S/N 2300-00232

A10-2884
Housing and Urban Development, Dept. of. Registry of Minority Contractors and Housing Professionals. Vol. 8, A Listing of Minority Construction Contractors and Housing Professionals for Cities in: Colorado, Montana, North Dakota, Utah and Wyoming. U.S. Gov't. Printing Office, 1973. 185 pp. $2.85. Order No. HH 1.2:M 66/4/v.8 S/N 2300-00233

A10-2885
Housing and Urban Development, Dept. of. Registry of Minority Contractors and Housing Professionals. Vol. 9A & 9B, A Listing of Minority Construction Contractors and Housing Professionals for Cities in: Arizona, California, Hawaii, and Nevada. U.S. Gov't. Printing Office, 1973. 2426 pp. $21.20. Order No. HH 1.2:M 66/4/v.9A & 9B S/N 2300-00234

A10-2886
Housing and Urban Development, Dept. of. Registry of Minority Contractors and Housing Professionals. Vol. 10, A Listing of Minority Construction Contractors and Housing Professionals for Cities in: Alaska, Idaho, Oregon, and Washington. U.S. Gov't. Printing Office, 1973. 226 pp. $3.35. Order No. HH 1.2:M 66/4/v.10 S/N 2300-00235

A10-2887
Housing and Urban Development, Dept. of. Services Available to Hud-related Business in International Trade. U.S. Gov't. Printing Office, 1975. 61 pp. $1.10. Order No. HH 1.40/5:Se 6 S/N 023-000-00301-5
This publication identifies various services available to U.S. businesses investigating international trade opportunities in the fields of housing and community development. The publication contains basic information about governmental and non-governmental organizations that can help businessmen find, consider, and undertake international business transactions.

A10-2888
Housing Research and Building Technology Activities of the Federal Government. U.S. Gov't. Printing Office, 1970. 117 pp. $1.45. Order No. PrEx 8.2:H 81/2 S/N 4106-00027

A10-2889
INBEX '71 Digest of Seminars--Industrialized Building Exposition and Congress. Cahners, 1972. 180 pp. $3.95
Experts, opinion-makers, and movers in the fastest-growing area of home building gathered at the Second Annual INBEX at Louisville, Kentucky. Over 150 talks are presented in a digested form.

A10-2890
Kelly, B. Design and the Production of Houses. McGraw-Hill, 1959. 428 pp. $12.50
This book surveys the process by which single new houses are designed and produced. Includes a discussion of the mobile home industry, labor, controls, descriptions of advances in housing and more.

A10-2891
Kelly, B. The Prefabrication of Houses. MIT Press, 1951. 166 pp. $12.50
A study of the postwar prefabrication industry in America, covering management, design, production, marketing, and technical problems, with an introductory historical survey.

A10-2892
Kelly, P. M. and Shillaber, C. International Bibliography of Prefabricated Housing. MIT Press, 1954. 85 pp. $1.00
This bibliography indicates the wide range of literature available in different languages for countries which produce or

make use of prefabricated housing. It covers the period from January, 1948 to June, 1954, although some significant earlier references have also been included. A list of prefabricating firms in various countries is given in the appendix.

A10-2893
Labor Statistics Bureau. Seasonality and Manpower in Construction. U. S. Gov't. Printing Office, 1970. 148 pp. $1.25. Order No. L2.3:1642

A10-2894
Lave, J. R. and Lave, L. B. The Hospital Construction Act: An Evaluation of the Hill-Burton Program. American Enterprise Institute, 1974. 41 pp. $3.00
The authors analyze the twenty-five year record and the current utility of this popular program. They conclude that Hill-Burton has fulfilled the goals of the original legislation and recommend its termination. The authors find no overall shortage of hospital beds in the United States today, despite overuse of in-patient facilities. Federal subsidies for construction are no longer needed and may indeed be counterproductive except in the case of hospitals with a high proportion of nonpaying patients.

A10-2895
Means, R. W. 1976 Labor Rates for the Construction Industry: City, State, National. R. W. Means, 1976. $17.50
Complete detailed wage rates, by trade, by city for more than 275 U.S. and Canadian cities, for up to 45 construction trades for each city. Also included are basic union hourly wage rates, plus fringe benefit packages and effective dates for negotiated Union contracts for 1976 and 1977.

A10-2896
National Research Council. Promotion of the Development and Use of the Subsystem Concept of Building Construction. National Academy of Sciences, 1973. 108 pp. $3.50
The report on the first of a two-phase project presents the results of an in-depth investigation into the desirability, feasibility, and means of implementing the concept of using dimensionally and functionally precoordinated subsystems as the basic building blocks in construction. Indicating that building with subsystems is the most promising alternative to the traditional process, the study proposes a program for stimulating the development, marketing, and widespread use of a broad range of precoordinated subsystems while minimizing legal and institutional constraints on their use.

A10-2897
Neutze, M. The Suburban Apartment Boom: Case Study of a Land Use Problem. Johns Hopkins Univ. Press, 1968. 170 pp. $5.00
In the 1950s apartment living seemed to be on the decline in the United States in relation to single family suburban housing. In the 1960s the emphasis has returned to apartments, most of

which are being built in the suburbs. Max Neutze assesses the dimensions of this development, inquires into the reasons for it, and analyzes its economic efficiency in terms of land use.

A10-2898
NHIC. Building Materials Technology and Selling Assignment Manual. National Home Improvement Council. $21.95 (Order from Distributive Education Instructional Mat'ls. Services, Univ. of Texas, Austin 78712)

For building material salesmen, home improvement salesmen and independent contractors. Product and selling information on lumber, paneling, ceilings, flooring, insulation, roofing, siding, gypsum, doors and windows, millwork, plywood, cement, masonry, paint, etc. The assignment manual, complete with study questions and projects stimulate motivation among trainees.

A10-2899
NICB. Economics of the Construction Industry. National Industrial Conference Board, 1969. 150 pp. $12.50

An examination of the relationship of contract construction to total construction activity. The industry's structure and its economic efficiency and performance are carefully scrutinized.

A10-2900
Organization for Social and Technical Innovation. Self-Help Housing in the U.S.A. National Technical Information Service, 1969. 269 pp. $3.00

This report is designed to provide an objective review of public and private self-help housing efforts and their present and potential contributions toward achieving national housing goals. The scope of work includes: factual analysis of nine specific self-help and mutual-help programs supported under the Section 207 program; identification and analysis of all other relevant HUD efforts in self and mutual-help housing. Identification and analysis of self and mutual-help housing efforts supported by other Federal departments and agencies, other levels of government and private organization as appropriate.

A10-2901
Paulus, V. Housing: A Bibliography, 1960-1972. Center for Urban Policy, Rutgers Univ., 1973. 350 pp. $25.00

A subject arrangement of nearly 4,000 citations including books, journal articles, government documents, technical reports, and theses, dealing with various aspects of housing. The period covered represents a time when housing problems and issues became a national priority and when a great deal of research in the housing field was published. The material selected includes such as economics, law, urban planning, sociology, and public policy.

A10-2902
Paulusci, L. Profitable Retailing of Building Supplies. Cahners, 1969. 124 pp. $12.95

Written for every building supply retailer who wants to

boost sales. Especially valuable to anyone thinking of opening a new store, enlarging an existing facility, or adding a new product category.

A10-2903
Rada, E. L. The Cal-Vet Program: A Study of State-Financed Housing in California. Univ. of California, Grad. School of Mgmt. 173 pp. $4.50

Presents an analysis of the financial management of the program and its role in California's housing and mortgage markets. The study includes original data on the characteristics of veterans who bought homes on Cal-Vet contracts and of the houses they acquired. The Cal-Vet program is appraised in terms of its benefits and costs.

A10-2904
Ragatz, R. L. Vacation Homes: An Analysis of the Market for Seasonal-Recreational Housing. Cornell Univ., Center for Urban Devel. Research. 388 pp. $5.00

The text begins with an overview of the vacation home market and effects of this land use. The factors contributing to its existence and growth are examined along with the study of the family characteristics of the owners. Statistics of the current supply of vacation housing along with annual starts, dollar volume, etc. are provided. Vacation housing production and financing are also studied.

A10-2905
Reid, M. Housing and Income. Univ. of Chicago Press, 1962. 415 pp. $7.50

This study of the relation of housing, its values and rents, to normal or expected long-run income, concludes that the proportion of income spent on housing rises as income increases. Using evidence from surveys and censuses covering the years 1918-60, the author assesses such factors as short-run income change, error in reporting income, cyclical change, composition of households, mobility of population, and market price of housing. She also shows that the rise in price of housing has a marked tendency to decrease housing consumption.

A10-2906
Reidelbach, J. A., Jr. Housing Compendiums. Vol. 1. Herman Publishing, Inc., 1972. 320 pp. $19.95

Guides reader to needed information in broad range of housing literature. References and abstracts. Six sections cover national goals and programs; legislation and laws; finance and fiscal policy; social concepts and environment; methods and materials.

A10-2907
Reidelbach, J. A., Jr. Modular Housing 1971--Facts and Concepts. Cahners, 1971. 237 pp. $7.95

A comprehensive survey of the modular housing industry, its impact on the American economy, and its growing place in the

housing industry are the subjects of this book by one of the top
authorities in the field.

A10-2908
Reidelbach, J. A., Jr. Modular Housing '72, Statistics & Specifics.
    Herman Publishing, Inc., 1972. 236 pp. $24.50
    Accurate report on today's modular housing industry and its
trends.

A10-2909
Ricks, R. B., ed. National Housing Models. Lexington Books,
    1973. 208 pp. $12.50
    This study brings together current work in the development
of housing-related models. Econometric analyses of residential
housing and construction include models developed for housing,
mortgages and housing starts. It presents an analytical summary
of the FRB-MIT-Penn model sectors covering financial intermediaries, mortgage market and housing. Discusses model breakdowns,
model-associated problems, the uses of models in forecasting and
in policy making.

A10-2910
Sandstrom, G. Man the Builder: Origins of Civil Engineering.
    McGraw-Hill, 1970. 280 pp. $24.95
    This unusual reference work recounts the engrossing history
of man's architectural and engineering feats from Stonehenge to
modern power dams. It is illustrated with more than 200 sketches,
engravings, maps, drawings and photographs.

A10-2911
Schmid, T. and Testa, C. Systems Building: An International
    Survey of Methods. Praeger, 1969. 240 pp. $20.00
    The authors survey the state of industrialized building today,
explaining new theories in modular building and the "reversed designing procedure." They also provide coverage of typical technical solutions in different building systems, using examples from
Western Europe, Scandinavia, England, the United States, and
Eastern European countries. A separate section evaluates present
methods and describes the conditions necessary for successful research.

A10-2912
Smith, D. L. How to Find Out in Architecture and Building--A
    Guide to Sources of Information. Pergamon Press. $6.50
    A source book for information on building, planning and architecture.

A10-2913
Sternlieb, G. and Listokin, D., eds. Housing 1973-74: An Anthology. Center for Urban Policy Research, Rutgers Univ.
    1974. 400 pp. $15.00
    The definitive annual compendium of significant research on

Construction Industry

housing finance, federal housing subsidies, residential segregation, new modes of housing development, and many other topics.

A10-2914
Sternlieb, G. and Paulus, V., eds. Housing, 1971-1972. Center for Urban Policy, Rutgers Univ., 1973. 500 pp. $15.00

A compendium of significant research published between July 1971 and December 1972 in journals, government documents, and technical reports. Included are materials dealing with housing market analysis, exclusionary zoning, mortgage finance, race and housing, fair share plans, scatter-site efforts, abandonment, and evaluation of federal subsidy programs.

A10-2915
Sternlieb, G. and Paulus, V., eds. Housing, 1971-1972. Vol. 2. AMS Press, 1974. 496 pp. $15.00

This second volume in an annual series of anthologies on contemporary housing problems includes twenty-nine major articles by thirty-eight contributors, published between June 1971 and June 1972. Covers current trends in housing, exclusionary zoning, and fair share, scatter-site, inner-city housing, economic analysis and housing markets, race and housing, housing finance, policy considerations.

A10-2916
Sternlieb, G. and Sagalyn, L. B., eds. Housing, 1970-1971. Vol. 1. AMS Press, 1972. 592 pp. $25.00

This first volume in an annual series of anthologies is designed for libraries and professionals interested in general housing problems. Thirty-four significant papers published in periodicals and journals are represented on such subjects as current trends in housing, economic analysis and housing markets, residential construction industry, housing and racial integration and changes in landlord-tenant relations.

A10-2917
Tavistock Institute of Human Relations. Interdependence and Uncertainty: A Study of the Building Industry. Tavistock 1966. $4.00 (available from Barnes and Noble)

A10-2918
U.S. Savings and Loan League. 1976 Savings and Loan Fact Book. U. S. Savings and Loan League, 1976. 144 pp. no charge

This is an annual publication that provides a comprehensive reference source on savings, home ownership, and residential construction and financing. The savings and loan business is analyzed as are the functions of the Federal Home Loan Bank System and the Federal Savings and Loan Insurance Corporation.

A10-2919
van den Broek. Habitation. Vol. 2, Belgium, Denmark, France, United Kingdom, Sweden, Switzerland and Czechosolvakia. North-Holland, 1959. 284 pp. $18.20

This series of three volumes on the world problem of housing was initiated at the Fourth Congress of the U.I.A. The programme section consists of a general survey of each country of the housing standards obtaining there, the legislation concerning the building of houses, and the manner in which housing is financed. In the section on design the various types of dwelling are treated, namely one-family houses, apartment houses, special forms of housing and district plans. The most important results achieved as regards normalization, standardization and modular coordination, development of standard ground-plans, standard houses and elements, and details in the field of design and equipment of dwellings, are surveyed.

A10-2920
van den Broek. Habitation. Vol. 3, USA, USSR, Finland, Greece, Hungary, Turkey and Yugoslavia. North-Holland, 1964. 350 pp. $29.10

A general review of each country is included in the Programme section, followed by an analysis of requirements, forms of legislation and financing. The design section is grouped according to types of dwelling; special forms of housing and district plans. The most notable results of normalization and standardization are recorded, along with ground plans, standard houses and elements, and details of household equipment. Traditional production methods, series production and an economic analysis are covered in the third section. Attention is given to the various forms of order and contract, and the influence of the methods of production on architecture.

A10-2921
Ward, J. W. Construction Information Source and Reference Guide. Construction Publications, 1973. 190 pp. $8.00

A listing of construction-related texts, manuals, handbooks, associations, societies, periodicals, publishers and book sources for use by engineers, architects, contractors and others.

A10-2922
Whitehead, C. M. E. The U. K. Housing Market: An Econometric Model. Lexington Books, 1973.

This study develops a model of the private housing market in the U.K. for the period 1955-1970. It examines the determinants of demand and supply taking account of changes in incomes, prices, costs, and the rate of inflation, and looks at the question of whether new housing can be regarded as an equilibrium market. The book also analyzes the effect of tightness in the money market on both the demand and the supply of new housing, and discusses the importance of public sector building and other government policies towards housing.

A10-2923
Wolman, H. L. Housing and Housing Policy in the U. S. and the U. K. Lexington Books, 1975. 144 pp. $13.00

Noting that cross-national comparison can provide fresh

alternatives to chronic problems, Wolman contrasts American and British housing problems and policies. The author examines housing policy in each country in its political, economic, and social aspects.

Section   A11

SELECTED TOPICS

       Drafting, Illustration, Blueprint Reading
       Exams for Professional Registration
       Fire Safety
       General Safety and Security

A11-2930
ACI. Manual of Standard Practice for Detailing Reinforced Concrete Structures. American Concrete Institute, 1974. 167 pp. $16.00 (for ACI members, $11.50)

This manual is the "bible" of design and drawing practice for reinforced concrete. It emphasizes techniques to reduce labor costs for forming and reinforcing and the use of waffle slabs. Prestressed and precast concrete details are also included as is a chapter on computer-aids to detailing.

A11-2931
AIA. Architectural Registration Handbook (NCARB). American Institute of Architects, 1973. $18.50 (for AIA members, less 10%)

A test guide for those taking the professional examination for architectural registration.

A11-2932
AIChE. Plant and Design Safety. American Institute of Chemical Engrs., 1965. 97 pp. $4.00 (for AIChE members, $2.00)

Twelve articles selected from Chemical Engineering Progress on evaluating and anticipating hazards; use of radioactive sources; handling flammable dusts, phosphorous, and other hazardous chemicals; others.

A11-2933
AISC. Problems and Solutions for Structural Steel Detailing. American Institute of Steel Construction. $5.00

This 2-unit set of exercise booklets has been prepared for use in technical schools or on-the-job training of structural steel draftsmen. The problems and their solutions are keyed to the basic textbook, the 2nd Edition of AISC's Structural Steel Detailing and to the 7th edition AISC Manual of Steel Construction. They provide supplementary exercises in the analysis, procedures, and calculations frequently required in detailing practice.

A11-2934
AISC. Structural Steel Detailing. American Institute of Steel Construction. 408 pp. $16.00

Reflects the many improvements that have taken place in the structural steel fabricating industry since publication of the first edition in 1966. Every facet of current drafting practice for structural steel fabrication is explained and illustrated by over 400 drawings showing details of beams, columns, trusses, and the miscellaneous members required in a completed building framework. Computations and design calculations that must be made by the structural detailer are presented throughout.

A11-2935
Alger, P. L., et al. Ethical Problems in Engineering. John Wiley, 1965. 299 pp. $9.95

Addresses ethical problems of consulting engineers, industrial engineers, governmental engineers, and construction engineers. Includes the ECPD Canons of Ethics and the NSPE and ASCE Codes.

A11-2936
Anderson, C. R. OSHA and Accident Control Through Training. Industrial Press, 1974.

The subjects covered include a discussion of the OSHA law and the consequences of non-compliance; establishing safety standards and regulations internally; the economic benefits of instituting a comprehensive safety program; the steps involved in initiating the safety training program; a description of how to determine safe operating procedures; the use of posters, multi-language signs, and incentives; case studies; accident reporting; inspection and presentation of statistics of management and trade unions.

A11-2937
Answers on Blueprint Reading. Howard Sams. 416 pp. $5.25

An instruction manual. Aids in understanding blueprints for machines and tools, electricity, and architecture. Gives short cuts and suggestions.

A11-2938
Apfelbaum, H. J. and Ottesen, W. O. Basic Engineering Sciences and Structural Engineering for Engineer-in-Training Examinations. Hayden, 1970. 424 pp. $13.95

This volume covers mathematics, statics and dynamics, electricity and electronics, physics, and all types of structural elements and materials. A nation-wide sampling of examination problems, and quick-reference arrangement of the material are featured.

A11-2939
Arco Editorial Board. Mechanical Engineer--Junior, Assistant and Senior Grades. Arco, 1974. 288 pp. $8.00

This review provides test preparation for a mechanical engineer career in civil service. Includes eleven actual previous exams with detailed answers and two specially constructed exams for predictive practice.

A11-2940
Arnell, A. Standard Graphical Symbols. McGraw-Hill, 1963. 536 pp. $24.50

Here is a precise, highly-illustrated reference book, full of symbols for use on drawings, plans, and specifications in different fields.

Selected Topics

A11-2941
Arnold, T. and Vaden, F. S. Invention Protection for Practicing Engineers. Cahners Books, 1971. 190 pp. $7.95 (paperback $4.95)

Fundamentals of the patent system, obtaining a patent, patent enforcement and litigation, and other patent-related matters are covered in this book.

A11-2942
ASCE. Definitions of Surveying and Associated Terms. American Society of Civil Engineers, 1972. 205 pp. $8.00 (for ASCE members, $4.00)

A glossary of terms has just been issued by ASCE and ACSM. This new publication, which replaces ASCE Manual 34, is intended to standardize terminology for national usage in surveying. The glossary includes current and historic land surveying terms, general surveying terms and selected terms in several fields closely related to surveying.

A11-2943
Associated Builders and Contractors. Construction Safety Guide. Associated Builders and Contractors, Inc., 1974. $40.00 (for ABC members, $22.00)

Eleven chapters deal with the requirements of the Occupational Safety and Health Act, organizing a safety program, safety inspection, accident investigation, first aid, fire protection and prevention, safety evaluation, training and record keeping for OSHA, and construction standards.

A11-2944
ASTM. Ignition, Heat Release, and Noncombustibility of Materials. American Society for Testing and Materials, 1972. 165 pp. $10.00 (for ASTM members, $8.00)

Ten papers cover methods of developing an understanding of fire problems defined in terms of characteristics of materials, assembly and conditions.

A11-2945
Austin, P. R. Design and Operation of Clean Rooms. Business News Publishing, 1970. 447 pp. $29.95

This book is important to those who design, operate or purchase components for clean rooms. Such subjects as air filtration, laminar flow, standards and monitoring are featured.

A11-2946
Bannister, A. and Raymond, S. Surveying. Pitman and Sons, 1974. 495 pp. £3.75 (paperback £2.50)

A text on the elements of surveying, written in SI units. The chapters on photogrammetry, further instruments and techniques, levelling, and traverse surveying, have been revised.

A11-2947
Barker, J. A. Reinforced Concrete Detailing. Oxford University

Press, 1967. 370 pp. $15.25
The process of reinforced concrete detailing is described. The book may be used by students or as a reference for architects, contractors and concrete manufacturers. The material included is in accord with the latest British standards.

A11-2948
Barry, B. A. Construction Measurements. John Wiley, 1973. 304 pp. $18.25
Contents include elements of construction measurements, difference of elevation, taping, angle measurements and layout, building construction layout, highway planning, shafts and tunnels, etc.

A11-2949
Barsukov, P. Building Construction Drawing. Beekman, 1976. 312 pp. $11.00
Translated from the Russian, the book deals with draftsman's materials, instruments, and accessories; the layout and title blocks of drawings; the geometrical constructions on a plane; exonometric protections; technical drawings; rectangular projections; and mechanical drafting.

A11-2950
Bellis, H. and Schmidt, W. Blueprint Reading for the Construction Trades. McGraw-Hill, 1968. 160 pp. $6.95
Studies blueprints related to all aspects of the construction trades, for carpenters, bricklayers, concrete workers, mechanics, plumbers, and electricians.

A11-2951
Bender, W. H. California Structural Engineer's License Review. Engineering Press, 1970. 290 pp. $12.00
The complete California Structural Engineering examinations, 1960 through 1964, are presented together with solutions to all problems in three of the five examinations.

A11-2952
Best's Safety Directory. A. M. Best, 1974. 1000 pp. $20.00
This annual directory combines an analysis of industrial safety, pollution control and security products with "how to do it" editorial material. Ideas, control techniques and technical advancements are available. Federal OSHA standards are presented and defined.

A11-2953
Bhatnager, V. M., ed. Advances in Fire Retardants, Part 1. Technomic, 1972. 200 pp. $25.00
Twelve studies detail recent developments in fire-retardant materials, fire-proofing and fire-retarding treatments, test methods and results, and flammability of materials. The materials examined include: plastics, wood, paint and other coatings, and textiles. Both practical applications and theory are presented.

Selected Topics

A11-2954
Bhatnager, V. M., ed. Advances in Fire Retardants, Part 2.
   Technomic, 1973. 200 pp. $25.00
   Contents include information on flame resistant phenolic fibers, fire retardation of cellulosic and plastic materials, fire-retardant textiles, wood, etc.

A11-2955
Bhatnager, V. M. Fire Retardant Formulations Handbook. Technomic, 1972. 245 pp. $20.00
   Here are 400 major fire-retardant formulations for the most widely used flammable materials. They apply to wood, paint, plastics, rubber, textiles, paper, and building materials. Original sources are given. The formulations reflect recent advances in fire retardance technology and trends in flammability regulations.

A11-2956
Bickmore, D., ed. Automatic Cartography and Planning Techniques in Automatic Cartography. Architectural Press, 1971.
   232 pp. £8.50
   This is a study conducted by the Cartography Research Unit at the Royal College of Art on the extent to which new techniques in computer-controlled cartography can be applied to improving methods of presenting planning data. It includes a catalogue of U.K. statistical material potentially suitable for mapping, a number of simple maps made to demonstrate methods of cartography that might be employed in handling planning problems and a section describing and comparing methods of automatic cartography.

A11-2957
BNA Editorial Staff. The Job Safety and Health Act of 1970.
   Bureau of National Affairs, 1971. $15.00
   This work explains the provisions of the Occupational Health and Safety (Williams-Steiger) Act and tells what it means to employers, employees, states, associations, and unions. Contains the text of the law, a legislative history, and documents which shed light on the intent of Congress.

A11-2958
Bond, H. NFPA Inspection Manual. National Fire Protection
   Assn., 1970. 352 pp. $6.75
   Tells what to look for in inspecting all types of properties. Describes common and special hazards, building features, exits, inspection of extinguishing equipment, detection systems, report writing and other essentials. Standard plan symbols in full color and many other illustrations.

A11-2959
Bouchard, H. and Moffitt, F. H. Surveying. Intext, 1965.
   754 pp. $9.95
   This 5th edition emphasizes the fundamental principles of surveying practice with basic information on special topics and

their applications. Introduces the student to automatic devices used in photogrammetry for obtaining earthwork quantities.

A11-2960
Breed, C. B. Surveying. John Wiley, 1971. 495 pp. $12.95
 Presents material on surveying from a practical point of view. Includes field exercises which provide a fundamental understanding of surveying techniques used in all branches of surveying. Modern methods and new equipment have been introduced in this edition.

A11-2961
Breed, C. B. and Hosmer, G. L. Principles and Practice of Surveying. Vol. 1, Elementary Surveying. John Wiley, 1966. 717 pp. $14.50
 Covers use, adjustment and care of instruments, surveying methods, computations, and plotting and electronic surveying.

A11-2962
Breed, C. B. and Hosmer, G. L. Principles and Practice of Surveying. Vol. 2, Higher Surveying. John Wiley, 1962. 543 pp. $13.25

A11-2963
Bretz, H. Sheet Metal Shop Drawing. Industrial Press, 1971. 301 pp. $15.00
 This book is a text for sheet metal draftsmen. It also provides shop drawing reference material and standards for sheet metal apprentices, journeymen, contractors, architects and engineers. The student is taught how to execute drawings for shop fabrication and field erection, coordination of work by various contractors, and for approval by the architect and engineer. Practical mathematics for sheet metal draftsmen included.

A11-2964
Brinker, R. C. Elementary Surveying. Intext, 1969. 620 pp. $10.50
 This treatment of modern surveying topics has been written for those who must know the fundamentals of surveying and apply them in their daily work. Part 1 covers fundamentals of elementary surveying; Part 2 covers advanced surveying topics and surveying applications.

A11-2965
Brinker, R. C. and Barry, B. A. Noteforms for Surveying Measurements. Intext, 1957. 93 pp. $1.50
 This set of sample field notes provides a guide for accepted recording practice. The cover and punched pages are easily separated for insertion in the standard looseleaf field book.

A11-2966
Brinker, R. C., et al. 1777 Review Problems from E.I.T. and Professional Registration Examinations: With Answers and

Typical Solutions. Intext, 1967. 524 pp. $9.50
This text is designed as a refresher course for the various E.I.T. and P.E. exams. Instructions explain the conditions under which the example examinations were given, so that the candidate can take one of these examinations under actual conditions.

A11-2967
Burden, E. Architectural Delineation: A Photographic Approach to Presentation. American Institute of Architects or McGraw-Hill, 1971. 310 pp. $21.95 (for AIA members, less 10%)
Mr. Burden has a new approach to architectural rendering. It enables quick determination of the ideal vantage point and ensures the rendering will be kept down to an optimum size. Includes basic concepts on camera and perspective; preparation of the layout; techniques of drawings; and presentation.

A11-2968
Capelle, F. Professional Perspective Drawing for Architects and Engineers. McGraw-Hill, 1969. 192 pp. $19.00
Written for professionals and students faced with the task of constructing accurate, appealing perspectives, this book presents an original, simple, and highly accurate technique for finding the size and location of the perspective between two vanishing points. Faster than conventional ones, this technique also reverses conventional procedures by enabling the designer to start with the size of the perspective.

A11-2969
Choate, C. Architectural Presentation in Opaque Watercolor. Van Nostrand Reinhold, 1961. 158 pp. $16.95
The author demonstrates the advantages of using opaque watercolor to create atmosphere, and how to indicate activity and function within the scene. He also analyzes the degree of psychological stimulus of colors and compositional devices, evaluating their effect on the viewer.

A11-2970
Coleman, R. M. Technical Drafting: Aerospace, Electrical-Electronic, Structural. Holt, Rinehart and Winston, 1971. 320 pp. $10.95
The text is divided into three main units of instruction: Aerospace, Electrical and Electronic, and Structural. Each unit is preceded by fundamentals of general drafting which include equipment and supplies, geometry of drafting, multiview projection, sectioning, auxiliary views, and pictorial representations.

A11-2971
Collins, S. P. Handbook of Accurate Surveying Methods. Pitman and Sons, 1973. 112 pp. £1.40
The scale and complexity of modern engineering construction projects demands precision and speed in their mapping and control by surveyor, and this book describes systems to achieve it.

Construction Information

A11-2972
Commerce Clearing House. Occupational Safety and Health Act of
 1970: Law and Explanation. Commerce Clearing House.
 128 pp. $3.50
 The full text plus a detailed explanation of the Occupational
Safety and Health Act of 1970. Contains explanations of the procedures the Secretary of Labor will follow in setting and enforcing safety standards, added responsibilities and temporary relief measures for employers and options open to employees and unions. Includes excerpts from controlling committee reports.

A11-2973
Conover, H. Grounds Maintenance Handbook. McGraw-Hill, 1958.
 538 pp. $22.50
 Practical, economical methods are stressed in this book which presents detailed information needed to supervise, maintain, and plan grounds from one to thousands of acres--public, private, or industrial.

A11-2974
Constance, J. Electrical Engineering for Professional Engineers'
 Examinations. McGraw-Hill, 1970. 456 pp. $17.00
 Designed to prepare candidates for the electrical engineering branch of the Professional Engineers' Examination, this book covers direct-current circuits, alternating-current circuits, electrical machinery, electrical distribution and transmission, and electrical controls.

A11-2975
Constance, J. How to Become a Professional Engineer. McGraw-
 Hill, 1966. 294 pp. $11.50
 Gives advice on what to do and how to do it when preparing for exams and filing application forms.

A11-2976
Constance, J. Mechanical Engineering for Professional Engineers'
 Examinations. McGraw-Hill, 1969. 492 pp. $17.00
 The updated edition of this popular book contains new material on computers, charts, cryogenics, gas dynamics, mechanical vibrations, gas compressors, etc. It includes a special section on licensing of foreign engineers. A helpful reference in preparing for state exams, the book is also useful for persons interested in civil service exams, design engineering, and plant engineering.

A11-2977
Coren, G. Mechanical Engineering. Arco, 1965. 288 pp. $5.00
 This study guide for the Mechanical Engineer Licensing Exam includes actual previous exams with detailed, worked-out answers.

A11-2978
Coulin, C. Step-By-Step Perspective Drawing: For Architects,
 Draftsmen, and Designers. Van Nostrand Reinhold, 1971.
 $13.95

Selected Topics

Every phase of perspective is covered with the aid of 43 full pages of descriptive geometry. Self-teaching is possible with the use of this book.

A11-2979
Croney, J. Anthropometrics for Designers. Reinhold. 176 pp. $7.95

Helps solve problems of environmental design through an account of man's dimensions and other physical data. Covers growth, figure typing, measurement, and human engineering.

A11-2980
CRSI. Reinforcing Bar Detailing. Concrete Reinforcing Steel Institute, 1970. 280 pp. $18.50

22 illustrated chapters covering materials, specifications, contract and placing drawings to current detailing practices and computer estimating and detailing. For technical and vocational schools, and industrial on-job training programs.

A11-2981
Cusset, F. English-French, French-English Technical Dictionary. Chemical Publishing Co., 1967. 436 pp. $9.00

Of particular use to engineers, chemists and metallurgists who must refer to French technical texts.

A11-2982
Dalzell, J. Plan Reading for Home Builders. McGraw-Hill, 1973. 172 pp. $9.95

Basic principles of reading construction plans are discussed in light of new advances. This second edition has also been updated as to the symbols used on drawings and the materials more widely used in house construction. Also included are specification changes needed because more appliances are being built into homes.

A11-2983
D'Amelio, J. Perspective Drawing Handbook. Tudor Publishing Co., $4.95

An instruction manual and reference guide on the fundamentals. Over 450 drawings and diagrams.

A11-2984
Davis, R., et al. Surveying--Theory and Practice. McGraw-Hill, 1966. 1120 pp. $16.00

A discussion of surveying that emphasizes the precision of measurements and computation.

A11-2985
Davis, R. E. and Kelly, J. W. Short Course in Surveying. McGraw-Hill, 1942. 330 pp. $11.00

Covers field and office work, errors, measurement of distance in elevation, differential and profile leveling, measurement of angles and directions, engineer's transit, the plane table, map plotting, and more.

A11-2986
Desch, H. E. Structural Surveying. Charles Griffin, 1970.
     269 pp. £3.50
     The author is aware of the various sources of timber decay, whether of a new council estate or a medieval college. He has surveyed many unusual, ancient, and large properties, and points out possible sources of trouble and the best way to make proposed improvements.

A11-2987
Doblin, J. Perspective: A New System for Designers. Watson-
     Guptill, 1956. 68 pp. $7.50
     For designers this text offers a simple method of visualizing any three-dimensional object accurately and quickly and eliminates complex mechanical drawing. For draftsmen it helps develop the freehand skill and judgment that any good student of perspective must have.

A11-2988
Dreyfuss, H. The Measure of Man: Human Factors in Design.
     Watson-Guptill, 1967. 20 pp. 32 charts $16.50
     Provides measurements of every part of the human body in standing or sitting positions, including sight lines, reach, and other design factors. The designs are based on three types of frames (small, medium, large) for men and women. Also included are charts on human strength, body clearances, climbing data, access openings.

A11-2989
Dudley, L. Architectural Illustration. Prentice-Hall, 1976.
     336 pp. $16.95
     Presents different types and uses of architectural illustration, leading techniques and procedures for learning and developing proficiencies in each of them. To-the-point topic coverage, approximately 400 illustrations (including diagrams, charts, and finished drawings), plus suggestions for inexpensive home-made drawing aids and equipment make this a practical book for practicing or aspiring architects.

A11-2990
Dunning, W. J. and Robin, L. P. Home Planning and Architec-
     tural Drawing. John Wiley, 1966. 81 pp. $5.25
     Contents include planning purposes and principles, the single line sketch, room analysis, roof plans, and elevations.

A11-2991
EJC. Thesaurus of Engineering and Scientific Terms. Engineers
     Joint Council, 1967. 696 pp. $25.00 (paperback $20.00)
     A standardized vocabulary reference for use in information storage and retrieval systems. Over 23,000 main entries include 18,000 preferred terms and 5,000 cross-references. The book is the result of a cooperative effort of EJC and the Department of Defense.

Selected Topics

A11-2992
Factory Mutual System. Handbook of Industrial Loss Prevention.
McGraw-Hill, 1967. 912 pp. $33.50
Gives practical details on how to protect industrial plants
and processes against damage by fire, explosions, lightning, wind,
and earthquakes. Offers recommendations for desirable construction, automatic sprinkling systems and water supplies, the safeguarding of special-hazard processes involving flammable materials,
the protection of high storage values, the anchoring of roofs against
wind uplift, and many other features.

A11-2993
Faires, V. M. and Richardson, J. O. E.I.T. Review. Prentice-
Hall, 1961. 256 pp. $12.40
Survey of the primary technical subjects of the four year
college engineering course. Presents an analytical study of mathematics, from Calculus to Thermodynamics and other scientific
laws, supplemented with illustrative material.

A11-2994
Fairweather, L. and Sliwa, J. AJ Metric Handbook. Architectural Press, 1971. 208 pp. £2.50
A guide to the change to metric for all connected with the
construction industry. The latest information on British Standards,
dimensional co-ordination, notation, product and building design
data, component sizes and environmental and structural design.

A11-2995
Farrall, A. W. Engineering for Dairy and Food Products. Avi
Publishing Co., 1963. 674 pp. $21.50
This book applies not only to the dairy industry but also
to the food industry in general, since it has been found that most
of the basic principles of engineering applicable to the dairy industry underlie the problems encountered in other branches of the food
industry. Presents basic material on food processes and information on materials, methods, and equipment such as pumps, vacuum
treatment, pasteurizing equipment, and drying and evaporation
among others.

A11-2996
Feldman, E. B. Building Design for Maintainability. McGraw-
Hill, 1976. 232 pp. $12.50
This book shows how to make building maintenance more
efficient by coordinating architectural and interior design with effective maintenance. The author uses numerous examples to show
how proper planning, both inside and outside the building, will yield
cost advantages over its life.

A11-2997
French, T. and Svensen, C. Mechanical Drawing. McGraw-Hill,
1966. $8.96

A11-2998
French, T. and Vierck, C. Manual of Engineering Drawing for
    Students and Draftsman. McGraw-Hill, 1966. 701 pp.
    $11.50
        Full details on all aspects of graphic communication. 1143
illustrations.

A11-2999
Fuller, D. Functional Drafting for Today. Cahners, 1965.
    350 pp. $9.95
        Explains the basic principles of functional drafting, establishes the acceptable level for free-hand drawing, demonstrates how to get maximum mileage from pressure-sensitive materials, and how to use transparencies and intermediates most effectively. Illustrated.

A11-3000
General Electric. Modern Drafting Practices and Standards.
    General Electric Co., 1976. 580 pp. $95.00
        This loose-leaf book, updated every six months, documents the information government, industrial and student draftsmen need to implement modern and uniform engineering documentation practices. The manual has 14 sections that cover everything from forms and routines to line conventions and indications of surface textures. Copious illustrations graphically portray design drawings, production drawings, printed wiring drawings-most of the styles and formats used today.

A11-3001
General Services Administration. Public Building Service International Conference on Fire Safety. U.S. Gov't. Printing Office, 1971. 240 pp. $2.30. Order No. GS6.2:F 51
    S/N 2204-00005
        Seventy fire safety experts participated in this conference to explore new or revised approaches to solving the problems of fire safety in high-rise buildings.

A11-3002
Giachino, J. W. and Beukema, H. J. American Technical Society's Freehand Sketching. 2nd ed. American Technical Society.
    120 pp. $3.25
        The skill of freehand sketching can be developed by following the course of study in this text.

A11-3003
Giachino, J. W. and Beukema, H. J. Drafting Technology. 2nd ed. American Technical Society, 1971. 458 pp. $8.50
        This second edition includes important changes in drafting practices now recommended by the National Standards Institute. A new chapter is included on pipe drafting; and information is added on electronics, dimensioning, fluid power, and latest techniques.

Selected Topics

A11-3004
Giachino, J. W. and Beukema, H. J. Everyday Sketching and Drafting. 2nd ed. American Technical Society, 1973. 172 pp. $4.50

This revised edition is a text-workbook. Each chapter is followed by a set of selfstudy-and-check questions and application-practice sheets. Has 90 ready-to-use basic drafting assignments covering a variety of drafting problems.

A11-3005
Giachino, J. W. and Beukema, H. J. Print Reading for Welders. American Technical Society, 1970. 133 pp. $5.25

This text, which also serves as a manual, helps welding operators become more proficient in reading welding prints. It introduces the welder to a few basic principles used by the draftsman in preparing shop drawings, and covers in detail the various welding symbols prescribed by the American Welding Society.

A11-3006
Gibby, J. C. Technical Illustration. 3rd ed. American Technical Society, 1969. 352 pp. $8.25

Instruction on inking, shading, and rendering, including airbrush. This is broad, coverage, planned to develop employable vocational skills.

A11-3007
Glidden, H. K. Reports, Technical Writing and Specifications. McGraw-Hill, 1964. 312 pp. $9.95

Explains the nature of a technical report and instructs on how to write one. Treats verbal, mathematical, chemical, and graphical languages.

A11-3008
Goodban, W. and Hayslett, J. Architectural Drawing and Planning. McGraw-Hill, 1972. 288 pp. $10.95

This book presents the procedures, methods of work, and standards of preparing sketches and working drawings for architectural drafting practice. It includes information on freehand sketching, approximate and projected perspective, devices for strengthening perspective, and procedures in developing working drawings from idea to print.

A11-3009
Gowan, J., ed. A Continuing Experiment: Learning and Teaching at the Architectural Association. Architectural Press, 1974. 128 pp. £2.00 (paperback £1.00)

Apart from being the oldest architectural school in Britain, the Architectural Association is one of the most interesting institutes of higher education in the world. This collection of essays and interviews sheds light on an institution whose architectural and educational influence is far greater and more internationally based than is often realized.

A11-3010
Grant, H. E. Engineering and Architectural Lettering. McGraw-
 Hill, 1960. 32 pp. $3.00
 Includes such topics as vertical and inclined engineering
lettering, lettering with instructions, pictoral lettering, title blocks,
engineering drawings, graphs, and architectural detailed drawings.

A11-3011
Grant, H. E. Engineering Drawing. McGraw-Hill, 1965. $7.95
 Practical in approach, the book is oriented towards modern
industrial applications. Consists of 380 line drawings and 40 half-
tones, a set of 60 drawing problems sheets, 24 pages of problems,
a work-plan, and references.

A11-3012
Hall, C. W., et al. Encyclopedia of Food Engineering. Avi Publishing Co., 1971. 755 pp. $52.00
 This volume provides A to Z (absolute pressure to zeolite)
information on food manufacturing and food plant planning, design
maintenance, and operation. Covered are those properties of the
products which relate to food engineering, aspects of utilities which
are important for plant operation, factors of plant design and operation, elements of food plant work, as distinguished from other types
of industrial plants, particularly sanitation, and microbiological
consideration. Basic information on food products, construction
materials for equipment and buildings, design procedures, raw materials, and operational procedures are included.

A11-3013
Halse, A. Architectural Rendering: The Techniques of Contemporary Presentation. McGraw-Hill, 1973. 336 pp. $24.50
 This new edition includes information on new techniques and
media. The chapters devoted to perspective, composition, color
theory, light, rendering entourage, equipment and general approach
have all been updated to include new material on purchasing supplies, using various mediums, and a close analysis of student renderings.

A11-3014
Healy, R. J. Design for Security. John Wiley, 1968. 309 pp.
 $13.50
 Covers facility layouts for security, electronic components,
lighting, vaults and locks, and the systems approach.

A11-3015
Heine, G. M., et al. How to Read Electrical Blueprints. American Technical Society. 320 pp. $5.25 (Study Guide $1.40)
 Covers electrical blueprint readings: architectural prints,
bell and signal wiring, house wiring, A.C. and D.C. motor winding diagrams, control diagrams, power station circuits, etc.

A11-3016
Hepler, D. and Wallach, P. Architectural Drafting and Design.

Selected Topics

McGraw-Hill, 1971. 568 pp. $10.28
A first course in architectural drafting and design which presents principles and practices specific to architectural drafting.

A11-3017
Hewitt, R., ed. Guide to Site Surveying. Architectural Press, 1972. 200 pp. £1.95
The book sets out a procedure to follow for site investigations and has special sections on climate and soil mechanics. Information is given on new and traditional techniques of measuring and leveling and on the latest surveying instruments.

A11-3018
Hicks, T. Professional Achievement for Engineers and Scientists. McGraw-Hill, 1963. 380 pp. $9.95
A book designed to help increase earning power and improve the quality of professional work by developing nontechnical skills.

A11-3019
Hilado, C. J. Flammability Test Method Handbook. Technomic, 1973. 120 pp. $25.00
Contents include: Theory and Definitions, Physical Classifications, Flammability Tests and Types of Tests, Tests for: Flammability of Liquids, Specific Flammability Characteristics, Ignition Characteristics, Surface Flame Spread, Smoke Evolution, Fibers and Fabrics, Foams, Building Construction Materials, Drapery Materials, Electrical Appliance Materials and Floor Covering Materials.

A11-3020
Hogarth, P. Drawing Architecture. Watson-Guptill, 1973. 192 pp. $14.50
Interesting architecture and how to draw it. Castles and fortresses; Art Nouveau townhouses of Moscow; the vernacular architecture of California and Florida and more.

A11-3021
Hooper, L. Introduction to Construction Drafting. Prentice-Hall, 1971. 247 pp. $13.75
Begins where a novice might start and proceeds through the skills and background required of competent draftsmen. Contains a variety of material to spark an unsuspected aptitude or interest in further study or specialization.

A11-3022
Hopf, P. S. Designer's Guide to OSHA. McGraw-Hill, 1975. 304 pp. $17.50
This book presents the structural details of safety-related items and design methods which are in accordance with the 1971 Occupational Safety and Health Act. This statute requires architects, engineers, contractors, and builders to be aware of safety considerations in the design and construction of occupational structures.

A11-3023
Hornung, W. J. Architectural Drafting. Prentice-Hall, 1971.
 293 pp. $11.88
 Organization follows actual order of planning and construction of a building. Covers dwellings and small industrial buildings and teaches construction principles for structural plumbing, heating, and electrical plans from footings to roofing.

A11-3024
Howarth, R. Building Craft Foremanship: A Manual for the Trainee Building Supervisor. David and Charles, 1973. $8.95
 This book is intended for mature foremen and students taking the Craft Foremanship Studies in the City and Guilds Full Technological Certificate Course in building crafts. The contents provide details of site management and supervision and there are chapters on the basic aspects of site management--elements of craft leadership, site and workshop organization, records, measurements, and joint industrial relations, etc.

A11-3025
Hubert, C. Preventive Maintenance of Electrical Equipment. McGraw-Hill, 1969. 448 pp. $13.95
 This work presents latest methods for preventive maintenance of all kinds of electrical equipment, provides methods by which common troubles can be identified, recommends emergency repairs that will keep equipment in operation until scheduled out of service, suggests operating procedures, and outlines inspection programs. It is designed for technicians, motor and industrial control inspectors, operating engineers, and maintenance engineers.

A11-3026
Hudson, R. L. Useful Formulae for the Surveyor: A Field Manual for Control and Land Surveying. Exposition Press, 1973. $20.00
 The book combines all the formulae, tables and other information needed in the field. It is intended for practicing surveyors, chiefs-of-party, and other members of the surveying profession who already have a basic knowledge of surveying terminology and techniques. It was conceived for control and land surveys, but can also be used for construction surveys.

A11-3027
International Atomic Energy Agency. Manual on Safety Aspects of The Design and Equipment of Hot Laboratories. International Publications, 1969. $3.00

A11-3028
Jacoby, H. New Architectural Drawings. Praeger, 1969. 96 pp. $14.00
 The author guides the reader through the various stages of a rendering, including such factors as the choice of a station point,

Selected Topics

angles of vision, perspective, light and shadow, color and tone, and environment.

A11-3029
Jacoby, H. New Techniques of Architectural Rendering. Praeger, 1971. 110 pp. $18.50
These renderings represent outstanding performances in modern techniques. The drawings are by architects themselves, their office collaborators, or freelance specialists. Particular attention has been paid to examples that offer detailed interpretations of architectural design.

A11-3030
Jensen, R., ed. Fire Protection Systems for the Design Professional. Cahners, 1975. 192 pp. $25.00
Designed to help professionals to perform in conformance with the current codes, standards, and rulings of enforcing authorities, the book features 23 chapters authored by specialists. Based on a series of articles originally published in Actual Specifying Engineer Magazine, these chapters are grouped under four sectional headings: codes, standards and approvals; water supply and water-related fire protection systems; special agent systems, extinguishers and HVAC systems; and fire alarm systems and other electrical fire protection problems.

A11-3031
Johnson, S. Deterioration, Maintenance and Repair of Structures. McGraw-Hill, 1965. 384 pp. $21.50
Describing types of structural deterioration, their causes, preventive measures, and repair methods, this book covers steel, concrete, and timber.

A11-3032
Jones, G. L. How to Prepare Professional Design Brochures. McGraw-Hill, 1976. 384 pp. $16.50
Here is a book for design professionals who do not have access to in-house graphics and publications department, or who want a better understanding of the functions of such departments. It describes the best ways to write, layout, produce, and distribute the promotional materials which are often the only means of reaching potential clients before the selection procedure begins. As an aid to decision making, the book includes a discussion of ethical and practical considerations, free idea sources, and dealing with printers.

A11-3033
Jones, L. D. Electrical Engineering License Review. Engineering Press, 1972. 186 pp. $6.95
The 11 chapters present the full range of problems in the California electrical engineering examinations, 1960-1971. Seventy-three problems are included with their solutions.

A11-3034
Kemmerich, C. Graphic Details for Architects. Praeger, 1968.
    172 pp. $7.50
    This book treats graphic accessories that fall between pure
technical drawing and illusionistic freehand drawing. It is printed
on only one side of the paper so that models can be traced directly. The examples are divided into trees in cross-section; trees
in outline; low-growing plants; ground coverings; people; vehicles;
and compass marks and proportion symbols. Each group varies
from very simplified, abstract figures to complicated forms, and
proportions range from 1:500 to 1:50.

A11-3035
Kemper, J. D. The Engineer and His Profession. Holt, Rinehart
    and Winston, 1967. 254 pp. $6.25
    Professor Kemper describes the career of engineering, pro-
viding information on ranges of salaries, professional registration,
professional societies, business organization, and patent procedures.

A11-3036
Kenney, J. Blueprint Reading for the Building Trades. McGraw-
    Hill, 1955. 120 pp. $6.95
    Describes what blueprints are, how they are drawn, and
how members of the building trades use them. Working drawings
are in blueprint form.

A11-3037
Kenney, J. and McGrail, J. Architectural Drawing for the Build-
    ing Trades. McGraw-Hill, 1949. 128 pp. $7.95
    This introduction to architectural drawing is a guide through
various stages of draftsmanship to the point where the user is considered competent in the use of tools and ready for the study of
design. Includes lessons in concrete, wall, brick, veneer wall,
solid brick, and wood frame construction.

A11-3038
Kissam, P. Surveying for Civil Engineers. McGraw-Hill, 1976.
    600 pp. $19.50
    This second edition, treating the full scope of engineering
surveying, contains many new items and gives the best available
procedures to meet today's surveying requirements. Major sections cover instruments and methods, operations, control procedures,
and serial mapping. Special features include a discussion of electronic measurement theory and its use, new material on laser alignment, and information on the application of the calculator and computer to surveying problems.

A11-3039
Kissam, P. Surveying: Instruments and Methods for Surveys of
    Limited Extent. McGraw-Hill, 1956. 482 pp. $11.95
    Discusses horizontal measurement, the transit, traverses,
the level and benchmark leveling, adjustment of level, optical tooling, aerial photography, and topographical surveying.

Selected Topics   501

A11-3040
Kissam, P. C.   Surveying Practice.   McGraw-Hill, 1971.   500 pp.
   $9.95
   Prepares the student to perform the duties of any member
of a surveying field party. Includes new material on electronic
distance measurement, D.M.D. method, stadia, photogrammetry,
a vertical curve in sag computed by direct method, volume computation, and property surveys.

A11-3041
Kurtz, M.   Engineering Economics for Professional Engineers' Examinations.   McGraw-Hill, 1959.   322 pp. $13.50
   This text reviews the simple mathematical principles that
constitute the foundation of economics covered by this license exam.
It discusses the mechanics of investment and explains financial
terms.

A11-3042
Kurtz, M.   Structural Engineering for Professional Engineers' Examinations.   McGraw-Hill, 1968.   384 pp. $15.00
   This book includes problems from recent exams to aid candidates preparing for a professional engineer's or architect's license.

A11-3043
Labor Department.   Construction Safety and Health Training Course, Instructor's Guide.   Manual 301.   Job Planning, Walking and Working Surfaces and Spaces, Fire Prevention, Electrical Hazards.   U.S. Gov't. Printing Office, 1972.   136 pp. $1.95.   Order No. L 35.8/2:301 S/N 029-015-00018-0
   Designed to train supervisory personnel and employees in safe work practices on construction sites.

A11-3044
Labor Department.   Construction Safety and Health Training Course, Instructor's Guide.   Manual 302.   Personal Protective Equipment, Hand Tools, Ladders and Scaffolds, Rigging.   U.S. Gov't. Printing Office, 1972.   120 pp. $1.85.   Order No. L 35.8/2:302 S/N 029-015-00019-8
   Designed to train supervisory personnel and employees in safe work practices on construction sites.

A11-3045
Labor Department.   Construction Safety and Health Training Course, Instructor's Guide.   Manual 303.   Powered Equipment, Material Handling, Welding, Flammable Liquids.   U.S. Gov't. Printing Office, 1972.   120 pp. $1.85.   Order No. L 35.8/2:303 S/N 029-015-00020-1
   Designed to train supervisory personnel and employees in safe work practices on construction sites.

A11-3046
Labor Department.   Construction Safety and Health Training Course,

Instructor's Guide. Manual 304. Blasting, Excavation, Traffic Control, Concrete Construction. U. S. Gov't. Printing Office, 1972. 112 pp. $1.80. Order No. L 35.8/2: 304 S/N 029-015-00021-0

Designed to train supervisory personnel and employees in safe work practices on construction sites.

A11-3047
Labor Department. Construction Safety and Health Training Course, Instructor's Guide. Manual 305. Marine Operations, Steelworking, Tunneling, Demolition, Special Hazards. U. S. Gov't. Printing Office, 1972. 120 pp. $1.85. Order No. L35.8/2:305 S/N 029-015-00022-8

Designed to train supervisory personnel and employees in safe work practices on construction sites.

A11-3048
LaLonde, W. S., Jr. Professional Engineering Examination Questions and Answers. McGraw-Hill, 1976. 640 pp. $15.00

This guide to the examinations for the Engineer-in-Training Certificate, the Professional Engineer's License, and the Land Surveyor's License has been updated to handle the increased scope of the examinations. Most questions and answers are a composite of problems that appeared on past examinations; with structural answers based on the latest ACI and AISC codes for concrete and steel construction.

A11-3049
Lang, T. Computer Programs for Mapping. Architectural Press, 1971. 204 pp. £20.00

A description of the computer mapping facilities developed by the Experimental Cartography Unit, the techniques of which are outlined in "Automatic Cartography and Planning." It shows how cartographic data are stored in computer accessible form and describes what the various programs do and how they are interrelated. The largest part of the book consists of instructions on how to operate the programs for mapping which have been developed.

A11-3050
Lawson, P. J. Perspective Charts. Van Nostrand Reinhold, 1940. $7.95

These eight charts aid when accurate perspective drawings of architecture, furniture, and industrial design are required. Eliminates the need for distant vanishing points, keeps each line in its true perspective direction.

A11-3051
Legault, A. R., et al. Surveying: An Introduction to Engineering Measurements. Prentice-Hall, 1956. 430 pp. $12.95

Importance of mensuration theory is developed through surveying principles, with emphasis on fundamentals as applied to all branches of engineering.

Selected Topics

A11-3052
Levinson, I. J. Preparing for the Professional Engineer's Examinations: A Review with Questions and Answers. Prentice-Hall, 1965. 349 pp. $12.95
   Review of the fundamentals of ten basic engineering fields through the presentation of graded problems, solutions, and answers. Aims at helping the reader pass the P.E. and E.I.T. examinations. Over 500 questions and problems in fourteen major branches of engineering typical of those asked in the professional portion of the exam. Solutions to over 650 problems.

A11-3053
Lewis, H. F., ed. Laboratory Planning for Chemistry and Chemical Engineering. Van Nostrand Reinhold, 1962. 536 pp. $27.50
   Covers all important aspects of the planning and design of chemical and chemical engineering laboratories for industry, government, and academic institutions.

A11-3054
Lie, T. T. Fire and Buildings. Applied Science, 1972. 276 pp. £6.00
   This book deals with various aspects of protection of buildings against spread of fire and collapse. Particular emphasis is placed on preventing the spread of fire by compartmentation of a building with fire-resisting constructions, and consideration is given to experimental and theoretical methods of determining the fire resistance of structural elements such as floors, walls, columns and beams.

A11-3055
Lockard, W. K. Drawing as a Means to Architecture. Van Nostrand Reinhold, 96 pp. $11.95
   A workbook describing the various ways architectural projects can be conceived, refined, and communicated through drawing.

A11-3056
Lyons, J. S. and Dublin, S. W. Electrical Engineering and Economics and Ethics for Professional Engineering Examinations. Hayden, 1970. 320 pp. $12.95
   This volume offers coverage of all areas of the field, including a broad range of actual engineering problems.

A11-3057
McCartney, T. Precision Perspective Drawing. McGraw-Hill, 1963. 256 pp. $14.00
   Points out the techniques for making precise perspective drawings. 200 illustrations.

A11-3058
McKibben, G. Looking Forward to a Career: Building Trades. Dillon, 1970. $3.95

A11-3059
McKinnon, G. P., ed. Fire Protection Handbook, 14th Edition.
National Fire Protection Assn., 1976. 1300 pp. $43.50
This is a comprehensive reference book on fire protection which has been updated to include data on special fire protection problems in areas of industrial, process, and transportation. Fire-hazard properties of materials are given with respect to their handling and storage.

A11-3060
Marchant, E. W. A Complete Guide to Fire and Buildings: A Comprehensive Survey of all Aspects of Fire Protection and Provision. Barnes & Noble, 1972. 268 pp. $17.50
This volume is based on a special course taught at the University of Edinburgh. Eight specialist authors discuss all the relevant problems: the cause of fires, escape routes, fire tests as required by the building regulations, fire protection techniques, behavior of building structures and the re-instatement of damaged buildings.

A11-3061
Morgan, S. W. Architectural Drawing. McGraw-Hill, 1950. 228 pp. $14.50
Covers all aspects of perspective drawing, including aerial perspective, special procedures, orthographic drawings, and symbols and definitions.

A11-3062
Morrison, J. W. EIT Examination Study Guide. Arco, 1976. 300 pp. $10.00
A review of the material covered by the Engineer-in-Training Examination is presented in the form of practice questions taken from actual exams given since 1971, and which are representative of the types of problems a candidate may encounter on future examinations. Questions cover mathematics, physics, chemistry, statics, dynamics, mechanics of materials, fluid mechanics, thermodynamics, electrical theory, materials science, and economic analysis.

A11-3063
Morrison, J. W. Professional Electrical Engineering Examination. Arco, 1976. 300 pp. $10.00
Electrical engineering problems with detailed solutions and explanatory test provide comprehensive preparation for the National Council of Engineering Examiners' Principles and Practice of Electrical Engineering Examination. Full information on the nature, intent, and scope of the examination is given, and a comprehensive bibliography on electrical engineering is provided for further study.

A11-3064
Morrow, L. Maintenance Engineering Handbook. McGraw-Hill, 1960. 1842 pp. $39.50
Prepared by 100 specialists, this reference work gives full-

scale coverage of responsibilities and procedures involved in the maintenance of equipment, buildings, and services in manufacturing industries, including technical how-to-do-it information.

A11-3065
Mott, L. C. Engineering Drawing and Construction. Oxford Univ. Press. Vol. 1, 1965. 174 pp. $4.50; Vol. 2, 1967. 124 pp. $4.50

Throughout these volumes the explanatory text and the full-page diagrams are on the facing pages. Several groups of exercises are included to help the user with the problems he is likely to meet in drawing office work.

A11-3066
Mulholland, J. R. Heating, Ventilation and Air Conditioning Plant: Planned Maintenance and Operation. Cahners, 1970. 154 pp. $13.50

A guide to the compilation of a planned maintenance system for a complex building installation. The book is the result of a seven year program of building up, operating and revising a planned maintenance system for the Millbank Tower development in London.

A11-3067
Muller, E. Architectural Drawing and Light Construction. Prentice-Hall, 1976. 480 pp. $13.95

Provides the beginner and the draftsman on the job with a complete guide to drawing homes and small buildings. Step-by-step, the author covers the fundamentals of drawing and continues through the various aspects of construction, planning, mechanical and electrical equipment, as well as writing specifications and building actual models of small buildings. Including actual architectural-style illustrations as examples of contemporary drafting, this new edition also features extensive treatment of perspective drawing, freehand drawing, and additional material on wood construction.

A11-3068
Muller, E. J. Reading Architectural Working Drawings. Prentice-Hall, 1971. 352 pp. $13.95

Introduces the world of construction by analyzing basic principles, related mathematics, technical graphics and architectural conventions. Explains symbols, abbreviations, principles of technical projection and reviews construction arithmetic and geometry. Contains information on the latest types of commercial building drawings.

A11-3069
Munson, A. E. Construction Design for Landscape Architects. McGraw-Hill, 1975. 212 pp. $12.50

This volume is a guide to the preparation of a building site for construction or landscaping. Such topics as surveying, layout, roads, dams, retaining walls, earthwork calculations, utilities, storm drainage, and construction materials are covered. The book was written for use as a rapid refresher for the practicing landscape architect as well as a handy reference guide to short-cut methods that will be of interest to the civil engineer doing site improvement plans.

A11-3070
Newman, C. D.   Engineer-in-Training License Review.   Engineering Press, 1971.   256 pp. $5.95
    This edition reflects the changes in the National Engineering Fundamentals Examination used by 47 states. The book is organized in the 11 subjects of the exam: mathematics, statics, dynamics, mechanics of materials, materials science, fluid mechanics, thermodynamics, electrical theory, economic analysis, chemistry, and nucleonics and wave phenomena.

A11-3071
Newman, D. G.   Civil Engineering License Review.   Engineering Press, 1970.   424 pp. $8.95
    This book presents the six recent California civil engineering examinations. In addition, the Part One problems from four additional examinations are included. Detailed step-by-step solutions are provided.

A11-3072
Newman, M.   Standard Structural Details for Building Construction.   R. S. Means or McGraw-Hill, 1968.   361 pp. $19.95
    This guide is divided into four large sections and covers all basic structural material: wood, concrete, masonry, and steel. Each detail is drawn to scale, fully described, and arranged in logical construction sequence. Using this book will help reduce the cost of producing structural drawings, and prevent misunderstandings between architects, engineers, draftsmen, and contractors.

A11-3073
Newman, O.   Crime Prevention Through Urban Design: Designing a House to Prevent Crime.   Macmillan, 1972. $7.95 (paperback $3.95)

A11-3074
Newman, O.   Defensible Space: Crime Prevention Through Urban Design.   Macmillan, 1972. $3.95

A11-3075
NFPA.   Designing Buildings for Fire Safety.   National Fire Protection Assn., 1975.   125 pp. $4.75
    Effective fire safety in buildings starts on the drafting board. Twenty-three articles from Fire Journal and Fire Technology magazines cover topics on good firesafe building design written by field experts.

A11-3076
NFPA.   Fire Protection by Sprinklers.   National Fire Protection

Assn., 1975. 100 pp. $4.75

Seventeen articles from Fire Journal and Fire Technology magazines to keep you updated on sprinklers. Reviews the past, present and future of sprinkler systems. Describes their performance in high-rise buildings and homes for the elderly and gives convincing reasons for consumer installation of sprinkler systems.

A11-3077
NFPA. <u>Fire Protection Guide on Hazardous Materials, Sixth Edition. National Fire Protection Assn., 1975. 1500 pp. $10.50</u>

Flash points of more than 8,800 trade name products are listed alphabetically as are the fire hazard properties of over 1,300 flammable substances. Data are included for about 388 chemicals on their fire, explosion, and toxicity hazards, including recommendations on storage and fire fighting. Information on 2,350 mixtures of 2 or more chemicals liable to cause fires, explosions, or detonations is given. The recommended system for identifying the severity of the health, flammability, and reactivity hazards of materials is explained.

A11-3078
NFPA. <u>Fire Protection Handbook, Fourteenth Edition. National Fire Protection Assn., 1976. 1296 pp. $43.50</u>

Authoritative information and reference data on all aspects of fire protection-the basic facts, plus explanations of the "whys" and "hows"! Designed to supplement--not duplicate--the National Fire Codes. Covers the subject in logical sequence so that it's ideal for students of the subject, while being so organized that the reference data needed by the experienced fire protection specialist is easy to find ... and use.

A11-3079
NFPA. <u>Fires in High-Rise Buildings. National Fire Protection Assn., 1974. 120 pp. $4.75</u>

Valuable lessons to be learned by architects, engineers, builders, fire officers from the twenty-three reprints from Fire Journal and Fire Technology in this book. Includes articles on automatic recall elevators, standards for refuse handling and problems faced in fighting high-rise fires in a variety of buildings.

A11-3080
NHIC. <u>Home Improvement Selling. National Home Improvement Council. $11.50 (order from Distributive Education Instructional Mats. Services, Univ. of Texas, Austin 78712)</u>

An instruction manual to aid in training personnel for the home improvement business. Some subjects covered are: estimating and pricing, drawing, blueprint reading, lumber measurement and arithmetic, and selling techniques.

A11-3081
Packer, M., ed. <u>Professional Engineer (Civil) State Board Examination Review. Arco, 1975. 544 pp. $10.00</u>

This manual outlines major engineering concepts and presents solutions to a broad range of practical problems typical of those found in actual practice and on the State Professional Engineering Examinations.

A11-3082
Perrott, S. W. Surveying for Young Engineers. John Wiley, 1970. 188 pp. $4.00

A11-3083
Perry, R. Engineering Manual. McGraw-Hill, 1967. 680 pp. $12.50

Condenses the fundamentals of the standard engineering disciplines into a single work; covers all areas of engineering--architectural, chemical, civil, electrical, mechanical, and nuclear--and considers advances made in each.

A11-3084
Pile, J., ed. Drawings of Architectural Interiors. Watson-Guptill, 172 pp. $29.50

This book contains a selection of sketches representing the work of 89 outstanding architects and illustrators, among them Le Corbusier, Florence Knoll Bassett, Frank Lloyd Wright, Walter Gropius, Mies van der Rohe, Richard Neutra, and I. M. Pei. The reader will discover new ways to make his own ideas visual, and present them with clarity and drama. 146 illustrations.

A11-3085
Piper, R. J. Opportunities in an Architecture Career. American Institute of Architects, 1970. $3.75 (paperback $1.95) (for AIA members, less 10%)

Intended primarily for students considering pursuit of an architectural career, this book presents an analysis of all occupational facets of the profession.

A11-3086
Plant Engineering. Plant Protection Library. Plant Engineering, 112 pp. $9.50

Three booklets in a loose-leaf folder cover such topics as fire protection, industrial security and safety for the plant engineer.

A11-3087
Polentz, L. Engineering Fundamentals for Professional Engineers' Examinations. McGraw-Hill, 1961. 394 pp. $14.00

This guide helps engineers prepare for Part 1 of this exam. Covers mathematics, statics, dynamics, fluid mechanics, thermodynamics, electricity, chemistry, metallurgy, and more.

A11-3088
Rathbone, R. R. Communicating Technical Information: A Guide to Current Uses and Abuses in Scientific and Engineering Writing. Addison-Wesley, 1966. 104 pp. $3.50

This book is intended as a primary text for short in-plant

Selected Topics

writing courses for engineers and scientists, and as a secondary text for college courses in engineering, science, and technical writing. It can serve as a self-help reference book for students and men on the job. Includes a chapter on editing, slated for the technical supervisor.

A11-3089
Rayner, W. H. and Schmidt, M. O. Fundamentals of Surveying. Van Nostrand Reinhold, 1969. 544 pp. $9.95

A11-3090
Renton, B. A. Electrical and Electronic Drafting. Hayden, 1971. 400 pp. $12.95
This course presents a cross-section of circuits, symbols, examples, and techniques, guiding the student from the simplest drafting skill to the most complex. Beginning with various types of electrical and electronic drawings, the book then covers graphic symbols, including both standard U.S. symbols and their Canadian equivalents.

A11-3091
Ripa, L. Surveying Manual. McGraw-Hill, 1964. 144 pp. $8.95
A manual of standard practice with instructions on the most commonly met field situations.

A11-3092
Royer, K. Applied Field Surveying. John Wiley, 1970. 205 pp. $9.75
A guide for surveying field exercises and methods for using surveying instruments. Qualifies student for work on a survey crew or gives construction workers additional skills in layout.

A11-3093
Sanders, M. and Dublin, S. W. Civil Engineering and Economics and Ethics for Professional Engineering Examinations. Hayden, 1973. 448 pp.
This volume presents theoretical and practical background and detailed solutions to a selection of recent exam problems. Individual chapters focus on structural analysis and design, soil mechanics and foundation design, surveying, transportation engineering, construction materials and methods, hydraulic and environmental engineering.

A11-3094
Schaarwachter, G. Perspectives for Architecture. Praeger, 1967. 120 pp. $10.00
Surveys fundamental theory and practice of perspective. Simple assignments with basic drafting instruments--such as a T-square, ruler, and arc protractor--lead by easy stages through ground plans and vertical projections to the use of axometer and perspectograph and a mastery of complex forms, such as spiral stairways and vaulted ceilings.

A11-3095
Schofield, W. Engineering Surveying: Theory and Examination Problems for Students. Butterworth, 1972. 256 pp. £3.50 (paperback £2.00)
Covers leveling, earthworks, the theodolite and its application, optical distance measurement, curves, and underground and hydrographic surveying. Uses SI units throughout.

A11-3096
Siders, R. A., et al. Computer Graphics: A Revolution in Design. American Management Assn., 1966. 160 pp. $9.00 (for AMA members, $7.00)
Discusses in nontechnical language how computers are producing graphic diagrams on cathode tubes to cut down time-consuming drafting problems in engineering design. Explains with drawings and photographs the functions of this new system and its economic advantages to the organization.

A11-3097
Simonds, J. Landscape Architecture. McGraw-Hill, 1961. 244 pp. $19.95
Every aspect of site planning is covered in detail in this book which outlines and analyzes the complete landscape process from site selection to completed project.

A11-3098
Smirnoff, M. V. Measurements for Engineering and Other Surveys. Prentice-Hall, 1961. 556 pp. $12.95
Art of surveying, dealing with the proper use of surveying instruments and equipment. Included are such topics as random errors in measurement, use of tape for measuring angles, traverses and distances, use of leveling rod and transit. Area computations and underground and topographic surveys are covered.

A11-3099
Snow, C. Electrical Drafting and Design. Prentice-Hall, 1975. 416 pp. $16.95
This book opens with a complete discussion of drafting procedures, from drafting materials through revising a given project. The book then proceeds to in-depth discussions of fundamentals of electricity, one-line diagrams, one-line and relay diagrams, elementary diagrams, riser diagrams, wire and cable schedules connection and interconnection wiring diagrams, grounding, raceway layouts, lighting layouts, power distribution and substations, system protection requirements and devices, application of system protective devices, sizing, wires, cables and conduits, and finally, equipment layouts, control panels and detail drawings.

A11-3100
Stamper, E. and Dublin, S. W. Mechanical Engineering and Economics and Ethics for Professional Engineering Examinations. Hayden Book Co., 1975. 416 pp. $17.20
This book is designed to familiarize the candidate for P.E.

Selected Topics

registration with the basics of mechanical engineering as required for the Part 3 examination. A second part of the book is devoted to economics and ethics which are usually part of the same examination. Sample problems enable the reader to understand the type of questions he will be expected to answer.

A11-3101
Steere, N. V., ed. CRC Handbook of Laboratory Safety. CRC Press, 1971. 854 pp. $26.00

Major topics include protective equipment, ventilation, fire hazards, chemical reactions, toxic hazards, radiation hazards, electrical and mechanical hazards, water supply and biological hazards. Included also are laboratory design and equipment.

A11-3102
Stegman, G. K. and Stegman, H. J. Architectural Drafting; Functional Planning and Creative Design. American Technical Society, 640 pp. $11.95

There are chapters on professional, architectural office practice, sketching, building material sizes plus coverage of metric measurements in the building industry. The chapter on light commercial building is expanded, covering medium-heavy construction. This second edition is slanted toward professional-level training, with the latest in architectural practices, building methods, and materials.

A11-3103
Strong, C. W. and Eidson, D. The Technical Writer's Handbook. Holt, Rinehart and Winston, 1971. 320 pp. $5.95

Future professional technical writers, journalists, scientists, administrators, and technicians, as well as undergraduates will need this combination text and reference. The text part covers style, format, types of reports and articles, techniques for preparing graphic ads, statistics, technical editing, and advertising.

A11-3104
Sundberg, E. W. Building Trades Blueprint Reading. Part 1, Fundamentals. American Technical Society, 1973. 184 pp. $4.95

This revised edition covers all the basics plus new information on: electrical, heating, air conditioning, concrete, masonry, and welding. The appendix features a simplified math review, metric conversion, decimal equivalents, and a glossary of terms.

A11-3105
Sundberg, E. W. Building Trades Blueprint Reading. Part 2, Residential and Light Construction. American Technical Society, 1973. 256 pp. $6.95

This text can be used by students who have some background in blueprint reading, for apprentice training, or as a reference. There are four complete sets of large blueprints covering residential and commercial buildings. Specifications, construction materials, and types of construction are also discussed.

A11-3106
Sundberg, E. W. Building Trades Blueprint Reading. Part 3, General Construction, Specifications, and Heavy Construction. American Technical Society, 1973. 197 pp. $7.25

This text covers the reading of working drawings beyond the light construction field. A townhouse and commercial building are studied to give general background in blueprint reading. One chapter deals with specifications as they relate to blueprint reading, another with heavy timber, structural steel, and reinforced concrete construction, and there is a study of the construction of a high-rise apartment building.

A11-3107
Taylor, J. Model Building for Architects and Engineers. McGraw-Hill, 1971. 256 pp. $18.50

This guide shows architects and engineers how to build models of architectural, engineering, and interior designs. It shows how to represent a large number of traditional and contemporary building materials and effects through model building.

A11-3108
Techniques in Automatic Cartography. Architectural Press, 1974. £6.00

This sequel to Automatic Cartography and Planning, describes the techniques that are currently available, discusses their potential and gives instructions on their application.

A11-3109
Thomas, T. Technical Illustration. McGraw-Hill, 1968. 352 pp. $10.95

Offering training in industrial methods and analyzing typical problems facing the technical illustrator, this book is of great value to commercial artists, technical illustrators, and industrial designers.

A11-3110
Thompson, N. J. Fire Behavior and Sprinklers. National Fire Protection Assn., 1964. 168 pp. $3.95

The fundamentals of fire behavior and the influence they have on automatic sprinkler protection. Concise explanations of the combustion process and factors governing sprinkler performances, including data on sprinkler capacities and limitations.

A11-3111
Turner, D. R. Civil Engineer, Senior and Supervising. Arco, 1974. 320 pp. $8.00

Preparation for government engineering positions. This book includes twelve previous exams with correct answers and official answer sheets.

A11-3112
Tuska, C. D. An Introduction to Patents for Inventors and Engineers. Dover. 192 pp. $2.00

Selected Topics 513

Former director of Patent Department of RCA explains patent law, procedures, precautions, pitfalls, with many case histories.

A11-3113
Walters, N. and Bromham, J. Principles of Perspective. Architectural Press, 1970. 128 pp. £2.25
A book which relates perspectives to other subjects such as solid geometry.

A11-3114
Weaver, G. Structural Detailing for Technicians. McGraw-Hill, 1974. 290 pp. $9.95
Written at the technician level and covers the basic principles and practices necessary for the preparation of structural drawings and details for the various segments of the construction industry. Each major structural material--wood, concrete, and steel--is introduced separately with a description of its history, and use, and then given a thorough examination in the light of applied detailing procedures for residential, commercial, and industrial projects.

A11-3115
[No entry]

A11-3116
Weidhaas, E. R. Architectural Drafting and Construction. Allyn and Bacon, 1974. $14.95
This text covers architectural drafting and construction of commercial and residential structures. The author covers timber, masonry, steel and concrete construction, and design as applied to residences. Sections on space frames, laminated timber construction, stressed-skin panel construction, glass masonry, cable roof structures, automatic welding, and many others.

A11-3117
Welling, R. The Technique of Drawing Buildings. Watson-Guptill, 1971. 160 pp. $10.95
This book shows how to draw buildings and other urban subjects--bridges, docks, boats, etc. Materials and tools, techniques and projects are other subjects covered.

A11-3118
Westinghouse Electrical Maintenance Hints. Westinghouse Electric Corp., 1974. 1450 pp. $15.00
This handbook is comprised of 46 chapters for use by industrial electrical maintenance personnel. Special emphasis is placed on motors and motor control, switchgear, transformers, and circuit breakers but general maintenance procedures are also covered.

A11-3119
Whyte, W. S. Basic Metric Surveying. Butterworth, 1969. 320 pp. £3.20 (paperback £2.25)
A review of metric surveying techniques for students and

practicing architects, engineers and surveyors. The book is intended for all who may require to carry out surveys of limited extent, but is aimed primarily at the needs of the construction industry.

A11-3120
Wilcox, C. and Snape, J. A.   Worked Examples in Measurement of Construction Work.   George Godwin.   Vol. 1, 1972. 180 pp.   £3.20 (paperback £2.00); Vol. 2, 1974.   185 pp. £4.00 (paperback £3.00)

These two volumes on the measurement of construction work use the group system of measurement and SMM phraseology. Volume 1 covers foundations, structural walls, roofs, floors, and partitions. Vol. 2 treats windows, doors, staircases, joinery fittings, and fixtures.

A11-3121
Woltjes, W.   Building Trades Blueprint Reading Examination Kit. American Technical Society. $2.15

Material which serves as a final examination, as a continuing project, and as a test to determine the need for additional training. 6 blueprints, 1 standard building specifications form, 8-page comprehensive examination.

A11-3122
Wyatt, W. E.   General Architectural Drawing.   Bennett Books, 1969.   556 pp. $11.04 (Student Guide $1.76)

This pre-vocational text goes into modular planning, prefabricated sections, windows and doors, and modern climate control. The basics are covered, with analyses of site selection, structure, function, materials, details of construction, techniques of planning and other factors, as well as work methods of the architectural draftsman.

APPENDIX

BOOK PUBLISHERS AND DISTRIBUTORS

Harry N. Abrams, Inc.: 110 E. 59 St., New York 10022
Academic Press, Inc.: 111 Fifth Ave., New York 10003
Addison-Wesley Publishing Co., Inc.: Reading, Mass. 01867
Adler's Foreign Books, Inc.: 162 Fifth Avenue, New York 10010
Aldine Publishing Co.: 529 S. Wabash Ave., Chicago, Ill. 60605
Allyn and Bacon: 470 Atlantic Ave., Boston, Mass. 02210
Aluminum Association: 750 Third Avenue, New York 10017
American Appraisal Associates, Inc.: 525 E. Michigan St., Milwaukee, Wis. 53201
American Association of Port Authorities: 1612 K St., N.W., Washington, D.C. 20006
American Ceramic Society, Inc.: 65 Ceramic Drive, Columbus, Ohio 43214
American Chemical Society: 1155 16 St., N.W., Washington, D.C. 20036
American Concrete Institute: 22400 W. Seven Mile Rd., Detroit, Mich. 48219
American Concrete Pipe Association: 1501 Wilson Blvd., Arlington, Va. 22209
American Conference of Government Industrial Hygienists: P.O. Box 1937, Cincinnati, Ohio 45201
American Elsevier Publishing Co., Inc.: 52 Vanderbilt Ave., New York 10017
American Enterprise Institute: 1150 Seventeenth St., N.W., Washington, D.C. 20036
American Hospital Association: 840 N. Lake Shore Dr., Chicago, Ill. 60611
American Institute of Architects: 1735 New York Ave., N.W., Washington, D.C. 20006
American Institute of Chemical Engineers: 345 E. 47 St., New York 10017
American Institute of Steel Construction, Inc.: 101 Park Ave., New York 10017
American Management Associations, Inc.: 135 W. 50 St., New York 10020
American Nuclear Society: 244 East Ogden Ave., Hinsdale, Ill. 60521
American Public Works Association: 1313 E. 60 St., Chicago, Ill. 60637

American Road Builders Association: ARBA Building, 525 School St., S.W., Washington, D.C. 20024
American School of Classical Studies at Athens: c/o Institute for Advanced Study, Princeton, N.J. 08540
American Society for Metals: 9533 Kinsman Rd., Metals Park, Ohio 44073
American Society for Nondestructive Testing: 914 Chicago Ave., Evanston, Ill. 60202
American Society for Testing and Materials: 1916 Race St., Philadelphia, Pa. 19103
American Society of Civil Engineers: 345 E. 47 St., New York 10017
American Society of Heating, Refrigerating and Air Conditioning Engineers: 345 E. 47 St., New York 10017
American Society of Mechanical Engineers: 345 E. 47 St., New York 10017
American Society of Planning Officials: 1313 E. 60 St., Chicago, Ill. 60637
American Society of Sanitary Engineering: 960 Illuminating Blvd., Cleveland, Ohio 44113
American Technical Society: 5608 Stony Island Ave., Chicago, Ill. 60637
American Welding Society: 2501 NW 7th St., Miami, Fla. 33125
AMS Press, Inc.: 56 E. 13 St., New York 10003
Anchor Neptune: available from Tailfeather, P.O. Box 1106, Moab, Utah 84532
Ann Arbor Science Publishers, Inc.: Drawer 1425, Ann Arbor, Mich. 48106
Applied Science Publishers Ltd.: Rippleside Commercial Estate, Barking, Essex, England
Architectural Book Publishing Co.: 10 E. 40 St., New York 10016
The Architectural Press: 9 Queen Anne's Gate, London SW1H 9BY, England
Arco Publishing Co., Inc.: 219 Park Ave. S., New York 10003
Arno Press, Inc.: 330 Madison Ave., New York 10017
Asia Publishing House, Inc.: 420 Lexington Ave., New York 10017
The Asphalt Institute: Asphalt Institute Building, College Park, Md. 20740
Associated Builders and Contractors, Inc.: P.O. Box 8733, Baltimore-Washington International Airport, Md. 21240
Atheneum Publishers: 122 E. 42 St., New York 10017
Aurora Publishers Inc.: 118 16th Ave. S., Nashville, Tenn 37219
Avi Publishing Co.: Box 831, Westport, Conn. 06880
Ballinger Publishing Co.: 17 Dunster St., Cambridge, Mass. 02138
A. S. Barnes and Co., Inc.: Forsgate Drive, Cranbury, N.J. 08512
Barnes and Noble Books: 10 E. 53 St., New York 10022
William L. Bauhan, Publisher: Dublin, N.H. 03444
Beekman Publishers, Inc.: 38 Hicks St., Brooklyn Heights, N.Y. 11201
Chas. A. Bennett Co., Inc.: 809 W. Detweiller Dr., Peoria, Ill. 61614

Book Publishers

A. M. Best Co.: 343 South Dearborn, Chicago, Ill. 60640
Billboard Publications, Inc.: 1719 West End Ave., Nashville, Tenn. 37203
Benjamin Blom, Inc.: 2521 Broadway, New York 10025
R. R. Bowker Co.: 1180 Ave. of the Americas, New York 10036
George Braziller, Inc.: One Park Avenue, New York 10016
Brick Institute of America: 1750 Old Meadow Rd., McLean, Va. 22101
British Book Centre, Inc.: 153 E. 78 St., New York 10021
Bruce Publishing Co.: orders to Macmillan Publishing Co., Inc.
Building Officials Code Administrators International: 1313 E. 60 St., Chicago, Ill. 60637
Building Research Institute: 2101 Constitution Ave., N.W., Washington, D.C. 20418
Bureau of National Affairs, Inc.: 1231 25th St., N.W., Washington, D.C. 20037
Business News Publishing Co.: P.O. Box 6000, Birmingham, Mich. 48012
The Butterworth Group: 4 and 5 Bell Yard, London WC2, England
Cahners Books: 89 Franklin St., Boston, Mass. 02110
Cahners Publishing Co., Inc.: 221 Columbus Ave., Boston, Mass. 02116
Cambridge University Press: 32 E. 57 St., New York 10022
Ceilings and Interior Systems Contractors Association: 1201 Waukegan Rd., Glenview, Ill. 60025
Cement and Concrete Association: Wexham Springs, Slough SL36PL, Bucks, England
Center for Urban Policy Research-Rutgers University: Building 4051, Kilmer Campus, New Brunswick, N.J. 08903
Central Book Co.: 850 DeKalb Ave., Brooklyn, N.Y. 11221
Chapman and Hall: distributed by Associated Book Publishers Ltd., Northway, Andover, Hampshire SP105BE, England
The Chatham Press, Inc.: 143 Sound Beach Ave., Old Greenwich, Conn. 06870
Chemical Publishing Co., Inc.: 155 W. 19 St., New York 10011
Chilton Book Co.: Chilton Way, Radnor, Pa. 19089
Chronicle Books: 870 Market St., San Francisco, Calif. 94102
Clarkson N. Potter, Inc.: 419 Park Avenue S., New York 10016
Colorado Associated University Press: 1424 15 St., Boulder, Colo. 80302
Columbia University Press: 562 W. 113 St., New York 10025
Commerce Clearing House: 4025 W. Peterson Ave., Chicago, Ill. 60646
Concrete Construction Publications, Inc.: P.O. Box 555, Elmhurst, Ill. 60126
Concrete Reinforcing Steel Institute: 180 North LaSalle St., Chicago, Ill. 60601
The Construction Press: Lynesdale House, Hornby, Lancaster LA2 8NB, England
Construction Publications: P.O. Box 15567, Phoenix, Ariz. 85060
Construction Publishing Co., Inc.: 2 Park Avenue, New York 10016
The Construction Specifications Institute: 1150 17th St., N.W., Washington, D.C. 20036

Cooling Tower Institute: 3003 Yale St., Houston, Texas 77018
Cornell Maritime Press, Inc.: Cambridge, Md. 21613
Cornell University Press: 124 Roberts Pl., Ithaca, N.Y. 14850
Council of Educational Facility Planners: 29 W. Woodruff Ave., Columbus, Ohio 43210
Craftsman Book Co.: 542 Stevens Ave., Solana Beach, Calif. 92075
Crane, Russak and Co., Inc.: 347 Madison Ave., New York 10017
CRC Press, Inc.: 18901 Cranwood Pkway., Cleveland, Ohio 44128
Crown Publishers, Inc.: 419 Park Ave. S., New York 10016
DaCapo Press, Inc.: 227 W. 17 St., New York 10011
The Darwin Press, Inc.: Box 2202, Princeton, N.J. 08540
Davey, Daniel and Co., Inc.: P.O. Box 97, Hartford, Conn. 06101
David and Charles, Inc.: North Pomfret, Vt. 05053
Marcel Dekker, Inc.: 270 Madison Ave., New York 10016
Design Data Center: P.O. Box 566, Lansdale, Pa. 19446
Dillon Press, Inc.: 510 S. Third St., Minneapolis, Minn. 55415
Dodd, Mead and Co.: 79 Madison Ave., New York 10016
Dorrance and Co., Inc.: 1617 J.F. Kennedy Blvd., Philadelphia, Pa. 19103
Dorsey Press: 1818 Ridge St., Homewood, Ill. 60430
Doubleday and Co., Inc.: 245 Park Ave., New York 10017
Dover Publications, Inc.: 180 Varick St., New York 10014
Dowden, Hutchinson and Ross: Box 699, 523 Sarah St., Stroudsburg, Pa. 18360
Drake Publishers, Inc.: 381 Park Avenue S., New York 10016
Earthquake Engineering Research Institute: 424 40th St., Oakland, Calif. 94609
Wm. B. Eerdmans Publishing Co.: 255 Jefferson Ave. S.E., Grand Rapids, Mich 49502
Elek Books: 2 All Saints St., London N1, England
Energy Research and Development Administration, U.S.: Washington, D.C. 20545
Engineering Press: P.O. Box 1142, San Jose, Calif. 95108
Engineers Joint Council: 345 E. 47 St., New York 10017
Equipment Guide-Book Co.: 2800 W. Bayshore Rd., P.O. Box 10113, Palo Alto, Calif. 94303
Exposition Press, Inc.: 50 Jericho Tpk., Jericho, N.Y. 11753
Frederick Fell Publishers, Inc.: 386 Park Ave., New York 10016
Fernhill House Ltd.: distributed by Humanities Press, Inc., Atlantic Highlands, N.J. 07716
Fifth Street Press: 1409 Fifth St., Berkeley, Calif. 94710
Foster Publishing Co.: 1602 Pattiz, Long Beach, Calif. 90815
G.T. Foulis see International Scholarly Book Service
Freeman, Cooper and Co.: 1736 Stockton St., San Francisco, Calif. 94133
Gale Research Co.: Book Tower, Detroit, Mich. 48226
Garden Way Publishing Co.: Charlotte, Vermont 05445
General Electric Co.: P.O. Box 43, Schenectady, N.Y. 12301
Glenwood Publishers: P.O. Box 194, Felton, Calif. 95018
George Godwin Ltd.: Aldwych House, 71/81/91 Aldwich, London WC2B3EX, England

Book Publishers

Goodheart-Wilcox Co.: 123 W. Taft Dr., S. Holland, Ill. 60473
Goodyear Publishing Co., Inc.: 15113 Sunset Blvd., Pacific Palisades, Calif. 90272
Gould Publications: 199 State St., Binghamton, N.Y. 13901
Stephen Greene Press: Box 1000, Brattleboro, Vt. 05301
Gregg International Publishers: 125 Spring St., Lexington, Mass. 02173
Charles Griffin and Co., Ltd.: c/o The Hafner Publishing Co., Inc., 866 Third Ave., New York 10022
Grosset and Dunlap, Inc.: 51 Madison Ave., New York 10010
Gulf Publishing Co.: Book Division, Box 2608, Houston, Texas 77001
Hacker Art Books, Inc.: 54 W. 57 St., New York 10019
Halsted Press: 605 Third Ave., New York 10016
Hammond Inc.: Maplewood, N.J. 07040
Harcourt Brace Jovanovich: 757 Third Ave., New York 10017
Harper and Row, Publishers: 10 E. 53 St., New York 10022
Hart Publishing Co., Inc.: 15 W. Fourth St., New York 10012
Harvard University Press: 79 Garden St., Cambridge, Mass. 02138
Hastings House, Publishers, Inc.: 10 E. 40th St., New York 10016
Hawthorn Books, Inc.: 260 Madison Ave., New York 10016
Hayden Book Co., Inc.: 50 Essex St., Rochelle Park, N.J. 07662
William S. Heinman: 1966 Broadway, New York 10023
Herman Publishing, Inc.: 45 Newbury St., Boston, Mass. 02116
Hoffman Publications: Sunrise Professional Bldg., Ft. Lauderdale, Fla. 33304
Holt, Rinehart and Winston: 383 Madison Ave., New York 10017
Horizon Press: 156 Fifth Ave., New York 10010
Hunter Associates: 792 Partridge Dr., Somerville, N.J. 08876
Illuminating Engineering Society: 345 E. 47 St., New York 10017
Industrial Press, Inc.: 200 Madison Ave., New York 10016
The Institute of Electrical and Electronics Engineers, Inc.: 345 E. 47 St., New York 10017
Institute of Real Estate Management: 155 E. Superior St., Chicago, Ill. 60611
International Conference of Building Officials: 5360 S. Workman Mill Rd., Whittier, Calif. 90601
International Publications Service: 114 E. 32 St., New York 10016
International Scholarly Book Services: Box 4347, Portland, Oregon 97208
International Union of Architects: Paris, France
The Interstate Printers and Publishers, Inc.: 19 N. Jackson St., Danville, Ill. 61832
Intext Publishers Group: 257 Park Avenue S., New York 10010
Iowa State University Press: Ames, Iowa 50010
Richard D. Irwin, Inc.: 1818 Ridge Rd., Homewood, Ill. 60430
Israel Program for Scientific Translations: P.O. Box 7145, Jerusalem, Israel
Johns Hopkins University Press: Baltimore, Md. 21218
Augustus M. Kelley, Publishers: 300 Fairfield Rd., Fairfield, N.J. 07006

Kennikat Press Corp.: 90 S. Bayles Ave., Port Washington, N.Y. 11050
Charles Knight and Co.: 25 New Street Square, London EC4A3JA, England
Alfred Knopf, Inc.: 201 E. 50 St., New York 10022
Know-How Publications: Box 7126, Landscape Sta., Berkeley, Calif. 94717
Lane Magazine and Book Co.: Menlo Park, Calif. 94025
Lefax Publishing Co.: 2867 E. Allegheny Ave., Philadelphia, Pa. 19134
Lenox Hill Publishing and Distributing Corp.: 235 E. 44 St., New York 10017
Lexington Books: 125 Spring St., Lexington, Mass. 02173
McGraw-Hill Book Co.: 1221 Ave. of the Americas, New York 10020
McGraw-Hill Information Systems: 1221 Ave. of the Americas, New York 10020
Machine Tool Publications see McGraw-Hill Book Co.
Macmillan, Inc.: 866 Third Ave., New York 10022
Manas Publications: Shore Tower 205, 1868 Shore Dr. S., St. Petersburg, Fla. 33707
Marble Institute of America: 1984 Chain Bridge Rd., McLean, Va. 22101
Masonry Institute of America: 2550 Beverly Blvd., Los Angeles, Calif. 90057
R. S. Means: Duxbury, Mass. 02332
Charles E. Merrill Publishing Co.: 1300 Alum Creek Dr., Columbus, Ohio 43216
Minnesota Electric Association: 515 S. 7th St., Minneapolis, Minn. 55415
The MIT Press: 28 Carleton St., Cambridge, Mass. 02142
Mobile Homes Manufacturers Association: P.O. Box 201, 14650 Lee Rd., Chantilly, Va. 22021
National Academy of Sciences: 2101 Constitution Ave., N.W., Washington, D.C. 20418
National Association of Corrosion Engineers: 2400 W. Loop, South, Houston, Texas 77027
National Association of Home Builders: 1625 L St., N.W., Washington, D.C. 20036
National Environmental Systems Contractors Association: 1501 Wilson Blvd., Arlington, Va. 22209
National Fire Protection Association: 470 Atlantic Ave., Boston, Mass. 02210
National Industrial Conference Board: 845 Third Ave., New York 10022
National Research Bureau, Inc.: 424 N. Third St., Burlington, Iowa 52601
National Society of Professional Engineers: 2029 K St., Washington, D.C. 20006
National Technical Information Service: U.S. Department of Commerce, 5285 Port Royal Rd., Springfield, Va. 22151
New York Graphic Society Ltd.: 140 Greenwich Ave., Greenwich, Conn. 06830

Book Publishers 521

New York Public Library: 5th Ave. and 42nd St., New York 10018
North Castle Books: 212 Bedford Rd., Greenwich, Conn. 06830
North-Holland: P.O. Box 211, Jan Van Galenstraat 335, Amsterdam, The Netherlands
W. W. Norton and Co., Inc.: 500 Fifth Ave., New York 10036
Noyes Data Corp.: Mill Rd. and Grand Ave., Park Ridge, N.J. 07656
Oceana Publications, Inc.: Dobbs Ferry, N.Y. 10522
Ottaviano Technical Services: 150 Broad Hollow Rd., Melville, N.Y. 11746
Oxford University Press, Inc.: 200 Madison Ave., New York 10016
Pelican Publishing Co., Inc.: 630 Burmaster St., Gretna, La. 70053
Pendell Publishing Co.: P.O. Box 1666, Midland, Mich. 48640
Pennsylvania State University Press: 215 Wagner Bldg., University Park, Pa. 16802
Peter Peregrinus see International Scholarly Book Services
Pergamon Press, Inc.: Maxwell House, Fairview Park, Elmsford, N.Y. 10523
Philips Books: available from Macmillan Publishers, Ltd., Little Essex St., London WC2R 3LF, England
Philosophical Library, Inc.: 15 E. 40 St., New York 10016
Pitman and Sons, Ltd.: Pitman House, Parker Square, Kingsway, London WC2B 5PB, England
Plant Engineering: 1301 South Grove Ave., Barrington, Ill. 60010
Plenum Publishing Corp.: 227 W. 17 St., New York 10011
Popular Science Publishing Co.: 380 Madison Ave., New York 10017
Portland Cement Association: Old Orchard Rd., Skokie, Ill. 60076
Potter see Clarkson N. Potter
Praeger Publishers, Inc.: 111 Fourth Ave., New York, N.Y. 10003
Prentice-Hall, Inc.: Englewood Cliffs, N.J. 07632
Prestressed Concrete Institute: 20 N. Wacker Dr., Chicago, Ill. 60606
Princeton University Press: Princeton, N.J. 08540
Public Administration Service: 1313 E. 60 St., Chicago, Ill. 60637
Purdue University Press: South Campus Courts-D West, Lafayette, Ind. 47907
Rand McNally and Co.: 8255 Central Park Ave., Skokie, Ill. 60076
Random House, Inc.: 201 E. 50 St., New York 10022
Regent Graphic Services: Box 8372, Swissvale, Pa. 15218
Reinhold Book Co.: 450 W. 33 St., New York 10001
Research and Education Association: 342 Madison Ave., New York 10017
Research Guide Publications: 139 W. Colorado Blvd., Pasadena, Calif. 91105
Resources for the Future, Inc.: 1755 Massachusetts Ave., N.W., Washington, D.C. 20036
Richardson Engineering Services, Inc.: 722 Genevieve St., P.O.

Box Y, Solana Beach, Calif. 92075
The Ronald Press Co.: 79 Madison Ave., New York 10016
Rowman and Littlefield: 81 Adams Dr., Totowa, N.J. 07512
Rutgers University Press: 30 College Ave., New Brunswick, N.J. 08901
Sage Publications, Inc.: 275 S. Beverly Dr., Beverly Hills, Calif. 90212
Howard W. Sams (Theodore Audel and Co.): 4300 W. 62 St., Indianapolis, Ind. 46286
Lee Saylor, Inc.: 1541 Palos Verdes Mall, Walnut Creek, Calif. 94596
Schenkman Publishing Co., Inc.: 3 Mt. Auburn Pl., Cambridge, Mass. 02138
Schocken Books, Inc.: 200 Madison Ave., New York 10016
Abner Schram: 1860 Broadway, New York 10023
Scott, Foresman and Co.: 1900 E. Lake Ave., Glenview, Ill. 60025
Charles Scribner's Sons: 597 Fifth Ave., New York 10017
Sheet Metal and Air Conditioning Contractors' National Association: 1611 N. Kent St., Arlington, Va. 22209
Sheppard's Citations: P.O. Box 1235, Colorado Springs, Colo. 80901
The Shoe String Press, Inc.: 995 Sherman Ave., Hamden, Conn. 06514
Sierra Club Books: 1050 Mills Tower, San Francisco, Calif. 94104
Society of American Value Engineers: 29551 Greenfield Rd., Southfield, Mich. 48076
Society of Mining Engineers of AIME: 345 E. 47 St., New York 10017
Southern Forest Products Association: 3525 N. Causeway Blvd., New Orleans, La. 70152
Springer-Verlag New York, Inc.: 175 Fifth Ave., New York 10010
State University of N.Y. Press: 99 Washington Ave., Albany, N.Y. 12210
Steel Structures Painting Council: 4400 Fifth Ave., Pittsburgh, Pa. 15213
Structures Publishing Co.: 24269 Indoplex Circle, Box 423, Farmington, Mich. 48024
Syentek Books: P.O. Box 277, Cotati, Calif. 94928
Syracuse University Press: Box 8, University Sta., Syracuse, N.Y. 13210
Tab Books: Blue Ridge Summit, Pa. 17214
Taplinger Publishing Co., Inc.: 200 Park Ave. S., New York 10003
Tavistock see Barnes and Noble
Technical Publishing Co.: Barrington, Ill. 60010
Technomic Publishing Co., Inc.: 265 W. State St., Westport, Conn. 06880
Theatre Arts Books: 333 6th Ave., New York 10014
Paul Theobald and Co.: 5 N. Wabash Ave., Chicago, Ill.
Trans-Tech Publications: 21330 Center Ridge Rd., Cleveland, Ohio 44116

Tudor Publishing Co.: 221 Park Ave. S., New York 10003
Twayne Publishing Co.: 70 Lincoln St., Boston, Mass. 02111
UNESCO Publications Center: 650 First Ave., New York 10016
Frederick Ungar Publishing Co., Inc.: 250 Park Ave. S., New York 10003
U.S. Atomic Energy Commission see Energy Research and Development Administration
U.S. Department of Housing and Urban Development: Washington, D.C. 20410
U.S. Environmental Protection Agency: Office of Air and Water Programs, Washington, D.C. 20460
U.S. Government Printing Office: Washington, D.C. 20402
U.S. Savings and Loan League: 111 E. Wacker Dr., Chicago, Ill. 60601
Universe Books: 381 Park Ave. S., New York 10016
University Microfilms: 300 N. Zeeb Rd., Ann Arbor, Mich. 48106
University of California, Graduate School of Management: Housing, Real Estate and Urban Land Studies, Los Angeles, Calif. 90024
University of California Press: 2223 Fulton St., Berkeley, Calif. 94720
University of Chicago Press: 5801 Ellis Ave., Chicago, Ill. 60637
University of Georgia Press: Athens, Ga. 30602
University of Illinois Press: Urbana, Ill. 61801
University of Iowa Press: Iowa City, Iowa 52242
University of Miami Press: Drawer 9088, Coral Gables, Fla. 33124
University of Michigan, Institute for Social Research: Ann Arbor, Mich. 48106
University of New Mexico Press: Albuquerque, N.M. 87131
University of Oklahoma Press: 1005 Asp Ave., Norman, Okla. 73069
University of Tennessee Press: Communications Bldg., Knoxville, Tenn. 37916
University of Texas: Distributive Education Instructional Materials, Austin, Texas 78712
University of Toronto Press: University of Toronto, Toronto, Canada M5S1A6
University Press of Virginia: P.O. Box 3608, University Sta., Charlottesville, Va. 22903
The Urban Land Institute: 1200 18 St., N.W., Washington, D.C. 20036
Van Nostrand Reinhold Co.: 450 W. 33 St., New York 10001
Vanderbilt University Press: Nashville, Tenn. 37235
The Viking Press, Inc.: 625 Madison Ave., New York 10022
Frank R. Walker Co.: 5030 N. Harlem Ave., Chicago, Ill. 60656
Watson-Guptill: 1 Astor Plaza, New York 10036
John Weatherhill, Inc.: 149 Madison Ave., New York 10016
Westinghouse Electric Corp.: Pittsburgh, Pa. 15235
John Wiley and Sons, Inc.: 605 Third Ave., New York 10016
George Wittenborn, Inc.: 1018 Madison Ave., New York 10021

Wolters-Noordhoff: Academic Book Services: Holland, P.O. Box 66, Groningen, The Netherlands

Yale University Press: 92A Yale Station, New Haven, Conn. 06520

INDEX

ACOUSTICS AND SOUND ABATEMENT

A04-0723	Beranek, Music, Acoustics and Architecture.
A04-0769	Building Construction and Design.
A04-0787	Canter and Lee, Psychology and the Built Environment.
A04-0844	Day, Building Acoustics.
A04-0858	Doelle, Environmental Acoustics.
A04-0878	Egan, Concepts in Architectural Acoustics.
A04-0907	Flynn and Segil, Architectural Interior Systems: Lighting, Air Conditioning, Acoustics.
A04-0938	Gilford, Acoustics for Radio and Television Studios.
A04-0978	Hamlin, Forms and Functions of Twentieth-Century Architecture. Vol. 1 The Elements of Building. Vol. 2 The Principles of Composition. Vol. 3 Building Types: Residence, Gatherings, Education, Government. Vol. 4 Building Types: Commerce and Industry, Public Health, Transportation, Social Welfare and Recreation.
A04-1038	Izenour, Theater Design.
A04-1046	Jones, Teach Yourself Acoustics.
A04-1068	Kinzey and Sharp, Environmental Technologies in Architecture.
A04-1072	Knudsen and Harris, Acoustical Designing in Architecture.
A04-1085	Kuttruff, Room Acoustics.
A04-1087	Lamb, The Dynamical Theory of Sound.
A04-1096	Lawrence, Architectural Acoustics.
A04-1104	Lindsay, Acoustics: Historical and Philosophical Development.
A04-1125	Mankovsky, Acoustics of Studios and Auditoria.
A04-1146	Moore, Design for Good Acoustics.
A04-1167	Negroponte, The Architecture Machine.
A04-1168	Negroponte, Soft Architecture Machines.
A04-1240	Rayleigh, The Theory of Sound.
A04-1248	Rettinger, Acoustic Design and Noise Control.
A04-1267	Sabine, Collected Papers on Acoustics.
A04-1286	Seto, Acoustics.
A04-1334	Suri, Acoustics: Design and Practice.
A04-1352	Vandenberg and Elder, Handbook of Building Enclosure.
A04-1385	Winckel, Music, Sound and Sensation: A Modern Exposition.

A04-1391	Wood, Acoustics.
A05-1604	Duerden, Noise Abatement.
A05-1665	Harris, Handbook of Noise Control.
A06-1973	ASME, AMD. Vol. 1 Isolation of Mechanical Vibration, Impact and Noise.
A06-1988	Beranek, Noise and Vibration Control.
A06-1989	Beranek, Noise Reduction.
A06-1991	Blake and Mitchell, Vibration and Acoustic Measurement Handbook.
A06-2010	Crocker, Noise and Vibration Control Engineering: Proceedings of the Purdue Noise Control Conference, 1971, Lafayette, Ind.
A06-2033	Flugge, Handbook of Engineering Mechanics.
A06-2035	Flynn and Segil, Architectural Interior Systems: Lighting, Air Conditioning, Acoustics.
A06-2048	Harris, Handbook of Noise Control.
A06-2099	McGuinness and Stein, Mechanical and Electrical Equipment for Buildings.
A06-2132	Petrusewicz and Longmore, Noise and Vibration Control.
A06-2163	Stephens, Sound: In Eight Languages.
A06-2164	Stoecker, Principles for Air Conditioning Practice.
A06-2186	Yerges, Sound, Noise and Vibration Control.
A08-2365	Ceiling Systems Handbook.
A08-2369	Close, Sound Control and Thermal Insulation of Buildings.
A08-2385	Diamant, Insulation of Buildings: Thermal and Acoustic.
A09-2619	The Building Research Establishment, Essential Data for Construction. Vol. 3 Services and Environmental Engineering.
A09-2622	Burgess, The Construction Industry Handbook.
A09-2787	Reiner, Methods and Materials of Construction: A Guide for Builders, Owners, Architects and Engineers.

ADHESIVES AND SEALANTS

A08-2350	Bodnar, Structural Adhesives Bonding.
A08-2358	Bucksch, Dictionary of Wood and Woodworking Practice.
A08-2364	Cagle, Handbook of Adhesive Bonding.
A08-2374	Cook, Construction Sealants and Adhesives.
A08-2380	Damusis, Sealants.
A08-2383	DeLollis, Adhesives for Metals: Theory and Technology.
A08-2424	Katz and Cagle, Adhesive Materials: Their Properties and Usage.
A08-2435	Lee and Neville, Handbook of Epoxy Resins.
A08-2451	May and Tanaka, Epoxy Resins: Chemistry and Technology.
A08-2519	Shields, Adhesives Handbook.
A08-2525	Skeist, Epoxy Resins.
A08-2526	Skeist, Handbook of Adhesives.
A08-2527	Skeist, Plastics in Building.
A08-2553	Watson, Construction Materials and Processes.

Index 527

AIRPORTS

A07-2211	Army Dept., Planning and Design of Roads, Airbases, and Heliports, in the Theater of Operations.
A07-2212	Army Engineers Corps., Roads and Airfields.
A07-2214	ASCE, Airport Terminal Facilities.
A07-2215	ASCE, Airports--Challenges of the Future.
A07-2216	ASCE, Airports--Key to the Air Transportation System.
A07-2224	Cedergren, Drainage of Highway and Airfield Pavements.
A07-2232	Hennes and Ekse, Fundamentals of Transportation Engineering.
A07-2235	Holland, Travellers' Architecture.
A07-2237	Horonjeff, Planning and Design of Airports.
A07-2238	Legault, Highway and Airport Engineering.
A07-2239	Li, Bibliography on Airport Engineering.
A07-2245	Navy Department, Design Manual, Airfield Pavements.
A07-2249	Paquette, Transportation Engineering--Planning and Design.
A07-2257	Sargious, Pavements and Surfacings for Highways and Airports.
A07-2258	Schriever and Seifert, Air Transportation 1975 and Beyond: A Systems Approach.
A08-2429	Koster, Expansion Joints in Bridges and Roads.
A09-2607	Blake, Civil Engineer's Reference Book.
A09-2819	Urquhart, Civil Engineering Handbook.

APPRENTICE TRAINING

A06-2119	NESCA, Basic Refrigeration, and Student Workbook for Basic Refrigeration.
A08-2365	Ceiling Systems Handbook.
A09-2583	Althouse, Modern Welding.
A09-2611	Bradzinski, Stair Layout (Design and Building).
A09-2612	Branden and Hartsell, Plastering Skill and Practice.
A09-2635	Chandler, Structures for Building Technicians.
A09-2647	Dalzell and Townsend, Concrete Block Construction for Home and Farm.
A09-2648	Dalzell and Townsend, Masonry Simplified, Vol. 1 Tools, Materials, Practice.
A09-2649	Dalzell, Masonry Simplified, Vol. 2 Practical Construction.
A09-2659	Durbahn, Fundamentals of Carpentry. Vol. 1 Tools, Materials, Practices.
A09-2684	Feirer and Hutchings, Carpentry in Building Construction.
A09-2690	Giachino, Welding Skills and Practices.
A09-2693	Groneman, General Woodworking.
A09-2694	Groneman and Glazener, Technical Woodworking.
A09-2703	Hodge, Brickwork for Apprentices.
A09-2709	Hutchinson and Barton, Physical Aspects of Building Science.
A09-2713	Inott, Carpentry & Joinery: Multi-Question Course.
A09-2726	Kratfel, Introduction to Modern Sheet Metal.

528   Construction Information

A09-2727   Labor Dept., Training and Entry into Union Construction.
A09-2729   Lair, Carpentry for the Building Trades.
A09-2751   Matthias, How to Design and Install Plumbing.
A09-2759   Navy Dept., Equipment Operator.
A09-2772   Plumbers and Pipe Fitters Library. Vol. 1 Materials, Tools, Calculations.
A09-2773   Plumbers and Pipe Fitters Library. Vol. 2 Drainage. Fittings, Fixtures.
A09-2774   Plumbers and Pipe Fitters Library. Vol. 3 Installation, Heating, Welding.
A09-2785   Reband, Related Mathematics for Carpenters.
A09-2798   Schofield, Mathematics for Construction Students.
A09-2800   Sedwell, Calculations for Bricklayers.
A09-2814   Thames Valley Group, Students Project Work in Construction.
A09-2829   Whittick, Questions and Answers on Carpentry and Joinery.
A11-3058   McKibben, Looking Forward to a Career: Building Trades.

ARCHITECTS, ENGINEERS

A01-0136   MacFadyen, Sir Ebenezer Howard and the Town Planning Movement.
A04-0654   AIA Membership Directory.
A04-0657   Alander, Viljo Revell: Works and Projects.
A04-0704   Aurenhammer, J. B. Fischer von Erlach.
A04-0709   Baldwin, Stanford White.
A04-0715   Barry, The Life and Works of Sir Charles Barry.
A04-0730   Blaser, Mies van der Rohe: The Art of Structure.
A04-0732   Boaga and Boni, The Concrete Architecture of Riccardo Morandi.
A04-0733   Boesiger, Le Corbusier.
A04-0734   Boesiger, Le Corbusier: Last Works.
A04-0735   Boesiger, Richard Neutra: Buildings and Projects.
A04-0736   Boesiger and Girsberger, Le Corbusier 1910-65.
A04-0742   Bowker, American Architects Directory.
A04-0747   Braham and Smith, Francois Mansart.
A04-0761   Brown, The Work of G. Rietveld, Architect.
A04-0788   Carter, Mies van der Rohe at Work.
A04-0809   Cole, From Tipi to Skyscraper: History of Women in Architecture.
A04-0812   Colvin, A Biographical Dictionary of English Architects, 1660-1840.
A04-0818   Connely, Louis Sullivan.
A04-0819   Conrads, Programs and Manifestoes on 20th Century Architecture.
A04-0821   Cook and Klotz, Conversations with Architects: Philip Johnson, Kevin Roche, Paul Rudolph, Bertrand Goldberg, Morris Lapidus, Louis Kahn, Charles Moore, Robert Venturi.
A04-0824   Cook, Archigram.

Index

A04-0826	Coope, Salomon de Brosse and the Development of the Classical Style in French Architecture from 1565 to 1630.
A04-0834	Creese, The Legacy of Raymond Unwin: A Human Pattern for Planning.
A04-0845	Deane, Constantinos Doxiadis, Master Builder for Free Men.
A04-0859	Downes, Hawksmoor.
A04-0863	Doxiadis, Architecture in Transition.
A04-0864	Drexler, Architecture of Skidmore, Owings & Merrill, 1963-1973.
A04-0873	Eaton, American Architecture Comes of Age: European Reaction to HH Richardson and Louis Sullivan.
A04-0874	Eaton, Two Chicago Architects and Their Clients: Frank Lloyd Wright and Howard Van Doren Shaw.
A04-0879	Eisenman, Five Architects: Eisenman, Graves, Gwathmey, Hejduk, Meier.
A04-0881	Elmes, Lectures on Architecture.
A04-0884	Ericsson, Sixty Years a Builder: The Autobiography of Henry Ericsson.
A04-0885	Etienne-Louis Boullee: Theoretician of Revolutionary Architecture.
A04-0887	Evenson, Le Corbusier: The Machine and The Grand Design.
A04-0888	Faber, Arne Jacobsen.
A04-0899	Ferrey, Recollections of A. W. N. Pugin and His Father, Augustus Pugin.
A04-0913	Francois Mansart.
A04-0931	Gebhard, Rudolph M. Schindler.
A04-0934	Geretsegger, Otto Wagner 1841-1918.
A04-0945	Glaeser, The Works of Frei Otto.
A04-0952	Gotch, Inigo Jones.
A04-0954	Grady, Architecture of Neel Reid in Georgia.
A04-0956	Granger, Charles Follen McKim: A Study of His Life and Work.
A04-0960	Gropius, Apollo in the Democracy.
A04-0965	Guinness and Sadler, Mr. Jefferson, Architect.
A04-0966	Gutheim, In the Cause of Architecture: Frank Lloyd for the Record.
A04-0983	Harbron, The Conscious Stone: The Life of Edward William Godwin.
A04-0984	Harker, Studio and Stage.
A04-0988	Hatch, Buckminster Fuller: At Home in the Universe.
A04-0995	Herrmann, The Theory of Claude Perrault.
A04-0999	Hildebrand, Designing for Industry: The Architecture of Albert Kahn.
A04-1001	Hitchcock, The Architecture of H. H. Richardson and His Times.
A04-1006	Hoffman, The Architecture of John Wellborn Root.
A04-1020	Huber and Steinegger, Jean Prouve: Prefabrication, Structures and Elements.
A04-1024	Hunt, Total Design: Architecture of Welton Becket and Associates.

A04-1026	Huxtable, Have You Kicked A Building Lately?
A04-1044	Jencks, Le Corbusier and the Tragic View of Architecture.
A04-1045	Joedicke, Architecture Since 1945 Sources and Directions.
A04-1053	Kaufman and Raeburn, Frank Lloyd Wright Writings and Buildings.
A04-1065	Kilham, Raymond Hood, Architect of Ideas.
A04-1076	Kornwolf, M. H. Baillie Scott and the Arts and Crafts Movement: Pioneers of Modern Design.
A04-1089	Landy, The Architecture of Minard Lafever.
A04-1100	Le Corbusier, Creation is a Patient Search.
A04-1107	Linstrum, Sir Jeffry Wyatville: Architect to the King.
A04-1121	McCoy, Five California Architects.
A04-1153	Mumford, The Brown Decades: A Study of the Arts in America 1865-1895.
A04-1157	Munz and Kunstler, Adolf Loos: Pioneer of Modern Architecture.
A04-1160	Museum of Modern Art, Ludwig Mies van der Rohe.
A04-1182	Olmsted and Kimball, Frederick Law Olmsted, Landscape Architect: 1822-1903.
A04-1183	O'Neal, The American Association of Architectural Bibliographers. Vol. 1
A04-1184	O'Neal, The American Association of Architectural Bibliographers. Vol. 2
A04-1185	O'Neal, The American Association of Architectural Bibliographers. Vol. 3
A04-1186	O'Neal, The American Association of Architectural Bibliographers. Vol. 4
A04-1187	O'Neal, The American Association of Architectural Bibliographers. Vol. 5
A04-1188	O'Neal, The American Association of Architectural Bibliographers. Vol. 6
A04-1189	O'Neal, The American Association of Architectural Bibliographers. Vol. 7
A04-1190	O'Neal, The American Association of Architectural Bibliographers. Vol. 8
A04-1191	O'Neal, The American Association of Architectural Bibliographers. Vol. 9
A04-1192	O'Neal, The American Association of Architectural Bibliographers. Vol. 10
A04-1196	Owen, Owen, Robert Dale, 1801-1877.
A04-1199	Palmes, Le Corbusier: My Work.
A04-1200	Papachristou, Marcel Breuer: New Buildings and Projects.
A04-1207	Pentagram, Pentagram: The Work of Five Designers.
A04-1214	Pevsner, Some Architectural Writers of the Nineteenth Century.
A04-1223	Pratt and Gunther, The Architecture of Sir Roger Pratt.
A04-1244	Reilly, McKim, Mead and White.
A04-1261	Roth, Monograph of the Work of McKim, Mead & White: 1879-1915.
A04-1268	Safdie, Beyond Habitat.

Index 531

A04-1279	Schwab, The Architecture of Paul Rudolph.
A04-1304	Smith, Builders in the Sun: Five Mexican Architects.
A04-1313	Soleri, Sketch Books of Paolo Soleri.
A04-1316	Spreiregen, On the Art of Designing Cities: Selected Essays of Elbert Peets.
A04-1324	Storrer, The Architecture of Frank Lloyd Wright: A Complete Catalog.
A04-1325	Street, Memoir of George Edmund Street, R. A. 1824-1881.
A04-1329	Sullivan, The Autobiography of an Idea.
A04-1337	Sweeney and Sert, Antoni Gaudi.
A04-1339	Tange and Kultermann, Kenzo Tange.
A04-1347	Torroja, The Structures of Eduardo Torroja.
A04-1353	Van Rensselaer, Henry Hobson Richardson and His Works.
A04-1356	Venture, Complexity and Contradiction in Architecture.
A04-1381	Wilson, Conversations with Lev Zetlin.
A04-1386	Wingler, The Bauhaus: Weimar Dessau Berlin Chicago.
A04-1392	Woods, Candilis-Josic-Woods: Building for People.
A04-1393	Wright, An American Architecture.
A04-1394	Wright, The Drawings of Frank Lloyd Wright.
A04-1395	Wright, The Early Work.
A04-1396	Wright, The Future of Architecture.
A04-1397	Wright, Genius and Mobocracy.
A04-1398	Wright, The Industrial Revolution Runs Away.
A04-1401	Wright, An Organic Architecture: The Architecture of Democracy.
A04-1402	Wright, A Testament.
A04-1403	Wright, The Work of Frank Lloyd Wright (The Wendingen Edition).
A04-1406	Worman and Feldman, The Notebooks and Drawings of Louis I. Kahn.
A05-1495	ASCE, Outstanding Papers of Thomas R. Camp.
A05-1829	Roland, Frei Otto: Tension Structures, Ideas and Experiments in Lightweight Construction.
A10-2851	ASCE, A Biographical Dictionary of American Civil Engineers.

ARCHITECTURAL DETAILS

A04-0663	Amery, Domestic Historic Buildings and Their Details.
A04-0671	Architects' Journal, Architect's Working Details. Vol. 1
A04-0672	Architects' Journal, Architect's Working Details. Vol. 2
A04-0673	Architects' Journal, Architect's Working Details. Vol. 3
A04-0674	Architects' Journal, Architect's Working Details. Vol. 4
A04-0675	Architects' Journal, Architect's Working Details. Vol. 5
A04-0676	Architects' Journal, Architect's Working Details. Vol. 6

A04-0677	Architects' Journal, <u>Architect's Working Details.</u> Vol. 7
A04-0678	Architects' Journal, <u>Architect's Working Details.</u> Vol. 8
A04-0679	Architects' Journal, <u>Architect's Working Details.</u> Vol. 9
A04-0680	Architects' Journal, <u>Architect's Working Details.</u> Vol. 10
A04-0681	Architects' Journal, <u>Architect's Working Details.</u> Vol. 11
A04-0682	Architects' Journal, <u>Architect's Working Details.</u> Vol. 12
A04-0683	Architects' Journal, <u>Architect's Working Details.</u> Vol. 13
A04-0684	Architects' Journal, <u>Architect's Working Details.</u> Vol. 14
A04-0719	Bayes and Franklin, <u>Designing for the Handicapped.</u>
A04-0781	Callender, <u>Time-Saver Standards.</u>
A04-0847	DeChiari and Callender, <u>Time Saver Standards of Building Types.</u>
A04-0853	Devereau, <u>Architects' Working Details Revisited.</u>
A04-0894	Faulkner, <u>Architecture and Color.</u>
A04-0902	Fink, <u>Adobes in the Sun: Portraits of a Tranquil Era.</u>
A04-0925	Gatz, <u>Detail; Contemporary Architectural Design.</u> Vol. 2
A04-0926	Gatz, <u>Detail; Contemporary Architectural Design.</u> Vol. 3
A04-0927	Gatz, <u>Detail; Contemporary Architectural Design.</u> Vol. 4
A04-0928	Gatz, <u>Detail; Contemporary Architectural Design.</u> Vol. 5
A04-0932	Geerlings, <u>Metal Crafts in Architecture.</u>
A04-0933	Geerlings, <u>Wrought Iron in Architecture.</u>
A04-0946	Gleis, <u>Kleinkirchenbau.</u>
A04-0948	Goldsmith, <u>Designing for the Disabled.</u>
A04-0955	Graf, <u>Don Graf's Data Sheets.</u>
A04-0972	Halfpenny and Halfpenny, <u>The Art of Sound Building.</u>
A04-0977	Hamlin, <u>Greek Revival Architecture in America.</u>
A04-1062	Kicklighter and Baird, <u>Architecture: Residential Drawing and Design.</u>
A04-1092	Langley, <u>The Builder's Director.</u>
A04-1093	Langley, <u>The Builder's Jewel.</u>
A04-1094	Langley, <u>The City and Country Builder's and Workman's Treasury of Designs.</u>
A04-1105	Link, <u>Residential Designs: How to Get the Most for Your Housing Dollar.</u>
A04-1112	Lloyd, <u>A History of English Brickwork.</u>
A04-1133	Meyer-Bohe, <u>Bauten fur die Jugend/Structures for Children.</u>
A04-1229	Pyne Press, American Historical Catalog Collection. Vol. 1 <u>Architectural Elements, 1850-1880.</u>
A04-1232	Ramsey and Sleeper, <u>Architectural Graphic Standards.</u>
A04-1233	Ramsey and Harvey, <u>Small Georgian Houses and Their</u>

Index 533

	Details: 1750-1820.
A04-1249	Reynolds, <u>Dutch Houses in the Hudson Valley Before 1776.</u>
A04-1293	Shipway and Shipway, <u>Decorative Design in Mexican Homes.</u>
A04-1302	Sleeper, Architectural Specifications.
A04-1303	Sleeper, <u>Building Planning and Design Standards for Architects, Engineers, Designers, Consultants, Building Committees, Draftsmen, and Students.</u>
A04-1335	Swan, A Collection of Design in Architecture.
A04-1336	Swan, <u>Upwards of One Hundred Fifty New Designs for Chimney Pieces.</u>
A04-1365	Walsh, <u>The Science of Daylight.</u>
A04-1375	Whittick, <u>Symbols for Designers.</u>
A05-1915	Wass, <u>Manual of Structural Details for Building Construction.</u>
A08-2351	Boyne and Wright, <u>Architects' Working Details.</u> Vol. 1-15
A08-2381	Danz, <u>Sun Protection.</u>
A08-2401	Gatz and Thierry, <u>Architect's Detail Library.</u> Vol. 1 <u>Wrought Iron Railings, Doors and Gates.</u>
A08-2402	Gatz and Thierry, <u>Architect's Detail Library.</u> Vol. 2 <u>Ceilings in Wood.</u>
A08-2403	Gatz and Thierry, <u>Architect's Detail Library.</u> Vol. 3 <u>Windows and Window Walls.</u>
A08-2404	Gatz and Thierry, <u>Architect's Detail Library.</u> Vol. 5 <u>Entrances and Staircases.</u>
A08-2405	Gatz and Thierry, <u>Architect's Detail Library.</u> Vol. 6 <u>Exterior Detailing in Concrete.</u>
A08-2420	Hornung, <u>Reinhold Data Sheets.</u>
A08-2447	Marsh, <u>Concrete as a Visual Material.</u>
A08-2508	Rostron, <u>Light Cladding of Buildings.</u>
A09-2617	The Building Research Establishment, <u>Essential Data for Construction.</u> Vol. 1 <u>Building Construction.</u>
A09-2620	The Building Research Establishment, <u>Essential Data for Construction.</u> Vol. 4 <u>Building Defects and Maintenance.</u>
A09-2636	Christian, <u>Negro Ironworkers in Louisiana.</u>
A09-2728	Lefeyer, <u>The Modern Builder's Guide.</u>
A09-2736	Lloyd, A <u>History of English Brickwork: With Examples and Notes of the Architectural Use and Manipulation of Brick from Medieval Times to the End of the Georgian Period.</u>
A11-3072	Newman, <u>Standard Structural Details for Building Construction.</u>
A11-3116	Weidhaas, <u>Architectural Drafting and Construction.</u>

ARCHITECTURE AND URBAN DESIGN

See all items in A04
A01-0033	Calsat and Sydler, <u>International Vocabulary of Town Planning and Architecture.</u>
A01-0056	Crosby, <u>Architecture: City Sense.</u>

A01-0057 Cullen, The Concise Townscape.
A01-0059 Dahinden, Urban Structures for the Future.
A01-0061 DeChiara and Koppelman, Planning Design Criteria.
A01-0063 deWolfe, Civilia: The End of Sub Urban Man.
A01-0070 Eckbo, The Landscape We See.
A01-0082 Gallion and Eisner, The Urban Pattern: City Planning and Design.
A01-0086 Gibberd, Town Design.
A01-0087 Godwin, Directory of Official Architecture and Planning.
A01-0131 Lewis, Urban Structure.
A01-0143 Mayer, The Urgent Future.
A01-0153 Newman, Defensible Space: People and Design in the Violent City.
A01-0158 Papageorgiu, Integration of the Past: Preservation in City Planning.
A01-0167 Pushkarev and Zupan, Urban Space for Pedestrians.
A01-0169 Redstone, The New Downtowns: Rebuilding Business Districts.
A01-0172 Reps, Monumental Washington: The Planning and Development of the Capital Center.
A01-0185 Sanoff and Cohn, EDRA 1: Proceedings of the 1st Annual Environmental Design Research Assn.
A01-0194 Sixta and Parr, Urban Structure.
A01-0195 Smithson and Smithson, Urban Structuring.
A01-0199 Spyer, Architect and Community.
A01-0214 Tandy, Handbook of Urban Landscape.
A01-0215 Tetlow and Gross, Homes, Towns, and Traffic.
A01-0238 Worskett, The Character of Towns.
A02-0280 Collison, The Developer's Dictionary and Handbook.
A02-0285 Costonis, Space Adrift - Landmark Preservation and the Marketplace.
A02-0360 Metcalf, Planning Academic and Research Library Buildings.
A02-0365 Nierstrasz, Building for the Aged.
A02-0382 Redstone, New Dimensions in Shopping Centers and Stores.
A02-0403 Weiss, Better Buildings for the Aged.
A03-0427 Aguilar, Systems Analysis and Design in Engineering, Architecture, Construction and Planning.
A03-0429 AIA, Comprehensive Architectural Services.
A03-0430 AIA, Creative Control of Building Costs.
A03-0435 Andersen, Financial Management for Architectural Firms - A Manual of Accounting Procedures.
A03-0439 Architectural Record, Techniques of Successful Practice for Architects and Engineers.
A03-0460 Brunton, Management Applied to Architectural Practice.
A03-0464 Case, The Economics of Architectural Practice.
A03-0465 Case, Methods of Compensation for Architectural Services.
A03-0466 Case, Profit Planning in Architectural Practice.
A03-0467 Caudill, Architecture by Team: A New Concept for the Practice of Architecture.
A03-0475 Cowgill and Small, Architectural Practice.

Index 535

A03-0476	Coxe, Marketing Architectural and Engineering Services.
A03-0488	Dibner, Joint Ventures for Architects and Engineers.
A03-0492	Dubin, Architectural Supervision of Modern Buildings.
A03-0505	Foxhall, Professional Construction Management and Project Administration.
A03-0506	Foxhall, Technique of Successful Practice for Architects and Engineers.
A03-0507	Freeth and Davey, The AJ Legal Handbook.
A03-0517	Green, The Architect's Guide to Running a Job.
A03-0518	Green, Architect's Guide to Site Management.
A03-0519	Griffin, Development Building: The Team Approach.
A03-0525	Harper, Computer Applications in Architecture and Engineering.
A03-0526	Harper, Financial Management Computer Users Manual.
A03-0527	Hauf, Building Contracts for Design and Construction.
A03-0528	Heery, Time, Cost, and Architecture.
A03-0533	Hunt, Total Design: Architecture of Welton Becket & Associates.
A03-0534	Hutton and Devonald, Value in Building.
A03-0543	Jones, How to Market Professional Design Services.
A03-0551	Lapidus, Architecture: A Profession and a Business.
A03-0552	Lawton, Planning and Managing Housing for the Elderly.
A03-0559	Macfarlane, Architectural Supervision on Site.
A03-0587	O'Brien, Value Analysis in Design and Construction.
A03-0593	Pilcher, Appraisal and Control of Project Costs.
A03-0606	Rosenfeld, Architect's Handbook of Professional Practice. 2 Volumes.
A03-0629	Walker, Legal Pitfalls in Architecture, Engineering, and Building Construction.
A03-0633	Willis and George, Architects in Practice.
A05-1872	Stuttgart Universität, Biology and Building I: Colloquium Report.
A05-1873	Stuttgart Universität, Biology and Building II: Soap Film Models.
A05-1874	Stuttgart Universität, Biology and Building III.
A05-1875	Stuttgart Universität, Convertible Roofs.
A05-1876	Stuttgart Universität, The Experimental Determination of Minimal Nets.
A05-1877	Stuttgart Universität, Project Study "City in the Arctic."
A05-1878	Stuttgart Universität, Shadow in the Desert.
A06-1981	Banham, Architecture of the Well-Tempered Environment.
A07-2207	APWA, Streetscape Equipment Sourcebook.
A07-2235	Holland, Travellers' Architecture.
A07-2240	Malt, Furnishing the City.
A08-2310	Anderson and Earle, Design and Aesthetics in Wood.
A08-2381	Danz, Sun Protection.
A08-2419	Hoffman, Building with Wood: Form Structural Design and Preservation.
A09-2605	Benjamin, The American Builder's Companion.
A09-2701	Hansen, Architecture in Wood.
A09-2718	Kidder and Parker, Architects' and Builders' Handbook.

A09-2741	McCue, Creating the Human Environment.
A10-2873	Guide to Architectural Information.
A10-2912	Smith, How to Find Out in Architecture and Building -- A Guide to Sources of Information.
A11-2967	Burden, Architectural Delineation: A Photographic Approach to Presentation.
A11-2979	Croney, Anthropometrics for Designers.
A11-2988	Dreyfuss, The Measure of Man: Human Factors in Design.
A11-2996	Feldman, Building Design for Maintainability.
A11-3009	Gowan, A Continuing Experiment: Learning and Teaching at the Architectural Association.
A11-3074	Newman, Defensible Space: Crime Prevention Through Urban Design.
A11-3077	NFPA, Designing Buildings for Fire Safety.
A11-3078	NFPA, Fires in High-Rise Buildings.
A11-3083	Perry, Engineering Manual.
A11-3085	Piper, Opportunities in an Architecture Career.

ARCHITECTURE - ASIA, AFRICA, AUSTRALIA

A04-0661	Altherr, Three Japanese Architects.
A04-0707	Badawy, Architecture in Ancient Egypt and the Near East.
A04-0737	Borras, Contemporary Japanese Architecture.
A04-0745	Boyd, New Directions in Japanese Architecture.
A04-0758	Brown, From Madina to Metropolis: Heritage and Change in the Near Eastern City.
A04-0759	Brown, Indian Architecture. Vol. 1 Buddhist and Hindu Periods.
A04-0760	Brown, Indian Architecture. Vol. 2 Islamic Period.
A04-0778	Bussagli, Oriental Architecture.
A04-0789	Casey, Early Melbourne Architecture 1840 to 1888.
A04-0796	Chambers, Designs of Chinese Buildings, Furniture, Dresses, Machines, and Utensils.
A04-0831	Cram, Impressions of Japanese Architecture and the Allied Arts.
A04-0836	Creswell, A Bibliography of the Architecture, Arts, and Crafts of Islam to 1st January, 1960.
A04-0917	Freeland, Architecture in Australia.
A04-0920	Fry and Drew, Tropical Architecture.
A04-0924	Garlake, The Early Islamic Architecture of the East African Coast.
A04-0950	Goodwin, A History of Ottoman Architecture.
A04-0969	Halfpenny and Halfpenny, Chinese and Gothic Architecture.
A04-0971	Halfpenny and Halfpenny, Rural Architecture in the Chinese Taste.
A04-0983	Harbron, The Conscious Stone: The Life of Edward William Godwin.
A04-0990	Hayashi, House Design in Today's Japan.
A04-1005	Hoag, Western Islamic Architecture.
A04-1033	Ishimoto and Ishimoto, Japanese House, Its Interior and

	Exterior.
A04-1034	Itoh, The Classic Tradition in Japanese Architecture Modern Versions of the Sukiya Style.
A04-1035	Itoh, The Elegant Japanese House: Traditional Sukiya Architecture.
A04-1036	Itoh, Kura, Design of Traditional Japanese Barns and Storehouses.
A04-1037	Itoh, Traditional Domestic Architecture of Japan.
A04-1081	Kuhnel, Islamic Art & Architecture.
A04-1083	Kultermann, New Directions in African Architecture.
A04-1147	Morgan and Gilbert, Early Adelaide Architecture 1836 to 1886.
A04-1150	Morse, Japanese Homes and Their Surroundings.
A04-1173	Nilsson, European Architecture in India, 1750-1850.
A04-1292	Shinohara, 16 Houses & Architectural Theory.
A04-1315	Sowden, Towards an Australian Architecture.
A04-1338	Tange and Kawazoe, Ise: Prototype of Japanese Architecture.
A04-1343	Tempel, New Japanese Architecture.
A04-1349	T'Serstevens, Living Architecture: Chinese.
A04-1350	Univ. Press of Africa, Homes for Kenya.
A04-1359	Vohwahsen, Living Architecture: India.
A04-1360	Vohwahsen, Living Architecture: Islamic Indian.
A04-1407	Yamamoto, Japan's Contemporary Houses.
A04-1409	Yoshida, The Japanese House and Garden.

ARCHITECTURE - CHURCH

A04-0655	AJ Editors, Church Buildings.
A04-0686	Architectural Journal, Church Buildings.
A04-0691	Architectural Record, Interior Spaces Designed by Architects.
A04-0764	Bruggink and Droppers, Christ and Architecture.
A04-0765	Bruggink and Droppers, When Faith Takes Form.
A04-0785	Cantacuzino, New Uses for Old Buildings.
A04-0800	Chiarelli, San Lorenzo and the Medici Chapels.
A04-0804	Clarke, Parish Churches of London.
A04-0847	DeChiari and Callender, Time Saver Standards of Building Types.
A04-0855	Dixon, Architectural Design Preview, U.S.A.
A04-0876	Ede, Canadian Architecture 1960/1970.
A04-0937	Gieselmann, New Churches.
A04-0946	Gleis, Kleinkirchenbau.
A04-0953	Gowans, Church Architecture in New France.
A04-0979	Hammond, Liturgy and Architecture.
A04-1078	Kubler, The Religious Architecture of New Mexico: In the Colonial Period and Since the American Occupation.
A04-1239	Rawlings, Virginia's Colonial Churches: An Architectural Guide Together with Their Surviving Books, Silver, and Furnishings.
A04-1319	Stanton, The Gothic Revival and American Church Architecture: An Episode in Taste, 1840-1856.
A04-1341	Taylor, Looking at Cathedrals.

A04-1347	Torroja, The Structures of Eduardo Torroja.
A09-2686	Fitchen, The Construction of Gothic Cathedrals: A Study of Medieval Vault Erection.

ARCHITECTURE - EUROPE

A04-0657	Alander, Viljo Revell: Works and Projects.
A04-0659	Allen, Stone Shelters.
A04-0662	Ambasz, Italy: The New Domestic Landscape: Achievements and Problems of Italian Design.
A04-0663	Amery, Domestic Historic Buildings and Their Details.
A04-0697	The Architecture of England.
A04-0705	Bachmann and Von Moos, New Directions in Swiss Architecture.
A04-0713	Barley, The House and Home: A Review of 900 Years of House Planning and Furnishing in Britain.
A04-0714	Barnard, The Decorative Tradition in Nineteenth Century Architecture.
A04-0721	Becher and Becher, Anonyme Sculpturen.
A04-0722	Benevolo, History of Modern Architecture.
A04-0725	Besset, New French Architecture.
A04-0740	Boudon, Lived-in Architecture, Le Corbusier's Pessac Revisited.
A04-0747	Braham and Smith, Francois Mansart.
A04-0749	Branner, Gothic Architecture.
A04-0761	Brown, The Work of G. Rietveld, Architect.
A04-0767	Brunskill, Illustrated Handbook of Vernacular Architecture.
A04-0774	Burchard, The Voice of the Phoenix: Postwar Architecture in Germany.
A04-0784	Candilis, Planning and Design for Leisure.
A04-0792	Catalogue of the Royal Institute of British Architects Library. 2 vols.
A04-0806	Clifton-Taylor, The Pattern of English Building.
A04-0812	Colvin, A Biographical Dictionary of English Architects, 1660-1840.
A04-0826	Coope, Salomon de Brosse and the Development of the Classical Style in French Architecture from 1565 to 1630.
A04-0840	Dannatt, The Architect's Year Book.
A04-0848	Defoe, A Tour Thro' London About the Year 1725.
A04-0852	Dercsenyi, Historical Monuments in Hungary, Restoration and Preservation.
A04-0859	Downes, Hawksmoor.
A04-0866	Dunbar, The Historic Architecture of Scotland.
A04-0867	Dunlop, The Chateaux of the Loire.
A04-0888	Faber, Arne Jacobsen.
A04-0889	Faber, New Danish Architecture.
A04-0901	Feuerstein, New Directions in German Architecture.
A04-0914	Frankl, Principles of Architectural History.
A04-0922	Fyodorov and Bartenev, North Russian Architecture.
A04-0923	Galardi, New Italian Architecture.
A04-0935	Germann, Gothic Revival in Europe and Britain Sources,

Index 539

	Influences, and Ideas.
A04-0941	Girbau, Contemporary Spanish Architecture.
A04-0942	Girouard, The Victorian Public House.
A04-0943	Girouard, The Victorian Country House.
A04-0947	Goldfinger, Villages in the Sun: Mediterranean Vernacular Architecture.
A04-0957	Gregotti, New Directions in Italian Architecture.
A04-0964	Guinness and Ryan, Irish Houses and Castles.
A04-1002	Hitchcock, Early Victorian Architecture in Britain.
A04-1004	Hitchcock, Rococo Architecture in Southern Germany.
A04-1007	Hoffmann, Row Houses and Cluster Houses: An International Survey.
A04-1010	Hogg, A Guide to English Country Houses.
A04-1039	Jackson, The Politics of Architecture.
A04-1051	Kaufmann, Architecture in the Age of Reason: Baroque and Post-Baroque in England, Italy, France.
A04-1055	Keller, The Renaissance in Italy.
A04-1064	Kidson and Murray, A History of English Architecture.
A04-1071	Knox, The Architecture of Poland.
A04-1077	Kraemer, One-Family Houses in Groups/Einfamilienhauser in der gruppe: A Collection of Examples.
A04-1088	Landau, New Directions in British Architecture.
A04-1102	Lindemann, A History of German Art: Painting, Sculpture, and Architecture.
A04-1108	Lissitzky, Russia: An Architecture for World Revolution.
A04-1109	Little, Birmingham Buildings.
A04-1110	Little, English Historic Architecture.
A04-1111	Lloyd, A History of the English Houses.
A04-1122	Macleod, Style and Society: Architectural Ideology in Britain, 1835-1914.
A04-1130	Maxwell, New British Architecture.
A04-1204	Pehnt, Expressionist Architecture.
A04-1205	Pehnt, German Architecture 1960-1970.
A04-1213	Petzsch, Architecture in Scotland.
A04-1226	Priestley, The English Home.
A04-1227	Proksch, Houses in the Alps/Maisons Dans les Alpes.
A04-1251	Richards, A Guide to Finnish Architecture.
A04-1270	Salokorpi, Modern Architecture in Finland.
A04-1272	Santini and Marini, Catalogo Bolaffi dell' Architettura Italiana, 1963-1966.
A04-1283	Senkevitch, Soviet Architecture, 1917-1962: A Bibliographic Guide to Source Material.
A04-1285	Service, Edwardian Architecture and Its Origins.
A04-1320	Stanton, Pugin.
A04-1342	Tempel, New Finnish Architecture.
A04-1373	Whiting, New Houses.
A04-1374	Whiting, New Single-Storey Houses.
A04-1389	Wolgensinger, Vacation Homes in the Sun/Ferienhaeuser in der Sonne.
A04-1390	Wolgensinger, Vacation Houses of Europe.
A04-1408	Yarwood, The Architecture of Europe.
A10-2919	van den Broek, Habitation. Vol. 2 Belgium, Denmark,

France, United Kingdom, Sweden, Switzerland and Czechoslovakia.
A10-2920 van den Broek, Habitation. Vol. 3 USA, USSR, Finland, Greece, Hungary, Turkey and Yugoslavia.

ARCHITECTURE - HISTORIC PERIODS AND STRUCTURES

A01-0017 Benevolo, The Origins of Modern Town Planning.
A01-0018 Beresford, New Towns of the Middle Ages.
A01-0148 Morris, History of Urban Form.
A01-0171 Reps, The Making of Urban America: A History of City Planning in the U.S.
A01-0172 Reps, Monumental Washington: The Planning and Development of the Capital Center.
A01-0183 Saalman, Haussmann: Paris Transformed.
A01-0232 Wiebenson, Tony Garnier: The Cite Industrielle.
A04-0651 Acton, Great Houses of Italy: The Tuscan Villas.
A04-0658 Allatios, The Newer Temples of the Greeks.
A04-0660 Allsopp, The Study of Architectural History.
A04-0664 Andrews, Architecture in America.
A04-0667 Andrews, Architecture in New England: A Photographic History.
A04-0668 Angus, The Old Stones of Kingston.
A04-0670 Architects' Emergency Committee, Great Georgian Houses of America.
A04-0697 The Architecture of England.
A04-0698 Argan, The Renaissance City.
A04-0700 Artistic Houses, Being a Series of Interior Views of a Number of the Most Beautiful and Celebrated Homes in the United States.
A04-0706 Bacon, Design of Cities.
A04-0708 Bailey, Pre-Revolutionary Dutch Houses and Families in Northern New Jersey and Southern New York.
A04-0713 Barley, The House and Home: A Review of 900 Years of House Planning and Furnishing in Britain.
A04-0714 Barnard, The Decorative Tradition in Nineteenth Century Architecture.
A04-0716 Bassi, The Convento della Carita.
A04-0718 Baumgart, A History of Architectural Styles.
A04-0722 Benevolo, History of Modern Architecture.
A04-0727 Bibiena, L'Architettura Civile.
A04-0731 Blunt, Neopolitan Baroque and Rococo Architecture.
A04-0743 Bowyer, History of Building.
A04-0749 Branner, Gothic Architecture.
A04-0763 Bruce and Grossman, Revelations of New England Architecture: People and Their Buildings.
A04-0782 Cambridge Historical Commission, Survey of Architectural History in Cambridge.
A04-0789 Casey, Early Melbourne Architecture 1840 to 1888.
A04-0791 Castedo, A History of Latin American Art and Architecture.
A04-0794 Chamberlain, A Tour of Old Sturbridge Village.
A04-0795 Chamberlain and Flynt, Historic Deerfield: Houses and

Index 541

	Interiors.
A04-0796	Chambers, Designs of Chinese Buildings, Furniture, Dresses, Machines, and Utensils.
A04-0797	Chambers, A Treatise on Civil Architecture.
A04-0803	Clapham, Romanesque Architecture.
A04-0807	Cobblestone, World Architecture: An Illustrated History.
A04-0808	Coffin and Holden, Brick Architecture of the Colonial Period in Maryland and Virginia.
A04-0817	Congdon, Old Vermont Houses, 1763-1850.
A04-0826	Coope, Salomon de Brosse and the Development of the Classical Style in French Architecture from 1565 to 1630.
A04-0828	Couperie, Paris Through the Ages.
A04-0865	Dulaney, The Architecture of Historic Richmond.
A04-0866	Dunbar, The Historic Architecture of Scotland.
A04-0868	Dunlop, Palaces and Progresses of Elizabeth I.
A04-0869	Dunlop, Versailles.
A04-0870	Duprey, Old Houses on Nantucket.
A04-0892	Farrar and Hines, Old Virginia Houses Along the Fall Line.
A04-0893	Farrar and Hines, Old Virginia Houses: The Northern Peninsulas.
A04-0895	Favero, The Villa Emo and Fanzolo.
A04-0904	Fitch, American Building: The Historical Forces that Shaped It.
A04-0906	Fletcher, A History of Architecture.
A04-0909	Forman, Maryland Architecture: A Short History from 1634 through the Civil War.
A04-0910	Forman, Old Buildings, Gardens and Furniture in Tidewater Maryland.
A04-0911	Forman, Virginia Architecture in the Seventeenth Century.
A04-0914	Frankl, Principles of Architectural History.
A04-0915	Frary, Early Homes of Ohio.
A04-0916	Fraser, Essays in the History of Architecture.
A04-0921	Furneaux, A Concise History of Western Architecture.
A04-0936	Gibbs, A Book of Architecture.
A04-0939	Gillon, Early Illustrations and Views of American Architecture.
A04-0942	Girouard, The Victorian Country House.
A04-0943	Girouard, The Victorian Public House.
A04-0950	Goodwin, A History of Ottoman Architecture.
A04-0968	Halfpenny and Halfpenny, The Art of Sound Building.
A04-0969	Halfpenny and Halfpenny, Chinese and Gothic Architecture.
A04-0970	Halfpenny and Halfpenny, Practical Architecture.
A04-0971	Halfpenny and Halfpenny, Rural Architecture in the Chinese Taste.
A04-0972	Halfpenny and Halfpenny, Useful Architecture.
A04-0977	Hamlin, Greek Revival Architecture in America.
A04-0994	Helick, Varieties of Human Habitation.
A04-0996	Hersey, High Victorian Gothic: A Study in Associationism.

A04-1000 Historic American Buildings Survey.
A04-1002 Hitchcock, Early Victorian Architecture in Britain.
A04-1004 Hitchcock, Rococo Architecture in Southern Germany.
A04-1009 Hofstatter, Living Architecture: Gothic.
A04-1013 Holmes, Elizabethan London.
A04-1019 Howells, Lost Examples of Colonial Architecture.
A04-1021 Hubert, The Carolingian Renaissance.
A04-1032 Isham and Brown, Early Connecticut Houses: An Historical and Architectural Study.
A04-1039 Jackson, The Politics of Architecture.
A04-1040 Jackson, Byzantine and Romanesque Architecture.
A04-1041 Jackson, Renaissance of Roman Architecture.
A04-1051 Kaufmann, Architecture in the Age of Reason: Baroque and Post-Baroque in England, Italy, France.
A04-1055 Keller, The Renaissance in Italy.
A04-1056 Kelly, Early Domestic Architecture of Connecticut.
A04-1064 Kidson and Murray, A History of English Architecture.
A04-1067 Kimball, Domestic Architecture of the American Colonies and of the Early Republic.
A04-1071 Knox, The Architecture of Poland.
A04-1086 Lafever, Beauties of Modern Architecture.
A04-1092 Langley, The Builder's Director.
A04-1093 Langley, The Builder's Jewel.
A04-1094 Langley, The City and Country Builder's and Workman's Treasury of Designs.
A04-1095 Langley and Langley, Gothic Architecture.
A04-1110 Little, English Historic Architecture.
A04-1112 Lloyd, A History of the English Houses.
A04-1113 Lockwood, Bricks and Brownstone the New York Row House, 1783-1929: An Architectural and Social History.
A04-1119 McC. Groff, New Jersey's Historic Houses: A Guide to Homes Open to the Public.
A04-1120 McCall, Old Philadelphia Houses on Society Hill, 1750-1840.
A04-1122 Macleod, Style and Society: Architectural Ideology in Britain, 1835-1914.
A04-1131 Mazzotti, Palladian and Other Venetian Villas.
A04-1137 Millon, Baroque and Rococo Architecture.
A04-1138 Millon, Key Monuments of the History of Architecture.
A04-1149 Morrison, Early American Architecture.
A04-1158 Murray, The Architecture of the Italian Renaissance.
A04-1159 Murray, History of Architecture Series. Vol. 1 Renaissance.
A04-1173 Nilsson, European Architecture in India, 1750-1850.
A04-1176 Norberg-Schulz, History of Architecture Series. Vol. 2 Baroque.
A04-1195 Overdyke, Louisiana Plantation Homes.
A04-1197 Paladio, The Four Books of Architecture.
A04-1212 Peterson, The Carpenters' Company 1786 Rule Book.
A04-1214 Pevsner, Some Architectural Writers of the Nineteenth Century.
A04-1216 Pierson, The Colonial and Neoclassical Styles.
A04-1219 Poor, Colonial Architecture of Cape Cod, Nantucket &

Index

	Martha's Vineyard.
A04-1221	Portoghesi, Roma Barocca: The History of an Architectonic Culture.
A04-1222	Pothorn, Architectural Styles.
A04-1223	Pratt and Gunther, The Architecture of Sir Roger Pratt.
A04-1229	Pyne Press, American Historical Catalog Collection. Vol. 1 Architectural Elements, 1850-1880.
A04-1233	Ramsey and Harvey, Small Georgian Houses and Their Details: 1750-1820.
A04-1241	Raymond, Early Domestic Architecture of Pennsylvania.
A04-1249	Reynolds, Dutch Houses in the Hudson Valley Before 1776.
A04-1263	Rubens, Pallazzi Di Genova.
A04-1264	Rubens, Pompa Introitus ... Ferdinandi Austriaci ... Cum Antiverpiam ... Adventu Suo Bearet, 15 Ka. Maii Anno 1665.
A04-1266	Saalman, Medieval Cities.
A04-1280	Scully, American Architecture and Urbanism.
A04-1285	Service, Edwardian Architecture and Its Origins.
A04-1310	Smith and Owens, The Majesty of Natchez.
A04-1311	Soc. of Mexican Architects, Four Thousand Years of Mexican Architecture.
A04-1322	Stevens, Restorations of Classical Buildings.
A04-1328	Sturgis, A Dictionary of Architecture and Building.
A04-1330	Summerson, The Classical Language of Architecture.
A04-1335	Swan, A Collection of Design in Architecture.
A04-1358	Vogt-Goknil, Living Architecture: Ottoman.
A04-1366	Waterman and Barrows, Domestic Colonial Architecture of Tidewater, Virginia.
A04-1367	Watterson, Architecture: A Short History.
A04-1372	Whiffen, American Architecture Since 1780: A Guide to the Styles.
A04-1377	Williams and Williams, A Guide to Old American Houses 1700-1900.
A04-1388	Wittkower, Gothic Vs. Classic: Architectural Projects in Seventeenth Century Italy.
A04-1408	Yarwood, The Architecture of Europe.
A07-2205	APWA, History of Public Works in the United States, 1776-1976.
A09-2686	Fitchen, The Construction of Gothic Cathedrals: A Study of Medieval Vault Erection.
A09-2712	Innocent, The Development of English Building Construction.
A09-2728	Lafeyer, The Modern Builder's Guide.
A09-2736	Lloyd, A History of English Brickwork: With Examples and Notes of the Architectural Use and Manipulation of Brick from Medieval Times to the End of the Georgian Period.

ARCHITECTURE - LATIN AMERICA

A04-0717	Bastlund, Jose Louis Sert: Architecture, City Planning, Urban Design.

A04-0720	Beacham, The Architecture of Mexico: Yesterday and Today.
A04-0772	Bullrich, New Directions in Latin American Architecture.
A04-0791	Castedo, A History of Latin American Art and Architecture.
A04-0839	Damaz, Art in Latin American Architecture.
A04-0898	Fernandez, Architecture in Puerto Rico.
A04-0920	Fry and Drew, Tropical Architecture.
A04-1054	Kelemen, Baroque and Rococo in Latin America.
A04-1256	Rojas, Art and Architecture of Mexico: 10,000 B.C. to the Present.
A04-1293	Shipway and Shipway, Decorative Design in Mexican Homes.
A04-1294	Shipway and Shipway, Houses of Mexico, Origins and Traditions.
A04-1295	Shipway and Shipway, Mexican Homes of Today.
A04-1296	Shipway and Shipway, The Mexican House: Old and New.
A04-1304	Smith, Builders in the Sun: Five Mexican Architects.
A04-1311	Soc. of Mexican Architects, Four Thousand Years of Mexican Architecture.

ARCHITECTURE - MODERN

A04-0657	Alander, Viljo Revell: Works and Projects.
A04-0710	Banham, Guide to Modern Architecture.
A04-0711	Banham, Modern Architecture: The Age of the Masters.
A04-0712	Banham, Theory and Design in the First Machine Age.
A04-0725	Besset, New French Architecture.
A04-0733	Boesiger, Le Corbusier.
A04-0734	Boesiger, Le Corbusier: Last Works.
A04-0740	Boudon, Lived-in Architecture, Le Corbusier's Pessac Revisited.
A04-0761	Brown, The Work of G. Rietveld, Architect.
A04-0774	Burchard, The Voice of the Phoenix: Postwar Architecture in Germany.
A04-0819	Conrads, Programs and Manifestoes on 20th Century Architecture.
A04-0820	Conrads and Sperlich, The Architecture of Fantasy Utopian Building and Planning in Modern Times.
A04-0823	Cook, Experimental Architecture.
A04-0841	Davern, Lewis Mumford: Architecture as a Home for Man.
A04-0887	Evenson, Le Corbusier: The Machine and the Grand Design.
A04-0888	Faber, Arne Jacobsen.
A04-0889	Faber, New Danish Architecture.
A04-0923	Galardi, New Italian Architecture.
A04-0951	Goody, New Architecture in Boston.
A04-0961	Gropius, The New Architecture and the Bauhaus.
A04-1006	Hoffmann, The Architecture of John Wellborn Root.
A04-1008	Hoffmann and Kultermann, Modern Architecture in Color.
A04-1026	Huxtable, Four Walking Tours of Modern Architecture

Index 545

	in New York City.
A04-1027	Huxtable, Have You Kicked a Building Lately?
A04-1034	Itoh, The Classic Tradition in Japanese Architecture Modern Versions of the Sukiya Style.
A04-1045	Joedicke, Architecture Since 1945 Sources and Directions.
A04-1075	Korn, Glass in Modern Architecture of the Bauhaus Period.
A04-1082	Kulski, Architecture in A Revolutionary Era.
A04-1101	Le Corbusier, Towards a New Architecture.
A04-1130	Maxwell, New British Architecture.
A04-1160	Museum of Modern Art, Ludwig Mies van der Rohe.
A04-1200	Papachristou, Marcel Breuer: New Buildings and Projects.
A04-1204	Pehnt, Expressionist Architecture.
A04-1207	Pentagram, Pentagram: The Work of Five Designers.
A04-1215	Pevsner, The Sources of Modern Architecture and Design.
A04-1231	Ragon, The Aesthetics of Contemporary Architecture.
A04-1252	Richards and Pevsner, The Anti-Rationalists.
A04-1255	Rogers, What's Up in Architecture: A Look at Modern Building.
A04-1262	Rowland, History of the Modern Movement: Art, Architecture, Design.
A04-1270	Salokorpi, Modern Architecture in Finland.
A04-1272	Santini and Marini, Catalogo Bolaffi dell' Architettura Italiana, 1963-1966.
A04-1279	Schwab, The Architecture of Paul Rudolph.
A04-1281	Scully, Modern Architecture.
A04-1289	Sharp, Twentieth Century Architecture.
A04-1290	Sharp, A Visual History of Twentieth Century Architecture.
A04-1299	Siegel, Structure and Form in Modern Architecture.
A04-1337	Sweeney and Sert, Antoni Gaudi.
A04-1386	Wingler, The Bauhaus: Weimar Dessau Berlin Chicago.

ARCHITECTURE - U.S. AND CANADA

A04-0652	AIA, Award Winning Architecture/USA.
A04-0665	Andrews, Architecture in Chicago and Mid-America: A Photographic History.
A04-0666	Andrews, Architecture in New York: A Photographic History.
A04-0667	Andrews, Architecture in New England: A Photographic History.
A04-0670	Architect's Emergency Committee, Great Georgian Houses of America.
A04-0691	Architectural Record, Interior Spaces Designed by Architects.
A04-0694	Architectural Record, Record Houses of 1970.
A04-0695	Architectural Record, Record Houses of 1973.
A04-0696	Architectural Record, Record Houses of 1976.
A04-0700	Artistic Houses, Being a Series of Interior Views of a

	Number of the Most Beautiful and Celebrated Homes in the United States.
A04-0708	Bailey, Pre-Revolutionary Dutch Houses and Families in Northern New Jersey and Southern New York.
A04-0709	Baldwin, Stanford White.
A04-0729	Blake, Rural Ontario.
A04-0738	Boston Society of Architects, Architecture Boston and Cambridge.
A04-0739	Boston Society of Architects, Boston/Architecture.
A04-0756	Brooks, The Prairie School.
A04-0757	Brown, New Haven: A Guide to Architecture and Urban Design.
A04-0762	Bruce and Aidala, The Great Houses of San Francisco.
A04-0763	Bruce and Grossman, Revelations of New England Architecture: People and Their Buildings.
A04-0766	Brumbaugh, Architecture of Middle Tennessee Historic American Buildings Survey.
A04-0788	Carter, Mies van der Rohe at Work.
A04-0794	Chamberlain, A Tour of Old Sturbridge Village.
A04-0795	Chamberlain and Flynt, Historic Deerfield: Houses and Interiors.
A04-0798	Chermayeff, Observations on American Architecture.
A04-0799	Chermayeff and Erwitt, Observations on American Architecture.
A04-0802	Christovich, New Orleans Architecture. Vol. 2 The American Sector.
A04-0808	Coffin and Holden, Brick Architecture of the Colonial Period in Maryland and Virginia.
A04-0814	Condit, Chicago, 1930-1970.
A04-0815	Condit, The Chicago School of Architecture.
A04-0816	Condit, Chicago Since 1910.
A04-1817	Congdon, Old Vermont Houses, 1763-1850.
A04-0855	Dixon, Architectural Design Preview, U.S.A.
A04-0856	Doane, A Book of Cape Cod Houses.
A04-0860	Downing and Scully, Architectural Heritage of Newport, Rhode Island, 1640-1915.
A04-0861	Downing, The Architecture of Country Houses.
A04-0864	Drexler, Architecture of Skidmore, Owings & Merrill, 1963-1973.
A04-0865	Dulaney, The Architecture of Historic Richmond.
A04-0870	Duprey, Old Houses on Nantucket.
A04-0873	Eaton, American Architecture Comes of Age: European Reaction to H. H. Richardson and Louis Sullivan.
A04-0876	Ede, Canadian Architecture 1960/70.
A04-0877	Edgell, The American Architecture of Today.
A04-0892	Farrar and Hines, Old Virginia Houses Along the Fall Line.
A04-0893	Farrar and Hines, Old Virginia Houses: The Northern Peninsulas.
A04-0898	Fernandez, Architecture in Puerto Rico.
A04-0900	Ferriss, Power in Buildings: An Artist's View of Contemporary Architecture.
A04-0902	Fink, Adobes in the Sun: Portraits of a Tranquil Era.

A04-0903	Fitch, American Building: The Environmental Forces that Shape It.
A04-0904	Fitch, American Building: The Historical Forces that Shaped It.
A04-0905	Fitch, Architecture and the Esthetics of Plenty.
A04-0909	Forman, Maryland Architecture: A Short History from 1634 through the Civil War.
A04-0910	Forman, Old Buildings, Gardens and Furniture in Tidewater Maryland.
A04-0911	Forman, Virginia Architecture in the Seventeenth Century.
A04-0915	Frary, Early Homes of Ohio.
A04-0939	Gillon, Early Illustrations and Views of American Architecture.
A04-0940	Gillon and Lancaster, Victorian Houses: A Treasury of Lesser Known Examples.
A04-0951	Goody, New Architecture in Boston.
A04-0954	Grady, Architecture of Neel Reid in Georgia.
A04-0958	Greiff, Great Houses from the Pages of Architecture.
A04-0959	Greiff, Princeton Architecture: A Pictorial History of Town and Campus.
A04-0965	Guinness and Sadler, Mr. Jefferson, Architect.
A04-0977	Hamlin, Greek Revival Architecture in America.
A04-0992	Helick, Merchant Built Houses in Western Pennsylvania.
A04-1000	Historic American Building Survey.
A04-1003	Hitchcock, Rhode Island Architecture.
A04-1006	Hoffmann, The Architecture of John Wellborn Root.
A04-1026	Huxtable, Four Walking Tours of Modern Architecture in New York City.
A04-1032	Isham and Brown, Early Connecticut Houses: An Historical and Architectural Study.
A04-1047	Jordy, Progressive and Academic Ideals at the Turn of 20th Century.
A04-1052	Kaufmann, The Rise of an American Architecture.
A04-1053	Kaufmann and Raeburn, Frank Lloyd Wright Writings and Buildings.
A04-1056	Kelly, Early Domestic Architecture of Connecticut.
A04-1057	Kemper, Drawings by American Architects.
A04-1061	Keyes, Nineteenth Century Home Architecture of Iowa City.
A04-1063	Kidney, The Architecture of Choice: Eclecticism in America 1880-1930.
A04-1066	Kimball, American Architecture.
A04-1067	Kimball, Domestic Architecture of the American Colonies and of the Early Republic.
A04-1073	Koeper, Illinois Architecture.
A04-1106	Linley, Architecture of Middle Georgia: The Oconee Area.
A04-1113	Lockwood, Bricks and Brownstone, The New York Row House 1783-1929: An Architectural and Social History.
A04-1118	Maass, The Victorian Home in America.
A04-1119	McC. Groff, New Jersey's Historic Houses: A Guide to Homes Open to the Public.

A04-1120 McCall, Old Philadelphia Houses on Society Hill, 1750-1840.
A04-1121 McCoy, Five California Architects.
A04-1123 Malo, Landmarks of Rochester and Monroe County: A Guide to Neighborhoods and Villages.
A04-1136 Miller, Great Houses of Washington, D. C.
A04-1149 Morrison, Early American Architecture.
A04-1155 Mumford, Sticks and Stones: A Study of American Architecture and Civilization.
A04-1166 Neff, Architecture of Southern California.
A04-1181 Olmstead, Here Today: San Francisco's Architectural Heritage.
A04-1195 Overdyke, Louisiana Plantation Homes.
A04-1198 Pallister, Pallister's Model Homes 1878.
A04-1206 Pelican Guide to Plantation Homes of Louisiana.
A04-1212 Peterson, The Carpenters' Company 1786 Rule Book.
A04-1216 Pierson, The Colonial and Neoclassical Styles.
A04-1217 Platt, America's Gilded Age: Its Architecture and Decoration.
A04-1219 Poor, Colonial Architecture of Cape Cod, Nantucket & Martha's Vineyard.
A04-1229 Pyne Press, American Historical Catalog Collection. Vol. 1 Architectural Elements, 1850-1880.
A04-1241 Raymond, Early Domestic Architecture of Pennsylvania.
A04-1247 Rettig, Guide to Cambridge Architecture: Ten Walking Tours.
A04-1249 Reynolds, Dutch Houses in the Hudson Valley Before 1776.
A04-1257 Rolleston, Historic Houses and Interiors in Southern Connecticut.
A04-1261 Roth, Monograph of the Work of McKim, Mead & White: 1879-1915.
A04-1278 Schuyler, American Architecture and Other Writings.
A04-1280 Scully, American Architecture and Urbanism.
A04-1298 Siegel, Chicago's Famous Buildings: A Photographic Guide to the City's Architectural Landmarks and Other Notable Buildings.
A04-1310 Smith and Owens, The Majesty of Natchez.
A04-1316 Spreiregen, On the Art of Designing Cities: Selected Essays of Elbert Peets.
A04-1321 Stern, New Directions in American Architecture.
A04-1324 Storrer, The Architecture of Frank Lloyd Wright: A Complete Catalog.
A04-1357 Venturi, Learning from Las Vegas.
A04-1362 Wagner, Great Houses for ... View Sites, Beach Sites, Sites in the Woods, Meadow Sites, Small Sites, Sloping Sites, Steep Sites, Flat Sites.
A04-1366 Waterman and Barrows, Domestic Colonial Architecture of Tidewater, Virginia.
A04-1377 Williams and Williams, A Guide to Old American Houses 1700-1900.
A04-1384 Wilson and Lemann, New Orleans Architecture. Vol. 1 The Lower Garden District.

Index 549

A04-1393 Wright, An American Architecture.
A04-1394 Wright, The Drawings of Frank Lloyd Wright.
A04-1398 Wright, The Industrial Revolution Runs Away.
A04-1402 Wright, A Testament.
A04-1403 Wright, The Work of Frank Lloyd Wright (The Wendingen Edition).

AUDITORIUMS, STADIUMS, THEATERS, FAIRGROUNDS

A02-0268 Auditorium/Arena/Stadium Guide and International Directory.
A02-0292 Deering, Auditoriums and Arenas.
A04-0696 Architectural Record, Interior Spaces Designed by Architects.
A04-0748 Braithwaite, Fairground Architecture: The World of Amusement Parks, Carnivals, and Fairs.
A04-0777 Burris-Meyer and Cole, Theatres and Auditoriums.
A04-0805 Clasen, Expositions, Exhibits, Industrial and Trade Fairs.
A04-0952 Gotch, Inigo Jones.
A04-0976 Ham, Theatre Planning.
A04-0978 Hamlin, Forms and Functions of Twentieth-Century Architecture. Vol. 1 The Elements of Building. Vol. 2 The Principles of Composition. Vol. 3 Building Types: Residence, Gatherings, Education, Government. Vol. 4 Building Types: Commerce and Industry, Public Health, Transportation, Social Welfare and Recreation.
A04-0984 Harker, Studio and Stage.
A04-1038 Izenour, Theater Design.
A04-1048 Joseph, Actor and Architect.
A04-1099 Leacroft, The Development of the English Playhouse.
A04-1135 Mielziner, The Shapes of Our Theatres.
A04-1151 Moynet, L'Envers du Theatre: Machines et Decorations.
A04-1152 Mullin, The Development of the Playhouse: A Survey of Theatre Architecture from the Renaissance to the Present.
A04-1165 Navy Dept. Design Manual: Community Facilities.
A04-1263 Rubens, Pompa Introitus ... Ferdinandi Austriaci ... Cum Antiverpiam ... Adventu Suo Bearet, 15 ka. Maii Anno 1665.
A04-1284 Serlio, The Book of Architecture.
A04-1288 Sharp, The Picture Palace: And Other Buildings for the Movies.
A04-1291 Sheringham and Laver, Design in the Theater.
A04-1300 Silverman and Bowman, Theatre Architecture: An Illustrated Survey and a Checklist of Publications, 1946-1964.
A04-1314 Southern, Proscenium and Sight Lines.
A04-1326 Structure, Space, Mankind; Expo '70.
A04-1346 Tidworth, Theatres: An Architectural and Cultural History.
A04-1347 Torroja, The Structures of Eduardo Torroja.

BLAST-RESISTANT DESIGN AND CONSTRUCTION

A05-1474 Allgood and Swihart, Design of Flexural Members for Static and Blast Loading.
A05-1541 Biggs, Introduction to Structural Dynamics.
A05-1627 Fertis, Dynamics and Vibrations of Structures.
A05-1682 Howells, Dynamic Waves in Civil Engineering.
A05-1685 Hurty and Rubinstein, Dynamics of Structures.
A06-1980 Baker, Use of Models and Scaling in Shock and Vibration.

BRIDGES

A04-0978 Hamlin, Forms and Functions of Twentieth-Century Architecture. Vol. 1 The Elements of Building. Vol. 2 The Principles of Composition. Vol. 3 Building Types: Residence, Gatherings, Education, Government. Vol. 4 Building Types: Commerce and Industry, Public Health, Transportation, Social Welfare and Recreation.
A04-1347 Torroja, The Structures of Eduardo Torroja.
A05-1589 Cusens and Pama, Bridge Deck Analysis.
A05-1638 Gaylord and Gaylord, Design of Steel Structures.
A05-1708 Komendant, Contemporary Concrete Structures.
A05-1732 Lothers, Advanced Design in Structural Steel.
A05-1733 Lothers, Design in Structural Steel.
A05-1740 McGuire, Steel Structures.
A05-1758 Merritt, Structural Steel Designers' Handbook.
A05-1785 Okamoto, Introduction to Earthquake Engineering.
A05-1830 Rolfe and Barsom, Fracture and Fatigue Control in Structures: Applications of Fracture Mechanics.
A05-1836 Sachs, Wind Forces in Engineering.
A05-1915 Wass, Manual of Structural Details for Building Construction.
A05-1922 White, Structural Engineering. Vol. 4 Design of Structures.
A07-2200 ACI, First International Symposium on Concrete Bridge Design.
A07-2201 ACI, Second International Symposium on Concrete Bridge Design.
A07-2233 Henry and Jerome, Modern British Bridges.
A07-2236 Hopkins, A Span of Bridges: An Illustrated History.
A07-2253 Ritter and Paquette, Highway Engineering.
A07-2256 Rowe, Concrete Bridge Design.
A07-2261 Steinman and Watson, Bridges and Their Builders.
A08-2295 ACI, Temperature and Concrete.
A08-2429 Koster, Expansion Joints in Bridges and Roads.
A08-2432 LaLonde and Janes, Concrete Engineering Handbook.
A09-2607 Blake, Civil Engineer's Reference Book.
A09-2652 deSousa Ferreira, Glossary of Building and Bridge Construction.
A09-2657 Dunham, Foundations of Structures.
A09-2722 Komendant, Contemporary Concrete Structures.

Index

BUILDING MATERIALS

See all entries in Section A08
See individual listings such as concrete, wood, metals, plastics, etc.

A02-0275	Building Design and Construction Specifying Guide and Directory.
A02-0276	Building Supply News Purchasing File Issue.
A04-0781	Callender, Time-Saver Standards.
A04-0854	Diamant, The Chemistry of Building Materials.
A04-0929	Gatz and Achterberg, Color and Architecture.
A04-0932	Geerlings, Metal Crafts in Architecture.
A04-0933	Geerlings, Wrought Iron in Architecture.
A04-0955	Graf, Don Graf's Data Sheets.
A04-1015	Hornbostel, Materials for Architecture: An Encyclopedic Guide.
A04-1075	Korn, Glass in Modern Architecture of the Bauhaus Period.
A04-1139	Mills, Planning: Vol. 1 Architects' Technical Reference.
A04-1229	Pyne Press, American Historical Catalog Collection. Vol. 1 Architectural Elements, 1850-1880.
A04-1235	Rapoport, House Form and Culture.
A04-1303	Sleeper, Building Planning and Design Standards for Architects, Engineers, Designers, Consultants, Building Committees, Draftsmen, and Students.
A04-1352	Vandenberg and Elder, Handbook of Building Enclosure.
A05-1509	ASME, Structures and Materials.
A05-1537	Benjamin, Structural Design with Plastics.
A05-1655	Gurfinkel, Wood Engineering.
A05-1757	Merritt, Standard Handbook for Civil Engineers.
A05-1803	Polakowski and Ripling, Strength and Structure of Engineering Materials.
A05-1834	Ruskin, Materials Considerations in Design.
A05-1901	Vinson and Chou, Composite Materials and Their Use in Structures.
A05-1915	Wass, Manual of Structural Details for Building Construction.
A06-2071	Javitz, Materials Science and Technology for Design Engineers.
A07-2265	Wallace and Martin, Asphalt Pavement Engineering.
A08-2384	Diamant, Chemistry of Building Materials.
A09-2586	Ambrose, Building Structures Primer.
A09-2607	Blake, Civil Engineer's Reference Book.
A09-2622	Burgess, The Construction Industry Handbook.
A09-2705	Huntington and Mickadeit, Building Construction: Materials and Types of Construction.
A09-2754	Merritt, Building Construction Handbook.
A09-2767	Parker, Materials and Methods of Architectural Construction.
A09-2787	Reiner, Methods and Materials of Construction: A Guide for Builders, Owners, Architects and Engineers.
A09-2797	Schmidt, Construction: Principles, Materials and Methods.

A09-2809	Taylor, Plastering.
A10-2898	NHIC, Building Materials Technology and Selling Assignment Manual.
A10-2902	Palusci, Profitable Retailing of Building Supplies.
A11-2944	ASTM, Ignition, Heat Release, and Noncombustibility of Materials.
A11-2952	Bhatnager, Advances in Fire Retardants, Part 1.
A11-2953	Bhatnager, Advances in Fire Retardants, Part 2.
A11-2954	Bhatnager, Fire Retardant Formulations Handbook.
A11-3012	Hall, Encyclopedia of Food Engineering.
A11-3069	Munson, Construction Design for Landscape Architects.
A11-3072	Newman, Standard Structural Details for Building Construction.
A11-3114	Weaver, Structural Detailing for Technicians.

BUILDING TRADES

See individual listings such as carpentry, plumbing, welding, etc.

A03-0450	Bayley, Building: Teamwork or Conflict.
A03-0457	BNA Editorial Staff, OSHA and the Unions; Bargaining on Job Safety and Health.
A03-0470	Clough, Construction Contracting.
A03-0495	Editorial Staff of "Construction Labor Report," Construction Craft Jurisdiction Agreements.
A03-0554	Lefkoe, The Crisis in Construction: There is an Answer.
A03-0596	Porter, Site Labour Guide.
A04-0776	Burke, Architectural and Building Trades Dictionary.
A04-0932	Geerlings, Metal Crafts in Architecture.
A04-0933	Geerlings, Wrought Iron in Architecture.
A04-1348	Trill, Teamwork in Building Construction.
A06-1978	AWS, Resistance Welding--Theory and Use.
A06-1979	Babbitt, Plumbing.
A06-1996	Bredahl, Home Wiring Manual.
A06-1998	Burch, Basic House Wiring.
A06-2012	Croft, American Electricians' Handbook.
A06-2023	Emerick, Troubleshooters' Handbook for Mechanical Systems.
A06-2038	Giachino, Welding Technology.
A06-2042	Graham, Industrial and Commercial Wiring.
A06-2043	Graham, Interior Electric Wiring--Residential.
A06-2073	Johnson, Electrical Wiring: Design and Construction.
A06-2086	Kurtz and Shoemaker, The Lineman's and Cableman's Handbook.
A06-2121	NFPA, Electrical Code for One and Two-Family Dwellings.
A06-2122	NFPA, National Electrical Code 1975.
A06-2143	Richter, Practical Electrical Wiring, 10th Ed.
A06-2176	Watt and Stetka, NFPA Handbook of the National Electrical Code.
A06-2177	Watt and Summers, NFPA Handbook of the National Electrical Code.
A08-2365	Ceiling Systems Handbook.

Index

A08-2370	Comber, Composition Flooring and Floorlaying.
A08-2412	Gross, Applications Manual for Paint and Protective Coatings: A Guide to Types of Coatings, Methods of Surface Preparation, and Hand Application Techniques.
A09-2583	Alerich, Electrical Construction Wiring.
A09-2584	Althouse, Modern Welding.
A09-2590	Anderson, How to Build a Wood-Frame House.
A09-2591	Anderson, Wood Frame House Construction.
A09-2592	Anderson, Wood Frame House Construction.
A09-2593	Army Department, Carpenter, U. S. Army Technical Manual. TM 5-551B.
A09-2601	Badzinski, Carpentry in Residential Construction.
A09-2611	Badzinski, Stair Layout (Design and Building).
A09-2612	Branden and Hartsell, Plastering Skill and Practice.
A09-2616	Builders Encyclopedia.
A09-2624	Burkhardt, Domestic and Commercial Oil Burners.
A09-2627	Carpenters and Builders Library. Vol. 1 Tools, Steel Square, Joinery.
A09-2628	Carpenters and Builders Library. Vol. 2 Builders Math, Plans, Specifications.
A09-2629	Carpenters and Builders Library. Vol. 3 Layouts, Foundations, Framing.
A09-2630	Carpenters and Builders Library. Vol. 4 Millwork, Power Tools, Painting.
A09-2631	Carpentry and Building.
A09-2636	Christian, Negro Ironworkers in Louisiana.
A09-2644	Dahl and Wilson, Cabinetmaking and Millwork: Tools, Materials, Layout, Construction.
A09-2648	Dalzell and Townsend, Masonry Simplified, Vol. 1 Tools, Materials, Practice.
A09-2649	Dalzell, Masonry Simplified. Vol. 2 Practical Construction.
A09-2658	Durbahn, Fundamentals of Carpentry. Vol. 1 Tools, Materials, Practices.
A09-2659	Eastwick-Field, Design and Practice of Joinery.
A09-2683	Feirer and Hutchings, Carpentry in Building Construction.
A09-2689	Giachino, Welding Skills and Practices.
A09-2690	Gladstone, Air Conditioning Testing and Balancing, A Field Practice Manual.
A09-2692	Groneman, General Woodworking.
A09-2693	Groneman and Glazener, Technical Woodworking.
A09-2695	Guild of Architectural Ironmongers, Timber Joinery and Applied Ironmongery.
A09-2702	Hodge, Brickwork for Apprentices.
A09-2703	House Wiring.
A09-2706	Hurst, Painting and Decorating.
A09-2712	Inott, Carpentry & Joinery: Multi-Question Course.
A09-2715	Jay and Hemsley, Electrical Services in Buildings.
A09-2716	Kaberlein, Air Conditioning Metal Layout.
A09-2720	Koenigsberger and Adair, Welding Technology.
A09-2725	Kratfel, Introduction to Modern Sheet Metal.
A09-2726	Labor Dept. Training and Entry into Union Construction.

A09-2728 Lair, Carpentry for the Building Trades.
A09-2733 Lindsey, Pipefitters Handbook.
A09-2742 Maguire, Carpentry in Commercial Construction.
A09-2747 Masons and Builders Library. Vol. 1 Concrete, Block, Tile, Terrazzo.
A09-2748 Masons and Builders Library. Vol. 2 Bricklaying, Plastering, Rock Masonry, Clay Tile.
A09-2749 Matthias, How to Design and Install Plumbing.
A09-2755 Morris, Air Conditioning Cutter's Ready Reference.
A09-2756 Mulligan, Handbook of Brick Masonry Construction.
A09-2757 Navy Dept., Equipment Operator.
A09-2759 Nichols, How to Operate Excavation Equipment.
A09-2760 Nichols, Moving the Earth (The Workbook of Excavation).
A09-2770 Plumbers and Pipe Fitters Library. Vol. 1 Materials, Tools, Calculations.
A09-2771 Plumbers and Pipe Fitters Library. Vol. 2 Drainage, Fittings, Fixtures.
A09-2772 Plumbers and Pipe Fitters Library. Vol. 3 Installation, Heating, Welding.
A09-2778 Purbahn and Sundberg, Fundamentals of Carpentry. Vol. 2 Practical Construction.
A09-2779 Putnam, Bricklaying Skill and Practice.
A09-2781 Questions and Answers for Plumbers Examinations.
A09-2782 Rampaul, Pipe Welding Procedures.
A09-2785 Reband, Related Mathematics for Carpenters.
A09-2787 Reiner, Methods and Materials of Construction: A Guide for Builders, Owners, Architects and Engineers.
A09-2792 Rose, Questions and Answers on Pipework and Pipe Welding.
A09-2794 Rufus, Practical Rafter Calculator.
A09-2800 Sedwell, Calculations for Bricklayers.
A09-2802 Siegele, Roof Framing.
A09-2803 SMACNA, Architectural Sheet Metal Manual.
A09-2809 Taylor, Plastering.
A09-2826 Weaver, English Leadwork: Its Art and History.
A09-2827 Welders Guide.
A09-2829 Whittick, Questions and Answers on Carpentry and Joinery.
A09-2830 Wilson, Practical House Carpentry.
A09-2831 Wilson and Werner, Simplified Roof Framing.
A09-2837 Zinngrabe, Mathematics for the Sheet Metal Technician, Part 1.
A10-2859 Construction Labor Report, Construction Craft Jurisdiction Agreements.
A10-2862 Dubinsky, Reform in Trade Union Discrimination in the Construction Industry.
A10-2866 Foster, Labor Supply in the Construction Industry: A Case Study of Upstate New York.
A10-2869 Gordon, Year-Round Employment in the Construction Industry: A Systems Analysis.
A10-2893 Labor Statistics Bureau, Seasonality and Manpower in Construction.
A10-2895 Means, 1976 Labor Rates for the Construction Industry:

Index

	City, State, National.
A11-3024	Howarth, Building Craft Foremanship: A Manual for the Trainee Building Supervisor.
A11-3058	McKibben, Looking Forward to a Career: Building Trades.
A11-3104	Sundberg, Building Trades Blueprint Reading. Part 1 Fundamentals.
A11-3105	Sundberg, Building Trades Blueprint Reading. Part 2 Residential and Light Construction.
A11-3121	Woltjes, Building Trades Blueprint Reading Examination Kit.

CARPENTRY

A09-2590	Anderson, How to Build a Wood-Frame House.
A09-2591	Anderson, Wood Frame House Construction.
A09-2592	Anderson, Wood Frame House Construction.
A09-2593	Army Dept., Carpenter, U.S. Army Technical Manual. TM 5-551B.
A09-2601	Badzinski, Carpentry in Residential Construction.
A09-2605	Benjamin, The American Builder's Companion.
A09-2611	Bradzinski, Stair Layout (Design and Building).
A09-2627	Carpenters and Builders Library. Vol. 1 Tools, Steel Square, Joinery.
A09-2628	Carpenters and Builders Library. Vol. 2 Builders Math, Plans, Specifications.
A09-2629	Carpenters and Builders Library. Vol. 3 Layouts, Foundations, Framing.
A09-2630	Carpenters and Builders Library. Vol. 4 Millwork, Power Tools, Painting.
A09-2631	Carpentry and Building.
A09-2644	Dahl and Wilson, Cabinetmaking and Millwork: Tools, Materials, Practices.
A09-2658	Durbahn, Fundamentals of Carpentry. Vol. 1 Tools, Materials, Practices.
A09-2659	Eastick-Field and Stillman, Design & Practice of Joinery.
A09-2683	Feirer and Hutchings, Carpentry in Building Construction.
A09-2692	Groneman, General Woodworking.
A09-2693	Groneman and Glazener, Technical Woodworking.
A09-2694	Guild of Architectural Ironmongers, Timber Joinery and Applied Ironmongery.
A09-2712	Inott, Carpentry & Joinery: Multi-Question Course.
A09-2727	Lafeyer, The Modern Builder's Guide.
A09-2728	Lair, Carpentry for the Building Trades.
A09-2736	Love, Stair Builders Handbook.
A09-2742	Maguire, Carpentry in Commercial Construction.
A09-2778	Purbahn and Sundberg, Fundamentals of Carpentry. Vol. 2 Practical Construction.
A09-2785	Reband, Related Mathematics for Carpenters.
A09-2794	Rufus, Practical Rafter Calculator.
A09-2802	Siegele, Roof Framing.
A09-2825	Wass and Sanders, Building Construction: Roof Framing.
A09-2829	Whittick, Questions and Answers on Carpentry and Joinery.
A09-2830	Wilson, Practical House Carpentry.
A09-2831	Wilson and Werner, Simplified Roof Framing.

CERAMICS AND GLASS

A08-2345	Battelle Memorial Institute, Engineering Properties of Selected Ceramic Materials.
A08-2352	Brady, Materials Handbook.
A08-2390	Evans, Selecting Engineering Materials for Chemical and Process Plant.
A08-2414	Guy, Essentials of Materials Science.
A08-2473	Norton, Fine Ceramics: Technology and Applications.
A08-2474	Norton, Refractories.
A08-2479	Parker, Materials Data Book.
A08-2787	Peter, Design with Glass Materials in Modern Architecture.
A08-2518	Shand, Glass Engineering Handbook.
A08-2548	Van Vlack, A Textbook of Materials Technology.
A08-2553	Watson, Construction Materials and Processes.

CHIMNEYS, FIREPLACES

A05-1836	Sachs, Wind Forces in Engineering.
A05-1915	Wass, Manual of Structural Details for Building Construction.
A05-1923	Wiegel, Earthquake Engineering.
A06-2027	Evans, Equipment Design Handbook for Refineries and Chemical Plants. Vol. 2
A06-2152	Sculthorpe, Design of High Pressure Steam and High Temperature Water Plants.
A09-2604	Barry, Construction of Building.
A09-2629	Carpenters and Builders Library. Vol. 3 Layouts, Foundations, Framing.
A09-2824	Wass, Methods and Materials of Residential Construction.

COMPOSITE AND SANDWICH MATERIALS

A08-2317	Ashton, Primer on Composite Materials: Analysis.
A08-2330	ASTM, Environmental Effects on Advanced Composite Materials.
A08-2346	Becker, U.S. Sandwich Panel Manufacturing/Marketing Guide.
A08-2350	Bodnar, Structural Adhesives Bonding.
A08-2371	Composite Materials in Engineering Design.
A08-2421	Jayne, Theory and Design of Wood and Fiber Composite Materials.
A08-2423	Jones, Mechanics of Composite Materials.
A08-2428	Knowles, Composite Steel and Concrete Construction.
A08-2456	Mohr, SPI Handbook of Technology and Engineering of Reinforced Plastics/Composites.
A08-2472	Nicholls, Composite Construction Materials Handbook.
A08-2482	Patton, Construction Materials.
A08-2483	Patton, Materials in Industry.
A08-2543	Tsai, Composite Materials Workshop.
A09-2655	Dietz, Dwelling House Construction.
A09-2719	Knowles, Composite Steel and Concrete Construction.

Index   557

COMPRESSED GAS

A06-2023   Emerick, Troubleshooters' Handbook for Mechanical Systems.
A06-2079   Katz, Handbook of Natural Gas Engineering.
A06-2123   NFPA, National Fuel Gas Code.
A06-2153   Segler, Gas Engineers Handbook.

COMPUTER-AIDED DESIGN AND PLANNING

A03-0436   Andree, Computer Programming: Techniques, Analysis, and Mathematics.
A03-0438   Antill and Woodhead, Critical Path Methods in Construction Practice.
A03-0448   Barrodale, Elementary Computer Applications: In Science, Engineering and Business.
A03-0500   Faulkner, Project Management with CPM.
A03-0506   Foxhall, Professional Construction Management and Project Administration.
A03-0525   Harper, Computer Applications in Architecture and Engineering.
A03-0526   Harper, Financial Management Computer Uses Manual.
A03-0532   Hovanessian and Pipes, Digital Computer Methods in Engineering.
A03-0539   James, Analog Computer Simulation of Engineering Systems.
A03-0540   James, Applied Numerical Methods for Digital Computation with FORTRAN.
A03-0541   Jamison, FORTRAN IV Programming: Based on the IBM System 1130.
A03-0547   Kelly, Handbook of Numerical Methods and Applications.
A03-0548   LaFara, Computer Methods for Science and Engineering.
A03-0558   McCracken, FORTRAN with Engineering Applications.
A03-0566   Martin and Perkins, Computers and Information Systems: An Introduction.
A03-0573   Moyle, Introduction to Computers for Engineers.
A03-0584   O'Brien, Scheduling Handbook.
A03-0598   Priluck and Hourihan, Practical CPM for Construction.
A03-0623   Swanson and Pazer, PERTSIM: Text and Simulation.
A04-0754   Broadbent, Design in Architecture.
A04-0783   Campion, Computers in Architectural Design.
A04-0872   Eastman, Spatial Synthesis in Computer-Aided Building Design.
A04-0886   Evans and Wheeler, Architectural Programming/Emerging Techniques.
A04-0985   Harper, Computer Applications in Architecture and Engineering.
A04-1126   March, The Architecture of Form.
A04-1144   Moore, Emerging Methods in Environmental Design and Planning.
A04-1145   Moore, Environmental Design and Planning.
A04-1238   Rattenbury and Van Eck, OSIRIS: Architecture & Design.
A05-1454   ACI, Computer Applications in Concrete Design & Tech.
A05-1460   ACI, Impact of Computers on the Practice of Structural Engineering in Concrete.

A05-1463	ACI, Realism in the Application of ACI Standard 214-65.
A05-1505	ASME, Computer Software in Structural Analysis.
A05-1506	ASME, Gen. Purpose Finite Element Computer Programs.
A05-1510	ASME, Thermal Structural Analysis Programs: A Survey and Evaluation.
A05-1511	ASME, Three Dimensional Continuum Computer Programs for Structural Analysis.
A05-1532	Bathe and Wilson, Numerical Methods in Finite Element Analysis.
A05-1533	Beaufait, Computer Methods of Structural Analysis.
A05-1542	Billington, Thin Shell Concrete Structures.
A05-1546	Borg and Gennaro, Modern Structural Analysis.
A05-1550	Bowles, Analytical and Computer Methods in Foundation Engineering.
A05-1595	Davies, Space Structures: A Study of Methods and Developments in Three-Dimensional Construction.
A05-1598	Desai and Abel, Introduction to the Finite Element Method.
A05-1623	Fenves, Computer Methods in Civil Engineering.
A05-1624	Fenves, Numerical and Computer Methods in Structural Mechanics.
A05-1639	Gaylord and Gaylord, Structural Engineering Handbook.
A05-1640	Gere and Weaver, Analysis of Framed Structures.
A05-1668	Harrison, Computer Methods in Structural Analysis.
A05-1670	Heins, Bending & Torsional Design in Structural Members.
A05-1674	Hodge, Computers for Engineers.
A05-1686	Institution of Structural Engineers, International Symposium on Computer-Aided Structural Design.
A05-1693	Jenkins, Matrix and Digital Computer Methods in Structural Analysis.
A05-1696	Jeppson, Analysis of Flow in Pipe Networks.
A05-1700	Johnston, Column Research Council Guide to Design Criteria for Metal Compression Members.
A05-1711	Kuzmanovic and Willems, Steel Design for Structural Engineers.
A05-1714	Laursen, Structural Analysis.
A05-1730	Litton, Automatic Computational Techniques in Civil and Structural Engineering.
A05-1739	McFarland, Analysis of Plates.
A05-1745	Majid, Non-Linear Structures: Matrix Methods of Analysis and Design by Computers.
A05-1757	Merritt, Standard Handbook for Civil Engineers.
A05-1780	Norris, Elementary Structural Analysis.
A05-1826	Rogers, Reinforced Concrete Design for Buildings.
A05-1832	Rubinstein, Matrix Computer Analysis of Structures.
A05-1864	Spillers, Automated Structural Analysis: An Introduction.
A05-1865	Spindel, Computer Applications in Civil Engineering.
A05-1866	Spunt, Optimum Structural Design.
A05-1869	Stephenson, Computer Simulation for Engineers.
A05-1870	Stevens, Finite Elements and Matrix Structural Analysis.
A05-1896	Ural, Finite Element Method: Basic Concepts and Applications.
A05-1897	Ural, Matrix Operations and Use of Computers in Structural Engineering.
A05-1908	Wang, Computer Methods in Advanced Structural Analysis.

Index

A05-1914	Wang, Numerical and Matrix Methods in Structural Mechanics with Applications to Computers.
A05-1916	Weaver, Computer Programs for Structural Analysis.
A05-1924	Wilby, Pre-Stressed Concrete Beams-Design and Logical Analysis.
A05-1928	Williams, Analysis of Indeterminate Structures.
A06-2060	Hovanessian and Pipes, Digital Computer Methods in Engineering.
A06-2078	Kasuda, Use of Computers for Environmental Engineering Related To Buildings.
A06-2110	Mischke, An Introduction to Computer-Aided Design.
A06-2144	Rogers and Connolly, Analog Computation in Engineering Design.
A06-2156	Sherratt, Air Conditioning System Design for Buildings.
A06-2182	Wolberg, Application of Computers Engineering Analysis.
A07-2206	APWA, Public Works Computer Applications.
A08-2284	ACI, Concrete Thin Shells.
A11-2930	ACI, Manual of Standard Practice for Detailing Reinforced Concrete Structures.
A11-2980	CRSI, Reinforcing Bar Detailing.
A11-3049	Lang, Computer Programs for Mapping.
A11-3096	Siders, Computer Graphics: A Revolution in Design.

CONCRETE, BRICK, MASONRY AND STONE

A02-0343	Kenny, Concrete Estimating Handbook.
A02-0344	Kenny, Masonry Estimating Handbook.
A02-0351	Le Jeune, Manual of Concrete Estimating.
A03-0423	ACI Manual of Concrete Inspection.
A04-1111	Lloyd, A History of English Brickwork.
A05-1452	Abeles, An Introduction to Prestressed Concrete.
A05-1454	ACI, Computer Applications in Concrete Design and Technology.
A05-1459	ACI, Fatigue of Concrete.
A05-1557	Bresler, Reinforced Concrete Engineering. Vol. 1 Materials, Structural Elements, Safety.
A05-1587	CRSI Handbook.
A05-1617	Eriksen, Theory and Practice of Structural Design Applied to Reinforced Concrete.
A05-1619	Everard and Tanner, Reinforced Concrete Design: Including 200 Solved Problems.
A05-1716	Lenczner, Elements of Loadbearing Brickwork.
A05-1757	Merritt, Standard Handbook for Civil Engineers.
A05-1810	Preece and Davies, Models for Structural Concrete.
A05-1815	Rao, Reinforced Cement Concrete--A Textbook for Polytechnic Students.
A05-1817	Reynolds, Basic Reinforced Concrete Design.
A05-1818	Rice and Hoffman, Structural Design Guide to the ACI Building Code.
A05-1851	Seelye, Data Book for Civil Engineers. Vol. 1 Design. Vol. 2 Specifications and Costs.
A05-1915	Wass, Manual of Structural Details for Building Construction.
A07-2259	Sharp, Concrete In Highway Engineering.

A08-2280	ACI, Behavior of Concrete Under Temperature Extremes.
A08-2281	ACI, Causes, Mechanism, and Control of Cracking in Concrete.
A08-2282	ACI, Cement and Concrete Terminology.
A08-2283	ACI, Commentary on Building Code Requirements for Reinforced Concrete.
A08-2285	ACI, Designing for Effects of Creep, Shrinkage and Temperature in Concrete Structures.
A08-2286	ACI, Durability of Concrete.
A08-2287	ACI, Epoxies with Concrete.
A08-2288	ACI, Expansive Concrete.
A08-2289	ACI, Fiber Reinforced Concrete.
A08-2290	ACI, Lightweight Concrete.
A08-2292	ACI, Polymers in Concrete.
A08-2294	ACI, Shotcreting.
A08-2295	ACI, Temperature and Concrete.
A08-2296	ACI and Natl. Concrete Masonry Assn., Menzel Symposium on High Pressure Steam Curing.
A08-2304	Akroyd, Concrete: Its Properties and Manufacture.
A08-2309	Amrhein, Reinforced Masonry Engineering Handbook.
A08-2312	Architectural Precast Concrete.
A08-2330	ASTM, Masonry: Past and Present.
A08-2337	ASTM, Significance of Tests and Properties of Concrete and Concrete Making Materials.
A08-2344	Bate, Handbook on the Unified Code for Structural Concrete.
A08-2347	BIA, Recommended Practice for Engineered Brick Masonry.
A08-2348	BIA, Reinforced Brick Masonry--Lateral Force Design.
A08-2349	Biczok, Concrete Corrosion and Concrete Protection.
A08-2355	Bricks: Their Properties and Use.
A08-2360	Bureau of Reclamation, Concrete Manual.
A08-2366	Childe, Concrete Finishes and Decoration.
A08-2367	Childe, Everymans Guide to Concrete Work.
A08-2372	Concrete Society, Polymers in Concrete.
A08-2375	Cordon, Freezing and Thawing of Concrete: Mechanisms and Controls.
A08-2377	Critchell, Joints and Cracks in Concrete.
A08-2379	Dalzell, Simplified Masonry Planning and Building.
A08-2384	Diamant, Chemistry of Building Materials.
A08-2390	Evans, Selecting Engineering Materials for Chemical and Process Plant.
A08-2391	Fibre Reinforced Cement and Concrete.
A08-2392	Fintel, Handbook of Concrete Engineering.
A08-2396	Gage, Guide to Exposed Concrete Finishes.
A08-2397	Gage and Kirkbride, Design in Blockwork.
A08-2398	Gage and Newman, Guide to Ready Mixed Concrete.
A08-2399	Gage and Newman, Specification and Use of Ready Mixed Concrete.
A08-2400	Garthwaite, Metric Handbook for Reinforced Concrete.
A08-2405	Gatz and Thierry, Architect's Detail Library. Vol. 6 Exterior Detailing in Concrete.
A08-2409	Glanville and Thomas, Explanatory Handbook on the BS

	Code of Practice for Reinforced Concrete CP 114:1957 with Metrix Appendix.
A08-2415	Guyon, Limit-State Design of Prestressed Concrete. Vol. 1 The Design of the Section.
A08-2418	Heinle and Bacher, Building in Visual Concrete.
A08-2422	Johnson, Designing, Engineering and Construction with Masonry Products.
A08-2425	Keyser, Materials Science in Engineering.
A08-2430	Krebs and Walker, Highway Materials.
A08-2431	Labahn and Kaminsky, Cement Engineers Handbook.
A08-2432	LaLonde and Janes, Concrete Engineering Handbook.
A08-2434	Lea, The Chemistry of Cement and Concrete.
A08-2436	Lesley, History of the Portland Cement Industry in the United States.
A08-2439	Love, Construction Manual: Concrete and Formwork.
A08-2440	Lydon, Concrete Mix Design.
A08-2441	McIntosh, Concrete and Statistics.
A08-2442	McMillan and Tuthill, Concrete Primer.
A08-2447	Marsh, Concrete as a Visual Material.
A08-2450	Masonry Institute of America, Masonry Design Manual.
A08-2453	MIA, Marble Design Manual.
A08-2455	Moffat, Plant Engineer's Handbook of Formulas, Charts and Tables.
A08-2463	Nesbit, Structural Lightweight-Aggregate Concrete.
A08-2465	Neville, Creep of Concrete: Plain, Reinforced, and Prestressed.
A08-2466	Neville, Hardened Concrete: Physical and Mechanical Aspects.
A08-2467	Neville, High Alumina Cement Concrete.
A08-2468	Neville, Properties of Concrete.
A08-2477	Orchard, Concrete Technology. Vol. 1 Properties of Materials.
A08-2478	Orchard, Concrete Technology. Vol. 2 Practice.
A08-2479	Parker, Materials Data Book.
A08-2483	Patton, Materials in Industry.
A08-2488	Petzold and Rohrs, Concrete for High Temperatures.
A08-2491	Plummer, Brick and Tile Engineering.
A08-2492	Pollack, Materials Science and Metallurgy.
A08-2493	Portland Cement Assn., Notes on ACI 318-71 Building Code Requirements with Design Applications.
A08-2494	Portland Cement Assn., Principles of Quality Concrete.
A08-2495	Portland Cement Assn., Proceedings of the PCA-ACI Teleconference on ACI 318-71 Building Code Requirements.
A08-2496	Portland Cement Assn., Special Concretes, Mortars and Products.
A08-2497	Powers, The Properties of Fresh Concrete.
A08-2498	Preston, Prestressed Concrete for Architects and Engineers.
A08-2499	Preston and Sollenberger, Modern Prestressed Concrete.
A08-2505	Rice, User's Guide to the ACI Building Code.
A08-2506	Robson, High Alumina Cements and Concretes.
A08-2509	Ryan, Gunite, A Handbook for Engineers.

A08-2510	Sahlin, Structural Masonry.
A08-2514	Scott, Explanatory Handbook of the BS Code of Practice for Reinforced Concrete CP 114 (1957) (including Amendment No. 1:1965).
A08-2516	Sementsov and Kameiko, Designer's Manual Masonry, Including Reinforced Masonry in Industrial, Residential and Communal Buildings and Structures.
A08-2517	Shacklock, Concrete Constituents and Mix Proportions.
A08-2520	Short and Kinniburgh, Lightweight Concrete.
A08-2523	Simpson and Horrobin, The Weathering and Performance of Building Materials.
A08-2528	Smith, Materials of Construction.
A08-2529	Snow, Formwork for Modern Structure.
A08-2536	Svec and Jeffers, Modern Masonry Panel Construction Systems.
A08-2537	Swenson, Performance of Concrete.
A08-2538	Taylor, Concrete Technology and Practice.
A08-2541	Tretyakov, Concrete and Concreting.
A08-2542	Troxell, Composition and Properties of Concrete.
A08-2546	UNESCO, Reinforced Concrete: An International Manual.
A08-2547	Van Amerongen, Dictionary of Cement: Manufacture and Technology. (German-English/English-German).
A08-2550	Waddell, Concrete Construction Handbook.
A08-2551	Waddell, Practical Quality Control for Concrete.
A08-2552	Walley and Bate, A Guide to the B.S. Code of Practice for Prestressed Concrete CP 115.
A08-2554	The Weathering and Performance of Building Materials.
A08-2555	Whitehurst, Evaluation of Concrete Properties from Sonic Tests.
A08-2558	Wilson, Exposed Concrete Finishes. Vol. 2 Finishes to Precast Concrete.
A08-2559	Woods, Durability in Concrete Construction.
A09-2581	ACI, Symposium on Concrete Construction in Aqueous Environments.
A09-2582	ACI Manual of Concrete Inspection.
A09-2604	Barry, Construction of Buildings.
A09-2607	Blake, Civil Engineer's Reference Book.
A09-2609	Boudreau, Making the Adobe Brick.
A09-2618	The Building Research Establishment, Essential Data for Construction. Vol. 2 Building Materials.
A09-2621	Building with Tilt-Up.
A09-2629	Carpenters and Builders Library. Vol. 3 Layouts, Foundations, Framing.
A09-2631	Carpentry and Building.
A09-2646	Dalzell, Simplified Concrete Masonry Planning and Building.
A09-2647	Dalzell and Townsend, Concrete Block Construction for Home and Farm.
A09-2648	Dalzell and Townsend, Masonry Simplified. Vol. 1 Tools, Materials, Practice.
A09-2649	Dalzell, Masonry Simplified. Vol. 2 Practical Construction.
A09-2702	Hodge, Brickwork for Apprentices.

Index

A09-2704	Huntington and Mickadeit, Building Construction: Materials and Types of Construction.
A09-2705	Hurd, Formwork for Concrete.
A09-2709	Illingworth, Movement and Distribution of Concrete.
A09-2735	Lloyd, A History of English Brickwork: With Examples and Notes of the Architectural Use and Manipulation of Brick from Medieval Times to the End of the Georgian Period.
A09-2747	Masons and Builders Library. Vol. 1 Concrete, Block, Tile, Terrazzo.
A09-2748	Masons and Builders Library. Vol. 2 Bricklaying, Plastering, Rock Masonry, Clay Tile.
A09-2756	Mulligan, Handbook of Brick Masonry Construction.
A09-2768	Peurifoy, Formwork for Concrete Structures.
A09-2773	Portland Cement Assn., Administrative Practices in Concrete Construction.
A09-2774	Portland Cement Assn., Basic Concrete Construction Practices.
A09-2775	Portland Cement Assn., Concrete Inspection Procedures.
A09-2776	Portland Cement Assn., Laboratory and Exercise Manual on Concrete Construction.
A09-2780	Putnam, Bricklaying Skill and Practice.
A09-2787	Reiner, Methods and Materials of Construction: A Guide for Builders, Owners, Architects and Engineers.
A09-2788	Reynolds, Concrete Construction.
A09-2789	Richardson, Formwork Notebook.
A09-2790	Richardson, Precast Concrete Production.
A09-2800	Sedwell, Calculations for Bricklayers.
A09-2806	Stewart, Design and Placing of High Quality Concrete.
A09-2818	Turner and Lakeman, Concrete Construction Made Easy.
A09-2819	Urquhart, Civil Engineering Handbook.
A09-2822	Waddell, Concrete Construction Handbook.
A09-2824	Wass, Methods and Materials of Residential Construction.
A11-2930	ACI, Manual of Standard Practice for Detailing Reinforced Concrete Structures.
A11-3072	Newman, Standard Structural Details for Building Construction.

CONCRETE AND MASONRY STRUCTURES

A04-1018	Howard, Structure--An Architect's Approach.
A04-1170	Nervi, Structures.
A04-1220	Portland Cement Assn., Modern Plans for Modern Farms.
A05-1451	Abbett, American Civil Engineering Practice.
A05-1455	ACI, Concrete Thin Shells.
A05-1457	ACI, Deflections of Concrete Structures.
A05-1461	ACI, Models for Concrete Structures.
A05-1462	ACI, Probabilistic Design of Reinforced Concrete Buildings.
A05-1465	ACI, Response of Multistory Concrete Structures to Lateral Forces.

A05-1497	ASCE, The Proceedings of the August 1972 International Conference on the Planning and Design of Tall Buildings.
A05-1542	Billington, Thin Shell Concrete Structures.
A05-1599	Design of Tilt-Up Buildings.
A05-1607	Dunham, Advanced Reinforced Concrete.
A05-1608	Dunham, Theory and Practice of Reinforced Concrete.
A05-1626	Ferguson, Reinforced Concrete Fundamentals.
A05-1628	Field, Lessons from Failures of Concrete Structures.
A05-1635	Forsyth, Unified Design of Reinforced Concrete Members.
A05-1639	Gaylord and Gaylord, Structural Engineering Handbook.
A05-1651	Granholm, A General Flexural Theory of Reinforced Concrete.
A05-1708	Komendant, Contemporary Concrete Structures.
A05-1719	Leonhardt, Prestressed Concrete Design and Construction.
A05-1757	Merritt, Standard Handbook for Civil Engineers.
A05-1767	Murashev, Design of Reinforced Concrete Structures.
A05-1792	Park and Paulay, Reinforced Concrete Structures.
A05-1793	Parker, Simplified Design of Reinforced Concrete.
A05-1795	Parker, Simplified Engineering for Architects and Builders.
A05-1806	Portland Cement Assn., Design of Multistory Reinforced Concrete Buildings for Earthquake Motions.
A05-1814	Ramaswamy, Design and Construction of Concrete Shell Roofs.
A05-1906	Walmer and Baron, Manual of Structural Design and Engineering Solutions.
A05-1922	White, Structural Engineering. Vol. 4 Design of Structures.
A05-1923	Wiegel, Earthquake Engineering.
A05-1930	Winter and Nilson, Design of Concrete Structures.
A07-2200	ACI, First International Symposium on Concrete Bridge Design.
A07-2201	ACI, Second International Symposium on Concrete Bridge Design.
A07-2256	Rowe, Concrete Bridge Design.
A08-2286	ACI, Durability of Concrete.
A08-2293	ACI, Precast Concrete Wall Panels.
A08-2392	Fintel, Handbook of Concrete Engineering.
A08-2406	Gerwick, Construction or Prestressed Concrete Structures.
A08-2432	LaLonde and Janes, Concrete Engineering Handbook.
A08-2539	Terrington and Turner, Design of Non-Planer Roofs.
A08-2553	Watson, Construction Materials and Processes.
A09-2580	ACI, Concrete Construction.
A09-2581	ACI, Symposium on Concrete Construction in Aqueous Environments.
A09-2585	Ambrose, Building Structures Primer.
A09-2684	Feld, Construction Failure.
A09-2701	Havers and Stubbs, Handbook of Heavy Construction.
A09-2704	Huntington and Mickadeit, Building Construction: Materials and Types of Construction.

Index 565

A09-2721	Komendant, Contemporary Concrete Structures.
A09-2722	Koncz, Manual of Precast Concrete Construction. Vol. 1
A09-2723	Koncz, Manual of Precast Concrete Construction. Vol. 2
A09-2724	Koncz, Manual of Precast Concrete Construction. Vol. 3
A09-2730	Leonhardt, Prestressed Concrete Design and Construction.
A09-2732	Libby, Modern Prestressed Concrete: Design Principles and Construction Methods.
A09-2744	Manning, Concrete Water Towers, Bunkers, Silos and Other Elevated Structures.
A09-2746	Manning, Reinforced Concrete Reservoirs and Tanks.
A09-2767	Perkins, Floors--Construction and Finishes.
A09-2774	Portland Cement Assn., Basic Concrete Construction Practices.
A09-2776	Portland Cement Assn., Laboratory and Exercise Manual on Concrete Construction.
A09-2788	Reynolds, Concrete Construction.
A09-2804	Smith, Principles and Practices of Heavy Construction.
A09-2818	Turner and Lakeman, Concrete Construction Made Easy.
A09-2819	Urquhart, Civil Engineering Handbook.
A09-2822	Waddell, Concrete Construction Handbook.
A11-2930	ACI, Manual of Standard Practice for Detailing Reinforced Concrete Structures.
A11-2947	Barker, Reinforced Concrete Detailing.
A11-3046	Labor Department, Construction Safety and Health Training Course, Instructor's Guide. Manual 304 Blasting, Excavation, Traffic Control, Concrete Construction.

CONSTRUCTION

See all entries in Section A09

A04-1352	Vandenberg and Elder, Handbook of Building Enclosure.
A05-1495	ASCE, Fourth American Conference on Soil Mechanics and Foundation Engineering. Vol. 1
A05-1496	ASCE, Fourth American Conference on Soil Mechanics and Foundation Engineering. Vol. 2
A05-1500	ASCE, Guide for Design of Steel Transmission Towers.
A05-1553	Boyes, Structural and Cut-Off Diaphragm Walls.
A05-1560	Brown, Permafrost in Canada.
A05-1595	Davies, Space Structures: A Study of Methods and Developments in Three-Dimensional Construction.
A05-1599	Design of Tilt-Up Buildings.
A05-1620	Fairhurst, Failure and Breakage of Rock: Proceedings, 8th Symposium on Rock Mechanics.
A05-1654	Grosvenor and Paulding, Status of Practical Rock Mechanics: Proceedings, 9th Symposium on Roch Mechanics.
A05-1669	Hart, Multi-Storey Buildings in Steel.
A05-1749	Manual of Steel Construction.
A05-1757	Merritt, Standard Handbook for Civil Engineers.
A05-1774	Navy Department, Design Manual, Cold Regions: Engineering.
A05-1846	Schwartz, Civil Engineering for the Plant Engineer.

Construction Information

A05-1851	Seeyle, Data Book for Civil Engineers. Vol. 1 Design. Vol. 2 Specifications and Costs.
A05-1877	Stuttgart Universität, Project Study "City in the Arctic."
A05-1884	Terzaghi and Peck, Soil Mechanics in Engineering Practice.
A05-1893	Tschebotarioff, Foundations, Retaining and Earth Structures: The Art of Design Construction and Its Scientific Basis in Soil Mechanics.
A05-1894	Tsytovich, Mechanics of Frozen Ground.
A05-1906	Walmer and Baron, Manual of Structural Design and Engineering Solutions.
A05-1915	Wass, Manual of Structural Details for Building Construction.
A05-1932	Woodward, Drilled Pier Foundations.
A05-1935	Zilly, Handbook of Environmental Civil Engineering.
A06-2086	Kurtz and Shoemaker, The Lineman's and Cableman's Handbook.
A07-2228	Cornick, Dock and Harbour Engineering. Vol. 4 Construction.
A07-2246	Navy Department, Design Manual, Dry-Docking Facilities.
A07-2259	Sharp, Concrete in Highway Engineering.
A07-2265	Wallace and Martin, Asphalt Pavement Engineering.
A07-2266	Wetteland and Bruun, Proceedings POAC-Norway 1971. Vols. 1 and 2
A08-2280	ACI, Behavior of Concrete Under Temperature Extremes.
A08-2294	ACI, Shotcreting.
A08-2348	BIA, Reinforced Brick Masonry--Lateral Force Design.
A08-2360	Bureau of Reclamation, Concrete Manual.
A08-2365	Ceiling Systems Handbook.
A08-2367	Childe, Everymans Guide to Concrete Work.
A08-2370	Comber, Composition Flooring and Floorlaying.
A08-2379	Dalzell, Simplified Masonry Planning and Building.
A08-2396	Gage, Guide to Exposed Concrete Finishes.
A08-2406	Gerwick, Construction or Prestressed Concrete Structures.
A08-2418	Heinle and Bacher, Building in Visual Concrete.
A08-2422	Johnson, Designing, Engineering and Construction with Masonry Products.
A08-2432	LaLonde and Janes, Concrete Engineering Handbook.
A08-2439	Love, Construction Manual: Concrete and Formwork.
A08-2450	Masonry Institute of America, Masonry Design Manual.
A08-2457	Moselle, Practical Lumber Computer.
A08-2478	Orchard, Concrete Technology. Vol. 2 Practice.
A08-2536	Svec and Jeffers, Modern Masonry Panel Construction Systems.
A08-2538	Taylor, Concrete Technology and Practice.
A08-2541	Tretyakov, Concrete and Concreting.
A08-2550	Waddell, Concrete Construction Handbook.
A08-2553	Watson, Construction Materials and Processes.
A10-2875	HUD, Bibliography on Housing Building and Planning.

CONSTRUCTION EQUIPMENT AND TOOLS

A02-0281	Construction Equipment Buyers Guide.
A02-0284	Cost Reference Guide.
A02-0349	Land, Machinery and Equipment Pricing Guide.
A02-0383	Rental Rate Blue Book.
A03-0445	APWA, Equipment Management Manual.
A03-0592	Peurifoy, Construction Planning, Equipment, and Methods.
A03-0627	Vancil, Leasing of Industrial Equipment.
A08-2294	ACI, Shotcreting.
A08-2529	Snow, Formwork for Modern Structure.
A09-2603	Barber, Builders' Plant and Equipment.
A09-2608	Bolz and Hagemann, Materials Handling Handbook.
A09-2615	Bucksch, Dictionary of Civil Engineering and Construction Machinery and Equipment.
A09-2627	Carpenters and Builders Library. Vol. 1 Tools, Steel Square, Joinery.
A09-2630	Carpenters and Builders Library. Vol. 4 Millwork, Power Tools, Painting.
A09-2642	Crimmins, Construction Rock Work Guide.
A09-2651	Day, Construction Equipment Guide.
A09-2655	Douglas, Construction Equipment Policy.
A09-2656	Drake, The Complete Handbook of Power Tools.
A09-2661	Equipment Guide-Book, Compaction Equipment.
A09-2662	Equipment Guide-Book, Construction Equipment Cost Reference Guide.
A09-2663	Equipment Guide-Book, Crawler Cranes. Vol. 1-3
A09-2664	Equipment Guide-Book, Electric Lift Trucks.
A09-2665	Equipment Guide-Book, Fork Lifts.
A09-2666	Equipment Guide-Book, Green Guide, Vols. 1-2
A09-2667	Equipment Guide-Book, Green Guide: Off-Highway Trucks and Trailers.
A09-2668	Equipment Guide-Book, Green Guide's Older Equipment.
A09-2669	Equipment Guide-Book, Hydraulic Cranes.
A09-2670	Equipment Guide-Book, Hydraulic Excavators.
A09-2671	Equipment Guide-Book, Lift Trucks Green Guide.
A09-2672	Equipment Guide-Book, Motor Graders.
A09-2673	Equipment Guide-Book, Motor Scrapers.
A09-2674	Equipment Guide-Book, Off-Highway Haul Units.
A09-2675	Equipment Guide-Book, OSHA Requirements for Construction. Vol. 1 Equipment.
A09-2676	Equipment Guide-Book, OSHA Requirements for Construction. Vol. 2 Operations Standards.
A09-2677	Equipment Guide-Book, Rental Rate Blue Book
A09-2678	Equipment Guide-Book, Trenching Equipment.
A09-2679	Equipment Guide-Book, Truck Cranes.
A09-2680	Equipment Guide-Book, Wheel & Crawler Loaders.
A09-2681	Equipment Guide-Book, Wheel & Crawler Tractors.
A09-2696	Hammond, Earthmoving and Excavating Plant.
A09-2697	Hammond, Mobile and Movable Cranes.
A09-2701	Havers and Stubbs, Handbook of Heavy Construction.
A09-2705	Hurd, Formwork for Concrete.

A09-2707 Huston, Hydraulic Dredging: Theoretical and Applied.
A09-2709 Illingworth, Movement and Distribution of Concrete.
A09-2740 McGregor, Drilling of Rock.
A09-2757 Navy Dept., Equipment Operator.
A09-2758 Nichols, Heavy Equipment Repair.
A09-2759 Nichols, How to Operate Excavation Equipment.
A09-2760 Nichols, Moving the Earth (The Workbook of Excavation).
A09-2766 Paxton, Manual of Civil Engineering Plant and Equipment.
A09-2768 Peurifoy, Formwork for Concrete Structures.
A09-2784 Rapp, Construction of Structural Steel Building Frames.
A09-2789 Richardson, Formwork Notebook.
A09-2793 Rossnagel, Handbook of Rigging.
A09-2804 Smith, Principles and Practices of Heavy Construction.
A09-2805 Smith, Principles and Practices of Light Construction.
A09-2820 Vallings, Mechanisation in Building.
A09-2822 Waddell, Concrete Construction Handbook.
A09-2824 Wass, Methods and Materials of Residential Construction.
A11-3044 Labor Department, Construction Safety and Health Training Course, Instructor's Guide. Manual 302 Personal Protective Equipment, Hand Tools, Ladders and Scaffolds, Rigging.
A11-3045 Labor Department, Construction Safety and Health Training Course, Instructor's Guide. Manual 303 Powered Equipment, Material Handling, Welding, Flammable Liquids.

CONSTRUCTION INDUSTRY

See all entries in Section A10
A01-0108 HUD, Housing in the Seventies.
A01-0109 HUD, 1972 Statistical Yearbook.
A01-0175 Ricks, National Housing Models.
A02-0275 Building Design and Construction Specifying Guide and Directory.
A02-0276 Building Supply News Purchasing File Issue.
A02-0301 Construction Labor Report's 1970-71 Wage Rate Guide.
A02-0336 HUD Condominium Cooperative.
A02-0396 Strickland, Reports, Specifications and Estimates of Public Works in the United States of America.
A03-0450 Bayley, Building: Teamwork or Conflict.
A03-0530 Hilton, Industrial Relations in Construction.
A03-0535 ICBO, Building Department Administration.
A03-0554 Lefkoe, The Crisis in Construction: There is an Answer.
A03-0582 O'Bannon, Building Department Administration.
A03-0612 Sanderson, Codes and Code Administration: An Introduction to Building Regulations in the United States.
A04-0813 Condit, American Building.
A04-0981 Handler, Systems Approach to Architecture.
A04-1350 Univ. Press of Africa, Homes for Kenya.
A07-2204 APWA, History of Public Works in the United States, 1776-1976.

Index

A07-2209	APWA Directory.
A07-2243	Moses, Public Works: A Dangerous Trade.
A08-2436	Lesley, History of the Portland Cement Industry in The United States.
A09-2623	Burgess, Progress in Construction Science and Technology.
A09-2739	McCue, Creating the Human Environment.
A09-2750	Mead and Mitchell, Plant Hire for Building and Construction.
A09-2783	Randall, History of the Development of Building Construction in Chicago.
A09-2791	Richie, Canada Builds, 1867-1967.
A09-2795	Salzman, Building in England.
A09-2796	Sandstrom, Man the Builder: Origins of Civil Engineering.
A09-2807	Stone, Building Economy: Design, Production and Organization--A Synoptic View.
A10-2850	AIA, Glossary of Construction Industry Terms.
A10-2852	Bender, Selected Technological Aspects of the American Building Industry. The Industrialization of Building.
A10-2855	Building Science Directory.
A10-2856	Bureau of the Census, New One-Family Homes Sold and For Sale: 1963-1967.
A10-2857	Burgess, The Construction Industry Handbook.
A10-2859	Construction Labor Report, Construction Craft Jurisdiction Agreements.
A10-2860	Davidson, Housing Demand: Mobile, Modular or Conventional?
A10-2861	Dietz, The Building Industry.
A10-2864	Engineering News Record, Probing the Future.
A10-2865	Europe Design Intelligence Unit, Europe's Building Industries United Kingdom.
A10-2866	Foster, Labor Supply in the Construction Industry: A Case Study of Upstate New York.
A10-2868	Godel, Guide to Information Sources in the Construction Industry.
A10-2869	Gordon, Year-Round Employment in the Construction Industry: A Systems Analysis.
A10-2871	Grebler, Large Scale Housing and Real Estate Firms: Analysis of a New Business Enterprise.
A10-2875	Housing and Urban Development, Dept. of, Bibliography on Housing, Building and Planning.
A10-2877	Housing and Urban Development, Dept. of, Registry of Minority Contractors and Housing Professionals. Vol. 1 A Listing for Cities in: Connecticut, Maine, Massachusetts, New Hampshire, Rhode Island, and Vermont.
A10-2878	Housing and Urban Development, Dept. of, Registry of Minority Contractors and Housing Professionals. Vol. 2 A Listing of Minority Construction Contractors and Housing Professionals for Cities in: New Jersey, New York, and Puerto Rico.
A10-2879	Housing and Urban Development, Dept. of, Registry of Minority Contractors and Housing Professionals. Vol.

	3 A Listing of Minority Construction Contractors and Housing Professionals for Cities in: Delaware, District of Columbia, Maryland, Pennsylvania, Virginia and West Virginia.
A10-2880	Housing and Urban Development, Dept. of, Registry of Minority Contractors and Housing Professionals. Vol. 4 A Listing of Minority Construction Contractors and Housing Professionals for Cities in: Alabama, Georgia, Florida, Kentucky, North Carolina, South Carolina, Tennessee, and Mississippi.
A10-2881	Housing and Urban Development, Dept. of, Registry of Minority Contractors and Housing Professionals. Vol. 5A & 5B A Listing of Minority Construction Contractors and Housing Professionals for Cities in: Illinois, Indiana, Michigan, Minnesota, Ohio, and Wisconsin.
A10-2882	Housing and Urban Development, Dept. of, Registry of Minority Contractors and Housing Professionals. Vol. 6 A Listing of Minority Construction Contractors and Housing Professionals for Cities in: Arkansas, Louisiana, Oklahoma, New Mexico and Texas.
A10-2883	Housing and Urban Development, Dept. of, Registry of Minority Contractors and Housing Professionals. Vol. 7 A Listing of Minority Construction Contractors and Housing Professionals for Cities in: Iowa, Kansas, Missouri, and Nebraska.
A10-2884	Housing and Urban Development, Dept. of, Registry of Minority Contractors and Housing Professionals. Vol. 8 A Listing of Minority Construction Contractors and Housing Professionals for Cities in: Colorado, Montana, North Dakota, Utah and Wyoming.
A10-2885	Housing and Urban Development, Dept. of, Registry of Minority Contractors and Housing Professionals. Vol. 9A & 9B A Listing of Minority Construction Contractors and Housing Professionals for Cities in: Arizona, California, Hawaii, and Nevada.
A10-2886	Housing and Urban Development, Dept. of, Registry of Minority Contractors and Housing Professionals. Vol. 10 A Listing of Minority Construction Contractors and Housing Professionals for Cities in: Alaska, Idaho, Oregon, and Washington.
A10-2887	Housing and Urban Development, Dept. of, Services Available to HUD-Related Business in International Trade.
A10-2888	Housing Research and Building Technology Activities of the Federal Government.
A10-2890	Kelly, Design and the Production of Houses.
A10-2893	Labor Statistics Bureau, Seasonality and Manpower in Construction.
A10-2895	Means, 1976 Labor Rates for the Construction Industry: City, State, National.
A10-2899	NICB, Economics of the Construction Industry.
A10-2902	Paulusci, Profitable Retailing of Building Supplies.
A10-2904	Ragatz, Vacation Homes: An Analysis of the Market for Seasonal-Recreational Housing.

A10-2907	Reidelbach, Modular Housing 1971--Facts and Concepts.
A10-2908	Reidelbach, Modular Housing '72, Statistics & Specifics.
A10-2910	Sandstrom, Man the Builder: Origins of Civil Engineering.
A10-2913	Sternlieb and Listokin, Housing 1973-74: An Anthology.
A10-2914	Sternlieb and Paulis, Housing, 1971-1972.
A10-2915	Sternlieb and Paulis, Housing, 1971-72. Vol. 2
A10-2916	Sternlieb and Sagalyn, Housing, 1970-1971. Vol. 1
A10-2917	Tavistock Institute of Human Relations, Interdependence and Uncertainty: A Study of the Building Industry.
A10-2922	Whitehead, The U.K. Housing Market: An Econometric Model.

CONTRACTING AND CONSTRUCTION MANAGEMENT

A02-0272	Benson, Building Contractor's and Home Builder's Handbook of Bidding, Surveying, and Estimating.
A02-0279	Colby, Practical Legal Advice for Builders and Contractors.
A02-0280	Collison, The Developer's Dictionary and Handbook.
A02-0290	Deatherage, Construction Estimating and Job Preplanning.
A02-0293	Dent, Construction Cost Appraisal.
A02-0337	Hunt, Creative Control of Building Costs.
A02-0352	Lowe, Critical Path Analysis by Bar Chart.
A02-0355	McKeever, Community Builders Handbook.
A02-0363	Minnesota Elec. Assn., Estimating-Accounting Manual for Electrical Contractor Dealers.
A02-0390	Rothschild, Construction Bonds and Insurance Guide.
A02-0393	Sokol, Contractor or Manipulator?
A02-0395	Stires and Wenig, Pert/Cost.
A02-0398	Turin, Aspects of the Economics of Construction.
A02-0399	Turner, Quantity Surveying.
A02-0401	Walker, The Building Estimator's Reference Book.
A02-0402	Wass, Building Construction Estimating.
A02-0411	Wynne, Building Estimating.
A03-0423	ACI Manual of Concrete Inspection.
A03-0425	Adrian, Quantitative Methods in Construction Contracting.
A03-0427	Aguilar, Systems Analysis and Design in Engineering, Architecture, Construction and Planning.
A03-0438	Antill and Woodhead, Critical Path Methods in Construction Practice.
A03-0440	ASCE, National Conference on Construction Contracts.
A03-0441	Ashley, Electrical Contracting.
A03-0444	Atkinson, Construction Management.
A03-0445	AWPA, Equipment Management Manual.
A03-0449	Battersby, Network Analysis for Planning and Scheduling.
A03-0450	Bayley, Building: Teamwork or Conflict.
A03-0451	Begley, Project Management for Construction Superintendents.
A03-0453	Benson, Critical Path Methods in Building Construction.
A03-0454	Berger and Godel, Estimating and Project Management for Small Construction Firms.
A03-0458	Bonny and Frein, Handbook of Construction Management

and Organization.
A03-0459 Brock, Cost Accounting Manual for Highway Contractors, A System for Cost Control.
A03-0461 Burman, Precedence Networks for Project Planning and Control.
A03-0467 Caudill, Architecture by Team: A New Concept for the Practice of Architecture.
A03-0469 Clark, Business Systems and Data Processing Procedures.
A03-0470 Clough, Construction Contracting.
A03-0472 Cohen, Public Construction Contracts and the Law.
A03-0473 Commerce Clearing House, 1972 Government Contracts Guide.
A03-0474 Coombs, Construction Accounting and Financial Management.
A03-0477 Critical Path Techniques for Construction.
A03-0480 Dand and Farmer, Purchasing in the Construction Industry.
A03-0482 Deatherage, Construction Company Organization and Management.
A03-0483 Deatherage, Construction Office Administration.
A03-0484 Deatherage, Construction Scheduling and Control.
A03-0485 Dell'Isola, Value Engineering in the Construction Industry.
A03-0487 Dent, Construction Cost Appraisal.
A03-0489 Douglas and Munger, Construction Management.
A03-0491 Dressel, Organisation and Management of A Construction Company.
A03-0492 Dubin, Architectural Supervision of Modern Buildings.
A03-0500 Faulkner, Project Management with CPM.
A03-0504 Forinton, Civil Engineering Contracts and Claims.
A03-0505 Foxhall, Professional Construction Management and Project Administration.
A03-0509 Geddes and Chrystal-Smith, Building and Public Works Administration, Estimating and Costing.
A03-0511 Gill, Systems Management Techniques for Builders and Contractors.
A03-0512 Gobourne, Cost Control in the Construction Industry.
A03-0514 Granof, How to Cost Your Labor Contract.
A03-0517 Green, The Architect's Guide to Running a Job.
A03-0518 Green, Architect's Guide to Site Management.
A03-0520 Grummitt, The Mechanics of Construction Management.
A03-0527 Hauf, Building Contracts for Design and Construction.
A03-0528 Heery, Time, Cost, and Architecture.
A03-0531 Hollins, Production and Planning Applied to Building.
A03-0545 Jurecka, Network Planning in the Construction Industry.
A03-0554 Lefkoe, The Crisis in Construction: There is an Answer.
A03-0557 Lucas, Accounting Guide for Construction Contractors.
A03-0559 Macfarlane, Architectural Supervision on Site.
A03-0560 McKaig, Field Inspection of Building Construction.
A03-0564 Marsh, Contracting for Engineering and Construction Projects.

Index 573

A03-0567	Martino, Critical Path Networks.
A03-0571	Merritt, Building Construction Handbook.
A03-0572	Miller, Successful Management for Contractors.
A03-0576	Nedved, Builder's Accounting.
A03-0577	Neidle, Electrical Contracting and Management.
A03-0579	NESCA, Liens and Claims Manual.
A03-0583	O'Brien, CPM in Construction Management, Project Management with CPM.
A03-0584	O'Brien, Scheduling Handbook.
A03-0585	O'Brien, Contractor's Management Handbook.
A03-0587	O'Brien, Value Analysis in Design and Construction.
A03-0588	Oppenheimer, Directing Construction for Profit: Business Aspects of Contracting.
A03-0590	Parker and Oglesby, Methods Improvement for Construction Managers.
A03-0591	Parris, Law and Practice of Arbitrations.
A03-0592	Peurifoy, Construction Planning, Equipment, and Methods.
A03-0593	Pilcher, Appraisal and Control of Project Costs.
A03-0594	Pilcher, Principles of Construction Management for Engineers and Managers.
A03-0596	Porter, Site Labour Guide.
A03-0597	Practical Accounting and Cost Keeping for Contractors.
A03-0598	Priluck and Hourihan, Practical CPM for Construction.
A03-0599	Radcliffe, Critical Path Method.
A03-0602	Ray-Jones and McCann, CI/SfB Project Manual.
A03-0603	Reiner, Handbook of Construction Management.
A03-0608	Rowe, Management Techniques for Civil Engineering Construction.
A03-0609	Royer, Desk Book for Construction Superintendents.
A03-0610	Rubey and Milner, Construction and Professional Management.
A03-0618	Seelye, Data Book for Civil Engineers. Vol. 3 Field Practice.
A03-0619	Shaffer, Critical Path Method.
A03-0620	Sheeran, Management Essentials for Public Works Administrators.
A03-0622	Stone, Building Design Evaluation.
A03-0623	Swanson and Pazer, PERTSIM: Text and Simulation.
A03-0624	Teets, Construction Management for the Subcontractor.
A03-0628	Volpe, Construction Management Practices.
A03-0629	Walker, Legal Pitfalls in Architecture, Engineering, and Building Construction.
A03-0631	Wass, Construction Management and Contracting.
A03-0635	Zehner, Builder's Guide to Contracting.
A04-1163	Nat'l. Aer. and Space Admin., Facilities Engineering Handbook.
A04-1202	Paul, Apartments: Their Design and Development.
A04-1348	Trill, Teamwork in Building Construction.
A04-1352	Vandenberg and Elder, Handbook of Building Enclosure.
A05-1757	Merritt, Standard Handbook for Civil Engineers.
A05-1851	Seelye, Data Book for Civil Engineers. Vol. 1 Design. Vol. 2 Specifications and Costs.

A05-1915	Wass, Manual of Structural Details for Building Construction.
A07-2207	APWA, Public Works Computer Applications.
A09-2604	Barry, Construction of Buildings. 4 Vols.
A09-2606	Berne, How to Make Money in the Remodeling Business.
A09-2607	Blake, Civil Engineer's Reference Book.
A09-2625	Bush, Handbook for the Construction Superintendent.
A09-2627	Carpenters and Builders Library. Vol. 1 Tools, Steel Square, Joinery.
A09-2628	Carpenters and Builders Library. Vol. 2 Builders Math, Plans, Specifications.
A09-2629	Carpenters and Builders Library. Vol. 3 Layouts, Foundations, Framing.
A09-2630	Carpenters and Builders Library. Vol. 4 Millwork, Power Tools, Painting.
A09-2631	Carpentry and Building.
A09-2642	Crimmins, Construction Rock Work Guide.
A09-2682	Federal Highway Admin., Construction Manual.
A09-2688	Geary, Work Study Applied to Building.
A09-2701	Havers and Stubbs, Handbook of Heavy Construction.
A09-2710	Imrie, Building Site Organization.
A09-2717	Kidder and Parker, Architects' and Builders' Handbook.
A09-2734	Lippsmeier, Building in the Tropics/Tropenbau.
A09-2750	Mead and Mitchell, Plant Hire for Building and Construction.
A09-2753	Merritt, Building Construction Handbook.
A09-2765	Parker, Materials and Methods of Architectural Construction.
A09-2773	Portland Cement Assn., Administrative Practices in Concrete Construction.
A09-2787	Reiner, Methods and Materials of Construction: A Guide for Builders, Owners, Architects and Engineers.
A09-2797	Schmidt, Construction: Principles, Materials and Methods.
A09-2799	Searles, Field Engineering.
A09-2807	Stone, Building Economy: Design, Production and Organization--A Synoptic View.
A09-2824	Wass, Methods and Materials of Residential Construction.
A10-2853	Berne, How to Make Money in the Remodeling Business.
A10-2896	National Research Council, Promotion of the Development and Use of the Subsystem Concept of Building Construction.
A10-2898	NHIC, Building Materials Technology and Selling Assignment Manual.
A10-2912	Smith, How to Find Out in Architecture and Building-- A Guide to Sources of Information.
A11-3022	Hopf, Designer's Guide to OSHA.
A11-3024	Howarth, Building Craft Foremanship: A Manual for the Trainee Building Supervisor.
A11-3080	NHIC, Home Improvement Selling.
A11-3120	Wilcox and Snape, Worked Examples in Measurement of Construction Work. Vols. 1-2

Index 575

CONTRACTS AND LAW

A02-0279	Colby, Practical Legal Advice for Builders and Contractors.
A02-0280	Collison, The Developer's Dictionary and Handbook.
A02-0362	Milne, Builder's Estimating Simply Explained.
A02-0405	Werbin, Practical Legal Library for Architects, Contractor and Engineers.
A03-0420	Abbett, Engineering Contracts and Specifications.
A03-0421	Abrahamson, Engineering Law and the ICE Contracts.
A03-0424	Acret, California Construction Law Manual.
A03-0428	AIA Building Construction Legal Citator: 1971.
A03-0431	AIChE, Engineering Construction Contracts, April 1970 Proceedings.
A03-0432	AIChE, Engineering Construction Contracts, August 1970 Proceedings.
A03-0433	AIChE, Engineering Construction Contracts.
A03-0434	AIChE, Engineering Construction Contracts.
A03-0437	Antill, Civil Engineering Management.
A03-0440	ASCE, National Conference on Construction Contracts.
A03-0444	Atkinson, Construction Management.
A03-0454	Berger and Godel, Estimating and Project Management for Small Construction Firms.
A03-0458	Bonny and Frein, Handbook of Construction Management and Organization.
A03-0463	Canfield and Bowman, Business, Legal and Ethical Phases of Engineering.
A03-0470	Clough, Construction Contracting.
A03-0472	Cohen, Public Construction Contracts and the Law.
A03-0473	Commerce Clearing House, 1972 Government Contracts Guide.
A03-0485	Dell'Isola, Value Engineering in the Construction Industry.
A03-0489	Douglas and Munger, Construction Management.
A03-0493	Dunham and Young, Contracts, Specifications and Law for Engineers.
A03-0498	Farmer, What You Should Know About Contracts.
A03-0504	Forinton, Civil Engineering Contracts and Claims.
A03-0506	Foxhall, Professional Construction Management and Project Administration.
A03-0507	Freeth and Davey, The AJ Legal Handbook.
A03-0514	Granof, How to Cost Your Labor Contract.
A03-0527	Hauf, Building Contracts for Design and Construction.
A03-0538	Jabine, Case Histories in Construction Law: A Guide for Architects, Engineers, Contractors, Builders.
A03-0542	Jessup and Jessup, Law and Specifications for Engineers and Scientists.
A03-0546	Kantor, Contractual Aspects of Value Engineering.
A03-0549	Laidlow, Engineering Law.
A03-0561	McKown, Comprehensive Guide to Factory Law.
A03-0563	Marks, Aspects of Civil Engineering Contract Procedure.
A03-0564	Marsh, Contracting for Engineering and Construction Projects.

A03-0575	NAHB, Land Development Law for the Builder and His Attorney.
A03-0578	NESCA, Liens and Claims Manual.
A03-0580	Nord, Legal Problems in Engineering.
A03-0588	Oppenheimer, Directing Construction for Profit: Business Aspects of Contracting.
A03-0591	Parris, Law and Practice of Arbitrations.
A03-0603	Reiner, Handbook of Construction Management.
A03-0606	Rosenfeld, Architect's Handbook of Professional Practice. 2 Vols.
A03-0625	Tomson and Coplan, Architectural and Engineering Law.
A03-0626	Turner, Building Contracts: A Practical Guide.
A03-0629	Walker, Legal Pitfalls in Architecture, Engineering, and Building Construction.
A03-0630	Walker-Smith, The Standard Forms of Building Contract.
A09-2777	Portland Cement Assn., Administrative Practices in Concrete Construction.
A09-2787	Reiner, Methods and Materials of Construction: A Guide for Builders, Owners, Architects and Engineers.
A11-2941	Arnold and Vaden, Invention Protection for Practicing Engineers.
A11-2980	CRSI, Reinforcing Bar Detailing.
A11-3112	Tuska, An Introduction to Patents for Inventors and Engineers.

COOLING TOWERS

A06-1961	AIChE, Cooling Towers.
A06-1972	ASHRAE, Equipment Volume 1972.
A06-1974	ASME, Dry and Wet/Dry Cooling Towers for Power Plants.
A06-2013	CTI, Cooling Tower Performance Curves.
A06-2027	Evans, Equipment Design Handbook for Refineries and Chemical Plants. Vol. 2
A06-2041	Gosling, Applied Air Conditioning and Refrigeration.
A06-2089	Laub, Air Conditioning and Heating Practice.
A06-2139	Power Magazine, Plant Energy Systems.
A06-2140	Price, Air Conditioning for Building Engineers and Managers: Operation and Maintenance.
A06-2164	Stoecker, Principles for Air Conditioning Practice.
A06-2165	Stoecker, Refrigeration and Air Conditioning.

COOPERATIVES, CONDOMINIUMS, APARTMENTS

A01-0132	Liblit, Housing--The Cooperative Way.
A01-0179	Rose, Landlords and Tenants: A Complete Guide to the Residential Rental Relationship.
A02-0262	Architectural Record, Apartments, Townhouses, and Condominiums.
A02-0277	Clurman and Hebard, Condominiums and Cooperatives.
A02-0300	Dombal, Residential Condominiums.
A02-0336	HUD Condominium Cooperative Study.
A02-0339	IREM, 1975 Income/Expense Analysis - Apartments,

Index   577

	Condominiums and Cooperatives.
A04-0687	Architectural Record, Apartments and Dormitories.
A04-0699	Arregger and Glaus, Highrise Building and Urban Design.
A04-0740	Boudon, Lived-In Architecture - Le Corbusier's Pessac Revisited.
A04-0847	DeChiari and Callender, Time Saver Standards of Building Types.
A04-1105	Link, Residential Designs: How to Get the Most for Your Housing Dollar.
A04-1202	Paul, Apartments: Their Design and Development.
A04-1323	Steyert, Economics of High-Rise Apartment Buildings of Alternate Design Configuration.
A10-2897	Neutze, The Suburban Apartment Boom: Case Study of a Land Use Problem.

COST ESTIMATES AND ANALYSIS

A01-0009	Bagby, Housing Rehabilitation Costs.
A02-0264	Aries and Newton, Chemical Engineering Cost Estimation.
A02-0265	Ashley, Electrical Estimating.
A02-0266	Atton, Estimating Applied to Building.
A02-0267	Atton, Introduction to Estimating.
A02-0269	Barton, Estimating for Heating and Ventilating.
A02-0270	Bauman, Fundamentals of Cost Engineering in the Chemical Industry.
A02-0272	Benson, Building Contractor's and Home Builder's Handbook of Bidding, Surveying, and Estimating.
A02-0273	Bifulco, How to Estimate Construction Costs of Electrical Power Substations.
A02-0278	Cohen, Electrical Estimating Handbook.
A02-0282	Cooper and Badzinski, Building Construction Estimating.
A02-0284	Cost Reference Guide.
A02-0288	Davis, Spon's Architects' and Builders' Price Book.
A02-0289	Davis, Spon's Mechanical and Electrical Services Price Book.
A02-0290	Deatherage, Construction Estimating and Job Preplanning.
A02-0293	Dent, Construction Cost Appraisal.
A02-0295	Dietz and Maloney, Builders' Estimating Fact Book.
A02-0297	Dodge, 1975 Guide for Estimating Public Works Construction Costs.
A02-0298	Dodge, 1976 Dodge Construction Systems Costs.
A02-0299	Dodge, 1976 Dodge Manual for Building Construction Pricing and Scheduling.
A02-0301	Construction Labor Report's 1970-71 Wage Rate Guide.
A02-0302	Edgerton Building Cost Calculator and Guide to Real Estate Valuation.
A02-0310	Foster, Construction Estimates from Take-Off to Bid.
A02-0311	Frost, Values for Money: Techniques of Cost Benefit Analysis.
A02-0313	Galeno, The Plumbing Estimating Handbook.
A02-0315	Geddes and Chrystal-Smith, Estimating for Building and

Construction Information

	Civil Engineering Works.
A02-0316	Gladstone, Mechanical Estimating Guidebook.
A02-0319	Grant and Ireson, Principles of Engineering Economy.
A02-0322	Guthrie, Process Plant Estimating, Evaluation and Control.
A02-0326	Hanford, Feasibility Study Guidelines.
A02-0328	Harberger, Benefit Cost Analysis 1971: An Aldine Annual.
A02-0330	Harry, Construction Cost Guide.
A02-0331	Heavy Construction File.
A02-0333	Hinrichs and Taylor, Systematic Analysis: A Primer on Benefit-Cost Analysis and Program Evaluation.
A02-0335	Hornung, Estimating Building Construction: Quantity Surveying.
A02-0337	Hunt, Creative Control of Building Costs.
A02-0343	Kenny, Concrete Estimating Handbook.
A02-0348	Kolstad, Rapid Electrical Estimating.
A02-0349	Land, Machinery and Equipment Pricing Guide.
A02-0350	Lee Saylor, Current Construction Costs 1976.
A02-0351	LeJeune, Manual of Current Estimating.
A02-0353	MBM, Building Cost File.
A02-0354	MBM, Design Cost File.
A02-0357	McNeill and Clark, Cost Estimating and Contract Pricing.
A02-0358	Means, Building Construction Cost Data.
A02-0359	Means Building Systems Cost Guide.
A02-0362	Milne, Builders' Estimating Simply Explained.
A02-0363	Minnesota Elec. Assn., Estimating-Accounting Manual for Electrical Contractor Dealers.
A02-0364	Moselle, National Construction Estimator.
A02-0367	Ottaviano, National Mechanical Estimator.
A02-0368	Ottaviano, National Mechanical Estimator.
A02-0369	Page, Estimator's Construction Man-Hour Manual.
A02-0370	Page, Estimator's Equipment Installation Man-Hour Manual.
A02-0371	Page, Estimator's Manual of Equipment and Installation Costs.
A02-0372	Page, Heating, Plumbing and Air-Conditioning Man-Hour Manual.
A02-0373	Page and Nation, Estimator's Electric Man-Hour Manual.
A02-0374	Page and Nation, Estimator's Piping Man-Hour Manual.
A02-0375	Parker, Planning and Estimating Dam Construction.
A02-0376	Parker, Planning and Estimating Underground Construction.
A02-0377	Paxton, National Repair and Remodeling Estimator.
A02-0379	Peurifoy, Estimating Construction Costs.
A02-0380	Pulver, Construction Estimates and Costs.
A02-0383	Rental Rate Blue Book.
A02-0384	Richardson, Fabricators and Erectors Estimating Standards.
A02-0388	Richardson, General Construction Estimating Standards.
A02-0386	Richardson, Light Construction Estimating Standards.
A02-0387	Richardson, Mechanical and Electrical Construction

Index

	Estimating Standards.
A02-0388	Richardson, Process Plant Construction Estimating Standards.
A02-0389	Roth, Architects, Contractors and Engineers Guide to Construction Costs.
A02-0394	Steinberg and Stempel, Estimating for the Building Trades.
A02-0396	Strickland, Reports, Specifications and Estimates of Public Works in the United States of America.
A02-0397	Thomas, How to Estimate Building Losses and Construction Costs.
A02-0399	Turner, Quantity Surveying.
A02-0401	Walker, The Building Estimator's Reference Book.
A02-0402	Wass, Building Construction Estimating.
A02-0410	Wood, Mechanical Estimators Handbook.
A02-0411	Wynne, Building Estimating.
A03-0430	AIA, Creative Control of Building Costs.
A03-0454	Berger and Godel, Estimating and Project Management for Small Construction Firms.
A03-0459	Brock, Cost Accounting Manual for Highway Contractors, A System for Cost Control.
A03-0468	Central Electricity Generating Board, Phraseology for Civil Engineering. Two Vols.
A03-0474	Coombs, Construction Accounting and Financial Management.
A03-0479	CSI, Uniform Construction Index.
A03-0485	Dell'Isola, Value Engineering in the Construction Industry.
A03-0487	Dent, Construction Cost Appraisal.
A03-0499	Fasal, Practical Value Analysis Methods.
A03-0501	First Annual Conference-Value Engrg. Assn.
A03-0502	Fletcher and Moore, Standard Phraseology for Bills of Quantities.
A03-0509	Geddes and Chrystal-Smith, Building and Public Works Administration, Estimating and Costing.
A03-0510	Gibson, Value Analysis--The Rewarding Infection.
A03-0512	Gobourne, Cost Control in the Construction Industry.
A03-0529	Heller, Value Management: Value Engineering and Cost Reduction.
A03-0534	Hutton and Devonald, Value in Building.
A03-0546	Kantor, Contractual Aspects of Value Engineering.
A03-0550	Langdon and Every, Model Descriptions for Engineering Services.
A03-0565	Marston, Engineering Valuation and Depreciation.
A03-0589	Oughton, Value Analysis and Value Engineering.
A03-0593	Pilcher, Appraisal and Control of Project Costs.
A03-0597	Practical Accounting and Cost Keeping for Contractors.
A03-0622	Stone, Building Design Evaluation.
A04-1323	Steyert, Economics of High Rise Apartment Buildings of Alternate Design of Configuration.
A05-1538	Benjamin and Cornell, Probability, Statistics and Decisions for Civil Engineering.
A05-1850	Seelye, Data Book for Civil Engineers. Vol. 1 Design.

	Vol. 2 Specifications and Costs.
A06-1970	ASHRAE, 1973 Systems Volume.
A06-2052	Herkimer, Cost Manual for Piping and Mechanical Construction.
A06-2089	Laub, Air Conditioning and Heating Practice.
A06-2135	Plant Engineering Electrical Library.
A06-2157	Sherratt, Integrated Environment in Building Design.
A09-2662	Equipment Guide-Book, Construction Equipment Cost Reference Guide.
A09-2666	Equipment Guide-Book, Green Guide, Vol. 1-2.
A09-2667	Equipment Guide-Book, Green Guide: Off Highway Trucks and Trailers.
A09-2668	Equipment Guide-Book, Green Guide's Older Equipment.
A09-2671	Equipment Guide-Book, Lift Trucks Green Guide.
A09-2677	Equipment Guide-Book, Rental Rate Blue Book.
A09-2686	Foster, Practical Tables for Building Construction.
A09-2832	Winslow, Construction Industry Production Manual.
A10-2895	Means, 1976 Labor Rates for the Construction Industry: City, State, National.
A11-3120	Wilcox and Snape, Worked Examples in Measurement of Construction Work. Vols. 1-2

DAMS

A02-0375	Parker, Planning and Estimating Dam Construction.
A04-1347	Torroja, The Structures of Eduardo Torroja.
A05-1596	Davis and Sorenson, Handbook of Applied Hydraulics.
A05-1649	Graf, Hydraulics of Sediment Transport.
A05-1757	Merritt, Standard Handbook for Civil Engineers.
A05-1785	Okamoto, Introduction to Earthquake Engineering.
A05-1884	Terzaghi and Peck, Soil Mechanics in Engineering Practice.
A07-2252	Reclamation Bureau, Design of Small Dams.
A07-2260	Sherard, Earth and Earth-Rock Dams: Engineering Problems of Design and Construction.
A07-2262	Thomas, The Engineering of Large Dams.
A08-2282	ACI, Temperature and Concrete.
A08-2328	Asphalt Institute, Asphalt in Hydraulic Structures.

DEMOLITION

A09-2638	Colby, Building Wrecking: The How and Why of an Important Business.
A09-2641	A Complete Guide to Demolition.
A09-2691	Gregory, Explosives for North American Engineers.

DICTIONARIES AND GLOSSARIES

A01-0033	Calsat and Sydler, International Vocabulary of Town Planning and Architecture.
A02-0274	Boyce, Real Estate Appraisal Terminology.
A02-0280	Collison, The Developer's Dictionary and Handbook.
A03-0606	Rosenfeld, Architect's Handbook of Professional Practice.

Index

	2 Vols.
A04-0684	Architect's Journal, Architect's Working Details.
A04-0776	Burke, Architectural and Building Trades Dictionary.
A04-0812	Colvin, A Biographical Dictionary of English Architects, 1660-1840.
A04-0829	Cowan, Dictionary of Architectural Science.
A04-0986	Harris, Dictionary of Architecture and Construction.
A04-0987	Harris and Lever, Illustrated Glossary of Architecture.
A04-1274	Saylor, Dictionary of Architecture.
A04-1327	Studio Dictionary of Design and Decoration.
A04-1328	Sturgis, A Dictionary of Architecture and Building.
A04-1363	Walker, Glossary of Art, Architecture and Design Since 1945.
A06-2066	IEEE Standard Dictionary of Electrical and Electronic Terms.
A06-2070	Jacobson, Plumbing Dictionary.
A06-2074	Jones and Schubert, Engineering Encyclopedia.
A06-2095	Lindeke, Technical Dictionary of Heating, Ventilation and Sanitary Engineering.
A08-2282	ACI, Cement and Concrete Terminology.
A08-2358	Bucksch, Dictionary of Wood and Woodworking Practice.
A08-2426	Kinniburgh, Dictionary of Building Materials.
A08-2547	van Amerongen, Dictionary of Cement: Manufacture and Technology. (German-English/English-German).
A08-2549	Volkart, Gypsum and Plaster Dictionary: German-English-French.
A08-2556	Whittington, Whittington's Dictionary of Plastics.
A09-2614	Brooks, Illustrated Encyclopedic Dictionary of Building and Construction Terms.
A09-2615	Bucksch, Dictionary of Civil Engineering and Construction Machinery and Equipment.
A09-2616	Builders Encyclopedia.
A09-2652	de Sousa Ferreira, Glossary of Building and Bridge Construction.
A09-2835	Zboinski and Tyszynski, Dictionary of Architecture and Building Trades: English/German/Polish/Russian.
A10-2850	AIA, Glossary of Construction Industry Terms.
A10-2851	ASCE, A Biographical Dictionary of American Civil Engineers.
A11-2942	ASCE, Definitions of Surveying and Associated Terms.
A11-2981	Cusset, English/French, French/English Technical Dictionary.
A11-2991	EJC, Thesaurus of Engineering and Scientific Terms.

DOORS AND WINDOWS

Also see Architectural Details

A08-2339	ASTM, Window and Wall Testing.
A08-2401	Gatz and Thierry, Architect's Detail Library. Vol. 1 Wrought Iron Railings, Doors and Gates.
A08-2403	Gatz and Thierry, Architect's Detail Library. Vol. 3 Windows and Window Walls.
A08-2553	Watson, Construction Materials and Processes.

A09-2617	The Building Research Establishment, Essential Data for Construction. Vol. 1 Building Construction.
A09-2704	Huntington and Mickadeit, Building Construction: Materials and Types of Construction.
A09-2787	Reiner, Methods and Materials of Construction: A Guide for Builders, Owners, Architects and Engineers.
A09-2826	Weaver, English Leadwork: Its Art and History.

DRAFTING, ILLUSTRATING, BLUEPRINT READING

A02-0282	Cooper and Badzinski, Building Construction Estimating.
A04-0775	Burden, Architectural Delineation: A Photographic Approach to Presentation.
A04-1062	Kicklighter and Baird, Architecture: Residential Drawing and Design.
A04-1233	Ramsey and Sleeper, Architectural Graphic Standards.
A04-1297	Shulman, Photographing Architecture and Interiors.
A04-1375	Whittick, Symbols for Designers.
A05-1739	MacGinley, Structural Steelwork Calculations and Detailing.
A06-2064	IEEE, Standard and American National Standard Graphic Symbols for Electrical and Electronics Diagrams.
A06-2158	Sherwood and Whistance, The Piping Guide.
A08-2312	Architectural Precast Concrete.
A08-2420	Hornung, Reinhold Data Sheets.
A09-2628	Carpenters and Builders Library. Vol. 2 Builders Math, Plans, Specifications.
A11-2930	ACI, Manual of Standard Practice for Detailing Reinforced Concrete Structures.
A11-2933	AISC, Problems and Solutions for Structural Steel Detailing.
A11-2934	AISC, Structural Steel Detailing.
A11-2937	Answers on Blueprint Reading.
A11-2940	Arnell, Standard Graphical Symbols.
A11-2947	Barker, Reinforced Concrete Detailing.
A11-2949	Barsukov, Building Construction Drawing.
A11-2950	Bellis and Schmidt, Blueprint Reading for the Construction Trades.
A11-2963	Bretz, Sheet Metal Shop Drawing.
A11-2967	Burden, Architectural Delineation: A Photographic Approach to Presentation.
A11-2968	Capelle, Professional Perspective Drawing for Architects and Engineers.
A11-2969	Choate, Architectural Presentation in Opaque Watercolor.
A11-2970	Coleman, Technical Drafting: Aerospace, Electrical-Electronic, Structural.
A11-2978	Coulin, Step-By-Step Perspective Drawing: For Architects, Draftsmen, and Designers.
A11-2979	Croney, Anthropometrics for Designers.
A11-2980	CRSI, Reinforcing Bar Detailing.
A11-2982	Dalzell, Plan Reading for Home Builders.
A11-2983	D'Amelio, Perspective Drawing Handbook.
A11-2987	Doblin, Perspective: A New System for Designers.

Index

A11-2988	Dreyfuss, The Measure of Man: Human Factors in Design.
A11-2989	Dudley, Architectural Illustration.
A11-2994	Fairweather and Sliwa, AJ Metric Handbook.
A11-2997	French and Svensen, Mechanical Drawing.
A11-2998	French and Vierck, Manual of Engineering Drawing for Students and Draftsman.
A11-2999	Fuller, Functional Drafting for Today.
A11-3000	General Electric, Modern Drafting Practices & Standards.
A11-3002	Giachino and Beukema, American Technical Society's Freehand Sketching.
A11-3003	Giachino and Beukema, Drafting Technology.
A11-3004	Giachino and Beukema, Everyday Sketching and Drafting.
A11-3005	Giachino and Beukema, Print Reading for Welders.
A11-3006	Gibby, Technical Illustration.
A11-3008	Goodban and Hayslett, Architectural Drawing and Planning.
A11-3010	Grant, Engineering and Architectural Lettering.
A11-3011	Grant, Engineering Drawing.
A11-3013	Halse, Architectural Rendering: The Technique of Contemporary Presentation.
A11-3015	Heine, How to Read Electrical Blueprints.
A11-3016	Hepler and Wallach, Architectural Drafting and Design.
A11-3020	Hogarth, Drawing Architecture.
A11-3021	Hooper, Introduction to Construction Drafting.
A11-3023	Hornung, Architectural Drafting.
A11-3028	Jacoby, New Architectural Drawings.
A11-3029	Jacoby, New Techniques of Architectural Rendering.
A11-3034	Kemmerich, Graphic Details for Architects.
A11-3036	Kenney, Blueprint Reading for the Building Trades.
A11-3037	Kenney and McGrail, Architectural Drawing for the Building Trades.
A11-3050	Lawson, Perspective Charts.
A11-3055	Lockard, Drawing as a Means to Architecture.
A11-3057	McCartney, Precision Perspective Drawing.
A11-3061	Morgan, Architectural Drawing.
A11-3065	Mott, Engineering Drawing and Construction.
A11-3067	Muller, Architectural Drawing and Light Construction.
A11-3068	Muller, Reading Architectural Working Drawings.
A11-3072	Newman, Standard Structural Details for Building Construction.
A11-3084	Pile, Drawings of Architectural Interiors.
A11-3090	Renton, Electrical and Electronic Drafting.
A11-3094	Schaarwachter, Perspectives for Architecture.
A11-3096	Siders, Computer Graphics: A Revolution in Design.
A11-3099	Snow, Electrical Drafting and Design.
A11-3102	Stegman and Stegman, Architectural Drafting; Functional Planning and Creative Design.
A11-3104	Sundberg, Building Trades Blueprint Reading. Part 1 Fundamentals.
A11-3105	Sundberg, Building Trades Blueprint Reading. Part 2 Residential and Light Construction.

Construction Information

A11-3106	Sundberg, Building Trades Blueprint Reading. Part 3 General Construction, Specifications, Heavy Construction.
A11-3109	Thomas, Technical Illustration.
A11-3113	Walters and Bromham, Principles of Perspective.
A11-3114	Weaver, Structural Detailing for Technicians.
A11-3116	Weidhaas, Architectural Drafting and Construction.
A11-3117	Welling, The Technique of Drawing Buildings.
A11-3121	Woltjes, Building Trades Blueprint Reading Examination Kit.
A11-3122	Wyatt, General Architectural Drawing.

ELECTRICAL COMPONENTS AND EQUIPMENT

A02-0265	Ashley, Electrical Estimating.
A02-0274	Bifulco, How to Estimate Construction Costs of Electrical Power Substations.
A02-0278	Cohen, Electrical Estimating Handbook.
A02-0289	Davis, Spon's Mechanical and Electrical Services Price Book.
A02-0348	Kolstad, Rapid Electrical Estimating and Pricing.
A02-0349	Land, Machinery and Equipment Pricing Guide.
A02-0363	Minnesota Elec. Assn., Estimating-Accounting Manual for Electrical Contractor Dealers.
A02-0373	Page and Nation, Estimator's Electric Man-Hour Manual.
A02-0387	Richardson, Mechanical and Electrical Construction Estimating Standards.
A03-0441	Ashley, Electrical Contracting.
A04-0883	ERDA, General Design Criteria Engineering Handbook-Appendix 6301.
A05-1500	ASCE, Guide for Design of Steel Transmission Towers.
A06-1977	ASTM, Metallic Electrical Conductors.
A06-1996	Bredahl, Home Wiring Manual.
A06-1998	Burch, Basic House Wiring.
A06-2012	Croft, American Electricians' Handbook.
A06-2020	Ebeling, Basic Guide to Electric Heating.
A06-2034	Flynn and Mills, Architectural Lighting Graphics.
A06-2039	Giles, Layout of E.H.V. Substations.
A06-2042	Graham, Industrial and Commercial Wiring.
A06-2043	Graham, Interior Electric--Wiring Residential.
A06-2062	IEEE, National Electrical Safety Code, 1973 Edition.
A06-2073	Johnson, Electrical Wiring: Design and Construction.
A06-2086	Kurtz and Shoemaker, The Lineman's and Cableman's Handbook.
A06-2099	McGuinness and Stein, Mechanical and Electrical Equipment for Buildings.
A06-2102	McPartland and Novak, Electrical Equipment Manual.
A06-2103	McPartland and Novak, Practical Electricity.
A06-2106	Marshall, Lightning Protection.
A06-2108	Merritt, Mechanical and Electrical Design of Buildings for Architects and Engineers.
A06-2121	NFPA, Electrical Code for One and Two Family Dwellings.

Index 585

A06-2122	NFPA, National Electrical Code 1975.
A06-2127	Pansini, Basic Electrical Power Distribution. Vol. 2
A06-2135	Plant Engineering Electrical Library.
A06-2136	Porges, The Design of Electrical Services for Buildings.
A06-2137	Porges, The Design of Electrical Services for Buildings.
A06-2143	Richter, Practical Electrical Wiring, 10th Ed.
A06-2170	Traister, Electrical Design for Building Construction.
A06-2176	Watt and Stetka, NFPA Handbook of the National Electrical Code.
A06-2177	Watt and Summers, NFPA Handbook of the National Electrical Code.
A08-2553	Watson, Construction Materials and Processes.
A09-2619	The Building Research Establishment, Essential Data for Construction. Vol. 3 Services and Environmental Engineering.
A09-2703	House Wiring.
A09-2715	Jay and Hemsley, Electrical Services in Buildings.
A09-2787	Reiner, Methods and Materials of Construction: A Guide for Builders, Owners, Architects and Engineers.
A11-3025	Hubert, Preventive Maintenance of Electrical Equipment.
A11-3043	Labor Department, Construction Safety and Health Training Course, Instructor's Guide. Manual 301 Job Planning, Walking and Working Surfaces and Spaces, Fire Prevention, Electrical Hazards.
A11-3099	Snow, Electrical Drafting and Design.
A11-3118	Westinghouse, Electrical Maintenance Hints.

ELECTRICAL ENGINEERING

A02-0273	Bifulco, How to Estimate Construction Costs of Electrical Power Substations.
A04-0883	ERDA, General Design Criteria Engineering Handbook - Appendix 6301.
A04-1068	Kinzey and Sharp, Environmental Technologies in Architecture.
A06-1987	Beeman, Industrial Power Systems Handbook.
A06-2004	Central Electricity Generating Board, Modern Power Station Practice. Vol. 1 Planning and Layout.
A06-2031	Fink and Carroll, Standard Handbook for Electrical Engineers.
A06-2055	Hill, The Science of Engineering Design.
A06-2062	IEEE, National Electrical Safety Code, 1973 Edition.
A06-2063	IEEE, Recommended Practice for Grounding of Industrial and Commercial Power Systems.
A06-2065	IEEE Recommended Practice for Protection and Coordination of Industrial and Commercial Power Systems.
A06-2066	IEEE, Standard Dictionary of Electrical and Electronics Terms.
A06-2073	Johnson, Electric Wiring: Design and Construction.
A06-2085	Kuljian, Nuclear Power Plant Design.
A06-2097	Loftness, Nuclear Power Plants: Design, Operating Experience, and Economics.
A06-2100	McPartland, How to Design Electrical Systems.

A06-2101	McPartland and Novac, Electrical Design Details.
A06-2103	McPartland and Novac, Practical Electricity.
A06-2106	Marshall, Lightning Protection.
A06-2108	Merritt, Mechanical and Electrical Design of Buildings for Architects and Engineers.
A06-2112	Morris, Essential Formulae for Electrical Engineers.
A06-2114	Morse, Power Plant Engineering.
A06-2122	NFPA, National Electrical Code 1975.
A06-2127	Pansini, Basic Electrical Power Distribution.
A06-2136	Porges, The Design of Electrical Services for Buildings.
A06-2137	Porges, The Design of Electrical Services for Buildings.
A06-2170	Traister, Electrical Design for Building Construction.
A06-2176	Watt and Stetka, NFPA Handbook of the National Electrical Code.
A06-2177	Watt and Summers, NFPA Handbook of the National Electrical Code.
A11-2974	Constance, Electrical Engineering for Professional Engineers' Examinations.
A11-3033	Jones, Electrical Engineering License Review.
A11-3056	Lyons and Dublin, Electrical Engineering and Economics and Ethics for Professional Engineering Examinations.
A11-3063	Morrison, Professional Electrical Engineering Examination.
A11-3083	Perry, Engineering Manual.

ELEVATORS AND ESCALATORS

A06-2099	McGuinness and Stein, Mechanical and Electrical Equipment for Buildings.
A06-2136	Porges, The Design of Electrical Services for Buildings.
A06-2137	Porges, The Design of Electrical Services for Buildings.
A08-2301	Adler, Vertical Transportation for Buildings.
A08-2311	Annett, Elevators.
A08-2455	Moffatt, Plant Engineer's Handbook of Formulas, Charts and Tables.
A08-2534	Strakosch, Vertical Transportation: Elevators and Escalators.
A09-2608	Bolz and Hagemann, Materials Handling Handbook.
A09-2787	Reiner, Methods and Materials of Construction: A Guide to Builders, Owners, Architects and Engineers.

ENERGY CONSERVATION

A06-2003	Caudill, A Bucket of Oil: The Humanistic Approach to Building Design for Energy Conservation.
A06-2029	Federal Energy Administration, Energy Conservation: Lighting and Thermal Operations.
A06-2030	Federal Energy Administration, Guidelines for Saving Energy in Existing Buildings.
A06-2044	Griffin, Energy Conservation in Buildings: Techniques for Economical Design.
A06-2058	HUD, Multifamily Housing, Final Report.
A06-2059	HUD, Single-Family Housing, Final Report.

Index 587

A06-2157 Sherratt, Integrated Environment in Building Design.

ENGINEERING MANAGEMENT

A02-0378 Peart, Design of Project Management Systems and Records.
A02-0391 Siddall, Analytical Decision Making in Engineering Design.
A02-0395 Stires and Wenig, Pert/Cost.
A03-0420 Abbett, Engineering Contracts and Specifications.
A03-0421 Abrahamson, Engineering Law and I.C.E. Contracts.
A03-0427 Aguilar, Systems Analysis and Design in Engineering, Architecture, Construction and Planning.
A03-0436 Andree, Computer Programming: Techniques, Analysis and Mathematics.
A03-0437 Antill, Civil Engineering Management.
A03-0439 Architectural Record, Techniques of Successful Practice for Architects and Engineers.
A03-0442 ASME, Managing for Improved Engineering Effectiveness.
A03-0448 Barrodale, Elementary Computer Applications: In Science, Engineering and Business.
A03-0449 Battersby, Network Analysis for Planning and Scheduling.
A03-0452 Benjamin and Cornell, Probability, Statistics and Decisions for Civil Engineering.
A03-0461 Burman, Precedence Networks for Project Planning and Control.
A03-0463 Canfield and Bowman, Business, Legal, and Ethical Phases of Engineering.
A03-0468 Central Electricity Generating Board, Phraseology for Civil Engineering.
A03-0469 Clark, Business Systems and Data Process Procedures.
A03-0476 Coxe, Marketing Architectural and Engineering Services.
A03-0477 Critical Path Techniques for Construction.
A03-0478 Cronstedt, Engineering Management and Administration.
A03-0479 CSI, Uniform Construction Index.
A03-0480 Dand and Farmer, Purchasing in the Construction Industry.
A03-0486 deNeufville and Stafford, Systems Analysis for Engineers and Managers.
A03-0488 Dibner, Joint Ventures for Architects and Engineers.
A03-0492 Dubin, Architectural Supervision of Modern Buildings.
A03-0493 Dunham and Young, Contracts, Specifications and Law for Engineers.
A03-0506 Foxhall, Techniques of Successful Practice for Architects and Engineers.
A03-0508 Fuller, Organizing, Planning and Scheduling Engineering Operations.
A03-0509 Geddes and Chrystal-Smith, Building and Public Works Administration, Estimating and Costing.
A03-0510 Gibson, Value Analysis-The Rewarding Infection.
A03-0513 Gothie, A Selected Bibliography on Applied Ethics in the Professions, 1950-1970.
A03-0514 Granof, How to Cost Your Labor Contract.

A03-0521	Haavind and Turmail, The Successful Engineer-Manager: A Practical Guide to Management Skills for Engineer and Scientist.
A03-0522	Hackney, Control and Management of Capital Projects.
A03-0523	Hajek, Project Engineering.
A03-0525	Harper, Computer Applications in Architecture and Engineering.
A03-0527	Hauf, Building Contracts for Design and Construction.
A03-0529	Heller, Value Management: Value Engineering and Cost Reduction.
A03-0532	Hovanessian and Pipes, Digital Computer Methods in Engineering.
A03-0539	James, Analog Computer Simulation of Engineering Systems.
A03-0540	James, Applied Numerical Methods for Digital Computation with Fortran.
A03-0541	Jamison, Fortran IV Programming: Based on the IBM System 1130.
A03-0543	Jones, How to Market Professional Design Services.
A03-0547	Kelly, Handbook of Numerical Methods and Applications.
A03-0548	LaFara, Computer Methods for Science and Engineering.
A03-0549	Laidlow, Engineering Law.
A03-0550	Langdon and Avery, Model Descriptions for Engineering Services.
A03-0553	Leech, Management of Engineering Design.
A03-0555	Lifson, Decision and Risk Analysis for Practicing Engineers.
A03-0556	Lock, Engineer's Guide to Management Techniques.
A03-0559	McCracken, Fortran with Engineering Applications.
A03-0567	Martino, Critical Path Methods.
A03-0570	Meredith, Design and Planning of Engineering Systems.
A03-0571	Merritt, Building Construction Handbook.
A03-0573	Moyle, Introduction to Computers for Engineers.
A03-0581	NSPE, Professional Index of Private Practice.
A03-0584	O'Brien, Scheduling Handbook.
A03-0587	O'Brien, Value Analysis in Design and Construction.
A03-0589	Oughton, Value Analysis and Value Engineering.
A03-0595	Popper, Modern Technical Management Techniques.
A03-0599	Radcliffe, Critical Path Method.
A03-0601	Rau, Optimization and Probability in Systems Engineering.
A03-0602	Ray-Jones and McCann, CI/SfB Project Manual.
A03-0607	Rossnagel, Checklists for Management, Engineering, Manufacturing, and Product Assurance.
A03-0611	Rubey, The Engineer and Professional Management.
A03-0613	SAVE, Conference Proceedings--1970.
A03-0614	SAVE, Conference Proceedings--1971.
A03-0615	SAVE, Conference Proceedings--1972.
A03-0616	SAVE, Conference Proceedings--1973.
A03-0617	SAVE, Conference Proceedings--1974.
A03-0619	Shaffer, Critical Path Method.
A03-0620	Sherran, Management Essentials for Public Health Administrators.

Index

A03-0621	Stanley, The Consulting Engineer.
A03-0622	Stone, Building Design Evaluation.
A03-0623	Swanson and Pazer, Pertsim: Text and Simulation.
A03-0634	Wilson, Concepts of Engineering System Design.
A05-1935	Zilly, Handbook of Environmental Civil Engineering.
A06-2094	Lewis, Facilities and Plant Engineering Handbook.
A11-2941	Arnold and Vaden, Invention Protection for Practicing Engineers.
A11-3018	Hicks, Professional Achievement for Engineers and Scientists.
A11-3035	Kemper, The Engineer and His Profession.
A11-3112	Tuska, An Introduction to Patents for Inventors and Engineers.

ETHICS AND CLIENT RELATIONS

A03-0463	Canfield and Bowman, Business, Legal and Ethical Phases of Engineering.
A03-0513	Gothie, A Selected Bibliography on Applied Ethics in the Professions, 1950-1970.
A11-2935	Alger, Ethical Problems in Engineering.
A11-3056	Lyons and Dublin, Electrical Engineering and Economics and Ethics for Professional Engineering Examinations.
A11-3093	Sanders and Dublin, Civil Engineering and Economics and Ethics for Professional Engineering Examinations.
A11-3100	Stamper and Dublin, Mechanical Engineering and Economics and Ethics for Professional Engineering Examinations.

EXAMINATIONS FOR PROFESSIONAL REGISTRATION

A11-2931	AIA, Architectural Registration Handbook (NCARB).
A11-2938	Apfelbaum and Ottesen, Basic Engineering Sciences and Structural Engineering for Engineer-in-Training Examinations.
A11-2939	Arco Editorial Board, Mechanical Engineer--Junior Assistant and Senior Grades.
A11-2951	Bender, California Structural Engineer's License Review.
A11-2966	Brinker, 1777 Review Problems from E.I.T. and Professional Registration Examinations: With Answers and Typical Solutions.
A11-2974	Constance, Electrical Engineering for Professional Engineers' Examinations.
A11-2975	Constance, How to Become a Professional Engineer.
A11-2976	Constance, Mechanical Engineering for Professional Engineers' Examinations.
A11-2977	Coren, Mechanical Engineering.
A11-2993	Faires and Richardson, E.I.T. Review.
A11-3033	Jones, Electrical Engineering License Review.
A11-3035	Kemper, The Engineer and His Profession.
A11-3041	Kurtz, Engineering Economics for Professional Engineers' Examinations.
A11-3048	LaLonde, Professional Engineering Examination Questions

A11-3052	and Answers. Levinson, Preparing for the Professional Engineer's Examinations: A Review with Questions and Answers.
A11-3056	Lyons and Dublin, Electrical Engineering and Economics and Ethics for Professional Engineering Examinations.
A11-3062	Morrison, EIT Examination Study Guide.
A11-3063	Morrison, Professional Electrical Engineering Examination.
A11-3070	Newman, Engineer-In-Training License Review.
A11-3071	Newman, Civil Engineering License Review.
A11-3081	Packer, Professional Engineer (Civil) State Board Examination Review.
A11-3087	Polentz, Engineering Fundamentals for Professional Engineers' Examinations.
A11-3093	Sanders and Dublin, Civil Engineering and Economics and Ethics for Professional Engineering Examinations.
A11-3100	Stamper and Dublin, Mechanical Engineering and Economics and Ethics for Professional Engineering Examinations.
A11-3111	Turner, Civil Engineer, Senior and Supervising.

EXCAVATION

See Tunnels and Underground Structures

A09-2594	ASTM, Performance Monitoring for Geotechnical Construction.
A09-2633	Carson, General Excavation Methods.
A09-2642	Crimmins, Construction Rock Work Guide.
A09-2691	Gregory, Explosives for North American Engineers.
A09-2696	Hammond, Earthmoving and Excavating Plant.
A09-2707	Huston, Hydraulic Dredging: Theoretical and Applied.
A09-2714	Intl. Soc. of Soil Mechanics and Foundation Engrg., Grouts and Drilling Muds in Engineering Practice.
A09-2740	McGregor, Drilling of Rock.
A09-2759	Nichols, How to Operate Excavation Equipment.
A09-2760	Nichols, Moving the Earth (The Workbook of Excavation).
A09-2787	Reiner, Methods and Materials of Construction: A Guide for Builders, Owners, Architects and Engineers.
A09-2804	Smith, Principles and Practices of Heavy Construction.
A09-2805	Smith, Principles and Practices of Light Construction.
A09-2810	Technical Group for Nuclear Explosives Engrg., Proceedings of the Special Session on Nuclear Excavation.
A09-2811	Teller, The Constructive Uses of Nuclear Explosives.
A09-2816	Tomlinson, Foundation Design and Construction.
A09-2819	Urquhart, Civil Engineering Handbook.
A09-2824	Wass, Methods and Materials of Residential Construction.
A09-2834	Yardley, Rapid Excavation--Problems and Progress: Proceedings of the Tunnel and Shaft Conference, 1968.

FANS AND BLOWERS

A06-1964	Alden and Kane, Design of Industrial Exhaust Systems.

Index

A06-1969	ASHRAE, Equipment Volume, 1972.
A06-2008	Church, Centrifugal Pumps and Blowers.
A06-2021	Eck, Fans: Design and Operation of Centrifugal, Axial Flow and Cross Flow Fans.
A06-2037	General Electric, Heat Transfer and Fluid Flow Data Books. Vol. 1, Heat Transfer.
A06-2126	Osborne, Fans.
A06-2139	Power Magazine, Plant Energy Systems.

FARM BUILDINGS

A04-0744	Boyd, Practical Farm Buildings: A Text and Handbook.
A04-0980	Hancocks, Animals and Architecture.
A04-1036	Itoh, Kura, Design of Traditional Japanese Barns and Storehouses.
A04-1132	Merrilees and Loveday, Pole Building Construction.
A04-1220	Portland Cement Assn., Modern Plans for Modern Farms.
A04-1309	Smith, The Design and Construction of Stables and Ancillary Buildings.
A04-1376	Wilkinson, Storage Specifics.
A04-1387	Witney, The Barn: A Vanishing Landmark in North America.
A09-2628	Carpenters and Builders Library. Vol. 2 Builders Math, Plans, Specifications.
A09-2737	Lytle, Farm Builders Handbook.

FINANCE

A01-0055	Crecine, Financing the Metropolis: Public Policy in Urban Economies.
A01-0103	Heroux and Wallace, Financial Analysis and the New Community Development Process.
A01-0105	Hoover, An Introduction to Regional Economics.
A01-0108	Housing and Urban Development Dept., Housing in the Seventies.
A01-0134	Listokin, The Dynamics of Housing Rehabilitation: Macro and Micro Analyses.
A01-0149	Morris, State Housing Finance Agencies: An Entrepreneurial Approach to Subsidized Housing.
A02-0271	Beasley, Fell's Guide to Buying, Building and Financing a Home.
A02-0281	Construction Equipment Buyers Guide.
A02-0306	Ellwood, Ellwood Tables for Real Estate Appraising and Financing. Part I--Explanatory Text.
A02-0307	Ellwood, Ellwood Tables for Real Estate Appraising and Financing. Part II--Tables.
A02-0308	Essex, Bonding Versus Pay-as-You-Go in the Financing of School Buildings.
A02-0317	Golemon, Financing Real Estate Development.
A02-0323	Halperin, Construction Funding: Where the Money Comes From.
A02-0324	Halsey, Borrowing Money for the Public Schools: A

	Study of Borrowing Practices in the Administration of Public Schools in Florida.
A02-0334	Hoagland and Stone, Real Estate Finance.
A02-0393	Sokol, Contractor or Manipulator?
A03-0435	Andersen, Financial Management for Architectural Firms-A Manual of Accounting Procedures.
A03-0474	Coombs, Construction Accounting and Financial Management.
A03-0519	Griffin, Development Building: The Team Approach.
A03-0526	Harper, Financial Management Computer Users Manual.
A04-1202	Paul, Apartments: Their Design and Development.
A10-2867	Ghosh, The Economics of Building Societies.
A10-2870	Gramlich and Jaffee, Savings Deposits, Mortgages and Housing.
A10-2914	Sternlieb and Listokin, Housing 1973-74: An Anthology.
A10-2918	U.S. Savings and Loan League, 1976 Savings and Loan Fact Book.

FIRE SAFETY

A06-1993	Bond, NFPA Inspection Manual - Third Edition.
A06-2083	King, Piping Handbook.
A06-2106	Marshall, Lightning Protection.
A06-2123	NFPA, National Fuel Gas Code.
A06-2136	Porges, The Design of Electrical Services for Buildings.
A06-2138	Porges, The Design of Electrical Services for Buildings.
A08-2300	Addleson, Materials for Building. Vol. 4
A08-2308	American Chemical Society, Fire Retardant Paints.
A08-2469	NFPA, Tentative Guide for Plastics in Building Construction.
A09-2787	Reiner, Methods and Materials of Construction: A Guide for Builders, Owners, Architects and Engineers.
A11-2932	AIChE, Plant and Design Safety.
A11-2944	ASTM, Ignition, Heat Release, and Noncombustibility of Materials.
A11-2953	Bhatnager, Advances in Fire Retardants, Part 1.
A11-2954	Bhatnager, Advances in Fire Retardants, Part 2.
A11-2955	Bhatnager, Fire Retardant Formulations Handbook.
A11-2958	Bond, NFPA Inspection Manual.
A11-2992	Factory Mutual System, Handbook of Industrial Loss Prevention.
A11-3001	General Services Administration, Public Building Service International Conference on Fire Safety.
A11-3019	Hilado, Flammability Test Method Handbook.
A11-3030	Jensen, Fire Protection Systems for the Design Professional.
A11-3054	Lie, Fire and Buildings.
A11-3059	McKinnon, Fire Protection Handbook, 14th Edition.
A11-3060	Marchant, A Complete Guide to Fire and Buildings: A Comprehensive Survey of all Aspects of Fire Protection and Provision.
A11-3075	NFPA, Designing Buildings for Fire Safety.
A11-3076	NFPA, Fire Protection by Sprinklers.

Index

A11-3077	NFPA, Fire Protection Guide on Hazardous Materials, Sixth Edition.
A11-3078	NFPA, Fire Protection Handbook, Fourteenth Edition.
A11-3079	NFPA, Fires in High-Rise Buildings.
A11-3110	Thompson, Fire Behavior and Sprinklers.

FLOOR, WALL, ROOF, CEILING SYSTEMS

A05-1798	Parker, Simplified Design of Roof Trusses for Architects and Builders.
A05-1875	Stuttgart Universität, Convertible Roofs.
A08-2313	Arnison, Floor and Structural Surfaces.
A08-2365	Ceiling Systems Handbook.
A08-2370	Comber, Composition Flooring and Floorlaying.
A08-2402	Gatz and Thierry, Architect's Detail Library.
A08-2411	Griffin, Manual of Built-Up Roof Systems.
A08-2483	Patton, Construction Materials.
A08-2489	Pindar, Engineering Wall and Partition Components for Pre-fabrication.
A08-2511	Salter, Floors and Floor Maintenance.
A08-2536	Svec and Jeffers, Modern Masonry Panel Construction Systems.
A08-2553	Watson, Construction Materials and Processes.
A09-2587	Amer. Institute of Architects, Manual of Built-Up Roofing Systems.
A09-2617	The Building Research Establishment, Essential Data for Construction. Vol. 1 Building Construction.
A09-2620	The Building Research Establishment, Essential Data for Construction. Vol. 4 Building Defects and Maintenance.
A09-2621	Building with Tilt-Up.
A09-2639	Collins, Manual of Tilt-Up Construction.
A09-2687	Gatz, Curtain Wall Construction.
A09-2695	Halperin, Building with Steel.
A09-2704	Huntington and Mickadeit, Building Construction: Materials and Types of Construction.
A09-2722	Koncz, Manual of Precast Concrete Construction. Vol. 1.
A09-2723	Koncz, Manual of Precast Concrete Construction. Vol. 2.
A09-2724	Koncz, Manual of Precast Concrete Construction. Vol. 3.
A09-2767	Perkins, Floors--Construction and Finishes.
A09-2787	Reiner, Methods and Materials of Construction: A Guide for Builders, Owners, Architects and Engineers.
A09-2804	Smith, Principles and Practices of Heavy Construction.
A09-2805	Smith, Principles and Practices of Light Construction.
A09-2825	Wass and Sanders, Building Construction: Roof Framing.
A09-2831	Wilson and Werner, Simplified Roof Framing.

GOVERNMENT LEGISLATION AND POLICY

A01-0005	Anderson, The Federal Bulldozer: A Critical Analysis of Urban Renewal, 1949-62.
A01-0008	Babcock and Bosselman, Exlusionary Zoning.
A01-0022	Berry, Housing: the Great British Failure.

A01-0024	Binns, Knight's Building Regulations.
A01-0043	Chartrand, Systems Technology Applied to Social and Community Problems.
A01-0048	Clapp, New Towns and Urban Policy.
A01-0055	Crecine, Financing the Metropolis.
A01-0062	DeChiara and Koppelman, Planning Design Criteria.
A01-0066	Downs, Federal Housing Subsidies: How Are They Working?
A01-0073	Everett and Johnston, Housing Series: Law and Contemporary Problems.
A01-0079	Frieden and Kaplan, The Politics of Neglect: Urban Aid from Model Cities to Revenue Sharing.
A01-0090	Golany, Strategy for New Community Development in the U.S.
A01-0094	Groberg, Urban Renewal Programs Assisted by Title 1 of the Housing Act of 1949.
A01-0108	HUD, Housing in the Seventies.
A01-0129	Levin, New Approaches to State Land Use Policies.
A01-0135	Listokin, Fair Share Housing Allocation.
A01-0138	McKown, Comprehensive Guide to Town Planning Law Procedures.
A01-0146	Mields, Federally Assisted New Communities.
A01-0150	Multiple Dwelling Law of New York State.
A01-0154	Nourse, The Effects of Public Policy on Housing Markets.
A01-0173	Richardson, The Cost of Environmental Protection: Regulating Housing Development in the Coastal Zone.
A01-0181	Rothblatt, Regional Planning: The Appalachian Experience.
A01-0182	Rubinowitz, Low-Income Housing: Suburban Strategies.
A01-0196	Solomon, Housing the Urban Poor.
A01-0208	Stevens, Impact of Federal Legislation and Programs on Private Land in Urban and Metropolitan Development.
A01-0213	Taggart, Low-Income Housing.
A01-0226	Washnis, Community Development Strategies in the Model Cities Program.
A01-0228	Wexler and Peck, Housing and Local Government.
A02-0305	Edson and Lane, A Practical Guide to Low and Moderate Income Housing.
A02-0334	Hoagland and Stone, Real Estate Finance.
A03-0515	Gray, Cases and Materials on Environmental Law.
A03-0561	McKown, Comprehensive Guide to Factory Law.
A06-1962	AIChE, Industrial Process Design for Pollution Control.
A09-2675	Equipment Guide-Book, OSHA Requirements for Construction. Vol. 1 Equipment.
A09-2676	Equipment Guide-Book, OSHA Requirements for Construction. Vol. 2 Operations Standards.
A10-2870	Gramlich and Jaffee, Savings Deposits, Mortgages and Housing.
A10-2874	Havighurst, Regulating Health Facilities Construction.
A10-2875	HUD, Services Available to HUD-Related Business in International Trade.
A10-2876	HUD, Operation Breakthrough, Vol. 3.

Index

A10-2894	Lave and Lave, The Hospital Construction Act: An Evaluation of the Hill-Burton Program.
A10-2900	Organization for Social and Technical Innovation, Self-Help Housing in the U.S.A.
A10-2903	Rada, The Cal-Vet Program: A Study of State-Financed Housing in California.
A10-2923	Wolman, Housing and Housing Policy in the U.S. and the U.K.
A11-2936	Anderson, OSHA and Accident Control Through Training.
A11-2943	Associated Builders and Contractors, Construction Safety Guide.
A11-2951	Best's Safety Directory.
A11-2957	BNA Editorial Staff, The Job Safety and Health Act of 1970.
A11-2972	Commerce Clearing House, Occupational Safety and Health Act of 1970, Law and Explanation.
A11-3022	Hopf, Designer's Guide to OSHA.

HARBORS AND PORTS

A02-0250	Adie, Marinas.
A02-0255	American Assn. of Port Authorities, Port Planning Design and Construction.
A02-0332	Hedden, Mission: Port Development.
A04-0701	ASCE, Analytical Treatment of Problems of Berthing and Mooring Ships.
A04-0702	ASCE, Small Craft Harbors.
A04-0849	DeMare, The Nautical Style.
A05-1479	Andersen, Substructure Analysis and Design.
A05-1649	Graf, Hydraulics of Sediment Transport.
A05-1682	Howells, Dynamic Waves in Civil Engineering.
A05-1690	Ippen, Estuary and Coastline Hydrodynamics.
A06-2149	Schenck, Introduction to Ocean Engineering.
A07-2223	Bruun, Port Engineering.
A07-2225	Cornick, Dock and Harbour Engineering. Vol. 1 The Design of Docks.
A07-2226	Cornick, Dock and Harbour Engineering. Vol. 2 The Design of Harbours.
A07-2227	Cornick, Dock and Harbour Engineering. Vol. 3 Buildings and Equipment.
A07-2228	Cornick, Dock and Harbour Engineering. Vol. 4 Construction.
A07-2241	Marcus, The Challenge of Deepwater Terminals.
A07-2242	Minikin, Winds, Waves and Maritime Structures.
A07-2244	Myers, Handbook of Ocean and Underwater Engineering.
A07-2246	Navy Department, Design Manual, Dry-Docking Facilities.
A07-2249	Paquette, Transportation Engineering--Planning and Design.
A07-2251	Quinn, Design and Construction of Ports and Marine Structures.
A07-2263	Thorn and Simmons, Sea Defence Works: Design, Construction and Emergency Works.
A07-2266	Wetteland and Bruun, Proceedings POAC-Norway 1971.

	Vols. 1 and 2.
A07-2267	Wiegel, Oceanographical Engineering.
A08-2334	ASTM, Materials Performance and the Deep Sea.
A09-2581	ACI, Symposium on Concrete Construction in Aqueous Environments.
A09-2597	ASTM, Underwater Soil Sampling, Testing and Construction Control.
A09-2607	Blake, Civil Engineer's Reference Book.
A09-2707	Huston, Hydraulic Dredging: Theoretical and Applied.

HARDWARE, ANCHORS, FASTENERS

A08-2291	ACI, Mechanical Fasteners for Concrete.
A08-2356	Brownell, Architectural Hardware Specifications Handbook.
A08-2433	Launchbury, AJ Handbook of Fixings and Fastenings.
A08-2448	Marsh and Beckett, Mechanical Fixing Devices in the Construction Industry.
A08-2553	Watson, Marble Design Manual.
A09-2704	Huntington and Mickadeit, Building Construction: Materials and Types of Construction.
A09-2787	Reiner, Methods and Materials of Construction: A Guide for Builders, Owners, Architects and Engineers.

HEATING, VENTILATION, AIRCONDITIONING

A02-0269	Barton, Estimating for Heating and Ventilating.
A02-0289	Davis, Spon's Mechanical and Electrical Services Price Book.
A02-0316	Gladstone, Mechanical Estimating Guide-Book.
A02-0367	Ottaviano, National Mechanical Estimator.
A02-0368	Ottaviano, National Mechanical Estimator.
A02-0370	Page, Estimator's Equipment Installation Man-Hour Manual.
A02-0372	Page, Heating, Plumbing and Air-Conditioning Man-Hour Manual.
A02-0388	Richardson, Mechanical and Electrical Construction Estimating Standards.
A03-0497	Emerick, Handbook of Mechanical Specifications for Buildings and Plants.
A03-0579	NESCA, Service Operation Manual.
A04-0883	ERDA, General Design Criteria Engineering Handbook-Appendix 6301.
A04-0907	Flynn and Segil, Architectural Interior Systems: Lighting, Air-Conditioning, Acoustics.
A04-0944	Givoni, Man, Climate and Architecture.
A04-1068	Kinzey and Sharp, Environmental Technologies in Architecture.
A04-1164	Natl. Aero. and Space Admin., NASA Facilities Engineering Handbook.
A04-1271	Salvadori, Mathematics in Architecture.
A04-1352	Vandenberg and Elder, Handbook of Building Enclosure.
A06-1960	ACGIH, Industrial Ventilation.

Index

A06-1963	Air Conditioning.
A06-1964	Alden and Kane, Design of Industrial Exhaust Systems.
A06-1966	Ambrose, Heat Pumps and Electric Heating: Residential, Commercial, Industrial Year-Round Air Conditioning.
A06-1967	Andrews, Architect's Guide to Mechanical Systems.
A06-1968	Angus, The Control of Indoor Climate.
A06-1969	ASHRAE, Equipment Volume 1972.
A06-1970	ASHRAE, 1974 Applications Volume.
A06-1971	ASHRAE, 1973 Systems Volume.
A06-1972	ASHRAE, 1972 Handbook of Fundamentals.
A06-1981	Banham, Architecture of the Well-Tempered Environment.
A06-1983	Barton, Domestic Heating and Hot Water Supply.
A06-1984	Baturin, Fundamentals of Industrial Ventilation.
A06-1990	Billington, Building Physics: Heat.
A06-1993	Bond, NFPA Inspection Manual - Third Edition.
A06-1999	Burkhardt, Residential and Commercial Air Conditioning.
A06-2002	Carrier Air Conditioning Co., Handbook of Air Conditioning System Design.
A06-2003	Caudill, A Bucket of Oil: The Humanistic Approach to Building Design for Energy Conservation.
A06-2006	Chalkley and Cater, Thermal Environment.
A06-2012	Croft, American Electricians' Handbook.
A06-2015	Davis, Plumbing, Heating, and Piping: Estimator's Guide.
A06-2018	Down, Heating and Cooling Load Calculations.
A06-2020	Ebeling, Basic Guide to Electric Heating.
A06-2022	Egan, Concepts in Thermal Comfort.
A06-2023	Emerick, Heating Handbook.
A06-2024	Emerick, Troubleshooters' Handbook for Mechanical Systems.
A06-2028	Fanger, Thermal Comfort: Analysis and Applications in Environmental Engineering.
A06-2029	Federal Energy Administration, Energy Conservation: Lighting and Thermal Operations.
A06-2030	Federal Energy Administration, Guidelines for Saving Energy in Existing Buildings.
A06-2035	Flynn and Segil, Architectural Interior Systems: Lighting, Air Conditioning, Acoustics.
A06-2036	Geiringer, High Temperature Water Heating: Its Theory and Practice for District and Space Heating Applications.
A06-2037	General Electric, Heat Transfer & Fluid Flow Data Books. Vol. 1, Heat Transfer. Vol. 2, Fluid Flow.
A06-2041	Gosling, Applied Air Conditioning and Refrigeration.
A06-2044	Griffin, Energy Conservation in Buildings: Techniques for Economical Design.
A06-2045	Gunther, Refrigeration, Air Conditioning, and Cold Storage: Principles and Applications.
A06-2046	Haines, Automatic Control of Heating and Air Conditioning.
A06-2047	Haines, Control Systems for Heating, Ventilating, and Air Conditioning.
A06-2050	Harris, Modern Air Conditioning Practice.

A06-2051	Hemeon, Plant and Process Ventilation.
A06-2056	Home Refrigeration and Air Conditioning.
A06-2061	Hutchinson, Design of Refrigeration Systems for Air Conditioning.
A06-2076	Kallen, Handbook of Instrumentation and Controls.
A06-2078	Kasuda, Use of Computers for Environmental Engineering Related to Buildings.
A06-2080	Kell and Martin, Faber and Kell's Heating and Air Conditioning of Buildings.
A06-2083	King, Piping Handbook.
A06-2087	Kut, Heating and Hot Water Services in Buildings.
A06-2088	Kut, Warm Air Heating.
A06-2089	Laub, Air Conditioning and Heating Practice.
A06-2090	Laube, How to Have Air Conditioning and Still Be Comfortable.
A06-2091	Lefax Pub. Co., Home Heating.
A06-2095	Lindeke, Technical Dictionary of Heating, Ventilation and Sanitary Engineering.
A06-2099	McGuinness and Stein, Mechanical and Electrical Equipment for Buildings.
A06-2108	Merritt, Mechanical and Electrical Design of Buildings for Architects and Engineers.
A06-2116	National Science Foundation, Demand Analysis, Solar Heating and Cooling of Buildings.
A06-2117	National Science Foundation, Proceedings of the Solar Heating and Cooling for Buildings Workshop, 1973, Part I, Technical Sessions.
A06-2118	National Science Foundation, Solar Cooling for Buildings.
A06-2125	Olivieri, How to Design Heating-Cooling Comfort Systems.
A06-2129	Patton, Solar Energy for Heating and Cooling of Buildings.
A06-2138	Porges and Porges, Handbook of Heating Ventilating and Air Conditioning.
A06-2139	Power Magazine, Plant Energy Systems.
A06-2140	Price, Air Conditioning for Building Engineers and Managers: Operation and Maintenance.
A06-2141	Ramsey, Tested Solutions to: Design Problems in Air Conditioning and Refrigeration.
A06-2145	Rogers, Thermal Design of Buildings.
A06-2148	Sauer and Howell, Environmental Control Principles.
A06-2152	Sculthorpe, Design of High Pressure Steam and High Temperature Water Plants.
A06-2155	Severns and Fellows, Air Conditioning and Refrigeration.
A06-2156	Sherratt, Air Conditioning System Design for Buildings.
A06-2157	Sherratt, Integrated Environment in Building Design.
A06-2159	Shields, Boilers.
A06-2164	Stoecker, Principles for Air Conditioning Practice.
A06-2165	Stoecker, Refrigeration and Air Conditioning.
A06-2166	Strock and Koral, Handbook of Air Conditioning, Heating and Ventilating.
A06-2168	Threlkeld, Thermal Environmental Engineering.
A06-2172	Van Straaten, Thermal Performance of Buildings.

A06-2178	Weaver and Kirkpatrick, Environment Control: Air Conditioning and Refrigeration Theory and Application.
A08-2381	Danz, Sun Protection.
A08-2455	Moffat, Plant Engineer's Handbook of Formulas, Charts and Tables.
A08-2553	Watson, Construction Materials and Processes.
A09-2598	Auslander, Domestic Oil Burners and Oil Heat.
A09-2619	The Building Research Establishment, Essential Data for Construction. Vol. 3 Services and Environmental Engineering.
A09-2622	Burgess, The Construction Industry Handbook.
A09-2624	Burkhardt, Domestic and Commercial Oil Burners.
A09-2637	Clarke, Installing Small Pipe Central Heating.
A09-2650	Daniels, Home Guide to Plumbing, Heating and Air Conditioning.
A09-2690	Gladstone, Air Conditioning Testing and Balancing, A Field Practice Manual.
A09-2716	Kaberlein, Air Conditioning Metal Layout.
A09-2725	Kratfel, Introduction to Modern Sheet Metal.
A09-2755	Morris, Air Conditioning Cutter's Ready Reference.
A09-2787	Reiner, Methods and Materials of Construction: A Guide for Builders, Owners, Architects and Engineers.
A11-3066	Mulholland, Heating, Ventilation and Air Conditioning Plant: Planned Maintenance and Operation.

HIGHWAY ENGINEERING

A07-2202	Antoniou, Environmental Management: Planning for Traffic.
A07-2207	APWA, Streetscape Equipment Sourcebook.
A07-2208	APWA, A Survey of Urban Arterial Design Standards.
A07-2210	ARBA Directory of Transportation and Agency Personnel.
A07-2211	Army Department, Planning and Design of Roads, Airbases, and Heliports, in the Theater of Operations.
A07-2212	Army Engineers Corps, Roads and Airfields.
A07-2213	Arnison, Roadwork Technology.
A07-2217	ASTM, An Analysis of the Literature on Tire-Road Skid Resistance.
A07-2218	ASTM, Skid Resistance of Highway Pavements.
A07-2219	ASTM, Surface Texture Versus Skidding.
A07-2220	Baerwald, Institute of Traffic Engineers Transportation and Traffic Engineering Handbook.
A07-2221	Barenberg, Pavement Distress Identification and Repair.
A07-2222	Baumann and Wilson, Urban Engineering and Transportation.
A07-2224	Cedergren, Drainage of Highway and Airfield Pavements.
A07-2229	Drew, Traffic Flow Theory and Control.
A07-2230	Halprin, Freeways.
A07-2231	Heggle, Transport Engineering Economics.
A07-2232	Hennes and Ekse, Fundamentals of Transportation Engineering.
A07-2234	Hickerson, Route Location and Design.

A07-2238   Legault, Highway and Airport Engineering.
A07-2240   Malt, Furnishing the City.
A07-2247   Odier, Low Cost Roads: Design, Construction and Maintenance.
A07-2248   Oglesby, Highway Engineering.
A07-2249   Paquette, Transportation Engineering--Planning and Design.
A07-2250   Pignataro, Traffic Engineering: Theory and Practice.
A07-2253   Ritter and Paquette, Highway Engineering.
A07-2254   Robinson, Highways and Our Environment.
A07-2255   Rodgers and Sands, Automobile Traffic Signal Control Systems.
A07-2257   Sargious, Pavements and Surfacings for Highways and Airports.
A07-2259   Sharp, Concrete in Highway Engineering.
A07-2265   Wallace and Martin, Asphalt Pavement Engineering.
A07-2268   Wohl and Martin, Traffic System Analysis for Engineers and Planners.
A07-2269   Woods, Highway Engineering Handbook.
A07-2270   Yang, Design of Functional Pavements.
A07-2271   Yoder and Witczak, Principles of Pavement Design.
A08-2429   Koster, Expansion Joints in Bridges and Roads.
A08-2430   Krebs and Walker, Highway Materials.
A08-2432   LaLonde and Janes, Concrete Engineering Handbook.
A08-2540   Tranxler, Asphalt: Its Composition, Properties and Uses.
A08-2560   Zaker, Asphalt.
A09-2607   Blake, Civil Engineer's Reference Book.
A09-2682   Federal Highway Administration, Construction Manual.
A09-2819   Urquhart, Civil Engineering Handbook.

HOSPITALS, NURSING HOMES

A02-0254   American Assn. of Hospital Consultants, Functional Planning of General Hospitals.
A02-0256   American Hospital Assn., Guide Issue--Hospitals.
A02-0263   Architectural Record, Hospitals, Clinics, and Health Centers.
A02-0294   H.E.W., Hill-Burton Project Register.
A02-0329   Harrell, Planning Medical Center Facilities for Education, Research and Public Service.
A02-0365   Nierstrasz, Building for the Aged.
A02-0407   Wheeler, Hospital Modernization and Expansion.
A04-0719   Bayes and Franklin, Designing for the Handicapped.
A04-0741   Bouwcentrum, General Hospitals: Functional Studies on the Main Departments.
A04-0855   Dixon, Architectural Design Preview, U.S.A.
A04-0948   Goldsmith, Designing for the Disabled.
A04-0978   Hamlin, Forms and Functions of Twentieth-Century Architecture.
A04-1016   Horrobin, Housing the Elderly.
A04-1022   Hudenberg, Planning the Community Hospital.
A04-1178   Nuffield Foundation, Children in Hospital: Studies in

Index 601

	Planning.
A04-1258	Rosenfield, Hospital Architecture and Beyond.
A04-1259	Rosenfield, Hospital Architecture: Integrated Components.
A04-1344	Thomson and Goldin, The Hospital: A Social and Architectural History.
A04-1371	Wheeler, Hospital Design and Function.
A05-1585	Corss and Noble, Handbook on Hospital Solid Waste Management.
A06-1969	ASHRAE, 1974 Applications Volume.
A10-2874	Havighurst, Regulating Health Facilities Construction.
A10-2894	Lave and Lave, The Hospital Construction Act: An Evaluation of the Hill-Burton Program.

HOUSE CONSTRUCTION

A02-0271	Beasley, Fell's Guide to Buying, Building and Financing a Home.
A02-0272	Benson, Building Contractor's and Home Builder's Handbook of Bidding, Surveying, and Estimating.
A02-0318	Goodkin, When Real Estate and Home Building Become Big Business.
A02-0409	Wicks, How to Plan, Buy or Build Your Leisure Home.
A04-0769	Building Construction and Design.
A04-1235	Rapoport, House Form and Culture.
A09-2590	Anderson, How to Build a Wood-Frame House.
A09-2591	Anderson, Wood Frame House Construction.
A09-2592	Anderson, Wood Frame House Construction.
A09-2601	Badzinski, Carpentry in Residential Construction.
A09-2602	Badzinski, Light Frame House Construction.
A09-2604	Barry, Construction of Buildings.
A09-2609	Boudreau, Making the Adobe Brick.
A09-2617	The Building Research Establishment, Essential Data for Construction.
A09-2628	Carpenters and Builders Library. Vol. 2 Builders Math, Plans, Specifications.
A09-2629	Carpenters and Builders Library. Vol. 3 Layouts, Foundations, Framing.
A09-2631	Carpentry and Building.
A09-2640	Collymore, Altering, Repairing and Extending Houses.
A09-2654	Dietz, Dwelling House Construction.
A09-2802	Siegele, Roof Framing.
A09-2812	Terry-Smith, Building Illustrated.
A09-2817	Trill and Bowyer, Problems in Building Construction: A Scientific Method Approach.
A09-2823	Walton, How to Build your Cabin or Modern Vacation Home.
A09-2824	Wass, Methods and Materials of Residential Construction.
A09-2825	Wass and Sanders, Building Construction: Roof Framing.
A09-2830	Wilson, Practical House Carpentry.
A09-2831	Wilson and Werner, Simplified Roof Framing.

HOUSE DESIGN

See Architectural Details

A02-0409	Wicks, How to Plan, Buy or Build Your Leisure Home.
A04-0670	Architects' Emergency Committee, Great Georgian Houses of America.
A04-0688	Architectural Record, The Architectural Record Book of Vacation Houses.
A04-0690	Architectural Record, Houses Architects Design for Themselves.
A04-0694	Architectural Record, Record Houses of 1970.
A04-0695	Architectural Record, Record Houses of 1973.
A04-0696	Architectural Record, Record Houses of 1976.
A04-0784	Candilis, Planning and Design for Leisure.
A04-0832	Crandall, They Chose to be Different: Unusual California Homes.
A04-0835	Creighton, Contemporary Houses: Evaluated by Their Owners.
A04-0856	Doane, A Book of Cape Cod Houses.
A04-0861	Downing, The Architecture of Country Houses.
A04-0876	Ede, Canadian Architecture 1960/70.
A04-0908	Ford and Creighton, Quality Budget Houses.
A04-0967	Habraken, Supports An Alternative to Mass Housing.
A04-0990	Hayashi, House Design in Today's Japan.
A04-0992	Helick, Merchant Built Houses in Western Pennsylvania.
A04-0993	Helick, The Regent System of Townhouse Variations.
A04-0994	Helick, Varieties of Human Habitation.
A04-1007	Hoffman, Row Houses and Cluster Houses: An International Survey.
A04-1016	Horrobin, Housing the Elderly.
A04-1017	Housing the Family.
A04-1050	Kaspar, Vacation Houses: An International Survey.
A04-1059	Kennedy, The House and the Art of its Design.
A04-1077	Kraemer, One-Family Houses in Groups/Einfamilienhauser in der gruppe: A Collection of Examples.
A04-1090	Lang, 101 Select Dream Houses.
A04-1105	Link, Residential Designs: How to Get the Most for Your Housing Dollar.
A04-1129	Mason and Koch, Cabins, Cottages and Summer Homes.
A04-1132	Merrilees and Loveday, Pole Building Construction.
A04-1142	Milstein and Walker, Designing Houses: An Illustrated Guide.
A04-1171	Neutra, Building with Nature.
A04-1227	Proksch, Houses in the Alps/Maisons Dans les Alpes.
A04-1233	Ramsey and Harvey, Small Georgian Houses and Their Details: 1750-1820.
A04-1253	Roberts, Your Engineered House.
A04-1293	Shipway and Shipway, Decorative Design in Mexican Homes.
A04-1296	Shipway and Shipway, The Mexican House: Old and New.
A04-1303	Sleeper, Building Planning and Design Standards for Architects, Engineers, Designers, Consultants, Building Committees, Draftsmen, and Students.

Index

A04-1331	Sunset Eds., Cabins and Vacation Homes.
A04-1332	Sunset Eds., Planning and Landscaping Hillside Homes.
A04-1333	Sunset Eds., Sunset Ideas for Planning Your New Home.
A04-1350	University Press of Africa, Homes for Kenya.
A04-1354	Vaux, Villas and Cottages.
A04-1362	Wagner, Great Houses for ... View Sites, Beach Sites, Sites in the Woods, Meadow Sites, Small Sites, Sloping Sites, Steep Sites, Flat Sites.
A04-1369	Wesdert, Private Homes: An International Survey.
A04-1373	Whiting, New-Houses.
A04-1374	Whiting, New Single-Storey Houses.
A04-1378	Wills, Better Houses for Budgeteers.
A04-1379	Wills, Houses for Good Living.
A04-1380	Wills, More Houses for Good Living.
A04-1389	Wolgensinger, Vacation Homes in the Sun/Ferienhaeuser in der Sonne.
A04-1390	Wolgensinger, Vacation Houses of Europe.
A04-1400	Wright, The Natural House.
A04-1404	Wurman, Various Dwellings Described in a Comparative Manner.
A04-1407	Yamamoto, Japan's Contemporary Houses.
A09-2634	Cassiday, Vacation Houses.
A10-2890	Kelley, Design and the Production of Houses.
A11-3073	Newman, Crime Prevention Through Urban Design: Designing a House to Prevent Crime.
A11-3074	Newman, Defensible Space: Crime Prevention Through Urban Design.

HOUSING

A01-0008	Babcock and Bosselman, Exclusionary Zoning: Land Use Regulation and Housing in the 1970s.
A01-0009	Bagby, Housing Rehabilitation Costs.
A01-0015	Becker, New Housing in Finland.
A01-0016	Bell, Urban Environments and Human Behavior.
A01-0021	Berry, Race and Housing: The Chicago Experience 1960-1975.
A01-0022	Berry, Housing: The Great British Failure.
A01-0032	Burchell and Hughes, Planned Unit Development: New Communities American Style.
A01-0034	Caminos, Urban Dwelling Environments: An Elementary Survey of Settlements for the Study of Design Determinants.
A01-0038	Case, Inner-city Housing and Private Enterprise.
A01-0042	Chapman, The History of Working-class Housing: A Symposium.
A01-0046	Chung, The Economics of Residential Rehabilitation: Social Life of Housing in Harlem.
A01-0047	Claire, Handbook on Urban Planning.
A01-0050	Clawson, Modernizing Urban Land Policy.
A01-0051	Clawson and Hall, Planning and Urban Growth: An Anglo-American Comparison.
A01-0053	Cooper and Guntermann, Real Estate and Urban Land

A01-0054 Analysis.
A01-0054 Crawford, A Decade of British Housing 1963-73.
A01-0061 DeChiara and Koppelman, Planning Design Criteria.
A01-0065 Dietz, Housing in Latin America.
A01-0066 Downs, Federal Housing Subsidies: How are They Working?
A01-0069 Eberhard, The Performance Concept: A Study of Its Application to Housing. Vol. 2
A01-0073 Everett and Johnston, Housing Series: Law and Contemporary Problems.
A01-0074 Feagin, Subsidizing the Poor: An Evaluation of the Boston Housing Authority Program.
A01-0088 Goetze, Building Neighborhood Confidence: A Humanistic Stragegy for Urban Housing.
A01-0091 Goldstein, Residential Mobility, Migration, and Metropolitan Change.
A01-0094 Groberg, Urban Renewal Programs Assisted by Title 1 of the Housing Act of 1949.
A01-0096 Gruen and Gruen, Low and Moderate Income Housing in the Suburbs: An Analysis for the Dayton, Ohio Region.
A01-0097 Haar and Iatridis, Housing the Poor in Suburbia: Public Policy at the Grass Roots.
A01-0102 Heilbrun, Real Estate Taxes and Urban Housing.
A01-0108 Housing and Urban Development Dept., Housing in the Seventies.
A01-0109 Housing and Urban Development Dept., 1972 HUD Statistical Yearbook.
A01-0111 Housing and Urban Development Dept., A Study of the Effects of Real Estate Property Tax Incentive Programs Upon Property Rehabilitation and New Construction.
A01-0113 Hughes, New Dimensions of Urban Planning: Growth Controls.
A01-0115 Ingram, The Detroit Prototype of the NBER Urban Simulation Model.
A01-0124 Lansing, New Homes and Poor People: A Study of Chains of Moves.
A01-0132 Liblit, Housing--The Cooperative Way.
A01-0134 Listokin, The Dynamics of Housing Rehabilitation: Macro and Micro Analyses.
A01-0145 Meyerson, Housing, People and Cities.
A01-0147 Mills, Urban Economics.
A01-0149 Morris, State Housing Finance Agencies: An Entrepreneurial Approach to Subsidized Housing.
A01-0152 Muth, Cities and Housing.
A01-0154 Nourse, The Effects of Public Policy on Housing Markets.
A01-0157 Page and Seyfried, Urban Analysis.
A01-0163 Peterson, Property Taxes, Housing, and the Cities.
A01-0166 Prichard, Housing and the Spatial Structure of the City: Residential Mobility and the Housing Market in an English City Since the Industrial Revolution.
A01-0168 Pynoos, Housing Urban America.
A01-0174 Richardson, Housing and Urban Spatial Structure.

Index

A01-0175	Ricks, National Housing Models.
A01-0176	Rodwin, Housing and Economic Progress: A Study of the Housing Experiences of Boston's Middle-Income Families.
A01-0182	Rubinowitz, Low-Income Housing: Suburban Strategies.
A01-0184	Sagalyn and Sternlieb, Zoning and Housing Costs: The Impact of Land-Use Controls on Housing Price.
A01-0186	Schafer, The Suburbanization of Multifamily Housing.
A01-0196	Solomon, Housing the Urban Poor: A Critical Analysis of Federal Housing Policy.
A01-0200	Stegman, Housing and Economics.
A01-0201	Stegman, Housing Investment in the Inner City: The Dynamics of Decline.
A01-0205	Sternlieb and Burchell, Residential Abandonment: The Tenement Landlord Revisited.
A01-0206	Sternlieb, The Affluent Suburb: Housing Needs and Attitudes.
A01-0207	Sternlieb, Housing Development and Municipal Costs: A Methodological Investigation.
A01-0209	Stone, Urban Development in Britain. Vol. 1 Population Trends and Housing.
A01-0211	Struyk and Marshall, Urban Home Ownership.
A01-0213	Taggart, Low-Income Housing: A Critique of Federal Aid.
A01-0228	Wexler and Peck, Housing and Local Government.
A01-0230	Whitehead, The U.K. Housing Market: An Econometric Model.
A02-0305	Edson and Lane, A Practical Guide to Low- and Moderate-Income Housing.
A02-0338	Huntoon, PUD: A Better Way for the Suburbs.
A02-0355	McKeever, Community Builders Handbook.
A02-0365	Nierstrasz, Building for the Aged.
A02-0392	Smith, Real Estate and Urban Development.
A02-0400	Urban Land Institute, PUD--A Better Way for the Suburbs.
A02-0403	Weiss, Better Buildings for the Aged.
A03-0422	Abrams and Blackman, Managing Low and Moderate Income Housing.
A03-0552	Lawton, Planning and Managing Housing for the Elderly.
A04-0740	Boudon, Lived-in Architecture Le Corbusier's Pessac Revisited.
A04-0841	Davern, Lewis Mumford: Architecture as a Home for Man.
A04-0967	Habraken, Supports, An Alternative to Mass Housing.
A04-1016	Horrobin, Housing the Elderly.
A04-1017	Housing the Family.
A04-1025	Hussain, Living Underwater.
A04-1115	Lynch, Site Planning.
A04-1154	Mumford, From the Ground Up: Observations on Contemporary Architecture, Housing, Highway Building, and Civic Design.
A04-1203	Pawley, Architecture Versus Housing.
A04-1276	Schmitt, Multistory Housing.

A04-1304	Smith, The American Endless Weekend.
A04-1392	Woods, Candilis-Josic-Woods: Building for People.
A10-2856	Bureau of the Census, New One-Family Homes Sold and For Sale: 1963-1967.
A10-2860	Davidson, Housing Demand: Mobile, Modular or Conventional?
A10-2870	Gramlich and Jaffee, Savings Deposits, Mortgages and Housing.
A10-2871	Grebler, Large Scale Housing and Real Estate Firms: Analysis of a New Business Enterprise.
A10-2873	Guide to Architectural Information.
A10-2875	Housing and Urban Development, Dept. of, Bibliography on Housing Building and Planning.
A10-2876	Housing and Urban Development, Dept. of, Operation Breakthrough. Volume 3
A10-2887	Housing and Urban Development, Dept. of, Services Available to HUD-related Business in International Trade.
A10-2888	Housing Research and Building Technology Activities of the Federal Government.
A10-2890	Kelly, Design and the Production of Houses.
A10-2891	Kelly, The Prefabrication of Houses.
A10-2900	Organization for Social and Technical Innovation, Self-Help Housing in the U.S.A.
A10-2901	Paulus, Housing: A Bibliography, 1960-1972.
A10-2903	Rada, The Cal-Vet Program: A Study of State-Financed Housing in California.
A10-2904	Ragatz, Vacation Homes: An Analysis of the Market for Seasonal-Recreational Housing.
A10-2905	Reid, Housing and Income.
A10-2906	Reidelbach, Housing Compendiums, Vol. 1.
A10-2909	Ricks, National Housing Models.
A10-2913	Sternlieb and Listokin, Housing 1973-74: An Anthology.
A10-2914	Sternlieb and Paulus, Housing, 1971-1972.
A10-2915	Sternlieb and Paulus, Housing, 1971-1972. Vol. 2
A10-2916	Sternlieb and Sagalyn, Housing, 1970-1971. Vol. 1
A10-2918	U.S. Savings and Loan League, 1976 Savings and Loan Fact Book.
A10-2919	van den Broek, Habitation. Vol. 2 Belgium, Denmark, France, United Kingdom, Sweden, Switzerland and Czechoslovakia.
A10-2920	van den Broek, Habitation. Vol. 3 USA, USSR, Finland, Greece, Hungary, Turkey and Yugoslavia.
A10-2922	Whitehead, The U.K. Housing Market: An Econometric Model.
A10-2923	Wolman, Housing and Housing Policy in the U.S. and the U.K.

HYDRAULICS

A05-1451	Abbett, American Civil Engineering Practice.
A05-1531	Bartlett, Pumping Stations for Water and Sewage.
A05-1554	Brater and King, Handbook of Hydraulics.
A05-1596	Davis and Sorenson, Handbook of Applied Hydraulics.

A05-1649	Graf, Hydraulics of Sediment Transport.
A05-1690	Ippen, Estuary and Coastline Hydrodynamics.
A05-1696	Jeppson, Analysis of Flow in Pipe Networks.
A05-1757	Merritt, Standard Handbook for Civil Engineers.
A05-1764	Morris and Wiggert, Applied Hydraulics in Engineering.
A05-1851	Seelye, Data Book for Civil Engineers.
A09-2607	Blake, Civil Engineer's Reference Book.

INDUSTRIAL BUILDINGS

A01-0086	Gibberd, Town Design.
A02-0322	Guthrie, Process Plant Estimating, Evaluation and Control.
A02-0355	McKeever, Community Builders Handbook.
A02-0366	Nutt, Functional Plant Planning, Layout, and Materials Handling.
A02-0384	Richardson, Process Plant Construction Estimating Standards.
A02-0385	Richardson, General Construction Estimating Standards.
A02-0408	Whitman and Schmidt, Plant Relocation.
A04-0669	Apple, Plant Layout and Materials Handling.
A04-0721	Becher and Becher, Anonyme Sculpturen.
A04-0876	Ede, Canadian Architecture 1960/70.
A04-0963	Grube, Industrial Buildings and Factories.
A04-0978	Hamlin, Forms and Functions of Twentieth-Century Architecture. Vol. 1 The Elements of Building. Vol. 2 The Principles of Composition. Vol. 3 Building Types: Residence, Gatherings, Education, Government. Vol. 4 Building Types: Commerce and Industry, Public Health Transportation, Social Welfare and Recreation.
A04-0998	High Rise Storage: Modern Materials Handling Guidebook.
A04-0999	Hildebrand, Designing for Industry: The Architecture of Albert Kahn.
A04-1139	Mills, The Changing Workplace: Modern Technology and The Working Environment.
A04-1156	Munce, Industrial Architecture.
A04-1161	Muther and Haganas, Systematic Handling Analysis.
A04-1209	Peters, Centers for Storage and Distribution.
A04-1210	Peters, Factories.
A04-1250	Richards, Functional Tradition in Early Industrial Buildings.
A04-1376	Wilkinson, Storage Specifics.
A06-2004	Central Electricity Generating Board, Modern Power Station Practice.
A09-2723	Koncz, Manual of Precast Concrete Construction. Vol. 2.
A09-2724	Koncz, Manual of Precast Concrete Construction. Vol. 3.

INDUSTRIALIZED, PREFABRICATED, PRE-ENGINEERED BUILDINGS

A01-0199	Spyer, Architect and Community.
A02-0361	MHMA, Mobile Home Site Planning Kit: Basic Information Concerning Construction Park Development.

A02-0381	Rabb and Rabb, Good Shelter: A Guide to Mobiles, Modulars, and Prefabricated Homes.
A04-1020	Huber and Steinegger, Jean Prouve: Prefabrication, Structures and Elements.
A04-1028	INBEX '70 Digest of Seminars--Industrialized Building Exposition and Congress.
A04-1029	INBEX '71 Digest of Seminars--Industrialized Building Exposition and Congress.
A04-1174	Nissen, Industrialized Building and Modular Design.
A05-1708	Komendant, Contemporary Concrete Structures.
A07-2222	Baumann and Wilson, Urban Engineering and Transportation.
A08-2489	Pindar, Engineering Wall and Partition Components for Prefabrication.
A08-2536	Svec and Jeffers, Modern Masonry Panel Construction Systems.
A09-2621	Building with Tilt-Up.
A09-2639	Collins, Manual of Tilt-Up Construction.
A09-2643	Cutler and Cutler, Handbook of Systems Housing for Designers and Developers.
A09-2653	Diamant, Industrialized Building.
A09-2654	Dietz, Dwelling House Construction.
A09-2721	Komendant, Contemporary Concrete Structures.
A09-2731	Lewicki, Building with Large Prefabricates.
A09-2732	Libby, Modern Prestressed Concrete: Design Principles and Construction Methods.
A09-2737	Lytle, Industrialized Builders Handbook.
A09-2754	Modular Building Standards Association, Modular Practice.
A09-2813	Testa, The Industrialization of Building.
A09-2821	Wachsmann, The Turning Point of Building.
A10-2852	Bender, Selected Technological Aspects of the American Building Industry. The Industrialization of Building.
A10-2854	Building Blocks: Design Potentials and Constraints.
A10-2858	Carreiro, The New Building Block: A Report on the Factory-Produced Dwelling Module.
A10-2860	Davidson, Housing Demand: Mobile, Modular or Conventional?
A10-2872	Grubb and Phares, Industrialization: A New Concept for Housing.
A10-2889	INBEX '71 Digest of Seminars--Industrialized Building Exposition and Congress.
A10-2890	Kelly, Design and the Production of Houses.
A10-2891	Kelly, The Prefabrication of Houses.
A10-2892	Kelly and Shillaber, International Bibliography of Prefabricated Housing.
A10-2896	National Research Council, Promotion of the Development and Use of the Subsystem Concept of Building Construction.
A10-2907	Reidelbach, Modular Housing 1971--Facts and Concepts.
A10-2908	Reidelbach, Modular Housing '72, Statistics & Specifics.
A10-2911	Schmid and Testa, Systems Building: An International Survey of Methods.

Index 609

INFORMATION SOURCES, DIRECTORIES

A01-0023	Bestor, City Planning Bibliography.
A01-0025	Branch, Comprehensive Urban Planning: A Selective Annotated Bibliography with Related Materials.
A01-0052	Clawson and Stewart, Land Use Information: A Critical Survey of U.S. Statistics, Including Possibilities for Greater Uniformity.
A01-0087	Godwin, Directory of Official Architecture & Planning.
A01-0089	Golany, New Towns Planning and Development: A Worldwide Bibliography.
A01-0109	Housing and Urban Development Dept., 1972 HUD Statistical Yearbook.
A01-0160	Passonneau and Wurman, Urban Atlas: 20 American Cities.
A01-0220	Urban Land Institute, New Towns Planning and Development: A Bibliography.
A01-0229	White, Sourcebook of Planning Information.
A01-0231	Whittick, Encyclopedia of Urban Planning.
A02-0268	Auditorium/Arena/Stadium Guide and International Directory.
A02-0275	Building Design and Construction Specifying Buying Guide and Directory.
A02-0276	Building Supply News Purchasing File Issue.
A02-0281	Construction Equipment Buyers Guide.
A02-0294	Dept. of Health, Education and Welfare, Hill-Burton Project Register.
A02-0296	Directory of Shopping Centers in the U.S. and Canada.
A03-0479	CSI, Uniform Construction Index.
A03-0502	Fletcher and Moore, Standard Phraseology for Bills of Quantities.
A03-0513	Gothie, A Selected Bibliography on Applied Ethics in the Professions, 1950-1970.
A04-0654	AIA Membership Directory.
A04-0742	Bowker, American Architects Directory.
A04-0792	Catalogue of the Royal Institute of British Architects Library.
A04-0812	Colvin, A Biographical Dictionary of English Architects, 1660-1840.
A04-0836	Creswell, A Bibliography of the Architecture, Arts, and Crafts of Islam to 1st January, 1960.
A04-1300	Silverman and Bowman, Theatre Architecture: An Illustrated Survey and a Checklist of Publications, 1946-1964.
A05-1489	ASCE, Bibliography on Bolted and Riveted Joints.
A05-1678	Hollis, Bibliography of Earthquake Engineering.
A06-2069	International Institute of Refrigeration (Paris), Bibliographic Guide to Refrigeration 1960-64.
A06-2130	Pesco, Solar Directory.
A07-2209	APWA Directory.
A07-2210	ARBA Directory of Transportation and Agency Personnel.
A07-2217	ASTM, An Analysis of the Literature on Tire-Road Skid Resistance.
A07-2239	Li, Bibliography on Airport Engineering.

A08-2346	Becker, U.S. Sandwich Panel Manufacturing/Marketing Guide.
A08-2368	Clauser, Encyclopedia of Engineering Materials and Processes.
A08-2376	Corrosion Abstracts Yearbooks.
A08-2461	NACE, 1960 Bibliographic Survey of Corrosion.
A08-2462	National Bureau of Standards, A Compilation and Evaluation of Mechanical, Thermal and Electrical Properties of Selected Polymers.
A08-2490	Plant Engineering Directory and Specifications Catalog.
A08-2507	Ross, Metallic Materials Specification Handbook.
A10-2855	Building Science Directory.
A10-2863	Electrical World Directory of Electric Utilities, 1975-1976.
A10-2868	Godel, Guide to Information Sources in the Construction Industry.
A10-2873	Guide to Architectural Information.
A10-2875	Housing and Urban Development, Dept. of, Bibliography on Housing and Planning.
A10-2888	Housing Research and Building Technology Activities of the Federal Government.
A10-2892	Kelly and Shillaber, International Bibliography of Prefabricated Housing.
A10-2901	Paulus, Housing: A Bibliography, 1960-1972.
A10-2906	Reidelbach, Housing Compendiums, Vol. 1.
A10-2912	Smith, How to Find Out in Architecture and Building-- A Guide to Sources of Information.
A10-2913	Sternlieb and Listokin, Housing 1973-74: An Anthology.
A10-2921	Ward, Construction Information Source and Reference Guide.
A11-2952	Best's Safety Directory.
A11-3012	Hall, Encyclopedia of Food Engineering.

INSPECTION, TESTING AND QUALITY CONTROL

A03-0423	ACI Manual of Concrete Inspection.
A03-0443	ASTM, Manual on Quality Control of Materials.
A03-0455	Betz, Principles of Magnetic Particle Testing.
A03-0456	Betz, Principles of Penetrants.
A03-0481	Davis, The Testing and Inspection of Engineering Materials.
A03-0489	Douglas and Munger, Construction Management.
A03-0503	Fordham, Non-destructive Testing Techniques.
A03-0536	ICBO, Plan Review Manual.
A03-0537	ICBO, A Training Manual in Field Inspection of Buildings and Structures.
A03-0544	Juran, Quality Control Handbook.
A03-0560	McKaig, Field Inspection of Building Construction.
A03-0562	McMaster, Nondestructive Testing Handbook.
A03-0586	O'Brien, Construction Inspection Handbook.
A03-0618	Seelye, Data Book for Civil Engineers. Vol. 3 Field Practice.
A05-1454	ACI, Computer Applications in Concrete Design and

A06-1976	Technology. ASSE, The Principles of Residential Plumbing Inspection.
A06-1993	Bond, NFPA Inspection Manual - Third Edition.
A06-2146	Rossi, Welding Engineering.
A08-2399	Gage and Newman, Specification and Use of Ready Mixed Concrete.
A08-2441	McIntosh, Concrete and Statistics.
A08-2477	Orchard, Concrete Technology. Vol. 1 Properties of Materials.
A08-2478	Orchard, Concrete Technology. Vol. 2 Practice.
A08-2550	Waddell, Concrete Construction Handbook.
A08-2551	Waddell, Practical Quality Control for Concrete.
A09-2580	ACI, Concrete Construction.
A09-2582	ACI Manual of Concrete Inspection.
A09-2595	ASTM, Sampling of Soil and Rock.
A09-2596	ASTM, Special Procedures for Testing Soil and Rock for Engineering Purposes.
A09-2597	ASTM, Underwater Soil Sampling, Testing and Construction Control.
A09-2699	Hanna, Foundation Instrumentation.
A09-2775	Portland Cement Association, Concrete Inspection Procedures.
A09-2776	Portland Cement Association, Laboratory and Exercise Manual on Concrete Construction.
A09-2797	Schmidt, Construction: Principles, Materials and Methods.
A09-2804	Smith, Principles and Practices of Heavy Construction.
A09-2822	Waddell, Concrete Construction Handbook.
A11-2958	Bond, NFPA Inspection Manual.
A11-2986	Desch, Structural Surveying.
A11-3019	Hilado, Flammability Test Method Handbook.

INSULATION, MOISTURE- AND WATERPROOFING

A06-1971	ASHRAE, 1972 Handbook of Fundamentals.
A06-2044	Griffin, Energy Conservation in Buildings: Techniques for Economical Design.
A06-2141	Ramsey, Tested Solutions to: Design Problems in Air Conditioning and Refrigeration.
A06-2172	Van Straaten, Thermal Performance of Buildings.
A08-2298	Addleson, Materials for Building. Vol. 2 Water and Its Effects-1.
A08-2300	Addleson, Materials for Building. Vol. 4.
A08-2328	Asphalt Institute, Asphalt in Hydraulic Structures.
A08-2331	ASTM, Heat Transmission Measurements in Thermal Insulations.
A08-2369	Close, Sound Control and Thermal Insulation of Buildings.
A08-2374	Cook, Construction Sealants and Adhesives.
A08-2378	Croome and Sherratt, Condensation in Buildings.
A08-2380	Damusis, Sealants.

A08-2385	Diamant, Insulation of Buildings: Thermal and Acoustic.
A08-2395	Frisch and Saunders, Plastic Foams.
A08-2410	Gratwick, Dampness in Buildings, 2nd ed.
A08-2444	Malloy, Thermal Insulation: Integrated Design Manual.
A08-2500	Probert, Thermal Insulation.
A08-2515	Seiffert, Damp Diffusion and Buildings.
A08-2523	Simpson and Horrobin, The Weathering and Performance of Building Materials.
A08-2528	Smith, Materials of Construction.
A08-2553	Watson, Construction Materials and Processes.
A09-2704	Huntington and Mickadeit, Building Construction: Materials and Types of Construction.

INTERIOR DESIGN

A04-0790	Casson, Inscape: The Design of Interiors.
A04-0894	Faulkner, Architecture and Color.
A04-0909	Forman, Maryland Architecture: A Short History from 1634 through the Civil War.
A04-0910	Forman, Old Buildings, Gardens and Furniture in Tidewater Maryland.
A04-0918	Friedmann, Interior Design: An Introduction to Architectural Interiors.
A04-0948	Goldsmith, Designing for the Disabled.
A04-0958	Greiff, Great Houses from the Pages of Antiques.
A04-0975	Halse, The Use of Color in Interiors.
A04-1117	Lynes, Principles of Natural Lighting.
A04-1169	Nelson, Problems of Design.
A04-1257	Rolleston, Historic Houses and Interiors in Southern Connecticut.
A04-1273	Saphier, Office Planning and Design.
A04-1295	Shipway and Shipway, Mexican Homes of Today.
A04-1310	Smith and Owens, The Majesty of Natchez.
A04-1327	Studio Dictionary of Designs and Decoration.
A04-1383	Wilson and Leaman, Color in Decoration.
A11-3084	Pile, Drawings of Architectural Interiors.

JOINTS - FASTENERS, WELDED AND BONDED

A05-1489	ASCE, Bibliography on Bolted and Riveted Joints.
A05-1570	Cernica, Strength of Materials.
A05-1632	Fisher and Struik, Guide to Design Criteria for Bolted and Riveted Joints.
A05-1638	Gaylord and Gaylord, Design of Steel Structures.
A05-1655	Gurfinkel, Wood Engineering.
A05-1656	Gurney, Fatigue of Welded Structures.
A05-1721	Levinson, Mechanics of Materials.
A05-1732	Lothers, Advanced Design in Structural Steel.
A05-1733	Lothers, Design in Structural Steel.
A05-1738	McCormac, Structural Steel Design.
A05-1740	McGuire, Steel Structures.
A05-1758	Merritt, Structural Steel Designers' Handbook.
A05-1766	Munse and Grover, Fatigue of Welded Steel Structures.

Index

A05-1767	Murashev, Design of Reinforced Concrete Structures.
A05-1786	Olsen, Strength of Materials.
A05-1795	Parker, Simplified Engineering for Architects and Builders.
A05-1797	Parker and Hauf, Simplified Design of Structural Steel.
A05-1838	Salmon and Johnson, Steel Structures: Design and Behavior.
A05-1852	Semerdjiev, Metal-to-Metal Adhesive Bonding.
A05-1929	Williams and Harris, Structural Design in Metals.
A05-1934	Yu, Cold-Formed Steel Structures.
A08-2350	Bodnar, Structural Adhesives Bonding.
A08-2364	Cagle, Handbook of Adhesive Bonding.
A08-2383	Delollis, Adhesives for Metals: Theory and Technology.
A08-2519	Shields, Adhesives Handbook.
A09-2588	American Institute of Timber Construction, Timber Construction Manual.
A09-2696	Halperin, Building with Steel.

LABORATORIES AND CLEAN ROOMS

A06-1964	Alden and Kane, Design of Industrial Exhaust Systems.
A06-1969	ASHRAE, 1974 Applications Volume.
A06-2051	Hemeon, Plant and Process Ventilation.
A11-2945	Austin, Design and Operation of Clean Rooms.
A11-3027	International Atomic Energy Agency, Manual of Safety Aspects of the Design and Equipment of Hot Laboratories.
A11-3053	Lewis, Laboratory Planning for Chemistry and Chemical Engineering.
A11-3101	Steere, CRC Handbook of Laboratory Safety.

LAND ACQUISITION AND USE

A01-0003	Altshuler, The City Planning Process: A Political Analysis.
A01-0011	Bannon, Leisure Resources: Its Comprehensive Planning.
A01-0028	Breese, The Impact of Large Installations on Near-by Areas: Accelerated Urban Growth.
A01-0032	Burchell and Hughes, Planned Unit Development: New Communities American Style.
A01-0041	Chaplin, Urban Land Use Planning.
A01-0049	Clawson, America's Land and Its Uses.
A01-0050	Clawson, Modernizing Urban Land Policy.
A01-0051	Clawson and Hall, Planning and Urban Growth: An Anglo-American Comparison.
A01-0052	Clawson and Stewart, Land Use Information: A Critical Survey of U.S. Statistics, Including Possibilities for Greater Uniformity.
A01-0061	DeChiara and Koppelman, Planning Design Criteria.
A01-0068	Ducsik, Shoreline for the Public. A Handbook of Social, Economic, and Legal Considerations Regarding Public Recreational Use of the Nation's Coastal Shoreline.
A01-0077	Flawn, Environmental Geology Conservation, Land-Use

	Planning, and Resource Management.
A01-0107	Housing and Urban Development Dept., Guidelines for Urban Renewal Land Disposition.
A01-0122	Keeble, Principles and Practice of Town and Country Planning.
A01-0129	Levin, New Approaches to State Land Use Policies.
A01-0133	Linowes and Allensworth, The Politics of Land Use: Planning, Zoning, and the Private Developer.
A01-0173	Richardson, The Cost of Environmental Protection: Regulating Housing Development in the Coastal Zone.
A01-0174	Richardson, Housing and Urban Spatial Structure.
A01-0180	Rose, The Transfer of Development Rights: A New Technique of Land Use Regulation.
A01-0187	Schmid, Converting Land From Rural to Urban Uses.
A01-0191	Shomon, Open Land for Urban America: Acquisition, Safekeeping, and Use.
A01-0192	Siegan, Land Use and Real Estate.
A01-0193	Siegan, Land Use Without Zoning.
A01-0208	Stevens, Impact of Federal Legislation and Programs on Private Land in Urban and Metropolitan Development.
A01-0234	Wingo, Cities and Space: The Future Use of Urban Land.
A01-0239	Yearwood, Land Subdivision Regulation: Policy and Legal Considerations for Urban Planning.
A02-0355	McKeever, Community Builders Handbook.
A03-0575	NAHB, Land Development Law for the Builder and His Attorney.
A04-1116	Lynch, Site Planning.

LANDSCAPING AND PARKS

A01-0011	Bannon, Leisure Resources: Its Comprehensive Planning.
A01-0016	Bell, Urban Environments and Human Behavior.
A01-0030	Brown and Sherrard, An Introduction of Town and Country Planning.
A01-0061	DeChiara and Koppelman, Planning Design Criteria.
A01-0083	Geddes, City Development: A Study of Parks, Gardens, and Culture-Institutes.
A04-0703	Ashihara, Exterior Design in Architecture.
A04-0779	Butler, Recreation Areas, Their Design and Equipment.
A04-0801	Choay, The Modern City: Planning in the 19th Century.
A04-0811	Colvin, Land and Landscape: Evolution, Design and Control.
A04-0910	Frigand, The New City: Architecture and Urban Renewal.
A04-0949	Goodman and Von Eckardt, Life for Dead Spaces: The Development of the Lacanburg Commons.
A04-0974	Halprin, The RSVP Cycles: Creative Processes in the Human Environment.
A04-1103	Lindley, Appreciation of Architecture--Landscape and Buildings.
A04-1139	Mills, Planning: Volume 1: Architects' Technical

A04-1182	Reference. Olmsted and Kimball, Frederick Law Olmsted, Landscape Architect: 1822-1903.
A04-1265	Rutledge, Anatomy of a Park: The Essentials of Recreation Area Planning and Design.
A04-1332	Sunset Eds., Planning and Landscaping Hillside Homes.
A04-1409	Yoshida, The Japanese House and Garden.
A07-2269	Woods, Highway Engineering Handbook.
A09-2787	Reiner, Methods and Materials of Construction: A Guide for Builders, Owners, Architects and Engineers.
A11-2973	Conover, Grounds Maintenance Handbook.
A11-3069	Munson, Construction Design for Landscape Architects.
A11-3097	Simonds, Landscape Architecture.

LIGHTING

A02-0363	Minnesota Electric Assn., Estimating-Accounting Manual for Electrical Contractor Dealers.
A04-0894	Faulkner, Architecture and Color.
A04-0907	Flynn and Segil, Architectural Interior Systems: Lighting, Air-Conditioning, Acoustics.
A04-1068	Kinzey and Sharp, Environmental Technologies in Architecture.
A04-1091	Lang, Designing for Human Behavior: Architecture and the Behavioral Sciences.
A04-1117	Lynes, Principles of Natural Lighting.
A04-1365	Walsh, The Science of Daylight.
A06-1965	Allphin, Primer of Lamps and Lighting.
A06-1981	Banham, Architecture of the Well-Tempered Environment.
A06-1986	Bean and Simons, Lighting Fittings-Performance and Design.
A06-1995	Boud, Lighting Design in Buildings.
A06-2003	Caudill, A Bucket of Oil: The Humanistic Approach to Building Design for Energy Conservation.
A06-2012	Croft, American Electricians' Handbook.
A06-2029	Federal Energy Administration, Guidelines for Saving Energy in Existing Buildings.
A06-2034	Flynn and Mills, Architectural Lighting Graphics.
A06-2035	Flynn and Segil, Architectural Interior Systems: Lighting, Air Conditioning, Acoustics.
A06-2044	Griffin, Energy Conservation in Buildings: Techniques for Economical Design.
A06-2057	Hopkinson and Kay, The Lighting of Buildings.
A06-2067	IES Lighting Handbook.
A06-2075	Kalff, The Creative Light.
A06-2092	Lefax Pub. Co., Illumination.
A06-2098	Lyons, Management Guide to Modern Industrial Lighting.
A06-2099	McGuinness and Stein, Mechanical and Electrical Equipment for Buildings.
A06-2108	Merritt, Mechanical and Electrical Design of Buildings for Architects and Engineers.
A06-2127	Pansini, Basic Electrical Power Distribution. Vol. 2.

A06-2134   Phillips, Lighting in Architectural Design.
A06-2135   Plant Engineering Electrical Library.
A06-2170   Traister, Electrical Design for Building Construction.
A06-2180   Westinghouse Lighting Handbook.
A08-2553   Watson, Construction Materials and Processes.
A09-2619   The Building Research Establishment, Essential Data for Construction. Vol. 3 Services and Environmental Engineering.
A09-2622   Burgess, The Construction Industry Handbook.

MECHANICAL, CHEMICAL, POWER ENGINEERING

A06-1975   ASME, ASME Handbook - Engineering Tables.
A06-1985   Baumelster and Marks, Standard Handbook for Mechanical Engineers.
A06-2001   Carmichael, Kent's Mechanical Engineers' Handbook.
A06-2005   Central Electricity Generating Board, Modern Power Station Practice. Vol. 2 Mechanical (Boilers, Fuel- and Ash-Handling Plant).
A06-2016   Den Hartog, Mechanics.
A06-2019   Draffin and Collins, Statics and Strength of Materials.
A06-2025   Eshbach and Souders, Handbook of Engineering Fundamentals.
A06-2026   Evans, Equipment Design Handbook for Refineries and Chemical Plants. Vol. 1.
A06-2027   Evans, Equipment Design Handbook for Refineries and Chemical Plants. Vol. 2.
A06-2033   Flugge, Handbook of Engineering Mechanics.
A06-2037   General Electric, Heat Transfer & Fluid Flow Data Books. Vol. 1 Heat Transfer. Vol. 2 Fluid Flow.
A06-2040   Goldstern, Steam Storage Installations.
A06-2053   Hicks, Standard Handbook of Engineering Calculations.
A06-2055   Hill, The Science of Engineering Design.
A06-2068   Inglis, Applied Mechanics for Engineers.
A06-2074   Jones and Schubert, Engineering Encyclopedia.
A06-2084   Kolousek, Dynamics in Engineering Structures.
A06-2085   Kuljian, Nuclear Power Plant Design.
A06-2093   Levinson, Introduction to Mechanics.
A06-2097   Loftness, Nuclear Power Plants: Design, Operating Experience, and Economics.
A06-2109   Miliaras, Power Plants with Air-Cooled Condensing Systems.
A06-2113   Morrison, Engineering Design: The Choice of Favorable Systems.
A06-2114   Morse, Power Plant Engineering.
A06-2115   Nadeau, Introduction to Elasticity.
A06-2120   Neville and Kennedy, Basic Statistical Methods for Engineers and Scientists.
A06-2128   Parrish, Mechanical Engineer's Reference Book.
A06-2131   Peters and Timmerhaus, Plant Design and Economics for Chemical Engineers.
A06-2147   Rothbart, Mechanical Design and Systems Handbook.
A06-2149   Schenck, Introduction to Ocean Engineering.

Index 617

A06-2152	Sculthorpe, Design of High Pressure Steam and High Temperature Water Plants.
A06-2159	Shields, Boilers.
A06-2160	Simmons, Wind Power 1975.
A06-2173	Vilbrandt and Dryden, Chemical Engineering Plant Design.
A10-2863	Electrical World Directory of Electric Utilities, 1975-1976.
A11-2976	Constance, Mechanical Engineering for Professional Engineers' Examinations.
A11-2977	Coren, Mechanical Engineering.
A11-2995	Farrall, Engineering for Dairy and Food Products.
A11-3012	Hall, Encyclopedia of Food Engineering.
A11-3083	Perry, Engineering Manual.
A11-3100	Stamper and Dublin, Mechanical Engineering and Economics and Ethics for Professional Engineering Examinations.

METAL PROPERTIES AND CORROSION

A05-1475	Almen and Black, Residual Stresses and Fatigue in Metals.
A05-1513	ASTM, Achievement of High Fatigue Resistance in Metals and Alloys.
A05-1515	ASTM, Effects of Environment and Complex Load History on Fatigue Life.
A05-1517	ASTM, Stress Analysis and Growth of Cracks.
A05-1747	Manson, Metal Fatigue Damage--Mechanism, Detection, Avoidance, and Repair.
A05-1766	Munse and Grover, Fatigue of Welded Steel Structures.
A05-1788	Osgood, Fatigue Design.
A05-1830	Rolfe and Barsom, Fracture and Fatigue Control in Structures: Applications of Fracture Mechanics.
A05-1834	Ruskin, Materials Considerations in Design.
A05-1858	Smith and Nicholson, Advances in Creep Design.
A06-1994	Bosich, Corrosion Prevention for Practicing Engineers.
A08-2299	Addleson, Materials for Building. Vol. 3 Water and Its Effects-2.
A08-2302	AISC, Iron and Steel Beams 1873 to 1952.
A08-2305	Aluminum Assn., Aluminum Finishing Seminar Papers.
A08-2306	Aluminum Assn., Aluminum Standards and Data.
A08-2318	ASM, Metals Handbook. Vol. 1 Properties and Selection.
A08-2319	ASM, Metals Handbook. Vol. 2 Heat Treating, Cleaning and Finishing.
A08-2320	ASM, Metals Handbook. Vol. 3 Machining.
A08-2321	ASM, Metals Handbook. Vol. 4 Forming.
A08-2322	ASM, Metals Handbook. Vol. 5 Forging and Casting.
A08-2324	ASM, Metals Handbook. Vol. 7 Atlas of Microstructures. Vol. 8 Metallography, Structures and Phase Diagrams. Vol. 9 Fractography and Atlas of Fractographs.
A08-2325	ASM, Metals Handbook. Vol. 10 Failure Analysis and

	Prevention.
A08-2326	ASME Handbook--Metals Engineering: Design.
A08-2327	ASME Handbook--Metals Properties.
A08-2329	ASTM, Corrosion in Natural Environments.
A08-2332	ASTM, Localized Corrosion--Cause of Metal Failure.
A08-2334	ASTM, Materials Performance and the Deep Sea.
A08-2335	ASTM, Metal Corrosion in the Atmosphere.
A08-2338	ASTM, Stress Corrosion Cracking of Metals--A State of the Art.
A08-2343	Barer and Peters, Why Metals Fail.
A08-2352	Brady, Materials Handbook.
A08-2353	Brasunas and Stansbury, Symposium on Corrosion Fundamentals.
A08-2354	Brick, Structure and Property of Alloys.
A08-2361	Burns and Bradley, Protective Coatings for Metals.
A08-2362	Butler and Ison, Corrosion and Its Prevention in Waters.
A08-2368	Clauser, Encyclopedia of Engineering Materials and Processes.
A08-2373	Conway, Fatigue, Tensile, and Relaxation Behavior of Stainless Steels.
A08-2376	Corrosion Abstracts Yearbooks.
A08-2384	Diamant, Chemistry of Building Materials.
A08-2386	Diamant, Prevention of Corrosion.
A08-2390	Evans, Selecting Engineering Materials for Chemical and Process Plant.
A08-2393	Fontana and Green, Corrosion Engineering.
A08-2414	Guy, Essentials of Materials Science.
A08-2416	Hanson and Parr, The Engineer's Guide to Steel.
A08-2425	Keyser, Materials Science in Engineering.
A08-2427	Kissin, Finishing of Aluminum.
A08-2438	Logan, The Stress Corrosion of Metals.
A08-2445	Mantell, Engineering Materials Handbook.
A08-2455	Moffat, Plant Engineer's Handbook of Formulas, Charts and Tables.
A08-2458	NACE, Control of Pipeline Corrosion.
A08-2459	NACE, Fundamental Aspects of Stress Corrosion Cracking.
A08-2461	NACE, 1960 Bibliographic Survey of Corrosion.
A08-2479	Parker, Materials Data Book.
A08-2480	Parker, Pipe Line Corrosion and Cathodic Protection.
A08-2483	Patton, Materials in Industry.
A08-2492	Pollack, Materials Science and Metallurgy.
A08-2502	Rabinowicz, Friction and Wear of Materials.
A08-2507	Ross, Metallic Materials Specification Handbook.
A08-2523	Simpson and Horrobin, The Weathering and Performance of Building Materials.
A08-2528	Smith, Materials of Construction.
A08-2530	Society of Manufacturing Engineers, Surface Preparation and Finishes for Metals.
A08-2544	Uhlig, Corrosion and Corrosion Control: An Introduction to Corrosion Science and Engineering.
A08-2545	Uhlig, Corrosion Handbook.
A08-2548	Van Vlack, A Textbook of Materials Technology.

Index

A08-2554 The Weathering and Performance of Building Materials.
A08-2557 Wilson and Oates, Corrosion and the Maintenance Engineer.
A09-2618 The Building Research Establishment, Essential Data for Construction. Vol. 2 Building Materials.
A09-2801 Shank, Control of Steel Construction to Avoid Brittle Failure.

MODELS, ARCHITECTURAL

A04-0871 Dutton, A Student's Guide to Model Making.
A04-1011 Hohauser, Architectural and Interior Models.
A04-1042 Janke, Architectural Models.
A04-1308 Smith and Hoppe, Building to Scale: A Manual for Model Home Construction.
A04-1340 Taylor, Model Building for Architects and Engineers.
A05-1582 Cowan, Models in Architecture.
A06-2004 Central Electricity Generating Board, Modern Power Station Practice.
A11-3107 Taylor, Model Building for Architects and Engineers.

MOTELS, HOTELS, RESTAURANTS

A04-0650 Abraben, Resort Hotels: Planning and Management.
A04-0656 AJ Editors, Principles of Hotel Design.
A04-0685 Architect's Journal, Hotels: Principles of Hotel Design.
A04-0692 Architectural Record, Motels, Hotels, Restaurants and Bars.
A04-0784 Candilis, Planning and Design for Leisure.
A04-0842 Davern and eds. of Architectural Record, Architectural Record Book of Places for People: Motels, Hotels, Restaurants, and Bars.
A04-0882 End, Interiors Book of Hotels and Motor Hotels.
A04-0884 Ericsson, Sixty Years a Builder: The Autobiography of Henry Ericsson.
A04-0896 Fengler, Restaurant Architecture and Design.
A04-0897 Fengler, Restaurant Architecture and Design: An International Survey of Eating Places.
A04-1080 Kuhne, New Restaurants: An International Survey.
A04-1097 Lawson, Designing Commercial Food Service Facilities.
A04-1098 Lawson, Hotel Planning and Design.
A04-1114 Lundberg, The Hotel and Restaurant Business.
A04-1218 Podd and Lesure, Planning and Operating Motels and Motor Hotels.
A04-1246 Restaurant Planning and Design.
A04-1368 Weisskamp, Hotels: An International Survey.
A04-1376 Wilkinson, Storage Specifics.

NEW TOWNS

A01-0010 Bailey, New Towns in America: The Design and Development Process.
A01-0014 Batty, Urban Modelling.

A01-0018	Beresford, New Towns of the Middle Ages.
A01-0029	Brooks, New Towns and Communal Values: Case Study of Columbia, Maryland.
A01-0035	Campbell, New Towns: Another Way to Live.
A01-0048	Clapp, New Towns & Urban Policy: Planning Metropolitan Growth.
A01-0061	DeChiara and Koppelman, Planning Design Criteria.
A01-0062	DeChiara and Koppelman, Planning Design Criteria.
A01-0072	Evans, New Towns: The British Experience.
A01-0089	Golany, New Towns Planning and Development: A Worldwide Bibliography.
A01-0103	Heroux and Wallace, Financial Analysis and the New Community Development Process.
A01-0112	Howard, Garden Cities of To-Morrow.
A01-0121	Kaplan, Urban Planning in the 1960s: A Design for Irrelevancy.
A01-0136	MacFadyen, Sir Ebenezer Howard and the Town Planning Movement.
A01-0146	Mields, Federally Assisted New Communities: New Dimensions in Urban Development.
A01-0155	Osborn, Green-Belt Cities.
A01-0156	Osborn and Whittick, The New Towns: The Answer to Megalopolis.
A01-0159	Pass, Vällingby and Farsta--From Idea to Reality: The New Community Development Process in Stockholm.
A01-0161	Perloff and Sandberg, New Towns: Why-And for Whom?
A01-0162	Perloff, Modernizing the Central City: New Towns Intown ... and Beyond.
A01-0190	Seeley, Planned Expansion of Country Towns.
A01-0197	Spiegel, New Towns in Israel: Urban and Regional Planning and Development.
A01-0202	Stein, Toward New Towns for America.
A01-0220	Urban Land Institute, New Towns Planning and Development: A Bibliography.
A01-0224	Von Hertzen and Spreiregen, Building A New Town: Finland's New Garden City, Tapiola.
A02-0277	Clurman and Hebard, Condominiums and Cooperatives.
A04-1269	Safdie, For Everyone a Garden.

OPEN HOUSING

A01-0021	Berry, Race and Housing.
A01-0050	Clawson, Modernizing Urban Land Policy.
A01-0227	Weaver, Dilemmas of Urban America.
A10-2913	Sternlieb and Listoken, Housing 1973-74: An Anthology.
A10-2914	Sternlieb and Paulus, Housing, 1971-1972.
A10-2915	Sternlieb and Paulus, Housing, 1971-1972.
A10-2916	Sternlieb and Sagalyn, Housing, 1970-1971.

PAINTS, COATINGS AND FINISHES

A08-2305	Aluminum Assn., Aluminum Finishing Seminar Papers.
A08-2308	American Chemical Society, Fire Retardant Paints.

Index   621

A08-2319   ASM, Metals Handbook. Vol. 2 Heat Treating, Cleaning and Finishing.
A08-2336   ASTM, Paint Testing Manual.
A08-2342   Banov, Paints and Coatings Handbook.
A08-2361   Burns and Bradley, Protective Coatings for Metals.
A08-2366   Childe, Everymans Guide to Concrete Work.
A08-2382   David Litter Labs, Paints and Protective Coatings.
A08-2385   Diamant, Chemistry of Building Materials.
A08-2412   Gross, Applications Manual for Paint and Protective Coatings: A Guide to Types of Coatings, Methods of Surface Preparation, and Hand Application Techniques.
A08-2449   Martens, The Technology of Paints, Varnishes and Lacquers.
A08-2460   NACE, Industrial Maintenance Painting.
A08-2475   Nylen and Sunderland, Modern Surface Coatings: A Textbook of the Chemistry and Technology of Paints, Varnishes and Lacquers.
A08-2531   Spring, Preparation of Metals for Painting.
A08-2532   Steel Structures Painting Council, Steel Structures Painting Manual. Vol. 1 Good Painting Practice.
A08-2533   Steel Structures Painting Council, Steel Structures Painting Manual. Vol. 2 Systems and Specifications.
A08-2553   Watson, Construction Materials and Processes.
A09-2620   The Building Research Establishment, Essential Data for Construction. Vol. 4 Building Defects and Maintenance.
A09-2630   Carpenters and Builders Library. Vol. 4 Millwork, Power Tools, Painting.
A09-2706   Hurst, Painting and Decorating.
A09-2787   Reiner, Methods and Materials of Construction: A Guide for Builders, Owners, Architects and Engineers.
A09-2809   Taylor, Plastering.

PARKING

A04-0753   Brierley, Parking of Motor Vehicles.
A04-1070   Klose, Metropolitan Parking Structures: A Survey of Architectural Problems and Solutions.
A04-1140   Mills, Planning: Vol. 1, Architect's Technical Reference.
A07-2220   Baerwald, Institute of Traffic Engineers Transportation and Traffic Engineering Handbook.
A08-2455   Moffat, Plant Engineer's Handbook of Formulas, Charts and Tables.

PILES, PILING, FOUNDATIONS

A05-1479   Andersen, Substructure Analysis and Design.
A05-1491   ASCE, Fourth American Conference on Soil Mechanics and Foundation Engineering.
A05-1492   ASCE, Fourth American Conference on Soil Mechanics and Foundation Engineering. Vol. 1.
A05-1493   ASCE, Fourth American Conference on Soil Mechanics

	and Foundation Engineering. Vol. 2.
A05-1512	Asplund, Structural Mechanics: Classical and Matrix Methods.
A05-1516	ASTM, Performance of Deep Foundations.
A05-1518	ASTM, Vibration Effects of Earthquakes on Soil and Foundations.
A05-1521	Baker, Raft Foundations.
A05-1530	Barkan, Dynamics of Bases and Foundations.
A05-1550	Bowles, Analytical and Computer Methods in Foundation Engineering.
A05-1552	Bowles, Foundation Analysis and Design.
A05-1568	Cassie, Fundamental Foundations.
A05-1572	Chellis, Pile Foundations.
A05-1599	Design of Tilt-Up Buildings.
A05-1633	Fletcher and Smoots, Construction Guide for Soils and Foundations.
A05-1639	Gaylord and Gaylord, Structural Engineering Handbook.
A05-1699	Johnson and Kavanagh, The Design of Foundations for Buildings.
A05-1702	Jumikis, Foundation Engineering.
A05-1715	Lee, An Introduction to Deep Foundations and Sheet Piling.
A05-1718	Leonards, Foundation Engineering.
A05-1757	Merritt, Standard Handbook for Civil Engineers.
A05-1764	Morris, Handbook of Structural Design.
A05-1800	Peck, Foundation Engineering.
A05-1811	Proceedings, Sixth International Conference on Soil Mechanics and Foundation Engineering.
A05-1819	Richart, Vibrations of Soils and Foundations.
A05-1846	Schwartz, Civil Engineering for the Plant Engineer.
A05-1847	Scott, An Introduction to Soil Mechanics and Foundations.
A05-1851	Seelye, Foundations: Design and Practice.
A05-1863	Spangler and Handy, Soil Engineering.
A05-1883	Teng, Foundation Design.
A05-1893	Tschebotarioff, Foundations, Retaining and Earth Structures: The Art of Design Construction and Its Scientific Basis in Soil Mechanics.
A05-1894	Tsytovish, Mechanics of Frozen Ground.
A05-1906	Walmer and Baron, Manual of Structural Design and Engineering Solutions.
A05-1918	Whitaker, The Design of Piled Foundations.
A05-1931	Wolfer, Elastically Supported Beams.
A05-1932	Woodward, Drilled Pier Foundations.
A05-1935	Zilly, Handbook of Environmental Civil Engineering.
A06-1982	Barkan, Dynamics of Bases and Foundations.
A08-2455	Moffat, Plant Engineer's Handbook of Formulas, Charts and Tables.
A09-2594	ASTM, Performance Monitoring for Geotechnical Construction.
A09-2607	Blake, Civil Engineer's Reference Book.
A09-2617	The Building Research Establishment, Essential Data for Construction. Vol. 1 Building Construction.
A09-2632	Carson, Foundation Construction.

A09-2657	Dunham, Foundations of Structures.
A09-2698	Hammond, Modern Foundation Methods.
A09-2699	Hanna, Foundation Instrumentation.
A09-2729	Leonards, Foundation Engineering.
A09-2745	Manning, Design and Construction of Foundations.
A09-2764	Parker and MacGuire, Simplified Site Engineering for Architects and Builders.
A09-2804	Smith, Principles and Practices of Heavy Construction.
A09-2816	Tomlinson, Foundation Design and Construction.
A09-2828	West, Piling Practice.
A09-2833	Woodward, Drilled Pier Foundations.
A09-2836	Zeevaert, Foundation Engineering for Difficult Subsoil Conditions.

PIPE, VALVES, PIPE FITTINGS

A02-0367	Ottaviano, National Mechanical Estimator.
A02-0374	Page and Nation, Estimator's Piping Man-Hour Manual/Revised Edition.
A02-0388	Richardson, Mechanical and Electrical Construction Estimating Standards.
A02-0410	Wood, Mechanical Estimators Handbook.
A06-1964	Alden and Kane, Design of Industrial Exhaust Systems.
A06-2007	Chasis, Plastic Piping Systems.
A06-2011	Crocker and King, Piping Handbook.
A06-2015	Davis, Plumbing, Heating, and Piping: Estimator's Guide.
A06-2024	Emerick, Troubleshooters' Handbook for Mechanical Systems.
A06-2027	Evans, Equipment Design Handbook for Refineries and Chemical Plants.
A06-2052	Herkimer, Cost Manual for Piping and Mechanical Construction.
A06-2081	Kellogg Company, Design of Piping Systems.
A06-2083	King, Piping Handbook.
A06-2096	Littleton, Industrial Piping.
A06-2142	Rase, Piping Design for Process Plants.
A06-2150	Schweitzer, Handbook of Corrosion Resistant Piping.
A06-2151	Schweitzer, Handbook of Valves.
A06-2152	Sculthorpe, Design of High Pressure Steam and High Temperature Water Plants.
A06-2158	Sherwood and Whistance, The Piping Guide.
A06-2179	Weaver, Process Piping Design.
A08-2458	NACE, Control of Pipeline Corrosion.
A08-2480	Parker, Pipe Line Corrosion and Cathodic Protection.
A08-2513	Schweitzer, Handbook of Corrosion Resistant Piping.
A09-2586	American Concrete Pipe Assn., Concrete Pipe Installation Manual.
A09-2733	Lindsey, Pipefitters Handbook.
A09-2782	Rampaul, Pipe Welding Procedures.
A09-2792	Rose, Questions and Answers on Pipework and Pipe Welding.

PLASTICS, RUBBER, ELASTOMERS

A05-1537	Benjamin, Structural Design with Plastics.
A05-1731	Long, Bearings in Structural Engineering.
A08-2314	Arnold, Introduction to Plastics.
A08-2315	ASCE, Structural Plastics--Properties and Possibilities.
A08-2341	Baer, Engineering Design for Plastics.
A08-2352	Brady, Materials Handbook.
A08-2357	Brydson, Plastics Materials.
A08-2368	Clauser, Encyclopedia of Engineering Materials and Processes.
A08-2380	Damusis, Sealants.
A08-2384	Diamant, Chemistry of Building Materials.
A08-2387	Dietz, Plastics for Architects and Builders.
A08-2388	DuBois and John, Plastics.
A08-2389	Duck, Plastics and Rubbers.
A08-2390	Evans, Selecting Engineering Materials for Chemical and Process Plant.
A08-2395	Frisch and Saunders, Plastic Foams.
A08-2407	Gibbs and Cox, Marine Design Manual for Fiberglass Reinforced Plastics.
A08-2408	Glanvill, Plastics Engineer's Data Book.
A08-2414	Guy, Essentials of Materials Science.
A08-2417	Harper, Handbook of Plastics and Elastomers.
A08-2425	Kayser, Materials Science in Engineering.
A08-2435	Lee and Neville, Handbook of Epoxy Resins.
A08-2443	Mallinson, Chemical Plant Design with Reinforced Plastics.
A08-2446	Mark, Encyclopedia of Polymer Science and Technology: Plastics, Resins, Rubbers, Fibers. Vol. 1 Ablative Polymers to Amino Acids.
A08-2451	May and Tanaka, Epoxy Resins: Chemistry and Technology.
A08-2452	Meinecke and Clark, Mechanical Properties of Polymeric Foams.
A08-2456	Mohr, SPI Handbook of Technology and Engineering of Reinforced Plastics/Composites.
A08-2462	National Bureau of Standards, A Compilation and Evaluation of Mechanical, Thermal and Electrical Properties of Selected Polymers.
A08-2464	Neumann and Bockhoff, Welding of Plastics.
A08-2469	NFPA, Tentative Guide for Plastics in Building Construction.
A08-2476	Ogorkiewicz, Engineering Properties of Thermoplastics.
A08-2481	Parkyn, Glass Reinforced Plastics.
A08-2482	Patton, Construction Materials.
A08-2484	Patton, Plastics Technology: Theory, Design and Manufacture.
A08-2486	Penn, Plastics-in-Building Handbook.
A08-2492	Pollack, Materials Science and Metallurgy.
A08-2501	Wuarmby, Plastics in Architecture.
A08-2504	Reboul and Mitchell, Plastics in the Building Industry.
A08-2512	Sarvetnick, Polyvinyl Chloride.

A08-2522	Simonds and Church, Concise Guide to Plastics.
A08-2523	Simpson and Horrobin, The Weathering and Performance of Building Materials.
A08-2525	Skeist, Epoxy Resins.
A08-2526	Skeist, Handbook of Adhesives.
A08-2527	Skeist, Plastics in Building.
A08-2528	Smith, Materials of Construction.
A08-2548	Van Vlack, A Textbook of Materials Technology.
A08-2554	The Weathering and Performance of Building Materials.
A08-2556	Whittington, Whittington's Dictionary of Plastics.
A09-2618	The Building Research Establishment, Essential Data for Construction. Vol. 2 Building Materials.

PLUMBING

A01-0126	Lent, Plumbing Code of New York City--Guide and Interpretation.
A02-0313	Galeno, The Plumbing Estimating Handbook.
A02-0367	Ottaviano, National Mechanical Estimator.
A02-0368	Ottaviano, National Mechanical Estimator.
A02-0372	Page, Heating, Plumbing and Air-Conditioning Man-Hour Manual.
A02-0374	Page and Nation, Estimator's Piping Man-Hour Manual/Revised Edition.
A02-0388	Richardson, Mechanical and Electrical Construction Estimating Standards.
A02-0410	Wood, Mechanical Estimators Handbook.
A04-1068	Kinzey and Sharp, Environmental Technologies in Architecture.
A04-1069	Kira, The Bathroom.
A06-1967	Andrews, Architect's Guide to Mechanical Systems.
A06-1976	ASSE, The Principles of Residential Plumbing Inspection.
A06-1979	Babbitt, Plumbing.
A06-1983	Barton, Domestic Heating and Hot Water Supply.
A06-1992	Blendermann, Design of Plumbing and Drainage Systems.
A06-2015	Davis, Plumbing, Heating, and Piping: Estimator's Guide.
A06-2070	Jacobson, Plumbing Dictionary.
A06-2083	King, Piping Handbook.
A06-2095	Lindeke, Technical Dictionary of Heating, Ventilation and Sanitary Engineering.
A06-2099	McGuinness and Stein, Mechanical and Electrical Equipment for Buildings.
A06-2104	Manas, National Plumbing Code Handbook.
A06-2105	Manas, National Plumbing Code Illustrated.
A06-2124	Nielsen, Standard Plumbing Engineering Design.
A06-2150	Schweitzer, Handbook of Corrosion Resistant Piping.
A06-2151	Schweitzer, Handbook of Valves.
A06-2166	Strock and Koral, Handbook of Air Conditioning, Heating and Ventilating.
A06-2179	Weaver, Process Piping Design.
A09-2650	Daniels, Home Guide to Plumbing, Heating and Air

	Conditioning.
A09-2733	Lindsey, Pipefitters Handbook.
A09-2749	Matthias, How to Design and Install Plumbing.
A09-2770	Plumbers and Pipe Fitters Library. Vol. 1 Materials, Tools, Calculations.
A09-2771	Plumbers and Pipe Fitters Library. Vol. 2 Drainage, Fittings, Fixtures.
A09-2772	Plumbers and Pipe Fitters Library. Vol. 3 Installation, Heating, Welding.
A09-2781	Questions and Answers for Plumbers Examinations.
A09-2787	Reiner, Methods and Materials of Construction: A Guide for Builders, Owners, Architects and Engineers.
A09-2792	Rose, Questions and Answers on Pipework and Pipe Welding.

POLLUTION AND ENVIRONMENTAL IMPACT

A01-0026	Branch, Planning Urban Environment.
A01-0084	George and McKinley, Urban Ecology: In Search of an Asphalt Rose.
A01-1225	Wagner, Environment and Man.
A02-0283	Corwin and Hefferman, Environmental Impact Assessment.
A02-0286	Cross and Simons, Industrial Plant Siting.
A02-0367	Ottaviano, National Mechanical Estimator.
A03-0447	Azad, Industrial Wastewater Management Handbook.
A03-0515	Gray, Cases and Materials on Environmental Law.
A05-1481	Anthrop, Noise Pollution.
A05-1485	APWA, Water Pollution Aspects of Urban Runoff.
A05-1601	Dober, Environmental Design.
A05-1613	Eckenfelder, Industrial Water Pollution Control.
A05-1648	Goodfriend, Noise Pollution.
A05-1709	Kulski, Architecture in A Revolutionary Era.
A05-1727	Liptak, Environmental Engineers' Handbook.
A05-1728	Liptak, Environmental Engineers' Handbook. Vol. 2 Air Pollution.
A05-1729	Liptak, Environmental Engineers' Handbook. Vol. 3 Land Pollution.
A05-1734	Lund, Industrial Pollution Control Handbook.
A05-1744	Magill, Air Pollution Handbook.
A05-1759	Metcalf and Eddy, Wastewater Engineering: Collection, Treatment, Disposal.
A05-1777	Noll and Duncan, Industrial and Air Pollution Control.
A05-1778	Nonhebel, Processes for Air Pollution Control.
A05-1804	Pollution Control Technology.
A06-1962	AIChE, Industrial Process Design for Pollution Control.
A06-2024	Emerick, Troubleshooters' Handbook for Mechanical Systems.
A06-2109	Miliaras, Power Plants with Air-Cooled Condensing Systems.
A07-2202	Antonious, Environmental Management: Planning for Traffic.
A09-2660	EPA, Processes, Procedures, and Methods to Control

Index 627

Pollution Resulting from all Construction Activity.

PRISONS

A04-0891 Fairweather, Prison Architecture.

PROPERTY MANAGEMENT, MAINTENANCE

A02-0327 Hanke, The Homes Association Handbook.
A02-0339 IREM, 1975 Income/Expense Analysis--Apartments, Condominiums & Co-operatives.
A02-0340 IREM, 1975 Office Building Experience Exchange Report.
A03-0422 Abrams and Blackman, Managing Low and Moderate Income Housing.
A03-0462 Calvert, Introduction to Building Management.
A03-0490 Downs, Principles of Real Estate Management.
A03-0516 Grear and Oxborough, Commercial Property Management: A Practical Guide to the Legal Ownership, Use, Sale, and Acquisition of Commercial Land and Property.
A03-0524 Hanford, Analysis & Management of Investment Property.
A03-0552 Lawton, Planning and Managing Housing for the Elderly.
A06-2140 Price, Air Conditioning for Building Engineers and Managers: Operation and Maintenance.
A07-2221 Barenberg, Pavement Distress Identification and Repair.
A08-2313 Arnison, Floor and Structural Surfaces.
A08-2511 Salter, Floors and Floor Maintenance.
A09-2620 The Building Research Establishment, Essential Data for Construction. Vol. 4 Building Defects and Maintenance.
A09-2713 Insall, The Care of Old Buildings Today: A Practical Guide.
A11-2973 Conover, Grounds Maintenance Handbook.
A11-2996 Feldman, Building Design for Maintainability.
A11-3025 Hubert, Preventive Maintenance of Electrical Equipment.
A11-3032 Johnson, Deterioration, Maintenance and Repair of Structures.
A11-3064 Morrow, Maintenance Engineering Handbook.
A11-3066 Mulholland, Heating, Ventilation and Air Conditioning Plant: Planned Maintenance and Operation.
A11-3118 Westinghouse, Electrical Maintenance Hints.

PUMPS

A05-1531 Bartlett, Pumping Stations for Water and Sewage.
A06-1972 ASHRAE, Equipment Volume 1972.
A06-2008 Church, Centrifugal Pumps and Blowers.
A06-2026 Evans, Equipment Design Handbook for Refineries and Chemical Plants.
A06-2054 Hicks and Edwards, Pump Application Engineering.
A06-2077 Karassik, Pump Handbook.
A06-2139 Power Magazine, Plant Energy Systems.
A06-2174 Walker, Pump Selection--A Consulting Engineer's Manual.

REAL ESTATE AND APPRAISAL

A01-0053	Cooper and Guntermann, Real Estate and Urban Land Analysis.
A01-0236	Wolff, The Unreal Estate.
A02-0251	American Appraisal Co., Boeckh Building Valuation Manual. Vol. 1 Residential and Agricultural.
A02-0252	American Appraisal Co., Boeckh Building Valuation Manual. Vol. 2 Commercial.
A02-0253	American Appraisal Co., Boeckh Building Valuation Manual. Vol. 3 Industrial and Institutional.
A02-0257	American Institute of Real Estate Appraisers, The Appraisal of Real Estate.
A02-0258	American Institute of Real Estate Appraisers, Condemnation Appraisal Practice. Vol. 1.
A02-0259	American Institute of Real Estate Appraisers, Condemnation Appraisal Practice. Vol. 2.
A02-0260	American Institute of Real Estate Appraisers, Problems in Urban Real Estate Appraisal.
A02-0261	Anderson, Basic Real Estate Tax Manual.
A02-0271	Beasley, Fell's Guide to Buying, Building and Financing a Home.
A02-0274	Boyce, Real Estate Appraisal Terminology.
A02-0280	Collison, The Developer's Dictionary and Handbook.
A02-0300	Dombal, Residential Condominiums: A Guide to Analysis and Appraisal.
A02-0302	Edgerton, Edgerton Building Cost Calculator and Guide to Real Estate Valuation.
A02-0306	Ellwood, Ellwood Tables for Real Estate Appraising and Financing. Part I--Explanatory Text.
A02-0307	Ellwood, Ellwood Tables for Real Estate Appraising and Financing. Part II--Tables.
A02-0314	Garrett, The Valuation of Shopping Centers.
A02-0317	Golemon, Financing Real Estate Development.
A02-0318	Goodkin, When Real Estate and Home Building Become Big Business.
A02-0320	Gross, Concise Desk Guide to Real Estate Practice and Procedure.
A02-0334	Hoagland and Stone, Real Estate Finance.
A02-0341	Johnson, The Instant Mortgage Equity Technique.
A02-0342	Johnson, Selling Real Estate by Mortgage Equity Analysis.
A02-0345	Kinnard, Income Property Valuation: Principles and Techniques of Appraising Income-Producing Real Estate.
A02-0346	Kinnard, Principles and Techniques of Real Property Appraising.
A02-0347	Knowles, Single-Family Residential Appraisal Manual.
A02-0392	Smith, Real Estate and Urban Development.
A02-0404	Wendt, Real Estate Appraisal: Review and Outlook.
A03-0490	Downs, Principles of Real Estate Management.
A04-0989	Hawkins, Appraisal Guide for School Facilities.
A10-2871	Grebler, Large Scale Housing and Real Estate Firms: Analysis of a New Business Enterprise.

REFRIGERATION

A06-1970	ASHRAE, 1974 Applications Volume.
A06-1971	ASHRAE, 1973 Systems Volume.
A06-1972	ASHRAE, 1972 Handbook of Fundamentals.
A06-2009	Commercial Refrigeration.
A06-2017	Dossat, Principles of Refrigeration.
A06-2026	Evans, Equipment Design Handbook for Refineries and Chemical Plants.
A06-2030	Federal Energy Administration, Guidelines for Saving Energy in Existing Buildings.
A06-2041	Gosling, Applied Air Conditioning and Refrigeration.
A06-2045	Gunther, Refrigeration, Air Conditioning, and Cold Storage: Principles and Applications.
A06-2056	Home Refrigeration and Air Conditioning.
A06-2069	International Institute of Refrigeration (Paris), Bibliographic Guide to Refrigeration 1960-64.
A06-2082	King, Piping Handbook.
A06-2119	NESCA, Basic Refrigeration, and Student Workbook for Basic Refrigeration.
A06-2139	Power Magazine, Plant Energy Systems.
A06-2168	Threlkeld, Thermal Environmental Engineering.
A06-2178	Weaver and Kirkpatrick, Environment Control: Air Conditioning and Refrigeration Theory and Application.
A06-2183	Woolrich, Handbook of Refrigerating Engineering. Vol. 1 Fundamentals.
A06-2184	Woolrich, Handbook of Refrigerating Engineering.
A06-2185	Woolrich and Hallowell, Cold and Freezer Storage Manual.

RENOVATION AND REHABILITATION

A01-0037	Cantacuzino, New Uses for Old Buildings.
A01-0100	Harvey, Conservation of Buildings.
A01-0137	McFarland and Vivrett, Residential Rehabilitation.
A02-0285	Costonis, Space Adrift--Landmark Preservation and the Marketplace.
A02-0291	Debaights, New Interiors for Old Houses.
A02-0303	Edgerton, How to Renovate a Brownstone.
A02-0304	Edgerton, Row House Renaissance, A Guide to Renovation.
A02-0356	McKenna, A House in the City.
A02-0377	Paxton, National Repair and Remodeling Estimator.
A02-0382	Redstone, New Dimensions in Shopping Centers and Stores.
A04-0770	Bullock, The Restoration Manual.
A04-0771	Bullock, The Restoration Manual.
A04-0785	Cantacuzino, New Uses for Old Buildings.
A04-0802	Christovich, New Orleans Architecture. Volume II The American Sector.
A04-0827	Costonis, Space Adrift: Landmark Preservation and the Marketplace.
A04-0837	Crosby, The Necessary Monument: Its Future in the

	Civilized City.
A04-0852	Dercsenyi, Historical Monuments in Hungary, Restoration and Preservation.
A04-1026	Huxtable, Have You Kicked a Building Lately?
A04-1030	Insall, The Care of Old Buildings Today: A Practical Guide.
A04-1345	Tibbs, King's College Chapel, Cambridge (England).
A04-1370	West, The Timber-Frame House in England.
A09-2606	Berne, How to Make Money in the Remodeling Business.
A09-2640	Collymore, Altering, Repairing and Extending Houses.
A09-2645	Dalzell, Repairing and Remodeling Guide for Home Interiors: Planning, Materials, Methods.
A09-2713	Insall, The Care of Old Buildings Today: A Practical Guide.
A10-2853	Berne, How to Make Money in the Remodeling Business.
A11-3031	Johnson, Deterioration, Maintenance and Repair of Structures.
A11-3060	Marchant, A Complete Guide to Fire and Buildings: A Comprehensive Survey of all Aspects of Fire Protection and Provision.
A11-3080	NHIC, Home Improvement Selling.

SAFETY AND SECURITY

A01-0153	Newman, Defensible Space: People and Design in the Violent City.
A03-0457	BNA, Editorial Staff, OSHA and the Unions: Bargaining on Job Safety and Health.
A06-2086	Kurtz and Shoemaker, The Lineman's and Cableman's Handbook.
A08-2455	Moffat, Plant Engineer's Handbook of Formulas, Charts and Tables.
A09-2784	Rapp, Construction of Structural Steel Building Frames.
A09-2787	Reiner, Methods and Materials of Construction: A Guide for Builders, Owners, Architects and Engineers.
A09-2808	Swearer, Installing and Servicing Electronic Protective Systems.
A09-2824	Wass, Methods and Materials of Residential Construction.
A11-2932	AIChE, Plant and Design Safety.
A11-2936	Anderson, OSHA and Accident Control Through Training.
A11-2943	Associated Builders and Contractors, Construction Safety Guide.
A11-2952	Best's Safety Directory.
A11-2957	BNA Editorial Staff, The Job Safety and Health Act of 1970.
A11-2972	Commerce Clearing House, Occupational Safety and Health Act of 1970 Law and Explanation.
A11-3014	Healy, Design for Security.
A11-3022	Hopf, Designer's Guide to OSHA.
A11-3027	International Atomic Energy Agency, Manual on Safety Aspects of the Design and Equipment of Hot Laboratories.

Index

A11-3043 Labor Department, Construction Safety and Health Training Course, Instructor's Guide. Manual 301 Job Planning, Walking and Working Surfaces and Spaces, Fire Prevention, Electrical Hazards.
A11-3044 Labor Department, Construction Safety and Health Training Course, Instructor's Guide. Manual 302 Personal Protective Equipment, Hand Tools, Ladders and Scaffolds, Rigging.
A11-3045 Labor Department, Construction Safety and Health Training Course, Instructor's Guide. Manual 303 Powered Equipment, Material Handling, Welding, Flammable Liquids.
A11-3046 Labor Department, Construction Safety and Health Training Course, Instructor's Guide. Manual 304 Blasting, Excavation, Traffic Control, Concrete Construction.
A11-3047 Labor Department, Construction Safety and Health Training Course, Instructor's Guide. Manual 305 Marine Operations, Steelworking, Tunneling, Demolition, Special Hazards.
A11-3073 Newman, Crime Prevention Through Urban Design: Designing a House to Prevent Crime.
A11-3074 Newman, Defensible Space: Crime Prevention Through Urban Design.
A11-3086 Plant Engineering, Plant Protection Library.
A11-3101 Steere, CRC Handbook of Laboratory Safety.

SCHOOLS, LIBRARIES, MUSEUMS, DORMITORIES

A01-0204 Sternberg and Sternberg, Community Centers and Student Unions.
A02-0308 Essex, Bonding Versus Pay-as-You-Go in the Financing of School Buildings.
A02-0321 Guide for Planning Educational Facilities.
A02-0324 Halsey, Borrowing Money for the Public Schools: A Study of Borrowing Practices in the Administration of Public Schools in Florida.
A02-0360 Metcalf, Planning Academic and Research Library Buildings.
A02-0406 What Went Wrong?
A04-0688 Architectural Record, Apartments and Dormitories.
A04-0689 Architectural Record, Campus Planning and Design.
A04-0724 Berriman and Harrison, British Public Library Buildings.
A04-0728 Birks, Building the New Universities.
A04-0750 Brawne, Libraries: Architecture and Equipment.
A04-0751 Brawne, The New Museum: Architecture and Display.
A04-0755 Brodshaug, Buildings and Equipment for Home Economics in Secondary Schools.
A04-0780 Byrne, Check List Materials for Public School Building Specifications, Covering the General Specifications.
A04-0793 Caudill, Toward Better School Design.
A04-0855 Dixon, Architectural Design Preview, U.S.A.
A04-0857 Dober, Campus Planning.
A04-0876 Ede, Canadian Architecture 1960/70.

A04-0880	Ellsworth, Academic Library Buildings: A Guide to Architectural Issues and Solutions.
A04-0978	Hamlin, Forms and Functions of Twentieth-Century Architecture. Vol. 1 The Elements of Building. Vol. 2 The Principles of Composition. Vol. 3 Building Types: Residence, Gatherings, Education, Government. Vol. 4 Building Types: Commerce and Industry, Public Health, Transportation, Social Welfare and Recreation.
A04-0989	Hawkins, Appraisal Guide for School Facilities.
A04-1128	Markus, Building Performance.
A04-1133	Meyer-Bohe, Bauten fur die Jugend/Structures for Children.
A04-1141	Mills and Kaylor, The Design of Polytechnic Institute Buildings.
A04-1148	Morisseau, The New Schools.
A04-1165	Navy Department, Design Manual: Community Facilities.
A04-1193	Orr, Designing Library Buildings for Activity.
A04-1201	Parsons, The Cornell Campus: A History of Its Planning and Development.
A04-1211	Peters, Libraries for Schools and Universities.
A04-1260	Roth, The New Schoolhouse.
A04-1282	Seaborne, The English School: Its Architecture and Organization.
A04-1355	Vazquez, The National Museum of Anthropology, Mexico.
A09-2724	Koncz, Manual of Precast Concrete Construction. Vol. 3.

SEISMIC, WINDSTORM DESIGN AND CONSTRUCTION

A05-1465	ACI, Response of Multistory Concrete Structures to Lateral Forces.
A05-1486	Architectural Institute of Japan, Design Essentials in Earthquake Resistant Buildings.
A05-1487	Architectural Institute of Japan, Design Essentials in Earthquake Resistant Buildings.
A05-1518	ASTM, Vibration Effects of Earthquakes on Soils and Foundations.
A05-1532	Bathe and Wilson, Numerical Methods in Finite Element Analysis.
A05-1541	Biggs, Introduction to Structural Dynamics.
A05-1576	Clough and Penzien, Dynamics of Structures.
A05-1582	Cowan, Models in Architecture.
A05-1599	Design of Tilt-Up Buildings.
A05-1610	Eagleman, Thunderstorms, Tornadoes and Building Damage.
A05-1611	Eaton, Wind Effects: Proceedings of the Fourth International Conference on Wind Effects on Buildings and Structures, September 1975.
A05-1615	EERI, Earthquake and Blast Effects on Structures.
A05-1616	EERI, First World Conference on Earthquake Engineering.
A05-1620	Fairhurst, Failure and Breakage of Rock: Proceedings, 8th Symposium on Rock Mechanics.

Index

A05-1627	Fertis, Dynamics and Vibrations of Structures.
A05-1645	Gill, Structures for Nuclear Power.
A05-1678	Hollis, Bibliography of Earthquake Engineering.
A05-1682	Howells, Dynamic Waves in Civil Engineering.
A05-1685	Hurty and Rubinstein, Dynamics of Structures.
A05-1687	International Atomic Energy Agency, Aseismic Design and Testing of Nuclear Facilities.
A05-1688	International Atomic Energy Agency, Earthquake Guidelines for Reactor Siting.
A05-1689	International Conference on Wind Effects on Buildings and Structures, 3rd, Tokyo, Sept. 1971.
A05-1702	Jumikis, Foundation Engineering.
A05-1717	Lenczner, Elements of Loadbearing Brickwork.
A05-1737	Macdonald, Wind Loading on Buildings.
A05-1741	McKaig, Building Failures.
A05-1750	Marshall and Thom, Wind Loads on Buildings and Structures: Building Science Series 30.
A05-1755	Meehan, Managua, Nicaragua Earthquake of Dec. 23, 1972, Reconnaissance Report.
A05-1769	National Bureau of Standards, Building Performance in the 1972 Managua Earthquake.
A05-1770	National Bureau of Standards, Design, Siting, and Construction of Low-Cost Housing and Community Buildings to Better Withstand Earthquakes and Windstorms.
A05-1771	National Research Council of Canada, Wind Effects on Buildings and Structures.
A05-1772	National Weather Service, Earthquake History of the United States.
A05-1773	National Weather Service, United States Earthquakes, 1971.
A05-1776	Newmark and Rosenblueth, Fundamentals of Earthquake Engineering.
A05-1785	Okamoto, Introduction to Earthquake Engineering.
A05-1792	Park and Paulay, Reinforced Concrete Structures.
A05-1806	Portland Cement Assn., Design of Multistory Reinforced Concrete Buildings for Earthquake Motions.
A05-1819	Richart, Vibrations of Soils and Foundations.
A05-1831	Rosman, Tables for the Internal Forces of Pierced Shear-Walls Subject to Lateral Loads.
A05-1836	Sachs, Wind Forces in Engineering.
A05-1923	Wiegel, Earthquake Engineering.
A06-2175	Waller and Atkins, Building on Springs.
A07-2242	Minikin, Winds, Waves and Maritime Structures.
A07-2266	Wetteland and Bruun, Proceedings POAC-Norway 1971. Vols. 1 and 2.
A08-2309	Amrhein, Reinforced Masonry Engineering Handbook.
A08-2348	BIA, Reinforced Brick Masonry--Lateral Force Design.
A08-1516	Sementsov and Kameiko, Designer's Manual Masonry, Including Reinforced Masonry in Industrial, Residential and Communal Buildings and Structures.
A09-2741	McKaig, Building Failures.

SHEET METAL WORK

A09-2716	Kaberlein, Air Conditioning Metal Layout.
A09-2725	Kratfel, Introduction to Modern Sheet Metal.
A09-2755	Morris, Air Conditioning Cutter's Ready Reference.
A09-2806	SMACNA, Architectural Sheet Metal Manual.
A09-2837	Zinngrabe, Mathematics for the Sheet Metal Technician.
A11-2963	Bretz, Sheet Metal Shop Drawing.

SHOPPING CENTERS, STORES, COMMERCIAL BUILDINGS

A01-0086	Gibberd, Town Design.
A02-0287	Darlow, Enclosed Shopping Centres.
A02-0296	Directory of Shopping Centers in the U.S. and Canada.
A02-0314	Garrett, The Valuation of Shopping Centers.
A02-0340	IREM, 1975 Office Building Experience Exchange Report.
A02-0355	McKeever, Community Builders Handbook.
A02-0382	Redstone, New Dimensions in Shopping Centers and Stores.
A04-0690	Architectural Record, Office Buildings.
A04-0699	Arregger and Glaus, Highrise Building and Urban Design.
A04-0846	Debaights, Shop Fronts.
A04-0978	Hamlin, Forms and Functions of Twentieth-Century Architecture. Vol. 1 The Elements of Building. Vol. 2 The Principles of Composition. Vol. 3 Building Types: Residence, Gatherings, Education, Government. Vol. 4 Building Types: Commerce and Industry, Public Health, Transportation, Social Welfare and Recreation.
A04-1012	Hohl, Office Buildings: An International Survey.
A04-1014	Hornbeck, Stores and Shopping Centers.
A04-1049	Kaspar, Shops and Showrooms: An International Survey.
A04-1060	Ketchum, Shops and Stores.
A04-1115	Lynch, Site Planning.
A04-1275	Schmertz, Office Building Design.
A11-2996	Feldman, Building Design for Maintainability.

SITE SELECTION

A01-0165	Pratt Guide to Planning and Renewal for New Yorkers.
A01-0219	Triton Foundation, Triton City: A Prototype Floating Community.
A02-0286	Cross and Simons, Industrial Plant Siting.
A02-0325	Hamer, Industrial Exodus from the Central City.
A02-0361	MHMA, Mobile Home Site Planning Kit: Basic Information Concerning Construction Park Development.
A02-0366	Nutt, Functional Plant Planning, Layout, and Materials Handling.
A02-0408	Whitman and Schmidt, Plant Relocation.
A04-1116	Lynch, Site Planning.
A05-1688	IAEA, Earthquake Guidelines for Reactor Siting.

Index 635

A09-2607	Blake, Civil Engineer's Reference Book.
A09-2660	EPA, Processes, Procedures, and Methods to Control Pollution Resulting from All Construction Activity.
A09-2787	Reiner, Methods and Materials of Construction: A Guide for Builders, Owners, Architects and Engineers.

SOLAR HEATING AND COOLING

A06-1970	ASHRAE, 1974 Applications Volume.
A06-1997	Brinkworth, Solar Energy for Man.
A06-2014	Daniels, Solar Homes and Sun Heating.
A06-2044	Griffin, Energy Conservation in Buildings: Techniques for Economical Design.
A06-2116	National Science Foundation, Demand Analysis, Solar Heating and Cooling of Buildings.
A06-2117	National Science Foundation, Proceedings of the Solar Heating and Cooling for Buildings Workshop, 1973.
A06-2118	National Science Foundation, Solar Cooling for Buildings.
A06-2129	Patton, Solar Energy for Heating and Cooling of Buildings.
A06-2130	Pesco, Solar Directory.
A06-2181	Williams, Solar Energy, Technology and Applications.

SOIL AND ROCK MECHANICS

A05-1451	Abbett, American Civil Engineering Practice.
A05-1491	ASCE, Fourth American Conference on Soil Mechanics and Foundation Engineering.
A05-1492	ASCE, Fourth American Conference on Soil Mechanics and Foundation Engineering. Vol. 1.
A05-1493	ASCE, Fourth American Conference on Soil Mechanics and Foundation Engineering. Vol. 2.
A05-1498	ASCE, Proceedings of the Conference on Performance of Earth and Earth-Supported Structures.
A05-1514	ASTM, Determination of the In Situ Modulus of Deformation of Rock.
A05-1516	ASTM, Performance of Deep Foundations.
A05-1530	Barkan, Dynamics of Bases and Foundations.
A05-1551	Bowles, Engineering Properties of Soils and Their Measurements.
A05-1552	Bowles, Foundation Analysis and Design.
A05-1559	British Geotechnical Society, Settlement of Structures.
A05-1560	Brown, Permafrost in Canada.
A05-1568	Cassie, Fundamental Foundations.
A05-1603	Dreyer, The Science of Rock Mechanics. Vol. 1 Strength Properties of Rocks.
A05-1618	Evdokimov and Sapegin, Stability, Shear and Sliding Resistance, and Deformation of Rock Foundations.
A05-1633	Fletcher and Smoots, Construction Guide for Soils and Foundations.
A05-1639	Gaylord and Gaylord, Structural Engineering Handbook.
A05-1654	Grosvenor and Paulding, Status of Practical Rock Mechanics: Proceedings, 9th Symposium on Rock Mechanics.

A05-1663	Harr, Foundations of Theoretical Soil Mechanics.
A05-1681	Hough, Basic Soils Engineering.
A05-1703	Jumikis, Introduction to Soil Mechanics.
A05-1757	Merritt, Standard Handbook for Civil Engineers.
A05-1783	Obert and Duvall, Rock Mechanics and the Design of Structures in Rock.
A05-1800	Peck, Foundation Engineering.
A05-1801	Perloff and Baron, Soil Mechanics--Principles and Applications.
A05-1807	Poulos and Davis, Elastic Solutions for Soil and Rock Mechanics.
A05-1811	Proceedings, Sixth International Conference on Soil Mechanics and Foundation Engineering.
A05-1819	Richart, Vibrations of Soils and Foundations.
A05-1837	Salencon, Applications of the Theory of Plasticity in Soil Mechanics.
A05-1845	Schroeder, Soils in Construction.
A05-1846	Schwartz, Civil Engineering for the Plant Engineer.
A05-1847	Scott, An Introduction to Soil Mechanics and Foundations.
A05-1848	Scott and Schoustra, Soil Mechanics and Engineering.
A05-1850	Seelye, Data Book for Civil Engineers.
A05-1859	Smith, Elements of Soil Mechanics for Civil and Mining Engineers.
A05-1860	Smith, Introduction to Matrix and Finite Elements in Civil Engineering.
A05-1863	Spangler and Handy, Soil Engineering.
A05-1883	Teng, Foundation Design.
A05-1884	Terzaghi and Peck, Soil Mechanics in Engineering Practice.
A05-1893	Tschebotarioff, Foundations, Retaining and Earth Structures: The Art of Design Construction and Its Scientific Basis in Soil Mechanics.
A05-1894	Tsytovich, Mechanics of Frozen Ground.
A05-1935	Zilly, Handbook of Environmental Civil Engineering.
A06-2149	Schenck, Introduction to Ocean Engineering.
A07-2266	Wetteland and Brunn, Proceedings POAC-Norway 1971.
A08-2455	Moffat, Plant Engineer's Handbook of Formulas, Charts, and Tables.
A09-2589	American Society of Civil Engrs., Underground Rock Chambers.
A09-2595	ASTM, Sampling of Soil and Rock.
A09-2596	ASTM, Special Procedures for Testing Soil and Rock for Engineering Purposes.
A09-2607	Blake, Civil Engineer's Reference Book.
A09-2657	Dunham, Foundations of Structures.
A09-2698	Hammond, Modern Foundation Methods.
A09-2699	Hanna, Foundation Instrumentation.
A09-2816	Tomlinson, Foundation Design and Construction.
A09-2836	Zeevaert, Foundation Engineering for Difficult Subsoil Conditions.

Index 637

SPECIFICATIONS AND TECHNICAL WRITING

A02-0276	Building Design and Construction Specifying Buying Guide and Directory.
A03-0420	Abbett, Engineering Contracts and Specifications.
A03-0446	Ayers, Specifications: For Architecture, Engineering and Construction.
A03-0454	Berger and Godel, Estimating and Project Management for Small Construction Firms.
A03-0468	Central Electricity Generating Board, Phraseology for Civil Engineering.
A03-0479	CSI, Uniform Construction Index.
A03-0493	Dunham and Young, Contracts, Specifications and Law for Engineers.
A03-0494	Eacott, Specification in the Construction Industry.
A03-0496	Edwards, Specifications.
A03-0497	Emerick, Handbook of Mechanical Specifications for Buildings and Plants.
A03-0502	Fletcher and Moore, Standard Phraseology for Bills of Quantities.
A03-0542	Jessup and Jessup, Law and Specifications for Engineers and Scientists.
A03-0568	Meier, Construction Specifications Handbook, Vol. 1.
A03-0569	Meier, Construction Specifications Handbook, Vol. 2.
A03-0571	Merritt, Building Construction Handbook.
A03-0600	Rathbun, Building Construction Specifications.
A03-0603	Reiner, Handbook of Construction Management.
A03-0604	Rosen, Construction Specifications Writing: Principles and Procedures.
A03-0605	Rosen, Principles of Specification Writing.
A03-0607	Rossnagel, Checklists for Management, Engineering, Manufacturing, and Product Assurance.
A03-0632	Watson, Specifications Writing for Architects and Engineers.
A04-0780	Byrne, Check List Materials for Public School Building Specifications, Covering the General Specifications.
A04-1232	Ramsey and Sleeper, Architectural Graphic Standards.
A04-1302	Sleeper, Architectural Specifications.
A05-1757	Merritt, Standard Handbook for Civil Engineers.
A05-1845	Schroeder, Soils in Construction.
A05-1850	Seelye, Data Book for Civil Engineers. Vol. 1 Design. Vol. 2 Specifications and Costs.
A05-1915	Wass, Manual of Structural Details for Building Construction.
A08-1312	Architectural Precast Concrete.
A08-2356	Brownell, Architectural Hardware Specifications Handbook.
A08-2490	Plant Engineering Directory and Specifications Catalog.
A08-2507	Ross, Metallic Materials Specification Handbook.
A08-2533	Steel Structures Painting Council, Steel Structures Painting Manual. Vol. 2 Systems and Specifications.
A09-2769	Plant Engineering Construction Library.
A09-2787	Reiner, Methods and Materials of Construction: A Guide

A11-3007   for Builders, Owners, Architects and Engineers.
A11-3007   Glidden, Reports, Technical Writing and Specifications.
A11-3032   Jones, How to Prepare Professional Design Brochures.
A11-3088   Rathbone, Communicating Technical Information: A Guide to Current Uses and Abuses in Scientific and Engineering Writing.
A11-3103   Strong and Eidson, The Technical Writer's Handbook.
A11-3105   Sundberg, Building Trades Blueprint Reading. Part 2 Residential and Light Construction.
A11-3106   Sundberg, Building Trades Blueprint Reading. Part 3 General Construction, Specifications, and Heavy Construction.

STANDARDS AND CODES

A01-0007   ASTM, 1975 Compilation of ASTM Standards in Building Codes.
A01-0024   Binns, Knight's Building Regulations.
A01-0061   DeChiara and Koppelman, Planning Design Criteria.
A01-0069   Eberhard, The Performance Concept: A Study of Its Application to Housing. Vol. 2.
A01-0076   Field and Rivkin, The Building Code Burden.
A01-0095   Group for Environmental Education, The Process of Choice.
A01-0126   Lent, Plumbing Code of New York City--Guide and Interpretation.
A01-0134   Listokin, The Dynamics of Housing Rehabilitation: Macro and Micro Analysis.
A01-0164   Pitt and Dufton, The Guide to The London Building (Constructional) Bylaws 1972 and Building Acts.
A03-0535   ICBO, Building Department Administration.
A03-0574   Nagarajan, Standards in Building.
A03-0582   O'Bannon, Building Department Administration.
A03-0612   Sanderson, Codes and Code Administration: An Introduction to Building Regulations in the United States.
A04-1140   Mills, Planning: Volume 1: Architects' Technical Reference.
A04-1164   Natl. Aero. and Space Admin., NASA Facilities Engineering Handbook.
A04-1303   Sleeper, Building Planning and Design Standards for Architects, Engineers, Designers, Consultants, Building Committees, Draftsmen, and Students.
A05-1463   ACI, Realism in the Application of ACI Standard 214-65.
A05-1468   ACI, Ultimate Strength Design Handbook. Vol. 1.
A05-1469   ACI, Ultimate Strength Design Handbook. Vol. 2 Columns.
A05-1793   Parker, Simplified Design of Reinforced Concrete.
A05-1846   Schwartz, Civil Engineering for the Plant Engineer.
A05-1922   White, Structural Engineering. Vol. 4 Design of Structures.
A06-1969   ASHRAE, Equipment Volume 1972.
A06-2012   Croft, American Electricians' Handbook.
A06-2062   IEEE, National Electrical Safety Code, 1973 edition.

A06-2063	IEEE, Recommended Practice for Grounding of Industrial and Commercial Power Systems.
A06-2064	IEEE, Standard and American National Standard Graphic Symbols for Electrical and Electronics Diagrams.
A06-2065	IEEE, Recommended Practice for Protection and Coordination of Industrial and Commercial Power Systems.
A06-2072	Johnson, Johnson's Guide to the Welding Section of the Los Angeles Building Code.
A06-2104	Manas, National Plumbing Code Handbook.
A06-2105	Manas, National Plumbing Code Illustrated.
A06-2121	NFPA, Electrical Code for One and Two-Family Dwellings.
A06-2122	NFPA, National Electrical Code 1975.
A06-2123	NFPA, National Fuel Gas Code.
A06-2143	Richter, Practical Electrical Wiring, 10th Ed.
A06-2176	Watt and Stetka, NFPA Handbook of the National Electrical Code.
A06-2177	Watt and Summers, NFPA Handbook of the National Electrical Code.
A07-2208	APWA, A Survey of Urban Arterial Design Standards.
A08-2283	ACI, Commentary on Building Code Requirements for Reinforced Concrete.
A08-2306	Aluminum Assn., Aluminum Standards and Data.
A08-2309	Amrhein, Reinforced Masonry Engineering Handbook.
A08-2344	Bate, Handbook on the Unified Code for Structural Concrete.
A08-2392	Fintel, Handbook of Concrete Engineering.
A08-2397	Gage and Kirkbride, Design in Blockwork.
A08-2409	Glanville and Thomas, Explanatory Handbook on the BS Code of Practice for Reinforced Concrete CP 114:1957 with Metric Appendix.
A08-2493	Portland Cement Assn., Notes on ACI 318-71 Building Code Requirements with Design Applications.
A08-2495	Portland Cement Assn., Proceedings of the PCA-ACI Teleconference on ACI 318-71 Building Code Requirements.
A08-2505	Rice, User's Guide to the ACI Building Code.
A08-2514	Scott, Explanatory Handbook of the BS Code of Practice for Reinforced Concrete CP 114 (1957) (including Amendment No. 1: 1965).
A08-2552	Walley and Bate, A Guide to the B.S. Code of Practice for Prestressed Concrete CP 115.
A09-2580	ACI, Concrete Construction.
A09-2583	Alerich, Electrical Construction Wiring.
A09-2703	House Wiring.
A09-2718	Knowles and Pitt, The History of Building Regulation in London 1189-1972.
A09-2753	Merritt, Building Construction Handbook.
A09-2787	Reiner, Methods and Materials of Construction: A Guide for Builders, Owners, Architects and Engineers.
A09-2797	Schmidt, Construction: Principles, Materials and Methods.
A11-2994	Fairweather and Sliwa, AJ Metric Handbook.

A11-3030	Jensen, Fire Protection Systems for the Design Professional.
A11-3078	NFPA, Fire Protection Handbook, Fourteenth Edition.

STEEL STRUCTURES

A02-0384	Richardson, Fabricators and Erectors Estimating Standards.
A04-1018	Howard, Structure--An Architect's Approach.
A05-1451	Abbett, American Civil Engineering Practice.
A05-1470	AISC, Plastic Design in Steel.
A05-1471	AISC, Plastic Design of Braced Multistory Steel Frames.
A05-1494	ASCE, Guide for Design of Steel Transmission Towers.
A05-1496	ASCE, Plastic Design in Steel-A Guide and Commentary.
A05-1497	ASCE, The Proceedings of the August 1972 International Conference on the Planning and Design of Tall Buildings.
A05-1524	Baker and Heyman, Plastic Design of Frames.
A05-1525	Baker, Steel Skeleton.
A05-1544	Bleich, Buckling Strength of Metal Structures.
A05-1563	Bryan, The Stressed Skin Design of Steel Buildings.
A05-1573	Chilver, Thin Walled Structures.
A05-1583	Crawley and Dillon, Steel Buildings: Analysis and Design.
A05-1592	Daniels, Inelastic Steel Structures.
A05-1638	Gaylord and Gaylord, Design of Steel Structures.
A05-1639	Gaylord and Gaylord, Structural Engineering Handbook.
A05-1643	Gibson, The Design of Cylindrical Shell Roofs.
A05-1656	Gurney, Fatigue of Welded Structures.
A05-1669	Hart, Henn, and Sonntag, Multi-Storey Buildings in Steel.
A05-1732	Lothers, Advanced Design in Structural Steel.
A05-1736	McCormac, Structural Steel Design.
A05-1739	MacGinley, Structural Steelwork Calculations and Detailing.
A05-1740	McGuire, Steel Structures.
A05-1749	Manual of Steel Construction.
A05-1757	Merritt, Standard Handbook for Civil Engineers.
A05-1766	Munse and Grover, Fatigue of Welded Steel Structures.
A05-1795	Parker, Simplified Engineering for Architects and Builders.
A05-1797	Parker and Hauf, Simplified Design of Structural Steel.
A05-1828	Rogers, Steel Columns Eccentrically Loaded.
A05-1836	Sachs, Wind Forces in Engineering.
A05-1906	Walmer and Baron, Manual of Structural Design and Engineering Solutions.
A05-1911	Wang, Matrix Methods of Structural Analysis.
A05-1915	Wass, Manual of Structural Details for Building Construction.
A05-1922	White, Structural Engineering. Vol. 4 Design of Structures.
A05-1923	Wiegel, Earthquake Engineering.
A05-1934	Yu, Cold-Formed Steel Structures.
A09-2607	Blake, Civil Engineer's Reference Book.

Index

A09-2610 Bouwcentrum, Modern Steel Construction in Europe.
A09-2684 Feld, Construction Failure.
A09-2695 Halperin, Building with Steel.
A09-2701 Havers and Stubbs, Handbook of Heavy Construction.
A09-2704 Huntington and Mickadeit, Building Construction: Materials and Types of Construction.
A09-2743 Makowski, Steel Space Structures.
A09-2762 Oppenheimer, Erecting Structural Steel.
A09-2784 Rapp, Construction of Structural Steel Building Frames.
A09-2801 Shank, Control of Steel Construction to Avoid Brittle Failure.
A09-2804 Smith, Principles and Practices of Heavy Construction.
A09-2819 Urquhart, Civil Engineering Handbook.
A11-2933 AISC, Problems and Solutions for Structural Steel Detailing.
A11-2934 AISC, Structural Steel Detailing.
A11-3072 Newman, Standard Structural Details for Building Construction.

STRUCTURAL ANALYSIS AND DESIGN

See Seismic, Windstorm Design and Construction
A04-0833 Crane, Architectural Construction: The Choice of Structural Design.
A04-0843 Davies and Petty, Building Elements.
A04-0883 ERDA, General Design Criteria Engineering Handbook-Appendix 6301.
A04-1018 Howard, Structure--An Architect's Approach.
A04-1163 Natl. Aero. and Space Admin., Facilities Engineering Handbook.
A04-1194 Otto, Tensile Structures.
A04-1271 Salvadori, Mathematics in Architecture.
A05-1450 Aalami and Williams, Thin Plate Design for Transverse Loading.
A05-1451 Abbett, American Civil Engineering Practice.
A05-1452 Abeles, An Introduction to Prestressed Concrete.
A05-1453 ACI, Analysis of Structural Systems for Torsion.
A05-1456 ACI, Cracking, Deflection, and Ultimate Load of Concrete Slab Systems.
A05-1457 ACI, Deflections of Concrete Structures.
A05-1458 ACI, Designing for Effects of Creep, Shrinkage and Temperature.
A05-1460 ACI, Impact of Computers on the Practice of Structural Engineering in Concrete.
A05-1461 ACI, Models for Concrete Structures.
A05-1462 ACI, Probabilistic Design of Reinforced Concrete Buildings.
A05-1463 ACI, Realism in the Application of ACI Standard 214-65.
A05-1464 ACI, Reinforced Concrete Design Handbook.
A05-1465 ACI, Response of Multistory Concrete Structures to Lateral Forces.
A05-1466 ACI, Shear in Reinforced Concrete.
A05-1467 ACI, Symposium on Reinforced Concrete Columns.

A05-1468	ACI, Ultimate Strength Design Handbook. Vol. 1.
A05-1469	ACI, Ultimate Strength Design Handbook. Vol. 2 Columns.
A05-1470	AISC, Plastic Design in Steel.
A05-1471	AISC, Plastic Design of Braced Multistory Steel Frames.
A05-1472	Allen, Safe-load Tables for Solid Slabs.
A05-1473	Allen, Analysis and Design of Structural Sandwich Panels.
A05-1474	Allgood and Swihart, Design of Flexural Members for Static and Blast Loading.
A05-1476	Ambartsumyan, Theory of Anisotropic Plates: Strength, Stability, Vibration.
A05-1478	Andersen, Statically Indeterminate Structures--Their Analysis and Design.
A05-1479	Andersen, Substructure Analysis and Design.
A05-1480	Andersen and Norby, Introduction to Structural Mechanics.
A05-1488	Argyris and Kelsey, Energy Theorems and Structural Analysis.
A05-1489	ASCE, Bibliography on Bolted and Riveted Joints.
A05-1490	ASCE, Design of Cylindrical Concrete Shell Roofs.
A05-1494	ASCE, Guide for Design of Steel Transmission Towers.
A05-1496	ASCE, Plastic Design in Steel-A Guide and Commentary.
A05-1497	ASCE, The Proceedings of the August 1972 International Conference on the Planning and Design of Tall Buildings.
A05-1499	ASCE, Safety and Reliability of Metal Structures.
A05-1502	ASCE, Survey of Current Structural Research.
A05-1503	Ashdown, The Design of Prismatic Structures.
A05-1504	Ashton and Whitney, Theory of Laminated Plates.
A05-1505	ASME, Computer Software in Structural Analysis.
A05-1506	ASME, General Purpose Finite Element Computer Programs.
A05-1510	ASME, Thermal Structural Analysis Programs: A Survey and Evaluation.
A05-1511	ASME, Three Dimensional Continuum Computer Programs for Structural Analysis.
A05-1512	Asplund, Structural Mechanics: Classical and Matrix Methods.
A05-1519	Au, Elementary Structural Mechanics.
A05-1520	Baker, Limit-State Design of Reinforced Concrete.
A05-1521	Baker, Limit-State Design of Reinforced Concrete.
A05-1522	Baker, Raft Foundations.
A05-1523	Baker and Rish, Structural Analysis of Shells.
A05-1524	Baker and Heyman, Plastic Design of Frames.
A05-1525	Baker, Steel Skeleton.
A05-1526	Baker, Similarity Methods in Engineering Dynamics: Theory and Practice of Scale Modeling.
A05-1527	Bakos, Structural Analysis for Engineering Technology.
A05-1528	Bares, Tables for the Analysis of Plates, Slabs and Diaphragms: Based on the Elastic Theory.
A05-1529	Bares and Massonet, Analysis of Beam Grids and Orthotropic Plates: By the Guyon-Massonnet-Bares Method.
A05-1532	Bathe and Wilson, Numerical Methods in Finite Element Analysis.

A05-1533	Beaufait, Computer Methods of Structural Analysis.
A05-1534	Beckett, Limit State Design of Reinforced Concrete Structures.
A05-1535	Beckett, The Ultimate Load Design of Continuous Concrete Beams.
A05-1536	Beedle, Plastic Design of Steel Frames.
A05-1537	Benjamin, Structural Design with Plastics.
A05-1539	Bennett, Structural Concrete Elements.
A05-1540	Bickerdike, Design Failures in Buildings.
A05-1543	Blaszkowiak and Kaczkowski, Iterative Methods in Structural Analysis.
A05-1544	Bleich, Buckling Strength of Metal Structures.
A05-1545	Borg, Fundamentals of Engineering Elasticity.
A05-1546	Borg and Gennaro, Modern Structural Analysis.
A05-1547	Borrego, Space Grid Structures and Stressed Skin Systems.
A05-1548	Borrego, Space Grid Structures: Skeletal Frameworks and Stressed Skin Systems.
A05-1549	Bowes and Russell, Stress Analysis by the Finite Element Method for Practicing Engineers.
A05-1550	Bowles, Analytical and Computer Methods in Foundation Engineering.
A05-1552	Bowles, Foundation Analysis and Design.
A05-1555	Brebbia and Connor, Fundamentals of Finite Element Techniques: For Structural Engineers.
A05-1556	Breneman, Strength of Materials.
A05-1557	Bresler, Reinforced Concrete Engineering. Vol. 1 Materials, Structural Elements, Safety.
A05-1558	Bresler, Design of Steel Structures.
A05-1561	Browne, Basic Theory of Structures.
A05-1562	Brush and Almroth, Buckling of Bars, Plates and Shells.
A05-1563	Bryan, The Stressed Skin Design of Steel Buildings.
A05-1564	Bull and Sved, Moment Distribution in Theory and Practice.
A05-1565	Byars and Snyder, Engineering Mechanics of Deformable Bodies.
A05-1566	Calcote, The Analysis of Laminated Composite Structures.
A05-1567	Carpenter, Structural Mechanics.
A05-1569	Castigliano, The Theory of Equilibrium of Elastic Systems and Its Applications.
A05-1570	Cernica, Strength of Materials.
A05-1571	Charlton, Model Analysis of Plane Structures.
A05-1573	Chilver, Thin Walled Structures.
A05-1574	Chou and Pagano, Elasticity: Tensor, Dyadic and Engineering Approaches.
A05-1576	Clough and Penzien, Dynamics of Structures.
A05-1577	Conner, Theory of Structural Member Systems.
A05-1578	Connolly, Design of Prestressed Concrete Beams.
A05-1579	Cook, Concepts and Applications of Finite Element Analysis: A Textbook for Beginning Courses in the Finite Element Method, as Used for the Analysis of

	Displacement, Strain and Stress.
A05-1580	Coull and Dyke, Fundamentals of Structural Theory.
A05-1581	Cowan, Architectural Structures: An Introduction to Structural Mechanics.
A05-1583	Crawley and Dillon, Steel Buildings: Analysis and Design.
A05-1584	Croll and Walker, Elements of Structural Stability.
A05-1586	CRSI Design Handbook--Working Stress Design.
A05-1587	CRSI Handbook.
A05-1588	CRSI Handbook--Ultimate Strength Design.
A05-1590	Czerniak, Reinforced Concrete Columns. Vol. 1 Working Stress Design for Concrete Columns.
A05-1591	Czerniak, Reinforced Concrete Columns. Vol. 2 Working Stress Design Charts for Spiral Columns.
A05-1592	Daniels, Inelastic Steel Structures.
A05-1593	Davies, Steel Concrete Composite Beams for Building.
A05-1594	Davies, Structural Concrete.
A05-1595	Davies, Space Structures: A Study of Methods and Developments in Three-dimensional Construction.
A05-1597	Den Hartog, Strength of Materials.
A05-1598	Desai and Abel, Introduction to the Finite Element Method.
A05-1600	Disque, Applied Plastic Design in Steel.
A05-1602	Dravid, Analysis of Continuous Beams and Rigid Frames.
A05-1605	Dugdale, Elements of Elasticity.
A05-1606	Dugdale and Ruiz, Elasticity for Engineers.
A05-1607	Dunham, Advanced Reinforced Concrete.
A05-1608	Dunham, Theory and Practice of Reinforced Concrete.
A05-1609	Durgar'yan, Theory of Shells and Plates.
A05-1612	Eckardt, Strength of Materials.
A05-1617	Eriksen, Theory and Practice of Structural Design Applied to Reinforced Concrete.
A05-1619	Everard and Tanner, Reinforced Concrete Design: Including 200 Solved Problems.
A05-1621	Fawcett, Timber Structures.
A05-1622	Feltham, Deformation and Strength of Materials.
A05-1625	Feodosyev, Strength of Materials.
A05-1626	Ferguson, Reinforced Concrete Fundamentals.
A05-1627	Fertis, Dynamics and Vibrations of Structures.
A05-1629	Filonenko-Borodich, Theory of Elasticity.
A05-1631	Fischer, Architectural Engineering New Structures.
A05-1635	Forsyth, Unified Design of Reinforced Concrete Members.
A05-1636	Gallagher and Zienkiewicz, Optimum Structural Design: Theory and Applications.
A05-1637	Gartner, Statically Indeterminate Structures.
A05-1638	Gaylord and Gaylord, Design of Steel Structures.
A05-1639	Gaylord and Gaylord, Structural Engineering Handbook.
A05-1640	Gere and Weaver, Analysis of Framed Structures.
A05-1641	Gerstle, Basic Structural Design.
A05-1642	Ghali and Neville, Structural Analysis: A Unified Classical and Matrix Approach.
A05-1645	Gill, Structures for Nuclear Power.

Index

A05-1646	Glushkov, Handbook of Formulas for the Analysis of Complex Frames and Arches.
A05-1647	Godden, Numerical Analysis of Beam and Column Structures.
A05-1650	Granet, Strength of Materials for Engineering Technology.
A05-1651	Granholm, A General Flexural Theory of Reinforced Concrete.
A05-1652	Griffel, Handbook of Formulas for Stress and Strain.
A05-1653	Griffel, Plate Formulas.
A05-1655	Gurfinkel, Wood Engineering.
A05-1657	Guyon, Limit-State Design of Prestressed Concrete, Vol. 1: The Design of the Section.
A05-1658	Guyon, Limit-State Design of Prestressed Concrete, Vol. 2: The Design of the Member.
A05-1659	Hahn, Structural Analysis of Beams and Slabs.
A05-1660	Hall and Woodhead, Frame Analysis.
A05-1661	Halperin, Statics and Strength of Materials for Technology.
A05-1666	Harris, Elementary Structural Design.
A05-1667	Harris, Strength of Materials.
A05-1670	Heins, Bending and Torsional Design in Structural Members.
A05-1671	Hendry, Elements of Experimental Stress Analysis.
A05-1672	Hermann, Dynamic Stability of Structures.
A05-1673	Heyman, Beams and Framed Structures.
A05-1675	Hodge, Plastic Analysis of Structures.
A05-1676	Hodgkinson, AJ Handbook of Building Structure.
A05-1677	Hoff, The Analysis of Structures: Based on the Minimal Principles and the Principle of Virtual Displacements.
A05-1679	Hopkins and Deere, Design Analysis of Shafts and Beams.
A05-1680	Horne and Merchant, The Stability of Frames.
A05-1683	Hsieh, Elementary Theory of Structures.
A05-1684	Hughes, Limit State Theory for Reinforced Concrete.
A05-1685	Hurty and Rubinstein, Dynamics of Structures.
A05-1691	Jaeger, Cartesian Tensors in Engineering Science.
A05-1692	Jaeger, Elementary Theory of Elastic Plates.
A05-1694	Jensen and Chenoweth, Applied Strength of Materials.
A05-1695	Jensen and Chenoweth, Statics and Strength of Materials.
A05-1697	Johnson, Structural Concrete.
A05-1698	Johnson, Composite Structures of Steel and Concrete. Vol. 1 Beams, Columns, Frames, and Applications in Building.
A05-1700	Johnston, Column Research Council Guide to Design Criteria for Metal Compression Members.
A05-1701	Joiner, Essentials of the Theory of Structures.
A05-1704	Juvinall, Engineering Considerations of Stress, Strain, and Strength.
A05-1705	Kardestuncer, Elementary Matrix Analysis of Structures.
A05-1706	Khachatupian and Gurfinkle, Prestressed Concrete.
A05-1707	Kobayashi, Experimental Techniques in Fracture Mechanics.

A05-1710	Kurtz, Comprehensive Structural Design Guide.
A05-1711	Kuzmanovic and Willems, Steel Design for Structural Engineers.
A05-1712	Lancaster and Mitchell, The Mechanics of Materials.
A05-1713	Large and Chen, Reinforced Concrete Design.
A05-1714	Laursen, Structural Analysis.
A05-1716	Lenczner, Elements of Loadbearing Brickwork.
A05-1717	Lenczner, Movement in Buildings.
A05-1719	Leonhardt, Prestressed Concrete Design and Construction.
A05-1720	Levinson, Mechanics of Materials.
A05-1721	Levinson, Statics and Strength of Materials.
A05-1722	Lightfoot, Moment Distribution.
A05-1723	Lin, Theory of Inelastic Structures.
A05-1724	Lin, Probabilistic Theory of Structural Dynamics.
A05-1731	Long, Bearings in Structural Engineering.
A05-1732	Lothers, Advanced Design in Structural Steel.
A05-1733	Lothers, Design in Structural Steel.
A05-1735	McCormac, Structural Analysis.
A05-1736	McCormac, Structural Steel Design.
A05-1738	McFarland, Analysis of Plates.
A05-1739	MacGinley, Structural Steelwork Calculations and Detailing.
A05-1740	McGuire, Steel Structures.
A05-1741	McKaig, Applied Structural Design of Buildings.
A05-1742	McKaig, Building Failures.
A05-1743	McMinn, Matrices for Structural Analysis.
A05-1746	Majid, Optimum Design of Structures.
A05-1748	Mantell and Marron, Structural Analysis.
A05-1749	Manual of Steel Construction.
A05-1751	Martin, Introduction to Matrix Methods of Structural Analysis.
A05-1752	Martin and Carey, Introduction to Finite Element Analysis.
A05-1753	Matheson, Hyperstatic Structures: An Introduction to the Theory of Statically Indeterminate Structures.
A05-1754	Maugh, Statically Indeterminate Structures.
A05-1756	Meek, Matrix Structural Analysis.
A05-1757	Merritt, Standard Handbook for Civil Engineers.
A05-1758	Merritt, Structural Steel Designers' Handbook.
A05-1760	Michalos and Wilson, Structural Mechanics and Analysis.
A05-1761	Miroloyubov, Aid to Solving Problems in Strength of Materials.
A05-1762	Morgan, Forces in Framed Structures.
A05-1763	Morgan, The Analysis of Single-Bay Frames.
A05-1765	Morris, Handbook of Structural Design.
A05-1767	Murashev, Design of Reinforced Concrete Structures.
A05-1768	Nash, Strength of Materials: Including 345 Solved Problems.
A05-1775	Neal, Structural Theorems and Their Applications.
A05-1779	Norrie and deVries, The Finite Element Method: Fundamentals and Applications.
A05-1780	Norris, Elementary Structural Analysis.

Index

A05-1781	Norris, Structural Design for Dynamic Loads.
A05-1784	Oden, Mechanics of Elastic Structures.
A05-1786	Olsen, Strength of Materials.
A05-1790	Pannell, Design Charts for Members Subjected to Biaxial Bending and Thrust.
A05-1791	Parcel and Moorman, Analysis of Statically Indeterminate Structures.
A05-1793	Parker, Simplified Design of Reinforced Concrete.
A05-1794	Parker, Simplified Design of Structural Timber.
A05-1795	Parker, Simplified Engineering for Architects and Builders.
A05-1796	Parker, Simplified Mechanics and Strength of Materials.
A05-1797	Parker and Hauf, Simplified Design of Structural Steel.
A05-1798	Parker, Simplified Design of Roof Trusses for Architects and Builders.
A05-1799	Parkes, Braced Frameworks: An Introduction to the Theory of Structures.
A05-1802	Pisani, Essentials of Strength of Materials.
A05-1803	Polakowski and Ripling, Strength and Structure of Engineering Materials.
A05-1805	Popko, Geodesics.
A05-1809	Prager and Hodge, Theory of Perfectly Plastic Solids.
A05-1810	Preece and Davies, Models for Structural Concrete.
A05-1812	Przemieniecki, Theory of Matrix Structural Analysis.
A05-1813	Pucher, Influence Surfaces of Elastic Plates.
A05-1815	Rao, Reinforced Cement Concrete--A Textbook for Polytechnic Students.
A05-1816	Raz, Analytical Methods in Structural Engineering.
A05-1817	Reynolds, Basic Reinforced Concrete Design.
A05-1818	Rice and Hoffman, Structural Design Guide to the ACI Building Code.
A05-1819	Richart, Vibrations of Soils and Foundations.
A05-1821	Robinson, Integrated Theory of Finite Element Methods.
A05-1822	Robinson, Structural Matrix Analysis for the Engineer.
A05-1823	Robinson, Mechanics of Materials.
A05-1824	Rockey, The Finite Element Method: A Basic Introduction.
A05-1825	Rogers and Causey, Mechanics of Engineering Structures.
A05-1826	Rogers, Reinforced Concrete Design for Buildings.
A05-1827	Rogers, Tables and Formulas for Fixed End Moments of Members of Constant Moment of Inertia and for Simply Supported Beams.
A05-1828	Rogers, Steel Columns Eccentrically Loaded.
A05-1829	Roland, Frei Otto: Tension Structures, Ideas and Experiments in Lightweight Construction.
A05-1833	Rubinstein, Structural Systems: Statics, Dynamics and Stability.
A05-1835	Rygol, Nomograms for the Analysis of Frames.
A05-1838	Salmon and Johnson, Steel Structures: Design and Behavior.
A05-1839	Salvadori and Heller, Structure in Architecture.
A05-1840	Salvadori and Levy, Structural Design in Architecture.
A05-1842	Sarkar, Structural Analysis-Influence Coefficient Method.

A05-1843	Save and Massonet, Plastic Analysis and Design of Plates, Shells and Disks.
A05-1849	Sechler, Elasticity in Engineering.
A05-1850	Seelye, Data Book for Civil Engineers. Vol. 1 Design. Vol. 2 Specifications and Costs.
A05-1853	Shaw, Virtual Displacements and Analysis of Structures.
A05-1854	Shepley, Continuous Beam Structures--A Degree of Fixity Method and the Moment of Distribution.
A05-1855	Shermer, Design in Structural Steel.
A05-1856	Simitses, An Introduction to the Elastic Stability of Structures.
A05-1857	Singer, Strength of Materials.
A05-1858	Smith and Nicholson, Advances in Creep Design.
A05-1860	Smith, Introduction to Matrix and Finite Elements in Civil Engineering.
A05-1861	Smolira, Analysis of Structures.
A05-1862	Smolira, Analysis of Tall Buildings by the Force Displacement Method.
A05-1866	Spunt, Optimum Structural Design.
A05-1867	Stanek, Stress Analysis of Circular Plates and Cylindrical Shells.
A05-1868	Stark and Nicholas, Civil Engineering Systems: Mathematical Techniques for Design.
A05-1873	Stuttgart Universität, Institute of Lightweight Structures, Biology and Building II: Soap Film Models.
A05-1875	Stuttgart Universität, Institute of Lightweight Structures, Convertible Roofs.
A05-1876	Stuttgart Universität, Institute of Lightweight Structures, The Experimental Determination of Minimal Nets.
A05-1877	Stuttgart Universität, Institute of Lightweight Structures, Project Study "City in the Arctic."
A05-1879	Sutherland and Reese, Reinforced Concrete Design.
A05-1880	Szilard, Theory and Analysis of Plates: Classical and Numerical Methods.
A05-1881	Takabeya, Calculation and Moment-Tables: The Methods of Cross, Kani, and Takabeya.
A05-1882	Tall, Structural Steel Design.
A05-1885	Thandani, Modern Methods in Structural Engineering.
A05-1886	Timoshenko, Strength of Materials. Part 1 Elementary Theory and Problems.
A05-1887	Timoshenko, Strength of Materials. Part 2 Advanced Theory and Problems.
A05-1888	Timoshenko and Gere, Mechanics of Materials.
A05-1890	Timoshenko and Young, Elements of Strength of Materials.
A05-1891	Timoshenko and Young, Theory of Structures.
A05-1892	Tonge, The Indeterminate Beam: Theory and Examples.
A05-1895	Tuma, Structural Analysis: Including 200 Solved Problems.
A05-1898	Vawter and Clark, Elementary Theory and Design of Flexural Members.
A05-1899	Venkatraman and Patel, Structural Mechanics with Introductions to Elasticity and Plasticity.

Index

A05-1900	Vinson, Structural Mechanics: The Behavior of Plates and Shells.
A05-1902	Vlasov and Leont'ev, Beams, Plates and Shells on Elastic Foundations.
A05-1903	Volterra and Gaines, Advanced Strength of Materials.
A05-1904	Wah and Calcote, Structural Analysis by Finite Difference Calculus.
A05-1905	Walker, Design and Analysis of Cold Formed Sections.
A05-1906	Walmer and Baron, Manual of Structural Design and Engineering Solutions.
A05-1907	Wang, Applied Elasticity.
A05-1909	Wang, Introductory Structural Analysis with Matrix Methods.
A05-1910	Wang, Matrix Methods of Structural Analysis.
A05-1911	Wang, Statically Indeterminate Structures.
A05-1912	Wang and Eckel, Elementary Theory of Structures.
A05-1913	Wang and Salmon, Reinforced Concrete Design.
A05-1917	Westergaard, Theory of Elasticity and Plasticity.
A05-1919	White, Structural Engineering. Vol. 1 Introduction to Design Concepts and Analysis.
A05-1920	White, Structural Engineering. Vol. 2 Indeterminate Structures.
A05-1921	White, Structural Engineering. Vol. 3 Behavior of Members and Systems.
A05-1924	Wilby, Pre-Stressed Concrete Beams-Design and Logical Analysis.
A05-1927	Willems and Lucas, Matrix Analysis for Structural Engineers.
A05-1928	Williams, Analysis of Indeterminate Structures.
A05-1929	Williams and Harris, Structural Design in Metals.
A05-1930	Winter and Nilson, Design of Concrete Structures.
A05-1931	Wolfer, Elastically Supported Beams.
A05-1933	Yamada and Gallagher, Theory and Practice in Finite Element Structural Analysis. Proceedings of the 1973 Tokyo Seminar on Finite Element Analysis.
A05-1934	Yu, Cold-Formed Steel Structures.
A06-2032	Fitzgerald, Strength of Materials.
A06-2084	Kolousek, Dynamics in Engineering Structures.
A06-2085	Kuljian, Nuclear Power Plant Design.
A08-2302	AISC, Iron and Steel Beams 1873 to 1952.
A08-2339	ASTM, Window and Wall Testing.
A08-2347	BIA, Recommended Practice for Engineered Brick Masonry.
A08-2348	BIA, Reinforced Brick Masonry--Lateral Force Design.
A08-2392	Fintel, Handbook of Concrete Engineering.
A08-2400	Garthwaite, Metric Handbook for Reinforced Concrete.
A08-2413	Gurfinkel, Wood Engineering.
A08-2415	Guyon, Limit-State Design of Prestressed Concrete. Vol. 1 The Design of the Section.
A08-2422	Johnson, Designing, Engineering and Construction with Masonry Products.
A08-2432	LaLonde and Janes, Concrete Engineering Handbook.
A08-2450	Masonry Institute of America, Masonry Design Manual.

A08-2472	Nicholls, Composite Construction Materials Handbook.
A08-2499	Preston and Sollenberger, Modern Prestressed Concrete.
A08-2510	Sahlin, Structural Masonry.
A08-2516	Sementsov and Kameiko, Designer's Manual: Masonry, Including Reinforced Masonry in Industrial, Residential and Communal Buildings and Structures.
A08-2539	Terrington and Turner, Design of Non-Planer Roofs.
A08-2546	UNESCO, Reinforced Concrete: An International Manual.
A09-2588	American Institute of Timber Construction, Timber Construction Manual.
A09-2607	Blake, Civil Engineer's Reference Book.
A09-2623	Burgess, Progress in Construction Science and Technology.
A09-2635	Chandler, Structures for Building Technicians.
A09-2719	Knowles, Composite Steel and Concrete Construction.
A09-2722	Koncz, Manual of Precast Concrete Construction. Vol. 1.
A09-2723	Koncz, Manual of Precast Concrete Construction. Vol. 2.
A09-2724	Koncz, Manual of Precast Concrete Construction. Vol. 3.
A09-2729	Leonards, Foundation Engineering.
A09-2730	Leonhardt, Prestressed Concrete Design and Construction.
A09-2732	Libby, Modern Prestressed Concrete: Design Principles and Construction Methods.
A09-2742	McKaig, Building Failures.
A09-2744	Manning, Concrete Water Towers, Bunkers, Silos and Other Elevated Structures.
A09-2745	Manning, Design and Construction of Foundations.
A09-2746	Manning, Reinforced Concrete Reservoirs and Tanks.
A09-2753	Merritt, Building Construction Handbook.
A09-2815	Timber Engineering Co., Timber Design and Construction Handbook.
A09-2816	Tomlinson, Foundation Design and Construction.
A09-2819	Urquhart, Civil Engineering Handbook.
A11-2930	ACI, Manual of Standard Practice for Detailing Reinforced Concrete Structures.
A11-2938	Apfelbaum and Ottesen, Basic Engineering Sciences and Structural Engineering for Engineer-in-Training Examinations.
A11-2994	Fairweather and Sliwa, AJ Metric Handbook.
A11-3022	Hopf, Designer's Guide to OSHA.
A11-3042	Kurtz, Structural Engineering for Professional Engineers' Examinations.

STRUCTURAL SHELLS

A05-1455	ACI, Concrete Thin Shells.
A05-1490	ASCE, Design of Cylindrical Concrete Shell Roofs.
A05-1523	Baker and Rish, Structural Analysis of Shells.
A05-1537	Benjamin, Structural Design with Plastics.
A05-1542	Billington, Thin Shell Concrete Structures.
A05-1562	Brush and Almroth, Buckling of Bars, Plates and Shells.
A05-1609	Durgar'yan, Theory of Shells and Plates.
A05-1624	Fenves, Numerical and Computer Methods in Structural

Index

	Mechanics.
A05-1630	Fischer, Theory and Practice of Shell Structures.
A05-1634	Flugge, Stresses in Shells.
A05-1639	Gaylord and Gaylord, Structural Engineering Handbook.
A05-1643	Gibson, The Design of Cylindrical Shell Roofs.
A05-1644	Gibson, Linear Elastic Theory of Thin Shells.
A05-1675	Hodge, Plastic Analysis of Structures.
A05-1782	Novozhilov, Thin Shell Theory.
A05-1783	Obert and Duvall, Rock Mechanics and the Design of Structures in Rock.
A05-1787	Olszak and Sawczuk, Inelastic Behavior in Shells.
A05-1789	Osipova and Tumarkin, Tables for the Computation of Toroidal Shells.
A05-1805	Popko, Geodesics.
A05-1814	Ramaswamy, Design and Construction of Concrete Shell Roofs.
A05-1832	Rubinstein, Matrix Computer Analysis of Structures.
A05-1833	Rubinstein, Structural Systems: Statics, Dynamics and Stability.
A05-1843	Save and Massonet, Plastic Analysis and Design of Plates, Shells and Disks.
A05-1844	Savin and Fleishman, Rib-Reinforced Plates and Shells.
A05-1867	Stanek, Stress Analysis of Circular Plates and Cylindrical Shells.
A05-1889	Timoshenko and Woinowsky-Krieger, Theory of Plates and Shells.
A05-1900	Vinson, Structural Mechanics: The Behavior of Plates and Shells.
A05-1902	Vlasov and Leont'ev, Beams, Plates and Shells on Elastic Foundations.
A05-1925	Wilby and Khwaja, Concrete Shell Roofs.
A05-1926	Wilby and Nagvi, Reinforced Concrete Conoidal Shell Roofs--Flexural Theory Design Tables.
A05-1930	Winter and Nilson, Design of Concrete Structures.
A08-2284	ACI, Concrete Thin Shells.
A09-2753	Merritt, Building Construction Handbook.

SURVEYING

A02-0272	Benson, Building Contractor's and Home Builder's Handbook of Bidding, Surveying, and Estimating.
A05-1757	Merritt, Standard Handbook for Civil Engineers.
A05-1846	Schwartz, Civil Engineering for the Plant Engineer.
A09-2607	Blake, Civil Engineer's Reference Book.
A09-2613	Brighty, Setting Out: Guide for Site Engineers.
A09-2819	Urquhart, Civil Engineering Handbook.
A11-2942	ASCE, Definitions of Surveying and Associated Terms.
A11-2946	Bannister and Raymond, Surveying.
A11-2948	Barry, Construction Measurements.
A11-2956	Bickmore, Automatic Cartography and Planning Techniques in Automatic Cartography.
A11-2959	Bouchard and Moffitt, Surveying.
A11-2960	Breed, Surveying.

A11-2961	Breed and Hosmer, Principles and Practice of Surveying. Vol. 1 Elementary Surveying.
A11-2962	Breed and Hosmer, Principles and Practice of Surveying. Vol. 2 Higher Surveying.
A11-2964	Brinker, Elementary Surveying.
A11-2965	Brinker and Barry, Noteforms for Surveying Measurements.
A11-2971	Collins, Handbook of Accurate Surveying Methods.
A11-2984	Davis, Surveying--Theory and Practice.
A11-2985	Davis and Kelly, Short Course in Surveying.
A11-3017	Hewitt, Guide to Site Surveying.
A11-3026	Hudson, Useful Formulae for the Surveyor: A Field Manual for Control and Land Surveying.
A11-3038	Kissam, Surveying for Civil Engineers.
A11-3039	Kissam, Surveying: Instruments and Methods for Surveys of Limited Extent.
A11-3040	Kissam, Surveying Practice.
A11-3048	LaLonde, Professional Engineering Examination Questions and Answers.
A11-3049	Lang, Computer Programs for Mapping.
A11-3051	Legault, Surveying: An Introduction to Engineering Measurements.
A11-3069	Munson, Construction Design for Landscape Architects.
A11-3082	Perrott, Surveying for Young Engineers.
A11-3089	Rayner and Schmidt, Fundamentals of Surveying.
A11-3091	Ripa, Surveying Manual.
A11-3092	Royer, Applied Field Surveying.
A11-3095	Schofield, Engineering Surveying: Theory and Examination Problems for Students.
A11-3098	Smirnoff, Measurements for Engineering and Other Surveys.
A11-3108	Techniques in Automatic Cartography.
A11-3119	Whyte, Basic Metric Surveying.

SWIMMING POOLS, SAUNAS

A02-0309	Fabian, Aquatic Buildings.
A02-0312	Gabrielsen, Swimming Pools: A Guide to Their Planning, Design and Operation.
A04-0779	Butler, Recreation Areas--Their Design and Equipment.
A04-0890	Fabian, Aquatic Buildings.
A04-1074	Konya and Burger, The International Handbook of Finnish Sauna.
A04-1165	Navy Dept., Design Manual: Community Facilities.
A06-1976	ASSE, The Principles of Residential Plumbing Inspection.
A06-2024	Emerick, Troubleshooters' Handbook for Mechanical Systems.

TRANSPORTATION PLANNING AND DESIGN

See all entries in Section A07
See Harbors and Ports

Index

A01-0061	DeChiara and Koppelman, Planning Design Criteria.
A01-0064	Dickey, Metropolitan Transportation Planning.
A01-0075	Federal Highway Administration, Coordination of Urban Development and the Planning and Development of Transportation Facilities.
A01-0114	Hutchinson, Principles of Urban Transport Planning.
A01-0215	Tetlow and Gross, Homes, Towns, and Traffic.
A01-0225	Wagner, Environment and Man.
A01-0229	White, Sourcebook of Planning Information.
A02-0392	Smith, Real Estate and Urban Development.
A04-0701	ASCE, Small Craft Harbors.
A04-0702	ASCE, Analytical Treatment of Problems of Berthing and Mooring Ships.
A04-0844	Day, Building Acoustics.
A04-1121	Malt, Furnishing the City.
A04-1154	Mumford, From the Ground Up: Observations on Contemporary Architecture, Housing, Highway Building, and Civic Design.
A04-1230	Ragette, Engineering and Architecture and the Future Environment of Man.
A05-1851	Seelye, Data Book for Civil Engineers. Vol. 1 Design. Vol. 2 Specifications and Costs.

TUNNELS AND UNDERGROUND STRUCTURES

A01-0006	ASCE, The Use of Underground Space to Achieve National Goals.
A02-0376	Parker, Planning and Estimating Under-Ground Construction.
A04-0753	Brierley, Parking of Motor Vehicles.
A04-0962	Gropius, Town Plan for the Development of Selb.
A05-1479	Andersen, Substructure Analysis and Design.
A05-1620	Fairhurst, Failure and Breakage of Rock: Proceedings, 8th Symposium on Rock Mechanics.
A05-1654	Grosvenor and Paulding, Status of Practical Rock Mechanics: Proceedings, 9th Symposium on Rock Mechanics.
A05-1785	Okamoto, Introduction to Earthquake Engineering.
A07-2203	APWA, Feasibility of Utility Tunnels in Urban Areas.
A07-2205	APWA, Proceedings, Engineering Utility Tunnels in Urban Areas.
A07-2264	Urban Mass Transportation Dept., Subway Environmental Design Handbook. Vol. 1 Principles and Application.
A09-2589	American Society of Civil Engrs., Underground Rock Chambers.
A09-2607	Blake, Civil Engineer's Reference Book.
A09-2632	Carson, Foundation Construction.
A09-2642	Crimmins, Construction Rock Work Guide.
A09-2691	Gregory, Explosives for North American Engineers.
A09-2696	Hammond, Earthmoving and Excavating Plant.
A09-2701	Havers and Stubbs, Handbook of Heavy Construction.
A09-2714	Intl. Soc. of Soil Mechanics and Foundation Engrg., Grouts and Drilling Muds in Engineering Practice.
A09-2740	McGregor, Drilling of Rock.

A09-2834 Yardley, Rapid Excavation--Problems and Progress: Proceedings of the Tunnel and Shaft Conference, 1968.

URBAN AND REGIONAL PLANNING

See all entries in Section A01

A02-0277 Clurman and Hebard, Condominiums and Cooperatives.
A02-0327 Hanke, The Homes Association Handbook.
A04-0717 Bastlund, Jose Louis Sert: Architecture, City Planning, Urban Design.
A04-0753 Brierley, Parking of Motor Vehicles.
A04-0774 Burchard, Bernini is Dead?: Architecture and the Social Purpose.
A04-0801 Choay, The Modern City: Planning in the 19th Century.
A04-0814 Condit, Chicago, 1930-1970.
A04-0816 Condit, Chicago Since 1910.
A04-0828 Couperie, Paris Through the Ages.
A04-0841 Davern, Lewis Mumford: Architecture as a Home for Man.
A04-0851 Dept. of Landscape Arch., U. of Ga., and NE Ga. Area Planning and Development Comm., Madison, A Community Civic Design Study.
A04-0863 Doxiadis, Ekistics: An Introduction to the Science of Human Settlements.
A04-0875 Eckbo, Urban Landscape Design.
A04-0887 Evenson, Le Corbusier: The Machine and The Grand Design.
A04-0912 Forrester, Urban Dynamics.
A04-0949 Goodman and Von Eckardt, Life for Dead Spaces: The Development of the Lavanburg Commons.
A04-0973 Halprin, Cities.
A04-0991 Hegemann and Peets, The American Vitruvius: An Architect's Handbook of Civic Art.
A04-0997 Hesselgren, The Language of Architecture.
A04-1058 Kennedy and Kennedy, Architects Year Book 14: The Inner City.
A04-1079 Kuenzlen, Playing Urban Games.
A04-1084 Kurtz, Wasteland, Building the American Dream.
A04-1103 Lindley, Appreciation of Architecture--Landscape and Buildings.
A04-1115 Lynch, The Image of the City.
A04-1124 Malt, Furnishing the City.
A04-1134 Michaelides, Hydra: A Greek Island Town, Its Growth and Form.
A04-1154 Mumford, From the Ground Up: Observations on Contemporary Architecture, Housing, Highway Building, and Civic Design.
A04-1162 Nairn, Britain's Changing Towns.
A04-1172 Neutra, Survival Through Design.
A04-1224 Preiser, Environmental Design Research. Vol. 1.
A04-1225 Preiser, Environmental Design Research. Vol. 2.
A04-1228 Proshansky, Environmental Psychology: Man and His Physical Setting.

Index

A04-1234	Ransom, The People's Architects.
A04-1237	Rasmussen, Towns and Buildings.
A04-1243	Regional Plan Assn., Urban Design: Manhattan.
A04-1266	Saalman, Medieval Cities.
A04-1281	Scully, Modern Architecture.
A04-1306	Smith, Amenity and Urban Planning: Origin and Role of the Aesthetic Element in Modern Practice.
A04-1312	Soleri, Arcology: The City in the Image of Man.
A04-1313	Soleri, Sketch Books of Paolo Soleri.
A04-1316	Spreiregen, On the Art of Designing Cities: Selected Essays of Elbert Peets.
A04-1317	Spreiregen, Urban Design: The Architecture of Towns and Cities.
A04-1318	Spyer, Architect and Community: Environmental Design in an Urban Society.
A04-1351	Unwin, Town Planning in Practice.
A04-1364	Wall, Visionary Cities: The Arcology of Paolo Soleri.
A04-1392	Woods, Candilis-Josic-Woods: Building for People.
A04-1399	Wright, The Living City.
A04-1411	Zucker, Town and Square From the Agora to the Village Green, 1959.
A05-1757	Merritt, Standard Handbook for Civil Engineers.
A07-2202	Antonious, Environmental Management: Planning for Traffic.
A07-2203	APWA, Feasibility of Utility Tunnels in Urban Areas.
A07-2222	Baumann and Wilson, Urban Engineering and Transportation.
A07-2231	Heggle, Transport Engineering Economics.
A07-2232	Hennes and Ekse, Fundamentals of Transportation Engineering.

URBAN DECAY AND RENEWAL

A01-0001	Abrams, The City is the Frontier.
A01-0005	Anderson, The Federal Bulldozer: A Critical Analysis of Urban Renewal, 1949-1962.
A01-0031	Browne, West End: Renewal of a Metropolitan Centre.
A01-0036	Cantacuzino, New Uses for Old Buildings.
A01-0046	Chung, The Economics of Residential Rehabilitation: Social Life of Housing in Harlem.
A01-0055	Crecine, Financing the Metropolis: Public Policy in Urban Economies.
A01-0067	Doxiadis, Urban Renewal and the Future of the American City.
A01-0078	Frieden, The Future of Old Neighborhoods: Rebuilding for a Changing Population.
A01-0081	Friendly, Benefit-Cost Applications in Urban Renewal: A Feasibility Study.
A01-0094	Groberg, Urban Renewal Programs Assisted by Title 1 of the Housing Act of 1949.
A01-0107	Housing and Urban Development Dept., Guidelines for Urban Renewal Land Disposition.
A01-0110	Housing and Urban Development Dept., A Study of

A01-0111 Property Taxes and Urban Blight. Housing and Urban Development Dept., A Study of the Effects of Real Estate Property Tax Incentive Programs Upon Property Rehabilitation and New Construction.
A01-0116 Jacobs, The Death and Life of Great American Cities.
A01-0120 Kaplan, Urban Renewal Politics: Slum Clearance in Newark.
A01-0132 Liblit, Housing--The Cooperative Way.
A01-0134 Listokin, The Dynamics of Housing Rehabilitation: Macro and Micro Analyses.
A01-0147 Mills, Urban Economics.
A01-0162 Perloff, Modernizing the Central City: New Towns In-town ... and Beyond.
A01-0163 Peterson, Property Taxes, Housing, and the Cities.
A01-0165 Pratt Guide to Planning and Renewal for New Yorkers.
A01-0169 Redstone, The New Downtowns: Rebuilding Business Districts.
A01-0178 Rose, Regent Park: A Study in Slum Clearance.
A01-0205 Sternlieb and Burchell, Residential Abandonment: The Tenement Landlord Revisited.
A01-0227 Weaver, Dilemmas of Urban America.
A01-0228 Wexler and Peck, Housing and Local Government.
A01-0233 Wilson, Urban Renewal: The Record and the Controversy.
A02-0303 Edgerton, How to Renovate A Brownstone.
A02-0304 Edgerton, Row House Renaissance, A Guide to Renovation.
A02-0392 Smith, Real Estate and Urban Development.
A04-0837 Crosby, The Necessary Monument: Its Future in the Civilized City.
A04-0919 Frigand, The New City: Architecture and Urban Renewal.
A04-1115 Lynch, The Image of the City.

VIBRATION

See Seismic Design and Construction, and Acoustics
A06-1970 ASHRAE, 1973 Systems Volume.
A06-1973 ASME, AMD Vol. 1 Isolation of Mechanical Vibration, Impact, and Noise.
A06-1980 Baker, Use of Models and Scaling in Shock and Vibration.
A06-1982 Barkan, Dynamics of Bases and Foundations.
A06-1988 Beranek, Noise and Vibration Control.
A06-1991 Blake and Mitchell, Vibration and Acoustic Measurement Handbook.
A06-2000 Burton, Vibration and Impact.
A06-2010 Crocker, Noise and Vibration Control Engineering: Proceedings of the Purdue Noise Control Conference, 1971, Lafayette, Ind.
A06-2033 Flugge, Handbook of Engineering Mechanics.
A06-2048 Harris, Handbook of Noise Control.
A06-2049 Harris, Shock and Vibration Handbook.

Index 657

A06-2084	Kolousek, Dynamics in Engineering Structures.
A06-2093	Levinson, Introduction to Mechanics.
A06-2107	Meirovitch, Elements of Vibration Analysis.
A06-2132	Petrusewicz and Longmore, Noise and Vibration Control.
A06-2154	Seto, Mechanical Vibrations: Including 225 Solved Problems.
A06-2161	Snowdon, Vibration and Shock in Damped Mechanical Systems.
A06-2162	Steidel, An Introduction to Mechanical Vibrations.
A06-2167	Thomson, Vibration Theory and Applications.
A06-2169	Timoshenko, Vibration Problems in Engineering.
A06-2175	Waller and Atkins, Building on Springs.
A06-2186	Yerges, Sound, Noise and Vibration Control.

WASTE MANAGEMENT, TREATMENT AND DISPOSAL

A01-0170	Regional Plan Association, Second Regional Plan Publications.
A01-0225	Wagner, Environment and Man.
A03-0447	Azad, Industrial Wastewater Management Handbook.
A03-0515	Gray, Cases and Materials on Environmental Law.
A04-1068	Kinzey and Sharp, Environmental Technologies in Architecture.
A05-1482	APWA, Control of Infiltration and Inflow into Sewer Systems.
A05-1484	APWA, Problems of Combined Sewer Facilities and Overflows.
A05-1500	ASCE, Sewage Treatment Plant Design.
A05-1507	ASME, 1970 National Incinerator Conference Proceedings.
A05-1508	ASME, Proceedings of 1968 National Incinerator Conference.
A05-1531	Bartlett, Pumping Stations for Water and Sewage.
A05-1585	Cross and Noble, Handbook on Hospital Solid Waste Management.
A05-1613	Eckenfelder, Industrial Water Pollution Control.
A05-1614	Eckenfelder, Water Quality Engineering for Practicing Engineers.
A05-1662	Hardenbergh and Rodie, Water Supply and Waste Disposal.
A05-1727	Liptak, Environmental Engineers' Handbook.
A05-1759	Metcalf and Eddy, Wastewater Engineering: Collection, Treatment, Disposal.
A05-1804	Pollution Control Technology.
A05-1841	Sanks, Land Treatment and Disposal.
A05-1871	Straub, Low-Level Radioactive Wastes: Their Handling, Treatment, and Disposal.
A05-1935	Zilly, Handbook of Environmental Civil Engineering.
A06-2024	Emerick, Troubleshooters' Handbook for Mechanical Systems.
A06-2095	Lindeke, Technical Dictionary of Heating, Ventilation and Sanitary Engineering.
A08-2485	Pawley, Garbage Housing.
A09-2607	Blake, Civil Engineer's Reference Book.

WATER SUPPLY AND DRAINAGE - HYDROLOGY

A04-1230   Ragette, Engineering and Architecture and the Future Environment of Man.
A05-1477   American Water Works Assoc., Water Quality and Treatment: A Handbook of Public Water Supplies.
A05-1482   APWA, Control of Infiltration and Inflow into Sewer Systems.
A05-1483   APWA, Practices in Detention of Urban Stormwater Runoff.
A05-1484   APWA, Problems of Combined Sewer Facilities and Overflows.
A05-1485   APWA, Water Pollution Aspects of Urban Runoff.
A05-1531   Bartlett, Pumping Stations for Water and Sewage.
A05-1575   Chow, Handbook of Applied Hydrology.
A05-1596   Davis and Sorenson, Handbook of Applied Hydraulics.
A05-1614   Eckenfelder, Water Quality Engineering for Practicing Engineers.
A05-1662   Hardenbergh and Rodie, Water Supply and Waste Disposal.
A05-1664   Harr, Groundwater and Seepage.
A05-1725   Linsley and Franzini, Water Resources Engineering.
A05-1726   Linsley, Hydrology for Engineers.
A05-1757   Merritt, Standard Handbook for Civil Engineers.
A05-1764   Morris and Wiggert, Applied Hydraulics in Engineering.
A05-1808   Powell, Water Conditioning for Industry.
A05-1935   Zilly, Handbook of Environmental Civil Engineering.
A06-1992   Blendermann, Design of Plumbing and Drainage Systems.
A06-2083   King, Piping Handbook.
A07-2224   Cedergren, Drainage of Highway and Airfield Pavements.
A09-2607   Blake, Civil Engineer's Reference Book.
A09-2626   Campbell and Lehr, Water Well Technology: Field Principles of Exploration and Drilling for Ground Water.
A09-2764   Parker and MacGuire, Simplified Site Engineering for Architects and Builders.
A09-2803   SMACNA, Architectural Sheet Metal Manual.
A09-2819   Urquhart, Civil Engineering Handbook.

WELDING, BRAZING, SOLDERING

A06-1978   AWS, Resistance Welding--Theory and Use.
A06-2038   Giachino, Welding Technology.
A06-2072   Johnson, Johnson's Guide to the Welding Section of the Los Angeles Building Code.
A06-2111   Morris, Welding Principles for Engineers.
A06-2133   Pfluger and Lewis, Weld Imperfections.
A06-2146   Rossi, Welding Engineering.
A06-2171   Udin, Welding for Engineers.
A08-2307   Alum. Assn. and American Welding Society, Aluminum Welding Seminar Papers.
A08-2323   ASM, Metals Handbook. Vol. 6 Welding and Brazing.
A09-2584   Althouse, Modern Welding.

Index

A09-2599	AWS, Brazing Manual.
A09-2600	AWS, Soldering Manual.
A09-2689	Giachino, Welding Skills and Practices.
A09-2720	Koenigsberger and Adair, Welding Technology.
A09-2752	Meleka, Electron-Beam Welding: Principles and Practice.
A09-2782	Rampaul, Pipe Welding Procedures.
A09-2792	Rose, Questions and Answers on Pipework and Pipe Welding.
A09-2827	Welders Guide.
A11-3005	Giachino and Beukema, Print Reading for Welders.

WOOD, MILLWORK, AND WOOD PRODUCTS

A08-2303	Akers, Particle Board and Hardboard.
A08-2310	Anderson and Earle, Design and Aesthetics in Wood.
A08-2316	ASCE Structural Div., Wood Structures.
A08-2340	Austin and Ueda, Bamboo.
A08-2352	Brady, Materials Handbook.
A08-2358	Bucksch, Dictionary of Wood and Woodworking Practice.
A08-2359	Building Research Establishment, The Strength Properties of Timber.
A08-2368	Clauser, Encyclopedia of Engineering Materials and Processes.
A08-2394	Forest Products Laboratory, Wood Handbook: Wood as an Engineering Material.
A08-2402	Gatz and Thierry, Architect's Detail Library. Vol. 2 Ceilings in Wood.
A08-2404	Gatz and Thierry, Architect's Detail Library. Vol. 5 Entrances and Staircases.
A08-2413	Gurfinkel, Wood Engineering.
A08-2419	Hoffman, Building with Wood: Form, Structural Design, and Preservation.
A08-2421	Jayne, Theory and Design of Wood and Fiber Composite Materials.
A08-2425	Keyser, Materials Science in Engineering.
A08-2437	Lloyd, Millwork--Principles and Practice.
A08-2457	Moselle, Practical Lumber Computer.
A08-2470	Nicholas, Wood Deterioration and Its Prevention by Preservative Treatments. Vol. 1 Degradation and Protection of Wood.
A08-2471	Nicholas, Wood Deterioration and Its Prevention by Preservative Treatments. Vol. 2 Preservatives and Preservative Systems.
A08-2479	Parker, Materials Data Book.
A08-2484	Patton, Materials in Industry.
A08-2492	Pollack, Materials Science and Metallurgy.
A08-2521	Siau, Flow in Wood.
A08-2523	Simpson and Horrobin, The Weathering and Performance of Building Materials.
A08-2524	Skaar, Water in Wood.
A08-2528	Smith, Materials of Construction.
A08-2553	Watson, Construction Materials and Processes.

	Construction Information
A08-2554	The Weathering and Performance of Building Materials.
A09-2588	American Institute of Timber Construction, Timber Construction Manual.
A09-2618	The Building Research Establishment, Essential Data for Construction. Vol. 2 Building Materials.
A09-2630	Carpenters and Builders Library. Vol. 4 Millwork, Power Tools, Painting.
A09-2644	Dahl and Wilson, Cabinetmaking and Millwork: Tools, Materials, Layout, Construction.
A09-2704	Huntington and Mickadeit, Building Construction: Materials and Types of Construction.
A09-2736	Love, Stair Builders Handbook.
A09-2824	Wass, Methods and Materials of Residential Construction.

WOODEN STRUCTURES

A04-0982	Hansen, Architecture in Wood: A History of Wood Building and Its Techniques in Europe and North America.
A04-1018	Howard, Structure--An Architect's Approach.
A04-1132	Merrilees and Loveday, Pole Building Construction.
A04-1245	Rempel, Building with Wood.
A04-1370	West, The Timber-Frame House in England.
A05-1621	Fawcett, Timber Structures.
A05-1655	Gurfinkel, Wood Engineering.
A05-1794	Parker, Simplified Design of Structural Timber.
A05-1795	Parker, Simplified Engineering for Architects and Builders.
A05-1906	Walmer and Baron, Manual of Structural Design and Engineering Solutions.
A05-1915	Wass, Manual of Structural Details for Building Construction.
A05-1922	White, Structural Engineering. Vol. 4 Design of Structures.
A08-2316	ASCE Structural Div., Wood Structures.
A08-2419	Hoffmann, Building with Wood: Form, Structural Design, and Preservation.
A09-2588	American Institute of Timber Construction, Timber Construction Manual.
A09-2607	Blake, Civil Engineer's Reference Book.
A09-2684	Feld, Construction Failure.
A09-2694	Guild of Architectural Ironmongers, Timber Joinery and Applied Ironmongery.
A09-2700	Hansen, Architecture in Wood.
A09-2704	Huntington and Mickadeit, Building Construction: Materials and Types of Construction.
A09-2711	Innocent, The Development of English Building Construction.
A09-2761	Oberg, Heavy Timber Construction.
A09-2804	Smith, Principles and Practices of Heavy Construction.
A09-2805	Smith, Principles and Practices of Light Construction.
A09-2815	Timber Engineering Co., Timber Design and Construction

Index

A09-2824	Handbook. Wass, Methods and Materials of Residential Construction.

ZONING

A01-0008	Babcock and Bosselman, Exclusionary Zoning: Land Use Regulation and Housing in the 1970s.
A01-0027	Branch, Urban Planning Theory.
A01-0030	Brown and Sherrard, An Introduction of Town and Country Planning.
A01-0061	DeChiara and Koppelman, Planning Design Criteria.
A01-0071	Erber, Urban Planning in Transition.
A01-0095	Group for Environmental Education, The Process of Choice.
A01-0097	Haar and Iatridis, Housing the Poor in Suburbia: Public Policy at the Grass Roots.
A01-0133	Linowes and Allensworth, The Politics of Land Use: Planning, Zoning, and the Private Developer.
A01-0165	Pratt Guide to Planning and Renewal for New Yorkers.
A01-0173	Richardson, The Cost of Environmental Protection: Regulating Housing Development in the Coastal Zone.
A01-0184	Sagalyn and Sternlieb, Zoning and Housing Costs: The Impact of Land-Use Controls on Housing Price.
A01-0192	Siegan, Land Use and Real Estate.
A01-0193	Siegan, Land Use Without Zoning.
A01-0239	Yearwood, Land Subdivision Regulation: Policy and Legal Considerations for Urban Planning.
A02-0277	Clurman and Hebard, Condominiums and Cooperatives.

Z
5851
G62

JUL 1 2 1978

RAYMOND H. FOGLER LIBRARY
DATE DUE